Key Challenges in Geography

EUROGEO Book Series

This book series addresses relevant topics in the wide field of geography, which connects the physical, human and technological sciences to enhance teaching, research, and decision making. Geography provides answers to how aspects of these sciences are interconnected and are forming spatial patterns and processes that have impact on global, regional and local issues and thus affect present and future generations. Moreover, by dealing with places, people and cultures, Geography explores international issues ranging from physical, urban and rural environments and their evolution, to climate, pollution, development and political economy.

Key Challenges in Geography is an initiative of the European Association of Geographers (EUROGEO), an organization dealing with examining geographical issues from a European perspective, representing European Geographers working in different professional activities and at all levels of education. EUROGEO's goal and the core part of its statutory activities is to make European Geography a worldwide reference and standard. The book series serves as a platform for members of EUROGEO as well as affiliated National Geographical Associations in Europe, but is equally open to contributions from non-members.

The book series addresses topics of contemporary relevance in the wide field of geography. It has a global scope and includes contributions from a wide range of theoretical and applied geographical disciplines.

Key Challenges in Geography aims to:

- present collections of chapters on topics that reflect the significance of Geography as a discipline;
- provide disciplinary and interdisciplinary titles related to geographical, environmental, cultural, economic, political, urban and technological research with a European dimension, but not exclusive;
- deliver thought-provoking contributions related to cross-disciplinary approaches and interconnected works that explore the complex interactions among geography, technology, politics, environment and human conditions;
- publish volumes tackling urgent topics to geographers and policy makers alike;
- publish comprehensive monographs, edited volumes and textbooks refereed by European and worldwide experts specialized in the subjects and themes of the books;
- provide a forum for geographers worldwide to communicate on all aspects of research and applications of geography, with a European dimension, but not exclusive.

All books/chapters will undergo a blind review process with a minimum of two reviewers.

An author/editor questionnaire, instructions for authors and a book proposal form can be obtained by contacting the Publisher.

The Key Challenges in Geography Series is now indexed in SCOPUS.

Laura Soares · Miguel Saraiva ·
Rafael de Miguel González
Editors

Compromised Geography

Spreading a New World

eurogeo

Springer

Editors
Laura Soares
Faculty of Arts and Humanities
Department of Geography
University of Porto
Porto, Portugal

Miguel Saraiva
Faculty of Arts and Humanities
Department of Geography
University of Porto
Porto, Portugal

Rafael de Miguel González
EUROGEO
Waardamme, Belgium

ISSN 2522-8420 ISSN 2522-8439 (electronic)
Key Challenges in Geography
ISBN 978-3-032-05200-1 ISBN 978-3-032-05201-8 (eBook)
https://doi.org/10.1007/978-3-032-05201-8

This Springer imprint is published by the registered company Springer Nature Switzerland AG
The registered company address is: Gewerbestrasse 11, 6330 Cham, Switzerland

Preface

Compromised Geography—Spreading a New World, the new volume of the EUROGEO Book Series *Key Challenges in Geography*, includes some of the best papers that were presented at the EUROGEO Conference 2024, which took place in Porto (Portugal) between May 30th and June 1st. With 125 participants from practically all corners of the world (Fig. 1), this edition of the EUROGEO conference was once again anchored in the multiculturalism and the "travelling" spirit of geographers, providing ample opportunities for networking and disseminating the latest in geographical research.

The conference was, we believe, a successful one. It included two keynote plenary sessions, 69 paper presentations, four workshops, one special session, the presentation of 13 posters, and two field trips: a shorter one to the Historic Centre of Porto, a World Heritage Site since 1996, and a more extend post-conference trip to the Douro Valley in northern Portugal.

The selected theme, *Compromised Geography—Spreading a New World*, calls for a geography committed to the contemporary challenges of our common reality; one that aims to find meaning among the "vertiginous" changes facing our daily lives, one that hones our knowledge and skills towards balanced and sustainable interventions within new territorial frameworks. Concomitantly, the "shrinking" of time and space brought about by technological innovations imposes new responsibilities on the research, the training and the professional practice of geographers. As is written in the International Charter of Geographical Education (2016, p. 4–5):

> Whether it is through appreciating the beauty of Earth, the immense power of Earth-shaping forces or the often ingenious ways in which people create their living in different environments and circumstances, studying geography helps people to understand and appreciate how places and landscapes are formed, how people and environments interact, the consequences that arise from our everyday spatial decisions, and Earth's diverse and interconnected mosaic of cultures and societies.

This means that there is a commitment that Geography makes with the constant reconfiguration of the world, at different scales of analysis and in different contexts—whether environmental, geopolitical, social, economic or, among others, educational.

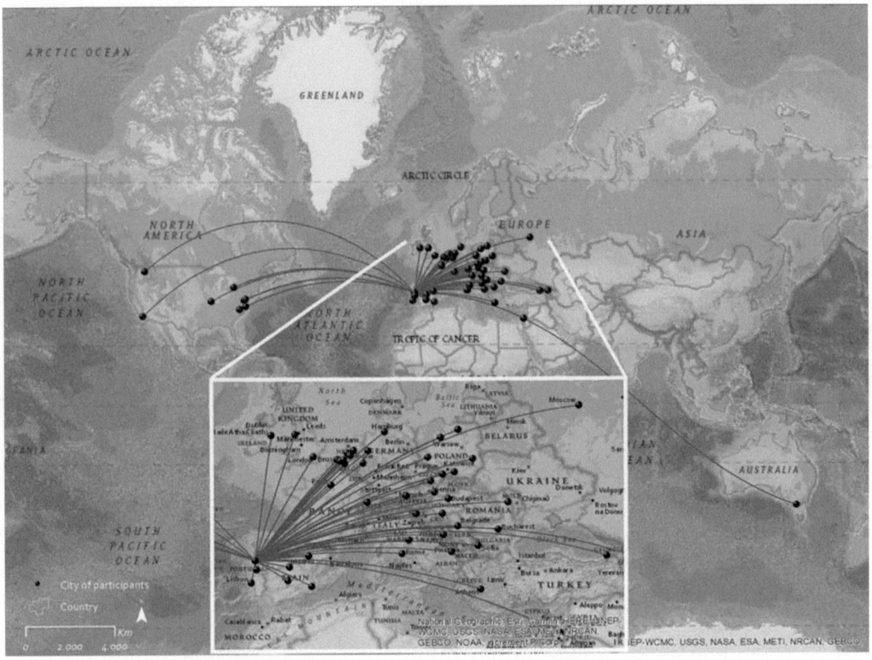

Fig. 1 Origin of participants in EUROGEO conference, Porto 2024

And that is the commitment made as well by the EUROGEO 2024 participants, who shared their research among the common ground of the great challenges ahead:

- What pathways are there for a (more) intelligible and practical geographical discourse?
- How important is the Noosphere in disseminating geographical knowledge?
- What are the implications of Artificial Intelligence (AI) for the evolution of geographical methodologies, research and practice?
- How is a new de-globalised world organised?
- How can we promote geographical research committed to the new world?

The volume we are now presenting shed insights into these themes, within six thematic areas:

I. Conferences;
II. Geography Education;
III. Geospatial Technologies and Artificial Intelligence;
IV. Human Geography;
V. Physical and Environmental Geography;
VI. Geoheritage and Tourism.

First Part is dedicated do the keynote conferences presented at EUROGEO 2024 by Miguel Bandeira (University of Minho—Portugal) and Nico Van de Weghe

(Ghent University, Belgium). **First Chapter**, entitled "Local Geography and the Irreplaceable Virtue of the Direct and Live", provides a critical analysis focused primarily on geographical education. The author defends the essential role of *face-to-face and perceptive interactions between educators and students*, prioritising direct and living relationships in the educational process. Bandeira further advocates the importance of engaging with the place where everything truly happens: *the place where we live* and can *exert our transformative influence.* **Second Chapter** discusses how generative AI is reshaping our world from a geographical perspective. It highlights the rapid spread of AI and its transformative effects on *how we work, learn and understand the world*. With a focus on education, the author explains the impact of *GenAI on geographical learning, the urgency of GenAI literacy and the need for critical spatial thinking*. Presenting some *practical applications of GenAI in geography education*, he also reminds us that, although *GenAI is here to stay, human judgement and values remain essential in shaping its role in society.*

Second Part (Geography Education) includes five chapters, where the common problem are the social and cultural issues related with the Educational System. In the **Third Chapter**, Ana Ferreira, Elsa Pacheco and Francisca Teixeira highlight the challenges faced by elementary and high schools due to an *increasing number of students from diverse geographical* contexts. They view schools as ideal spaces for promoting *behavioural changes and shaping the attitudes of future generations*, so they analyse the perceptions of geography teachers and students regarding the *integration* of international students. In the **Fourth Chapter**, Elsa Pacheco, Laura Soares and Paulo Lemos begin with a challenging question—*Teachers for what?* They answer it by examining the interplay between *desires, supply and demand*, based on the students who choose geography as an undergraduate course. They further debate why is there so much resistance to change essential learnings, if desires change with the times, and how can schools be transformed in response to such contextual changes. In the **Fifth Chapter**, Hélder Santos and colleagues uncover Portuguese socio-spatial inequalities in education, emphasising that it *is essential to combat existing territorial vulnerabilities in the education system* and necessary to understand the fragilities *of the territory to develop strategies that ensure a more integrated, inclusive and equitable development of the education system.* The authors use a methodology *based on a multidimensional territorial approach*, organised into six analytical dimensions and 29 indicators, to reach distinct territorial educational profiles in Portugal. In the **Sixth Chapter**, Javier Álvarez-Otero and Isabel María Gómez-Trigueros discuss training Geography teachers using AI apps to *manage and represent geographical data, as well as to visualise and interpret phenomena at local and global scales.* The authors believe that these tools *facilitate an understanding of the environmental, social, and economic dimensions of sustainability*. However, they call attention *to the ethical essential use of AI* and engage in *critical reflection on its implications*. Finally, in the **Seventh Chapter**, Stefana Matović, Suzana Lović and Darko Vuković discuss *Sustainable Development Goal 4.2*, namely in addressing the challenges of monitoring early childhood education, a life phase that plays a fundamental role in shaping children's cognitive, social and emotional development. The authors highlight the *disparities between European countries, with a focus on*

Serbia, appealing to the *importance of innovative teaching methods and data-driven policies to ensure that all children have access to high-quality early education.*

Third Part (Geospatial Technologies and Artificial Intelligence) includes four chapters, the first two explicitly concerning AI. In the **Eighth Chapter**, Simona Epasto explores the dual sides of AI *as a tool for enhanced territorial control and surveillance and as a source of new vulnerabilities and instabilities in the geopolitical landscape.* Analysing critical case studies (the Israel-Hamas conflict, the Russia-Ukraine war, and technological rivalry between USA-China), the author reveals how *AI-driven technologies, such as drones and predictive analytics, are transforming conflict dynamics*, increasing territorial inequalities, and worsening geopolitical tensions. In the **Ninth Chapter**, Ivana Vyslúžilová focuses on the Impact of Artificial Intelligence on urban transformation, especially in Smart Cities. Considering that digital transformation leverages smart infrastructure, the primary approach of innovative urban solutions *can lead to several benefits*, which can be seen *through reduced pollution and more efficient management of energy, water, and waste.* Data should be used to *make more correct decisions and to support more efficient operations*, however, it is crucial to strike a balance between the risks and opportunities of AI services, which can significantly impact the future of technological trends. In the **Tenth Chapter**, Suzana Lović Obradović and colleagues use geospatial technologies to understand cardiovascular mortality patterns, *the leading cause of death in Serbia.* Using geographically weighted regression, the authors highlight how their spatial distribution *is shaped by a complex interaction of demographic, socioeconomic, health, and environmental factors, reflecting deep regional disparities.* Lastly, in the **Eleventh Chapter**, Laura Soares, Elsa Pacheco and Paulo Lemos apply geospatial technologies to address landscape as heritage and the role of teacher-learning. They present work projects developed by Portuguese Geography students from three levels of education, using *geospatial technologies adapted to each level*, including, e.g. Wikiloc or Google Earth (elementary and high school) and ArcGIS 3D Analyst (undergraduate).

Fourth Part (Human Geography) integrates five chapters. In the **Twelfth Chapter**, Andreia Monteiro, Miguel Saraiva and Teresa Sá Marques present a *theoretical reflection on issues of injustice and inequality in transport and mobility, which compromises the development of adequate policies.* The authors revisit *some of the classic theories of justice to identify how they can or have been translated into the (geographical) challenges of current transport and mobility planning.* In the **Thirteenth Chapter**, María Zúñiga Antón looks at the role of Cognitive Cartography in the cartographic design process, by exploring *the theoretical foundations and current developments* and *focusing on its implications for map design and user interaction.* Allowing *for a deeper understanding of how users cognitively engage with maps*, this chapter *contributes to the development of evidence-based strategies that improve spatial understanding and communication.* In the **Fourteenth Chapter**, Monika Wesołowska and Anita Kulawiak discuss the urban-rural migration in Poland, analysing if they are 'an opportunity or a threat for rural areas'. Based*on field observations*, the authors *identify and define the types of people who* *decide to move to the countryside*, as well as *the profile of the Polish counterurbanite*. In the **Fifteenth Chapter**, Pedro Vaz Serra and Cláudia Seabra tell us about three UNESCO Creative Cities located in the interior of Portugal. *Due to their location, these cities* have suffered a systematic decline *in residents, progressive ageing, and increased difficulty in maintaining and retaining younger people*. Using a SWOT analysis, the authors reflect on how these cities *have incorporated the status of Creative Cities into everyday life, economic activity and cultural programming*. Finally, in the **Sixteenth Chapter**, Teresa Sá Marques and colleagues consider *the need to develop integrated territorial approaches that can support the design of public policies* about unworthy housing in Portugal. For that, they *develop a methodological approach based on multivariate analysis to identify territorial profiles based on different types of unworthy housing*.

Fifth Part (Physical and Environmental Geography) comprises two chapters. In the **Seventeenth Chapter**, Roman Maisuradze, Tamar Khardziani and Gia Chilingarashvili *examine the distribution of megalithic and rock-cut structures in Samtskhe-Javakheti (Georgia), analysing their spatial and temporal characteristics*. Thematic classifiers were developed to group landscapes based on different criteria, allowing for further exploration of this region's rich heritage. In the **Eighteenth Chapter**, Tamar Khardziani analyses *riparian forest fragmentation*, also in Georgia, *using case studies from protected and unprotected areas*. The Landscape Ecological Assessment undertaken includes *remote sensing and geospatial analysis to assess current fragmentation*, while historical fragmentation *has been investigated through old literary and cartographic sources*. The author analyses the impact on biodiversity and the *regional extinctions of particular fauna*.

Lastly, **Sixth Part** (Geoheritage and Tourism) consists of four chapters. In the **Nineteenth Chapter**, Ana Milanović Pešić, Jovana Brankov, and Željko Bjeljac warn about the fact that *tourism and other economic activities can contribute to the degradation of cultural heritage sites and pollute* the surrounding environment. The authors interviewed 52 municipalities in Serbia to address this relationship between cultural heritage and the environmental state, and how it can be managed. In the **Twentieth Chapter**, Jovana Brankov, Ana Milanović Pešić, and Željko Bjeljac discuss Serbian*traditional sports and games*, emphasising their significance as a vital part of cultural heritage that must be preserved and protected. A survey identifies the non-inclusion in the National Register of Intangible Cultural Heritage, *the lack of knowledge transfer and weak support from the local community* as significant obstacles to the affirmation traditional sports and games. In the **Twenty-First Chapter**, Laura Comănescu and Alexandru Nedelea, from Romania, discuss *legislative, educational, and structural measures* to protect *valuable elements that hold scientific, educational, cultural, aesthetic, and touristic*. The authors aim to create a national inventory of geoheritage*and propose measures for its enhanced management* and protection. Finally, in the **Twenty-Second Chapter**, Putkaradze Mera and Putkaradze Landa highlight the urgent challenges of sustainable development in Georgia's mountainous regions, particularly Ajara, which has faced *demographic and economic crises in recent years*. The authors employ various methods to assess the mountain's potential for tourism and recreational purposes, as well as how tourism influences the structure of the local economy, thus identifying tourism priorities and development pathways.

Porto, Portugal
2025

Laura Soares
Miguel Saraiva
Rafael de Miguel González

Contents

Geospatial Technologies and Artificial Intelligence

Human Geography

Physical and Environmental Geography

Geoheritage and Tourism

Conferences

Local Geography and the Irreplaceable Virtue of the Direct and Live

Miguel Sopas de Melo Bandeira

to my teacher, Conceição Coelho Ferreira

Abstract Geography is one of the oldest and most perennial disciplines of knowledge taught and learned in the global education system. It is practically omnipresent in the curriculum of the history of European education. For better or for worse, it has expressed the challenges of our contemporary societies, going beyond the strict training of younger generations or a particular type of scientific knowledge. Geography has been more about common sense than the specificity that nobles the many and most recent scientific disciplinary affirmations. What it has achieved as a "liberal art," a design that has always had difficulty imposing, has faded into epistemological identity. Of course, this is irrelevant outside a corporate framework. Knowledge advances regardless of its regimentation. Whatever meaning it has, or whatever we want to give it, geography has, over time, been a field of civic education. There is, however, a semantic and operational heritage that needs to be reconstructed and valued, not because it can be a guarantee of professional continuity and identity for geographers, but because of its potential to demonstrate a way of being, above all, in the educational process and decision support. It is necessary to return to reflecting and exercising a critical spirit, with a philosophical, social, and political scope, as a scientific-pedagogical objective, handling the corporeality of the territory rather than being dazzled by permanent innovation or by all the technological, uncritical, and subservient followings, so often only to justify disciplinary survival. We will discuss the resurgence of locally based geography, articulated in the perennial mechanism that gives it its raison d'être: scale and orientation. In valuing direct and live interaction, realising the exploratory vocation and itinerary attitude it contains. Just as sociology does not make sense without social intervention, economics without finance, history without archives, there is no geography without being on the ground. Geography teachers are the officiants of democratically recognised geography, those who give it its true raison d'être in the epistemological broth that is boiling today.

M. S. de M. Bandeira (✉)
University of Minho, Education Institute, Braga, Portugal
e-mail: bandeira@ie.uminho.pt

University of Minho, Rector's Office, Braga, Portugal

L. Soares et al. (eds.), *Compromised Geography*, Key Challenges in Geography,
https://doi.org/10.1007/978-3-032-05201-8_1

Keywords Geography · Geographic education · Professional identity · Local analysis

Geography is, More Than Anything, Geographic Education

Geography is one of the oldest and most enduring subjects taught and learned in the global education system. It is one of the few subjects that warranted a manual from UNESCO (1965) for its worldwide teaching (Fig. 1). Geography is practically omnipresent in the curriculum of European education history. For better or worse, it has always reflected the contemporary challenges faced by different generations of students. This is why its formative dimension was so intuitively affirmed by its civic vocation. Beyond being merely, a subject for the young—or seen as a science in the articulation of modern knowledge—geography, as we understand it today, has been perceived more as a domain of common sense rather than a liberal profession. This perception has empowered several contemporary scientific domains.

In the meantime, what it has managed to obtain as a "liberal art," a designation it has always struggled to impose, has faded into an epistemological identity among its peers. Of course, this is irrelevant outside a corporation professional framework. Knowledge advances regardless of its categorisation, colours and identities that

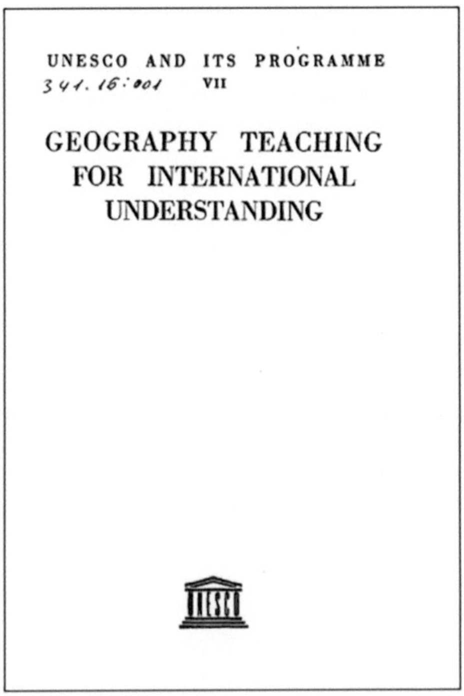
UNESCO AND ITS PROGRAMME
341.16:001 VII
GEOGRAPHY TEACHING FOR INTERNATIONAL UNDERSTANDING

Fig. 1 UNESCO programme for teaching Geography (1951). *Source* https://unesdoc.unesco.org/ark:/48223/pf0000128311

certify it. Whatever meaning it has, or whatever meaning we want to assign to it, geography has, over time, been fundamentally recognised as a field of collective education. There is, however, a semantic and operational heritage that needs to be reconstructed and valued, not because it guarantees professional continuity and identity for geographers, those "who live in the country of geography," as E. Soja (Benach et al. 2010, p. 34) put it, but because of its potential to demonstrate a specific way of being, especially in the educational and decision-support process. This potential is imperative to safeguard democratic societies, regardless of their designation.

It is high time to reflect and exercise a critical spirit with a philosophical, social, and political perspective as a scientific and pedagogical objective in itself. It involves dealing with the corporeality of the territory rather than the amazement at technological development or the permanent innovation or, even more, the devotion to technological paraphernalia that fosters an uncritical and subservient way of being to the capital. This often merely justifies a disciplinary survival and preservation of the *status quo*. Therefore, let's discuss didactics of territory, open and of interest to the public sphere, to quote J. Habermas (1990), reaffirming a locally based Geography, articulated in the perennial mechanism that gives it its *raison d'être*: scale and orientation. By valuing the interaction of the physical presence that induces a face-to-face way of being defined by live and direct, we realise the exploratory vocation and itinerant attitude it contains. Itinerant, literally, in the physical sense! Just as sociology makes no sense without social intervention, economics without resources, and history without archives, there is no geography without an on-site presence. As Orlando Ribeiro—one of Portugal's most known geographers—used to wittily tell his students, "science enters through the soles of the boots"!

Geography teachers are the legitimate professionals of democratically recognised geography, those who develop its potential and meaning, especially in today's epistemological turbulence. This is a current subject of democratic debate, because it is citizens who support public higher education.

Geography is mainly a subject of collective self-knowledge due to its potential to help us understand ourselves better, which is, after all, what matters. Our understanding of "selves" in an organised—spatialized society—An understanding that is not limited to our inner being but dedicated to the community.

Science is either open and accessible, or it makes no sense. Today, we have a consensual commitment to a sustainable environment, a just society, and a circular economy. In other words, the Sustainable Development Goals (SDGs) are commitments to our global citizenship. This triad of circumstances calls us all together, regardless of our way of being, even more so as it presents an active and unavoidable perspective. In Western democracies, we citizens should have the right to access science, but this is only possible in a society where people are well educated and adequately qualified.

Geography belongs to society more than any other area of knowledge, geography must affirm itself not according to a kind of science separated from society, but in conjunction with society.

Fig. 2 Meet our experienced typical hispanic geography teacher in middle school[1]—generate stunning images and art with AI (24JAN2024). Licence: Free to use with a backlink to Easy-Peasy.AI

The Geography of Our Discontent

We must ask ourselves why the image of geography among the general public, or the media, continues to be mainly associated with anthroponymy, even though we know that this has not been the case for a long time. But it is worth asking why this image is so strong and lasting?

What makes these ideas subsist? Have geographers really changed over the years the way they present the image and the main concepts of geography to the public? Do we feel uncomfortable that geography is still associated with the state and the colonial period? (Fig. 2).

[1] *A geography teacher of Hispanic descent, aged around 50, walking through the halls of a middle school. She's not smiling and there are students around her. She's bearing an air of confidence, is not thin and has an elongated physique. She's dressed in jeans with a black blazer and white sneakers. Her short, straight hair, cut in a neat bob with a straight fringe that hides her forehead, is grey. Her complexion is olive with subtle wrinkles. She has large, golden green eyes. She's holding a laptop in her hand.*

Despite being fundamental pillars of social and epistemological affirmation in the past, *scale*, *position* and *direction* are geographical archetypes that have been almost eradicated from geography. Its instrumental refinement has now migrated to other specific fields of knowledge. Who knows if, sooner rather than later, the basic concepts of geography will return to the agenda? Due to imminent conflicts? It is certainly not for a good reason.

But who today can deny that such an unconditional imperative might result in the urgency of having to recover basic elements abandoned in the citizen formation? Like having to develop the necessary resources to understand the terrain, stimulate how to choose where to go, the times of departure and arrival, the recognition of the weather conditions, and knowing how to read the human, physical, and temporal surroundings, which today we call the environment (Fig. 3). For example, having to draw up an itinerary, make a sketch, progress through the terrain, take shelter, conserve energy—in other words, value vital perceptions and sensitivities towards the territory. Perhaps even having to relearn basic survival skills. Undoubtedly, these are all skills for which geography is an essential field …

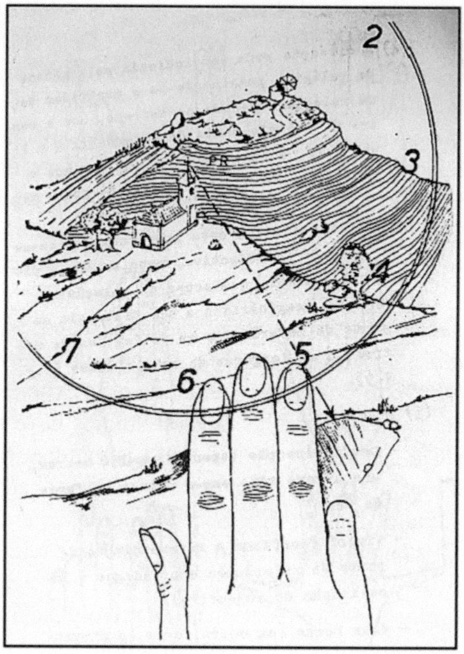

Fig. 3 Do we still know-how to read and sketch *the human, physical, and temporal surroundings*? *Source* MTIC-EP (1983)

… and, However, Geographic Education

There is a whole set of ideas that need to be overcome. Geography has lost peer prestige with the globalisation of mass media and digital multimedia technologies. It has been reconfigured according to the immediate availability of information rather than the production of knowledge, leading to a dispersive detachment that has forgotten the wisdom and respectability of an *old lady*. While the world has become smaller and lifestyles have become standardised into an indistinct uniformity, geography has lost the clarity of its identity. We should accept that Geography knowledges earth's space, and territory intervention are not exclusive to geography. These competencies remain inalienable to geography education. The larger the audience of geography, the more it justifies the recovery of old branches of knowledge, such as political geography, cultural geography, and landscape studies. From functions that a few years ago were considered moribund, today, the teaching of geography emerges, convening lived local experiences. A trail that, more than any other wonder, guarantees its unity and future through education.

Aldo Tobler (1970), credited with formulating the audacious "first law of Geography", posited that "everything is related to everything else, but near things are more related than distant things." Despite the current trend towards *glocalism*, this concept highlights the preponderance of proximity in fostering intense interactions that drive initiative and decision-making. In the recent years, the local scale has gained increasing importance in education, especially in geography teaching. A curriculum lexicon now integrates with the democratisation of education, beginning with the introduction of social sciences and later environmental studies, emphasising an integrative approach to society and the environment. Citizenship education plays a crucial role in stimulating participation through active pedagogies.

The *window of our school*, once a staple teaching resource, has lost its prominence in today's educational landscape. Instead, there is a growing preference for *blackbox* classrooms (Fig. 4), interactive environments, and immersive digitalisation, which offer seductive and engaging experiences but also pose risks to the educator-student relationship. Several studies have highlighted the potential pitfalls of this shift, suggesting that it could jeopardise the essence of schooling as we know it. Therefore, our thesis naturally leads us to advocate for the reclamation of the inherent value of face-to-face and perceptual interactions between educators and students. This entails prioritising direct and live relationships in the educational process.

Didactics, again. In line with the teachings of Brazilian pedagogue Paulo Freire (1996), education is not merely the transmission of knowledge but the creation of conditions for its reconstruction, fostering social awareness through the recognition of cultural identity. Students build their knowledge based on their experiences and the concepts they already have, adds our colleague H. Cachinho (2019). Therefore, it is necessary to reinforce confidence in what is known, and recognise the support for future knowledge. This forms the foundation for a didactic approach to geography grounded in proximity.

Fig. 4 Cyberpunk classroom in the dark concept art—image generated by AI, Freepik.[2] Acess 4.05.2025

Given the transformative impact of artificial intelligence, which is destroying the spontaneous process of knowledge acquisition and personal development, geography education emerges as a vital field for inclusion among vast sectors threatened with exclusion and irrelevance. In this context, we draw upon the theory of multiple intelligences to broaden educational horizons, leveraging face-to-face interactions and local contexts. By prioritising the corporeal relationship between the senses and the environment, we emphasise the value of kinaesthetic and emotional intelligence. Through the cultivation of balance, orientation, perceptions, and sensibilities, geography education can enhance acuity across all age groups, fostering a deeper understanding of the world from the local to the global scale.

We believe that rescuing geography's identity transcends mere preservation of cultural heritage, which risks becoming a metaphor. Instead, geography remains an indispensable branch of common knowledge, constituting the heritage of humanity itself. Its educational significance surpasses all others, as geography inherently embodies a predisposition towards exploration and an itinerary attitude, mirroring the rhythm of human existence.

Geography's value lies not only in the individual contributions of its educators but also in its capacity for mediation, the cognitive bridges it constructs, and the connections it fosters with the general public. It serves as a guiding beacon of knowledge, honing sensitivity to detect significant deviations from prevailing trends and offering the tools to navigate corrective trajectories. In today's interconnected and hybrid world, geography stands as a synthesis or a crossroads, facilitating transversal understanding and adaptation.

---

[2] https://fr.freepik.com/images-ia-premium/salle-classe-cyberpunk-dans-art-conceptuel-sombre_31764543.htm.

However, unlike half a century ago, when the prophetic revelation of the *new geography* exalted the liberal arts along with various radicalities, this supposed contradiction quickly became irrelevant because, ultimately, geography did not detach itself from its disciplinary matrix, that is, is, politics, war and diplomacy.

Let us never forget the geographic essence found in classical works that constitute the fundamental structure of different cultures. These texts possess a timeless philosophical dimension where certain historical currents bear geographical significance, offering didactic and comprehensive value for exploring fresh interpretations of history and culture. However, it is intended to be a process of revisiting the past, integrated into its historical context, detached from the attribution of penalties and the search for reparations that only serve only to foment resentment.

The didactic value of geography remains pertinent today, particularly through its inherent travel-oriented dimension. From leisurely strolls through the neighbourhood to embarking on long-distance journeys, geography fosters exploration. Yet, we lament the loss of storytellers, leisurely observers, and cultivators of *ekphrasis*—masters of the art of words whose captivating narratives once brought audiences together (Fig. 5). Unfortunately, the prevailing educational context often remains disciplinary and restrictive, leaving students feeling confined within the classroom. Increasingly, their attention is drawn solely to the allure of mobile phones, offering an escape from the confines of four walls.

Having exhausted the advancements of the recent past, wherein we brought the world into the classroom through technological mediums, primarily reliant on imagery, the present challenge lies in venturing outside the confines of the classroom for extended periods, fostering unforgettable experiences that embody practical know-how and engage participants' abilities and senses. It is necessary to recover the spirit of the *Grand Tour*, now from a perspective of democratic access.

We currently inhabit a sedentary society, despite our increasing movement through physical space. Our bodies have become "cockpitized", as if confined in an airplane (Fig. 6). Conditioned to passive and uniform behaviour, we navigate through life in

Fig. 5 *Geography fosters exploration… Source webpage* Fundação de Serralves. Acess: 24J.01.2024

Fig. 6 *Our bodies have become "cockpitized", as if confined in an airplane.* Cockpit—graphic essay—Mai2025

orderly fashion, guided by a standard crew monitoring computer system with a myriad of instruments at their disposal. Yet, amidst this routine, we still harbour the promise of discovering new and unfamiliar destinations, perpetually seeking novelty. We are becoming obese, struck with the illusion of movement, disconnected from our natural biorhythms, blurring the distinction between day and night and diminishing our awareness of the changing seasons. We find ourselves living increasingly segmented lives, confined to narrow mental channels. We frequent the same places, embark on long-distance travels to destinations that increasingly resemble one another, merely to confirm what we have seen before, "collecting countries" to the rhythm of selfies.

To Continue ...

Once again, the realm of geographic education presents boundless possibilities. Hence, it is crucial to foster connections between places, serving as advocates for in-person exploration. There is an indelible value in direct, first-hand experiences (Fig. 7).

In a way, it is a yearning to return home—to the dwelling place where everything originates and everything converges—a sentiment echoed in various fields such as economics, ecology, ecumene, and ekistics (the study of human settlements). To quote E. Soja (Benach et al. 2010, p. 243–244), synecism (synoikismós), highlighting the unique conditions of interdependence, creativity and efficiency that arise from community life. This inclination serves as a response to the standardised society marked by ISO quality standards, individual exhaustion, the infinity of acronyms

Fig. 7 "The eye of geography"—CEG. IGOT-UL. *(…) it's crucial to foster connections between places, serving as advocates for in-person exploration*

and syncretic language removed from life. Every day, the litigious nature of society asserts itself through exhaustion, similar to dictatorships, instilling fear, as in the past. And now, also, through the arbitrariness of cancellation, threats and exclusion.

The challenges of organising field trips are manifold, from navigating terms of responsibility, obtaining authorisations and insurance, to ensuring adequate supervision amidst the proliferation of fake news and the decline of the nuclear family structure. It is baffling to witness the persistence of the "uniform and ready-to-wear curriculum" in education, as J. Formosinho said (1987).

At its core, there remains a space for the indispensable, unwavering researcher, and the dedicated educator who is deeply ingrained in their community—present, sensitive, and creative. They navigate the intricate web of scales with a dynamic ontological purpose, as eloquently articulated by Milton Santos (2000), aiming to bring "totality to the place". The place where everything really happens. It is the place where we reside and exert our transformative influence—reminiscent of Voltaire's notion of cultivating our garden. Live and direct!

References

Benach N, Albet A, Soja EW (2010) La perspectiva postmoderna de un geógrafo radical. Icaria, Espacios críticos, Barcelona

Cachinho H (2019) Desafios da formação em geografia e na educação geográfica, Conhecimento poderoso e conceitos liminares. Rev. Elet. Educação Geográfica em Foco. Ano 3, Nº6, especial 2º ELG, outubro/2019

Formosinho J (1987) O currículo uniforme pronto a vestir de tamanho único. Cadernos de Análise Social da Educação: O Insucesso Escolar em Questão

Freire P (1996) Pedagogia da autonomia: saberes necessários à prática educativa. (Coleção Leitura), São Paulo: Paz e Terra

Habermas J (1990) O Discurso Filosófico da Modernidade. Publicações Dom Quixote, Lisboa

Santos M (2000) Por uma outra globalização: do pensamento único à consciência universal. Record; São Paulo

Tobler WR (2004) Sobre a Primeira Lei da Geografia: Uma Resposta. Anais da Associação Americana de Geógrafos 94(2):304–310

UNESCO (1951). Geographical teaching for international understanding. https://unesdoc.unesco.org/ark:/48223/pf0000128311. Acess 4.05.2025

UNESCO (1965). Manual da Unesco Para o Ensino da Geografia. Editorial Estampa. Lisboa 1978

How Generative AI is Changing Our World: Reflections from a Geographical Perspective

Nico Van de Weghe

Abstract Generative artificial intelligence (GenAI) is rapidly transforming how we work, learn, and understand the world. This chapter examines the rise of GenAI through a geographical lens. We begin by explaining what GenAI is and why it has proliferated so quickly. We then explore its societal impacts, from changes in professional practice to concerns about bias, misinformation, and environmental costs. A significant focus is placed on education: how GenAI is influencing geographical learning, the urgency of GenAI literacy, and the need for critical spatial thinking. Along the way, we introduce the concept of vibe writing to describe a new way of writing in collaboration with GenAI, where the writer steers the process rather than crafting every sentence. We also discuss practical applications of GenAI in geography and spatial education, alongside the ethical imperatives of human oversight ("human-in-the-loop"). Throughout, we highlight the balance between opportunity and responsibility—illustrating how geographers and educators can harness GenAI for open, inclusive, and critical engagement with the world. The chapter concludes with reflections on the future of AI in geography, emphasizing that while GenAI is here to stay, human judgment and values remain paramount in shaping its role in society.

Keywords Generative artificial intelligence (GenAI) · Geography education · GenAI literacy · Ethical AI · Spatial thinking

Introduction

In the recent years, generative artificial intelligence (GenAI) has emerged from research labs into mainstream society with astonishing speed. Tools like OpenAI's ChatGPT have demonstrated AI's capability to produce human-like text, images, code, audio, and even video content. GenAI refers to algorithms and models that

N. Van de Weghe (✉)
Ghent University, Department of Geography, GeoAI Research Center, Ghent, Belgium
e-mail: nico.vandeweghe@ugent.be

L. Soares et al. (eds.), *Compromised Geography*, Key Challenges in Geography,
https://doi.org/10.1007/978-3-032-05201-8_2

can create new content—e.g., writing an essay, composing music, or generating a map—often based on patterns learned from vast datasets.

The rise of GenAI has been remarkably fast. By early 2023, ChatGPT, a conversational large language model (LLM), reached an estimated 100 million users within two months of launch, becoming the fastest-growing consumer application in history (Hu 2023). For comparison, it took popular apps like Instagram and TikTok many months or years to achieve similar user bases. This explosive adoption underlines the public's keen interest and curiosity—people are experimenting with AI chatbots to draft emails, solve coding problems, translate text, plan lessons, and much more. Such widespread use of GenAI across sectors suggests we are at the cusp of a broader digital transformation, one that demands our critical attention as geographers and educators.

From a geographical perspective, the advent of GenAI raises important questions. Geography as a discipline is fundamentally concerned with space, place, and human–environment interactions, as well as the tools (like maps and spatial data) we use to understand our world. How will GenAI influence the way we interpret spatial information or engage with global issues? Can it enhance geographical education and promote global citizenship, or will it inadvertently introduce new biases and inequalities?

This chapter examines how GenAI is changing our world, with a special focus on implications for geography and education. We first provide a grounding in what GenAI is and how it works, in accessible terms. Next, we outline GenAI's societal impacts—both positive transformations and critical concerns. We then look into into GenAI's role in education, particularly geographic education and spatial learning. Issues of GenAI literacy, academic integrity, and the evolving role of educators and students are discussed. In a subsequent section, we consider GenAI's intersection with geographical thinking and geospatial technologies: how might it aid in spatial analysis or map-based reasoning, and what are the limitations? Throughout these sections, we interweave ethical considerations such as bias, misinformation (including so-called AI hallucinations), environmental sustainability, and global equity in AI access. A recurring theme is the "human-in-the-loop" principle—the idea that human oversight and critical thinking must remain central even as we leverage AI tools (Shneiderman 2020). This principle resonates strongly with the need for responsible engagement in the face of digital transformation, a core value for the geographical community.

This chapter aims to provide a comprehensive overview for geographers, educators, and concerned citizens. GenAI is often described in polarized terms—some hail it as revolutionary, others worry it heralds a kind of digital doom. In geography too, one might ask: Is GenAI a tool for empowerment and innovation, or does it compromise the integrity of geographical knowledge? We argue that such binary thinking is less productive than a critical, nuanced engagement. Generative AI undoubtedly offers powerful capabilities that can accelerate learning and research, but it also challenges us to rethink notions of creativity, authorship, and authenticity in the context of place and space. The task ahead is to integrate GenAI into our practices in thoughtful ways—leveraging its strengths (e.g., efficiency, accessibility) while mitigating its weaknesses (e.g., bias, energy use)—such that it supports the larger goals of geographical education and global understanding.

What is Generative Artificial Intelligence (GenAI)?

The Mechanics of GenAI

GenAI encompasses a variety of models and techniques, but a common thread is the use of machine learning, particularly deep learning, to generate new data that mimics the patterns of training data (Brown et al. 2020). A prominent example is the large language model (LLM)—an AI trained on massive text corpora to predict and produce human-like language (Vaswani et al. 2017). Models like GPT (by OpenAI), Gemini (by Google), Claude (by Anthropic), and LLaMA (by Meta) are LLMs with hundreds of billions of parameters (internal settings adjusted during training). These parameters encode probabilistic relationships between words and phrases learned from reading the Internet, books, and other text sources. When prompted with a question or statement, an LLM like ChatGPT generates a continuation by selecting likely next words one after another, a process akin to autocomplete but at an advanced level (Yang et al. 2023). For instance, if the model has seen the phrase "They are eating pizza" frequently during training, it might predict "pizza" as a likely next word after "They are eating …" with a high probability. The model doesn't "know" in a human sense, but it leverages statistical patterns—it has essentially learned that "eating pizza" is a common collocation in English. By stringing together many such predictions with contextual awareness, the model can produce coherent sentences and paragraphs on a given topic.

Modern generative models are not limited to text. Generative adversarial networks (GANs) and diffusion models can create realistic images (e.g., generating a photorealistic picture of a city skyline that never existed) or even videos (Rombach et al. 2022). Other GenAI systems compose music or synthesize speech in anyone's voice. Crucially, these systems are generative in that they produce content that did not exist before (unlike a search engine that only retrieves existing documents). In the geography domain, we see emerging generative applications such as AI-produced maps or simulated environmental data (Janowicz et al. 2020). For example, researchers have shown that satellite imagery can be "deepfaked"—using GANs to generate fake but visually convincing maps of landscapes (Zhao et al. 2021). This demonstrates both the power and the potential peril of generative tech in geography: we can create hypothetical scenarios or fill gaps in data, but we could also be fooled by fabricated geospatial information if proper checks are not in place.

Several factors have converged to propel GenAI's rapid development. One is computational power—specialized hardware like graphics processing units (GPUs) and tensor processing units (TPUs) allow researchers to train larger and more complex models than ever before. Another factor is the availability of massive datasets: the internet itself has become a training ground, providing text (webpages, books, social media) and images in the billions for AI to learn from. Additionally, improved algorithms and architectures (transformer models, in particular) have dramatically boosted performance of AI in generating coherent language and high-quality images (Raffel et al. 2020). Finally, significant investment and research interest in AI has created a competitive environment—companies and institutions are pouring resources into creating ever more capable generative models, leading to a fast-paced "AI race" in the tech sector.

The Allure and Utility of GenAI

Why has GenAI captured public imagination and adoption so quickly? One reason is its apparent versatility. A single well-trained model can perform a wide array of tasks: answer questions, draft essays, translate languages, write computer code, even engage in dialog that feels surprisingly human (Yang et al. 2023). This opens up countless applications. In everyday life, people use ChatGPT or similar tools to summarize documents, brainstorm ideas, or get help with writing. For example, a student might ask for a summary of a lengthy article to save time, or a professional might have an AI draft an email or report which they then refine. The ability to generate content on demand acts like a productivity booster.

In fields like computer programming, AI assistants can generate code snippets or help with debugging—effectively engaging in metaprogramming, where developers guide the AI rather than writing everything from scratch. This shifts the human role from direct content creator to a higher-level editor or decision-maker. Recently, this has also been referred to as vibe coding—a term introduced informally by Andrej Karpathy. It describes a new way of programming in which developers no longer write code line by line but instead describe in plain language what they want. An AI model then translates this description into functional code.

A similar shift can be observed in writing since the introduction of ChatGPT and similar tools. Instead of carefully crafting every sentence, users describe what they want to express—the message, the structure—and let generative AI assist in turning it into clear and coherent text. Until now, I have referred to this approach as meta writing. However, inspired by the notion of vibe coding, I propose to call it vibe writing. To the best of my knowledge, this term has not yet been formally introduced in the literature. The idea is straightforward: the writer shifts from composing every word to steering the writing process and reviewing the result. The focus moves to the underlying ideas, while the AI supports the formulation.

Another attraction of GenAI is its ability to stimulate creativity. For those who face writer's block or need inspiration, generative models can provide immediate suggestions and draft ideas. For example, when asked to help design a research experiment, a GenAI tool can propose possible research questions, variables to consider, and even suggest a basic methodology. This illustrates how GenAI can function as a brainstorming partner, helping users to structure their ideas and get started more efficiently. Rather than replacing human thinking, it can accelerate the initial stages of a creative or academic process, leaving more time for critical reflection and deeper analysis (Kasneci et al. 2023).

Moreover, GenAI tools are increasingly user-friendly, often accessible through simple chat interfaces. This lowers the barrier to entry for non-experts. One does not need to know how the AI works internally to benefit from it. This democratization of AI has profound implications: advanced computational power is now at the fingertips of anyone with an internet connection (Roy et al. 2024). In educational contexts, for instance, even a secondary school geography teacher with minimal AI background can experiment with ChatGPT to generate a quiz or a lesson outline on climate change, just by typing a request in plain language (Kasneci et al. 2023).

However, the ease and speed of GenAI also come with pitfalls that contribute to a sense of unease. The idea that an AI can draft a competent essay in seconds raises concerns about plagiarism, learning integrity, and the very nature of creative work. There is an uncanny quality to some AI outputs—text that reads correctly but feels soulless, or images that look real but aren't.

In summary, GenAI represents a major milestone in computing, characterized by its content-creating ability, broad utility, and rapid uptake. Its emergence sets the stage for significant societal changes. In the next section, we examine GenAI's societal impacts, ranging from the way we work to broad ethical challenges. These issues form the backdrop against which the geography and education communities must position themselves when engaging with AI.

Societal Impacts of GenAI

The adoption of GenAI is already influencing various sectors of society. In this section, we explore some key impacts: changes in workplaces and professional practice, the proliferation of AI-generated content (and associated concerns of truth and authenticity), and broader ethical and environmental issues. We also discuss how these impacts can vary geographically, potentially exacerbating global inequalities.

Transforming Work and Creative Practice

One of the clearest societal effects of GenAI is its impact on work and productivity. In many knowledge-based professions, tasks that once took hours can now be completed in minutes with AI assistance. For example, content writers can use tools like ChatGPT to draft articles or marketing copy, software developers employ GitHub's Copilot (powered by OpenAI's Codex LLM) to generate code functions, and graphic designers experiment with image generators to prototype designs (Brown et al. 2020). Rather than making human jobs obsolete overnight, what we see is a shift in the nature of work—a process sometimes described as automation and augmentation. Routine or boilerplate aspects of tasks (writing boilerplate code, composing template-like reports) can be offloaded to AI, freeing humans to focus on higher-level planning, quality control, and creative refinement. In programming, this has been termed metaprogramming, where "AI generates code and programmers guide the process" (Shneiderman 2022). Instead of manually writing every line, a programmer might instruct the AI, review its output, and iteratively improve it—effectively collaborating with the AI.

In geography, we can imagine a similar augmentation. A researcher might ask a GenAI to summarize trends from a dataset or to produce a literature review draft on a topic like urban migration. The AI could output a structured overview within seconds or minutes, albeit with possible errors or gaps (Janowicz et al. 2020). The human expert then assumes the role of editor and fact-checker, verifying the AI's points against sources and adding the nuanced insight that the AI lacks. This mode of working requires a new skill set: not just knowing one's domain, but also having GenAI literacy—understanding what the GenAI can and cannot do reliably, and how to prompt or steer it effectively (Kasneci et al. 2023).

The creative industries are also being transformed. Architects might use generative design algorithms to produce innovative building concepts. Musicians remix AI-generated melodies. Importantly, GenAI can serve as a great equalizer in creativity by lowering skill barriers (Roy et al. 2024). A person with no training in drawing can prompt an image model to create a concept illustration; someone who doesn't code can still build a simple app by instructing an AI assistant. This democratization of creation tools is empowering, but it also stirs debate about the value of human craftsmanship. If an AI can produce a passable painting or a decent essay in seconds, what motivates humans to spend years mastering these skills? Some fear a decline in human creativity or a devaluation of expert labor (Floridi and Cowls 2019). However, historical perspective reminds us that new tools often provoke such questions—photography was once thought to threaten painting, calculators to threaten arithmetic skills, etc. In practice, humans often recalibrate and find new forms of expression; for instance, photography became its own art form instead of rendering painting obsolete.

That said, certain jobs will undoubtedly be disrupted. Clerical and administrative tasks are already being automated with AI (e.g., AI chatbots handling customer service inquiries). Some forecasts suggest roles in journalism, translation, and even law (drafting contracts, for example) could be significantly affected by GenAI. In the context of geography, consider fields like cartography and GIS: AI might soon help in automating map generation or in the writing of geographic reports ("write a summary of the demographics of region X") (Janowicz et al. 2020). These changes raise critical questions about job training and re-skilling. Geography graduates, e.g., may need to be trained not only in GIS or statistics, but also in how to work effectively with AI tools to analyze spatial data.

From a geographical perspective on work, another consideration is how different regions of the world will be affected unevenly. Highly developed economies with tech industries are driving GenAI innovation and will likely reap productivity benefits early, while lower-income regions might lag in access to these technologies. Moreover, within countries, the digital divide means not everyone can equally leverage GenAI. This can worsen inequalities if not addressed.

Truth, Misinformation, and the Challenge of AI-Generated Content

As GenAI floods society with content, distinguishing fact from fiction is increasingly challenging. AI can produce text that reads confidently on any topic—but it does not guarantee truth or accuracy (Floridi and Cowls 2019). One well-documented problem is the generation of fabricated information or references, often termed "hallucinations" in AI research (Zhao et al. 2021). For instance, ChatGPT might produce a scholarly sounding paragraph with citations that look real but are entirely made-up (e.g., referencing a non-existent article). Early versions of these models frequently produced false bibliographies, which alarmed academics. While newer systems and tools (like the browser-based AI Perplexity or plugins connected to real databases) are improving the reliability of references, the risk of misinformation remains real.

This is akin to an information pollution problem. If students, journalists, or policymakers start using AI-generated content without verification, errors can propagate rapidly. Already, there have been cases of AI-generated news articles containing inaccuracies. Deepfake images and videos—another form of generative output—can spread disinformation; for example, fictitious satellite photos could be used propagandistically to mislead viewers about events on the ground. The geographic community is concerned with geospatial truth—for instance, authentic mapping of environmental phenomena. The advent of "deepfake geography" (Zhao et al. 2021) suggests we'll need robust methods to authenticate data and imagery. Encouragingly, parallel research is looking at deepfake detection (including for satellite imagery), but it is an arms race between generation and detection techniques.

In education and research, the issue of academic integrity arises. If a student uses GenAI to write an essay, is it plagiarism? If a researcher has AI draft parts of a paper, who is the author? Different institutions are grappling with these questions. Leading academic journals have responded by updating policies: some prohibit listing AI as a co-author (since it cannot bear responsibility), and many require that any use of AI in writing or analysis be disclosed. There is a growing consensus that transparency is vital—readers or evaluators should know if and how AI contributed.

But beyond formal policies, there's the practical question: can AI-generated work be detected reliably? Some teachers have attempted to use AI detection software on student assignments, with claims that they can tell if an essay is "too good" (or not in the student's voice). However, detection tools have proven very unreliable, often flagging false positives. One educator likened AI plagiarism detection to a cat-and-mouse game akin to doping tests in sports—as soon as a detection method is known, people find ways to evade it. In fact, one AI company's detector was discontinued in

2023 due to poor accuracy and bias. We claim that "recognizing an AI text because it's too fluent" is misguided. Good writing is not evidence of cheating per se—one of GenAI's effects is that it elevates the baseline writing fluency, making it harder to distinguish average student work from polished AI-assisted work on style alone.

The more sustainable approach, as many suggest, is to adapt our teaching and evaluation methods rather than rely on detection. This might include more oral examinations, presentations, and interactive assessments where students must explain or defend what is written (hence exposing if they don't truly understand content that an AI might have generated). In professional settings, it means instituting verification steps—for instance, a journalist using AI to draft an article must thoroughly fact-check every claim before publishing. Essentially, critical thinking and source verification become even more crucial in the age of GenAI. Rather than banning the technology (which is arguably neither feasible nor beneficial in the long term), society will need to cultivate skills to manage and audit AI outputs.

Ethical Considerations: Bias, Fairness, and Environmental Cost

GenAI systems inherit the biases and blind spots of their training data. If predominantly English and Western data are used, the model's outputs will reflect those cultural perspectives, potentially marginalizing other viewpoints (Bender et al. 2021). Efforts are underway to train models on a more diverse linguistic base—including projects focusing on African, Asian, and indigenous languages—to preserve cultural uniqueness and prevent a one-size-fits-all AI that overlooks local context (Joshi et al. 2020). For the geography community, which values place-based knowledge and diversity, this is critical. Geographical phenomena (like place names, cultural landscape descriptions) must be handled accurately by AI, which means incorporating local data and languages.

Bias isn't only linguistic. These models can also reflect social biases (gender, racial, etc.) present in their training text. For instance, an AI might generate a scenario where a "scientist" is assumed male or a "nurse" is assumed female, if those stereotypes were common in its training data (Bolukbasi et al. 2016). Or it might produce content with a Western-centric narrative when asked about world history (Sheng et al. 2021). Such biases in AI outputs can perpetuate stereotypes and misinformation. There is also the concern of political bias or censorship: models like OpenAI's ChatGPT have certain content filters (to avoid hate speech, etc.), and different jurisdictions are deploying AI under different rules. A well-known example is China's "DeepSeek" model, that avoids or gives censored answers about sensitive topics like Tiananmen Square, aligning with state guidelines (Roberts et al. 2020). This shows that GenAI can be shaped by power structures—raising questions about who controls the narrative. The flip side is that societies may demand such control to avoid harmful outputs, but it is a fine line between moderation and overreach.

Fairness in AI also touches on who benefits and who might be harmed. If GenAI automates certain jobs, are we equipped to support workers who are displaced? (Frey and Osborne 2017) Will the economic gains from AI (increased productivity) be shared broadly or concentrated among a few tech companies? These are open questions that policymakers and scholars, including geographers interested in economic geography and development, need to examine. The concept of a universal basic income (UBI) has been floated by tech leaders as a remedy if AI takes over large swaths of work—freeing people from jobs but providing income to live. This remains speculative, but it underscores that AI's societal impact could be radical and may require equally radical policy responses to ensure social stability and equity.

Another ethical dimension is the environmental impact of GenAI. Training large models is resource-intensive. A recent analysis estimated that training GPT-3 (175 billion parameters) consumed about 1287 MWh of electricity and led to approximately 500 metric tons of CO_2 emissions (Patterson et al. 2021). This is roughly equivalent to the annual CO_2 emissions of around 100 gasoline cars (Strubell et al. 2019). And GPT-3 is only one model; countless models of similar or even larger size are being trained and deployed. Data centers running AI services already contribute significantly to the IT sector's carbon footprint, which is estimated at 2–4% of global emissions (Dhar 2020). Beyond the training phase, the widespread use of GenAI also brings ongoing environmental costs. Each time a user interacts with a GenAI system—for instance, by asking a question in a chatbot, requesting a text summary, or generating an image—the model has to process the input and produce an output. This process, known as inference, also requires substantial computing power. Especially the newer reasoning models, which can perform multi-step problem-solving or complex language tasks, rely on longer and more energy-intensive inference chains. Some researchers warn that as GenAI becomes deeply embedded in everyday applications, the cumulative energy demand for inference could rise dramatically. One study estimated that a single AI-enhanced search query may require up to five times more computing resources than a traditional query (Schwartz et al. 2020). With billions of queries processed daily, the energy footprint of GenAI's inference phase could soon surpass that of its already costly training phase, raising concerns about the long-term sustainability of large-scale AI deployment.

These concerns are particularly relevant in the context of climate change—a global challenge that geographers and earth scientists are deeply engaged in. The good news is that there is growing awareness and effort to develop greener GenAI. Approaches include using renewable energy for data centers, designing more energy-efficient algorithms, and even reusing waste heat from AI computations (Dodge et al. 2022). Some tech companies have committed to carbon neutrality for their cloud services by the mid-2020s. Research competitions are pushing for training AI on smaller datasets. These initiatives align with the concept of sustainable digitalization—ensuring that digital transformation (AI included) does not compromise environmental sustainability.

In summary, the societal impacts of GenAI are multifaceted: it changes how we work and create, challenges our ability to discern truth, and raises new ethical questions about bias, fairness, and sustainability. From a geographical standpoint, these issues manifest unevenly across space and communities. There is both opportunity and risk. The next sections turn to one of the domains most affected by GenAI—education—and then to the specific interplay between GenAI and geographical knowledge. In those contexts, we will also consider how society (through educators, institutions, and policies) can respond to the challenges outlined here, e.g., by fostering GenAI literacy and establishing frameworks for responsible use.

GenAI in Education: Opportunities and Challenges

Education has quickly become a frontline for GenAI's impacts. The introduction of accessible AI tools in late 2022 and 2023—epitomized by ChatGPT—sent shockwaves through schools and universities. Educators were confronted with both exciting possibilities and new uncertainties (Chan 2023). This section examines how GenAI is being applied in educational settings (with an emphasis on geography education), the benefits it can bring, and the challenges it poses for teaching, learning, and assessment. A key concept in this discussion is GenAI literacy—the ability of learners and educators to use GenAI tools responsibly, effectively, and critically. It is important to distinguish this from the broader concept of AI literacy. AI literacy refers to understanding how artificial intelligence systems work, including knowledge of concepts such as neural networks, deep learning, and algorithmic decision-making. In our opinion, GenAI literacy focuses on knowing how to interact with GenAI systems in practice. Becoming GenAI literate does not require advanced technical knowledge of how AI models function internally. Teachers and students do not need to be AI experts to use GenAI effectively in learning and teaching contexts. However, the reverse is also true: there are many people who are technically AI literate but who lack the practical skills and critical awareness needed to engage with GenAI tools wisely.

AI as a Tool for Learning and Teaching

GenAI can serve as a powerful educational assistant for both teachers and students. Numerous use cases have emerged.

Content creation and lesson planning: Teachers can save time by having AI generate draft lesson plans, lecture outlines, or teaching materials. For example, a geography teacher could prompt an AI: "Write an introductory lesson plan about GIS for a 50-min class." The AI might produce a structured outline with key concepts (definitions of GIS, examples of applications), suggested activities or case studies, and even discussion questions. While the generated plan might not be perfect or tailored to a specific class's needs, it provides a solid starting point (Kasneci et al. 2023). The teacher can then adjust the content, incorporate local examples, or ensure alignment with curriculum standards. This use of AI automates the first draft of planning, leaving the creative and reflective refinements to the human expert. Future AI systems might improve in personalizing such content to a teacher's style and the students' level, especially if they can learn from feedback over time.

Summarizing and explaining: Students often need to digest large amounts of information, such as textbook chapters or research articles. GenAI is effective at summarization; for instance, a student could ask, "Summarize Chapter 3 of the world geography textbook focusing on urbanization" and receive a concise overview. Whether this is beneficial, however, depends on the learning objectives. If the goal is to quickly review or get an overview, automated summarization can be helpful. But if students are expected to engage deeply with the full text, summarization may be counterproductive. The process of reading, selecting key points, and structuring a summary is itself an essential part of learning. By outsourcing this step, students risk missing a crucial part of the learning process. Another scenario is explaining difficult concepts. AI can provide explanations in simpler terms or even use analogies. For instance, if a student doesn't understand plate tectonics, they might prompt the AI to explain it "like I'm 12 years old," and receive a more accessible description. Tools like the Khan Academy's Khanmigo (an AI tutor built on GPT-4) are leveraging this capability—it can respond to freeform student questions and guide them through problems step by step (Escobar et al. 2024). Khanmigo, in particular, is designed not just to give answers, but to ask the learner questions back, simulating the Socratic method to prompt deeper thinking (). For example, in an interactive session about inflation (an economics concept also relevant to human geography), the AI tutor might explain basic terms then ask the student follow-up questions ("Why do you think prices rise?") to ensure understanding. Early trials of such AI tutors are promising, suggesting they can provide personalized learning experiences at scale.

Writing improvement and feedback: As discussed earlier, AI can act as a writing coach. In educational contexts, this means students can use AI to refine their essays and reports. For example, a student writing a geography term paper can ask the AI to check grammar, improve clarity, or even suggest how to structure an argument. As an illustration, one could improve a rough abstract about a data science project: the original text is rambling and unclear, but after asking AI to rephrase and correct errors, the output can be far more professional and coherent. Using AI in this manner can teach students about the revision process. They might compare their draft and the AI's suggestions to learn how to write more clearly. Importantly, instructors can encourage this as long as students remain in control of the content—the goal is not to have AI write the essay, but to use it like a spellchecker or style advisor. Some universities have indeed updated their academic integrity guidelines to state that using AI for proofreading or copy-editing is acceptable (similar to using Grammarly or Google Translate), whereas using it to generate substantive content without attribution is not (Perkins et al. 2023).

Brainstorming and research guidance: Students and researchers can consult GenAI as a sounding board for ideas. If one is designing a research project, AI can generate potential hypotheses or methods to consider. It might list relevant variables or previous studies (one must verify the references, of course). This can broaden a researcher's perspective, perhaps bringing up interdisciplinary angles they hadn't thought of (Liang et al. 2023). Some have described AI as an "idea generator" that can provoke new questions. For instance, an AI might suggest exploring a certain dataset or coupling two theories in geography that sparks a novel approach. The human then evaluates those ideas, keeping some and discarding others. In teaching, educators can use AI to generate multiple examples or case studies to discuss in class, enriching the material.

Assessment and evaluation support: Grading and creating assessments can be time-consuming. AI is beginning to assist here as well. One can prompt: "Create a rubric to assess a group project in a GIS course" and get a reasonable draft rubric listing criteria like teamwork, analytical accuracy, presentation clarity, etc. This can be refined by the teacher. AI can also generate quiz questions or even exam questions on the fly. For example, given a chapter of material, the AI can produce a set of multiple-choice and short-answer questions covering key points. An instructor should vet these questions for correctness and appropriateness, but it's a quick way to get a test bank started (Zawacki-Richter et al. 2019). Students, on the other hand, might use the same functionality to create practice quizzes for themselves: "Give me 5 practice questions on population geography and then provide the answers." This turns the AI into a study buddy. Such use arguably enhances learning—it encourages students to test themselves and engage with the content actively.

Given these examples, it is evident that GenAI can be a multifaceted asset in education. Particularly in geography education, which spans factual knowledge (like capitals, landforms, concepts of spatial analysis) and skills (like GIS usage, critical thinking about human–environment relations), AI can help with memorization (through flashcards, quizzes), with concept explanation (through analogies, e.g., "explain monsoons using a simple analogy"), and even with technical tasks (perhaps assisting in writing a GIS script in Python). However, harnessing these opportunities comes with significant pedagogical and ethical considerations.

Rethinking Teaching and Assessment in the AI Era

The immediate reaction of many educational institutions to GenAI was one of alarm—concerns about cheating and erosion of learning skills (Cotton et al. 2023). Some schools moved to ban GenAI usage outright on assignments; others began revamping their assessments. As the dust settles, a more balanced view is emerging (Newton and Xiromeriti 2023). It is becoming clear that education must evolve in response to AI, much as it has in response to calculators or the internet in the past (Zawacki-Richter et al. 2019). Key areas of adaptation include assessment design, curriculum focus, and cultivating GenAI literacy.

Assessment design: If AI can write essays or solve problem sets, then traditional take-home assignments lose their effectiveness as a measurement of student learning. Educators are exploring alternatives that are more "AI-resistant" or that incorporate AI use transparently (Cotton et al. 2023). One approach is to emphasize in-class work and oral components. For example, rather than grading solely a written essay on urban sustainability, a teacher might have students submit a draft (which could be AI-assisted) and then conduct an in-person discussion or defense of that essay with each student. The student's understanding, beyond what's written, becomes part of the evaluation. Another approach is giving assignments that require personal reflection or local context, which an AI would not easily have. For instance, an assignment might be: "Relate the concept of sustainable cities to your own city's initiatives and propose a recommendation." Since this requires personal input, an AI's generic answer would not suffice without the student's contribution. Some educators are integrating AI use into the assignment itself. Instead of prohibiting it, they ask students to, say, produce an initial answer with AI and then critique it (Kasneci et al. 2023). A geography class might pose: "Use ChatGPT to generate a brief report on the impacts of deforestation in the Amazon. Then, in your submission, identify at least three errors or biases in that report and correct them using course concepts and external sources." This way, students exercise critical thinking and fact-checking—crucial skills in the AI era—and the presence of AI is acknowledged and leveraged for learning, not hidden as cheating.

Curriculum focus: With rote tasks being automated, education can put greater emphasis on higher-order skills (Mollick and Mollick 2023). I. In geography education, this means more focus on interpretation, synthesis, and critical evaluation of information. For example, knowing all capital cities by memory is less important when an AI or any digital tool can retrieve that instantly. But analyzing why a capital's location matters (politically, environmentally, spatially) or debating policy options for a city's development—those are human intellectual activities that education will double down on. The value of critical thinking is amplified. One might say that AI is a tool and not a replacement for critical thinking and self-study. Instructors might incorporate more project-based learning, where students have to navigate complex tasks possibly with AI as one tool among many (Zawacki-Richter et al. 2019).

GenAI literacy for students and teachers: Both need to learn how to use AI appropriately (Foltýnek et al. 2020). For students, GenAI literacy entails knowing how to prompt effectively, how to cross-verify AI outputs, understanding issues like bias and hallucination, and applying ethical judgment about when use is allowed (Sullivan et al. 2023). It also involves knowing how to cite AI contributions if relevant (e.g., "This section was generated with the assistance of ChatGPT and then edited by the author" might be a note in an assignment if permitted by the teacher). Teachers, on the other hand, need training on AI tools to enhance their teaching practice (Newton and Xiromeriti 2023). They should be aware of the latest capabilities (for instance, new features like multi-modal input where students could ask questions about an image, diagram or map to an AI). Professional development programs in education are starting to include AI workshops (Kasneci et al. 2023). Some universities have issued guidance documents for instructors on incorporating GenAI in their courses (e.g., suggesting syllabus statements about AI use, designing assignments differently, etc.). In the context of geography, educators can share specific strategies, like using AI to generate different regional case studies for class comparisons, or pitfalls like AI's lack of actual map-reading ability.

AI in Geographical Education and Spatial Learning

Geography, as a discipline taught from primary school through university, has its opportunities and challenges with GenAI. Let's consider a few.

Geo-facts versus geo-thinking: Geography involves a lot of factual knowledge (place names, statistics, etc.) and conceptual understanding (why those facts matter, patterns and processes). AI can supply facts quickly, so educators might shift focus to interpretation. However, factual knowledge can't be ignored entirely—it is the foundation for higher reasoning. One strategy is using AI to handle factual queries (like a student asking, "What is the population of Lagos?" and getting an answer) so that class time can be spent discussing implications (urban challenges in mega-cities, for example). Teachers could assign students to query the AI for data then bring that data to class for analysis. This promotes a form of active learning where AI is a data provider and the student is an analyst (Bednarz and Bednarz 2015).

Spatial analysis and GIS: At present, text-based GenAI like ChatGPT has limitations with purely spatial tasks (it cannot natively read maps or know coordinates unless specifically trained or connected to tools). However, there are emerging integrations: some GIS software companies are exploring AI assistants that can take natural language and perform GIS operations (e.g., "Show me a heat map of population density in Europe" could trigger a GIS routine). In a sense, this is moving toward natural language GIS, where instead of mastering complex software interfaces, users might just ask for geospatial analysis in plain terms. For education, that could lower technical barriers and let students focus more on geographic reasoning than on software mechanics (Kerski 2015). But it also means we need to ensure students still understand how the analysis is done—otherwise, they may take AI-produced maps at face value without comprehension (akin to blindly trusting a GPS without map-reading skills).

Global perspectives and case studies: Geography curriculum often tries to include examples from around the world to foster global citizenship. AI can assist by providing quick case briefs from various countries. Ask about "urban agriculture examples in Africa, Asia, and Latin America"—it might list a few for each (again, to be checked for accuracy). This could save instructors time in collecting diverse case studies, ensuring that the discussion isn't narrowly focused on well-known Western examples (Tuck and McKenzie 2015). However, if the AI is primarily trained on English sources, some regions might be underrepresented or misrepresented. Teachers should strive to verify and supplement AI-provided cases with local sources or knowledge where possible.

Language translation and inclusivity: Multilingual context in education can benefit from AI translation (like DeepL). A geography teacher could translate materials or even student essays between languages with high quality. This fosters inclusion for students who are non-native speakers of the instruction language (García and Li 2014). It also encourages cross-cultural exchange: classes in different countries could communicate via AI translation tools—for instance, Belgian and Finnish students sharing local geographic issues in their own language and reading each other's reports translated.

A recent survey of educators framed GenAI in education as having paradoxical qualities—it can potentially diminish certain skills (like basic writing or memorization) if misused, yet it can also significantly enhance personalized learning and creative inquiry if used well (Lim et al. 2023). The net outcome will depend on how we manage this transition. Responsible integration, rather than blanket bans or uncritical embrace, is key.

GenAI in Geographical Thinking and Spatial Practice

Beyond education, GenAI has implications for the practice of geography itself—how geographers conduct research, analyze spatial data, and communicate findings. This section briefly considers the current and potential future roles of GenAI in geographical research and geospatial technology.

Geographical research often involves handling diverse data sources: quantitative data (e.g., census statistics, satellite imagery), qualitative data (e.g., interviews, ethnographies), and literature from various fields. GenAI tools can assist geographers in several ways.

Literature review and synthesis: A researcher exploring a topic like climate-induced migration might ask a GenAI to summarize key findings from a set of papers or to outline theoretical perspectives (e.g., pull main arguments from human geography vs. economics literature on migration). The AI can produce a synopsis that, while needing verification, can help the researcher see the forest for the trees (Xu et al. 2023). It can also suggest connections between concepts, potentially sparking interdisciplinary ideas. There are specialized AI-powered literature review tools now that ingest PDFs and allow querying (such as Google's NotebookLM). A geographer could upload a set of reports, ask "What are the main drivers of urban sprawl according to these documents?" and get a collated answer referencing each document. Notably, NotebookLM explicitly cites the source passages, which is a positive step for academic trustworthiness (Chen and Park 2024). This kind of tool could greatly speed up the literature digestion phase of research.

Data analysis suggestions: Suppose a geographer has a rich dataset—say, a CSV of environmental measurements across regions and years. AI can aid in exploratory data analysis by suggesting statistical methods or even executing AI can write the code (in R or Python) to do this (Xu et al. 2023). This doesn't eliminate the need for analytical skills; rather it accelerates the process and opens advanced techniques to those who might not have coding expertise. For instance, an AI might generate a regression analysis or a clustering without the researcher writing a line of code, enabling the geographer to focus on interpreting results. Again, human oversight is crucial—the researcher must ensure the analysis makes sense and is correctly applied.

Geospatial analysis and AI integration: Traditional GIS operations can be complemented by AI. While a pure language model lacks spatial reasoning, integrated systems can overcome this (Li and Wang 2024). One approach is coupling LLMs with geo-databases. For example, an LLM can receive structured outputs from a database (like spatial query results) and then summarize or reason about them in text. There is research on spatially aware language models that incorporate location as part of their context. In practice, geographers might soon use voice or text to direct analyses: "AI, compare the land cover changes in these two satellite images and describe the differences." Some early tools can detect changes in images and an LLM can draft a human-readable report of those changes. This would expedite tasks like environmental monitoring reports or damage assessments after disasters. However, ensuring accuracy is challenging—interpreting imagery correctly still benefits from expert validation.

Conclusion

GenAI is transforming our world in ways that are both promising and challenging. From the perspective of geography—a field that straddles the natural and social sciences and engages deeply with questions of place, space, and society—GenAI's rise offers new tools for exploration and understanding, while also demanding vigilant critique and adaptation.

In this chapter, we have traversed the landscape of GenAI's impacts, guided by themes from the EUROGEO 2024 keynote "How GenAI is changing our world". We began by clarifying what GenAI is: an unprecedented class of AI algorithms capable of creating text, images, code, and more, by learning patterns from vast datasets. We noted the explosive growth and accessibility of these tools—exemplified by ChatGPT's record-breaking user adoption—which signposts that GenAI is not a distant prospect but an immediate reality in our daily lives and professions.

Moving through societal impacts, we saw that GenAI is reshaping how work is done, enabling new forms of collaboration between humans and machines (from writing and coding to brainstorming research) and raising productivity in many domains. At the same time, it challenges existing norms around authenticity and skill. We face a world where content is abundant, but its truth value can be questionable if produced by AI without oversight. Misinformation and fabricated outputs (hallucinations) pose risks that require us to double down on critical thinking and verification. Ethically, we grapple with AI's embedded biases and uneven representation of different cultures, and we cannot ignore the environmental footprint of this digital revolution, as highlighted by the significant energy and carbon cost of training large models. These are not reasons to halt progress, but clarion calls to guide it conscientiously—to develop mitigation strategies (from bias audits to green computing) and policies (like the EU AI Act and UNESCO's guidance on GenAI in education) that ensure AI develops in alignment with human values and sustainability.

In the realm of education, GenAI has quickly become both a disruptor and a catalyst for innovation. We discussed how educators are harnessing AI for lesson planning, content creation, and personalized tutoring, with early experiments like Khan Academy's GPT-4 powered tutor indicating the potential for more individualized learning support. Students, too, are using AI as a study aid, a writing coach, and an information resource. Rather than fight a losing battle against the use of AI, educational institutions need to rethink pedagogy: emphasizing GenAI literacy, redesigning assessments to focus on understanding and original thought, and teaching students to use AI ethically and effectively rather than clandestinely. Geography education can leverage AI for diverse case studies, real-time data analysis, and multilingual resources, enriching the learning experience and preparing students for a world where spatial thinking and digital literacy go hand in hand. AI is a tool, not a replacement for critical thinking; a sentiment that encapsulates the educational approach we must adopt. Students need to learn with AI, but also beyond AI—to question its outputs and cross-check against reality (for instance, using field observations or authoritative data in geography to validate AI-generated conclusions).

The chapter also explored into the intersection of GenAI with geographical practice and research. We illustrated that geographers can use GenAI to expedite literature reviews, parse data, generate hypotheses, and communicate findings. These augmentations could accelerate research cycles and perhaps democratize some aspects of analysis (e.g., enabling a researcher without advanced programming skills to nonetheless perform complex spatial data analysis via AI assistance). However, maintaining the human-in-the-loop is indispensable. In geography, context is king—and AI lacks context unless we provide it. The human geographer must guide the AI, ensuring that outputs make sense in light of real-world geospatial knowledge, and that ethical safeguards (like respect for local knowledge and privacy of data) are observed.

A key reflection from a geographical perspective is that GenAI is here to stay. The pace of AI development is exponential, and this means geographers (and all professionals) have to cultivate adaptability and a mindset of continuous, lifelong learning. We can't focus solely on "what AI cannot do" as if those limitations are permanent; rather, we should focus on "what it currently cannot do" and be proactive in anticipating the next steps. For example, if today's models struggle with real-time spatial reasoning, tomorrow's might integrate live GIS feeds—are we ready to handle that?

Yet, amidst rapid change, some fundamentals remain. Critical thinking, ethical reasoning, and human creativity are more important than ever. This has always been the case when new technologies emerged. The arrival of the printing press, the computer, the internet, search engines like Google, and later the smartphone all transformed how we access and process information, but they never replaced the need for human judgment, reflection, and creativity. They are our compass in the AI-enhanced world. Geography as a discipline offers a critical lens—geographers examine how phenomena vary across space and affect different communities, which is crucial when assessing an AI's impact (be it the digital divide or cultural bias in outputs). Geographers also emphasize sustainable and equitable development, which informs how we should direct AI advancements to serve all of humanity, not just a privileged subset.

Our reflections suggest that GenAI, if left unguided, could compromise aspects of geographical inquiry (for instance, by flooding information channels with unchecked outputs, or by entrenching dominant narratives from data biases). However, if approached with the spirit of open, critical, and inclusive engagement, GenAI can also enhance geography—making it more accessible, more interactive, and even more impactful in addressing real-world problems through simulations and data-driven insights.

The future of AI in geography, education, and society at large will be shaped by how we act now. Embracing GenAI's opportunities while upholding our commitment to truth, equity, and sustainability will ensure that this powerful technology becomes a boon rather than a bane. As with any tool, the value ultimately lies in the user. By training the next generation of "GenAI-literate" geographers and global citizens, and by advocating for policies that promote open data and ethical AI, the geographical community can help steer GenAI toward enriching our understanding of the world rather than compromising it.

The journey is just beginning. AI will continue to evolve—we might soon discuss AI agents that autonomously perform complex tasks, or world models that incorporate not just text but the full sensorium of data about Earth systems. Geographers should be at the table when these developments happen, to provide insight into spatial contexts and human–environment relationships. After all, GenAI might compute and generate, but it is our human geography perspective that will interpret and give meaning to what is generated, keeping it grounded in the real world we all share.

References

Bednarz SW, Bednarz RS (2015) Spatial thinking: the key to success in using geospatial technologies in the social studies classroom. J Geogr 114(6):238–251. https://doi.org/10.1080/00221341.2015.1061585

Bender EM, Gebru T, McMillan-Major A, Shmitchell S (2021) On the dangers of stochastic parrots: can language models be too big? In: Proceedings of the 2021 ACM conference on fairness, accountability, and transparency, pp 610–623. https://doi.org/10.1145/3442188.3445922

Bolukbasi T, Chang K-W, Zou J, Saligrama V, Kalai A (2016) Man is to computer programmer as woman is to homemaker? Debiasing word embeddings. Adv Neural Inf Process Syst 29. https://doi.org/10.48550/arXiv.1607.06520

Brown TB, Mann B, Ryder N, Subbiah M, Kaplan J, Dhariwal P, Neelakantan A, Shyam P, Sastry G, Askell A, Agarwal S, Herbert-Voss A, Krueger G, Henighan T, Child R, Ramesh A, Ziegler DM, Wu J, Winter C, Amodei D et al (2020) Language models are few-shot learners. Adv Neural Inf Process Syst 33:1877–1901. https://papers.nips.cc/paper/2020/file/1457c0d6bfcb4967418bfb8ac142f64a-Paper.pdf

Chan CKY (2023) A comprehensive AI policy education framework for university teaching and learning. arXiv. https://arxiv.org/abs/2305.00280

Chen Y, Park J (2024) Societal, economic, political and environmental: a review of benchmarks and AI-assisted systematic literature review of impact of open government data. Retrieved from: https://www.semanticscholar.org/paper/95a9f7ebd186bc843a19c1d2723a8beb8acfaf22

Cotton DRE, Cotton PA, Shipway JR (2023) Chatting and cheating: ensuring academic integrity in the era of ChatGPT. Innov Educ Teach Int. https://doi.org/10.1080/14703297.2023.2190148

Dhar P (2020) The carbon impact of artificial intelligence. Nature Mach Intell 2(8):423–425. https://doi.org/10.1038/s42256-020-0219-9

Dodge J, Sap M, Marasović A, Agnew W, Ilharco G, Groeneveld D, Gardner M (2022) Documenting large webtext corpora: a case study on the colossal clean crawled corpus. Trans Assoc Comput Linguistics 10:1176–1192. https://doi.org/10.1162/tacl_a_00520

Escobar AO, Suárez Jasso E, Anota MM (2024) Comparative analysis of generative language tools for higher education. In: 2024 IEEE international conference on engineering veracruz (ICEV), pp 1–6

Floridi L, Cowls J (2019) A unified framework for five principles for AI in society. Harvard Data Sci Rev. https://doi.org/10.1162/99608f92.8cd550d1

Foltýnek T, Dlabolová D, Linkeschová D (2020) Academic integrity in the age of AI: challenges and opportunities. J Acad Ethics 18(2):123–137. https://doi.org/10.1007/s10805-019-09342-4

Frey CB, Osborne MA (2017) The future of employment: how susceptible are jobs to computerisation? Technol Forecast Soc Chang 114:254–328. https://doi.org/10.1016/j.techfore.2016.08.019

García O, Li W (2014) Translanguaging: language, bilingualism and education. Palgrave Macmillan. https://doi.org/10.1057/9781137385765

Hu K (2023) ChatGPT sets record for fastest-growing user base—analyst note. Reuters. https://www.reuters.com/technology/chatgpt-sets-record-fastest-growing-user-base-analyst-note-2023-02-01/

Janowicz K, Gao S, McKenzie G, Hu Y, Bhaduri B (2020) GeoAI: spatially explicit artificial intelligence techniques for geographic knowledge discovery and beyond. Int J Geogr Inf Sci 34(4):625–636. https://doi.org/10.1080/13658816.2019.1684500

Joshi P, Santy S, Budhiraja A, Bali K, Choudhury M (2020) The state and fate of linguistic diversity and inclusion in the NLP world. In: Proceedings of the 58th annual meeting of the association for computational linguistics, pp 6282–6293. https://doi.org/10.18653/v1/2020.acl-main.560bid

Kasneci E, Seßler K, Küchemann S, Bannert M, Dementieva D, Fischer F, Gasser U, Groh G, Günnemann S, Hüllermeier E, Krusche S, Kutyniok G, Michaeli T, Nerdel C, Pfeffer J, Poquet O, Sailer M, Schmidt A, Seidel T, Kasneci G et al (2023) ChatGPT for good? On opportunities and challenges of large language models for education. Learn Individ Differ 103:102274. https://doi.org/10.1016/j.lindif.2023.102274

Kerski JJ (2015) Geo-awareness, geo-enablement, geotechnologies, citizen science, and storytelling: geography in the 21st century. J Geogr 114(3):109–119. https://doi.org/10.1080/00221341.2015.1004355

Liang W, Yuksekgonul M, Mao Y, Wu E, Zou J (2023) GPT detectors are biased against non-native English writers. arXiv. https://arxiv.org/abs/2304.02819

Lim WM, Gunasekara A, Pallant JL, Pallant JI, Pechenkina E (2023) Generative AI and the future of education: Ragnarök or reformation? A paradoxical perspective from management educators. Int J Manage Educ 21(2):100790. https://doi.org/10.1016/j.ijme.2023.100790

Li H, Wang Q (2024) AI-assisted landscape design optimization and innovation research based on AI assistance. Retrieved from https://www.semanticscholar.org/paper/8b6347eefc65859644a9316c29e5591ee5892064

Mollick ER, Mollick L (2023) Using AI to implement effective teaching strategies in classrooms: five strategies, including prompts. SSRN. https://doi.org/10.2139/ssrn.4391243

Newton PM, Xiromeriti M (2023) ChatGPT performance on MCQ-based exams. Assess Eval High Educ. https://doi.org/10.1080/02602938.2023.2213169

Patterson D, Gonzalez J, Le Q, Liang C, Munguia L-M, Rothchild D, So D, Texier M, Dean J (2021) Carbon emissions and large neural network training. arXiv. https://doi.org/10.48550/arXiv.2104.10350

Raffel C, Shazeer N, Roberts A, Lee K, Narang S, Matena M, Zhou Y, Li W, Liu PJ (2020) Exploring the limits of transfer learning with a unified text-to-text transformer. J Mach Learn Res 21(140):1–67. https://jmlr.org/papers/volume21/20-074/20-074.pdf

Roberts H, Cowls J, Morley J, Taddeo M, Wang V, Floridi L (2020) The Chinese approach to artificial intelligence: an analysis of policy, ethics, and regulation. AI & Soc 36(1):59–77. https://doi.org/10.1007/s00146-020-00992-2

Rombach R, Blattmann A, Lorenz D, Esser P, Ommer B (2022) High-resolution image synthesis with latent diffusion models. In: Proceedings of the IEEE/CVF conference on computer vision and pattern recognition (CVPR), pp 10684–10695. https://openaccess.thecvf.com/content/CVPR2022/papers/Rombach_High-Resolution_Image_Synthesis_With_Latent_Diffusion_Models_CVPR_2022_paper.pdf

Roy PC, Dey N, Kumar M, Moses GA (2024) Critical analysis of the technological affordances, challenges and future directions of Generative AI in education: a systematic review. Educ Inf Technol 29:6143–6169. https://doi.org/10.1007/s10639-023-12096-7

Tuck E, McKenzie M (2015) Place in research: theory, methodology, and methods. Routledge. https://doi.org/10.4324/9781315775352

Schwartz R, Dodge J, Smith NA, Etzioni O (2020) Green AI. Commun ACM 63(12):54–63. https://doi.org/10.1145/3381831

Sheng E, Chang K-W, Natarajan P, Peng N (2021) "Nice try, kiddo": investigating ad hominems in dialogue responses. In: Proceedings of the 2021 conference of the North American chapter of the association for computational linguistics: human language technologies, pp 750–767. https://doi.org/10.18653/v1/2021.naacl-main.60

Shneiderman B (2020) Human-centered artificial intelligence: reliable, safe & trustworthy. Int J Human-Comp Interact 36(6):495–504. https://doi.org/10.1080/10447318.2020.1741118

Strubell E, Ganesh A, McCallum A (2019) Energy and policy considerations for deep learning in NLP. In: Proceedings of the 57th annual meeting of the association for computational linguistics, pp 3645–3650. https://doi.org/10.18653/v1/P19-1355

Xu Z, Zhang Y, Zhang X (2023) Natural language generation and understanding of big code for AI-assisted programming: a review. arXiv. https://doi.org/10.48550/arXiv.2307.02503

Yang J, Jin H, Tang R, Han X, Feng Q, Jiang H, Yin T, Bai Q, Wu B, He Y, Hu X (2023) Harnessing the power of LLMs in practice: a survey on ChatGPT and beyond. arXiv preprint arXiv:2304.13712. https://arxiv.org/pdf/2304.13712.pdf

Zawacki-Richter O, Marín VI, Bond M, Gouverneur F (2019) Systematic review of research on artificial intelligence applications in higher education. Int J Educ Technol High Educ 16(1). https://doi.org/10.1186/s41239-019-0171-2

Zhao B, Zhang S, Xu C, Yifan S, Deng C (2021) Deep fake geography? When geospatial data encounter Artificial Intelligence. Cartogr Geogr Inf Sci. https://doi.org/10.1080/15230406.2021.1910075

Geography Education

Integration and Multiculturalism: Essays on EBS Fontes Pereira de Melo

Ana Ferreira, Elsa Pacheco, and Francisca Teixeira

Abstract One of the most interesting consequences of the intensification of population flows on an international scale is the promotion of multicultural societies. As responsible agents for the education of young people in Portugal, we cannot fail to note the challenge facing primary and secondary schools which, in the recent years, have been confronted with an increasing number of students from very diverse geographical backgrounds. Developing a set of activities that highlight the differences that characterise the diversity and cultural richness of our planet, the geography teacher develops attitudes and promotes behaviours that value multiculturalism at school and in society that shape the values of future citizens. In fact, in a global world in which human interaction is increasingly a reality (and our country is no exception), welcoming communities that are more and more diverse in linguistic and cultural terms, the school as a privileged space for human interaction is the ideal stage to change behaviours and shape attitudes of the next generations. This study aims to analyse the views of geography teachers and pupils on the 'integration' of foreign pupils at school and to demonstrate the added value of this heterogeneity of geographical origins for the teaching and learning of Geography. To achieve this goal, we used a qualitative and quantitative work methodology in the case of the Fontes Pereira de Melo Primary and Secondary School (EBSFPM), in Porto, whose results suggest that, with in-depth work on multicultural education, the obvious richness of being able to teach and learn from difference enhances the role of the school in contemporary society.

A. Ferreira (✉)
Escola Superior de Hotelaria e Turismo (ESHT), Geography and Spatial Planning Research Centre (CEGOT); Centre for Tourism Research, Development and Innovation (CiTUR), Porto, Portugal
e-mail: anaferreira@esht.ipp.pt

E. Pacheco
Faculty of Arts and Humanities of University of Porto, Transdisciplinary Research Centre for Culture, Space and Memory (CITCEM), University of Porto, Porto, Portugal
e-mail: elsap@letras.up.pt

F. Teixeira
Agrupamento de Escolas Fontes Pereira de Melo, Porto, Portugal
e-mail: francisca.teixeira@aefontespmelo.pt

L. Soares et al. (eds.), *Compromised Geography*, Key Challenges in Geography,
https://doi.org/10.1007/978-3-032-05201-8_3

Keywords Multicultural education · Geography · School · Multiculturality

Introduction

The intensification of international flows, paired with the growing attraction that Portugal has for different multinational communities, has made Portuguese society a multicultural society. Inevitably, the issue of cultural diversity has repercussions at educational level, not only because the school reflects society, but also because the school is one of the favoured spaces for promoting cultural integration.

In fact, although the school—as Viana and Silva (2020) points out—is more comfortable with homogenisation and standardisation, today it faces the growing challenge of opening to diversity since classes integrate a diverse range of national and international students. As a multicultural space, the school, as a stage for training citizens, is the ideal place to discuss multiculturalism, particularly in the subject of Geography, where, as Cavalcanti (2012; Viana and Silva 2020, p. 210) notes, 'places that are different, like diverse cultures, can be points of reference in schoolwork'. In other words, the geography teacher should take advantage of the diversity of symbols, values, attitudes, feelings, expectations and knowledge of the various students who are in the classroom, as an opportunity to broaden geographical knowledge education and training for citizenship.

To do this, schools must change their methodological strategies, since it is not just a question of overcoming language barriers but also of adopting pedagogical practices that highlight cultural diversity within the classroom and other school spaces.

This article aims to show how geography can make a decisive contribution to truly multicultural teaching, using the Fontes Pereira de Melo primary and secondary school as a case study. In this school, in the 2023–2024 academic year, 254 students of 31 different nationalities shared the school space and interacted daily with the remaining students of Portuguese nationality.

Literature Review

Multicultural Education and the Teacher's Role

Multiculturalism, which in a simplified way can be understood as a set of cultures in contact without merging, is now a fact in Portuguese society and requires a redefinition of attitudes and behaviours to adopt an integrative perspective which, while admitting diversity, emphasises differences and accepts the heterogeneous (Canclini 2004).

As one of the fundamental supports of national culture, from which it spreads, the school has until now played a unifying role, homogenising local differences to reinforce a national identity. However, aware of the impact of current migratory flows and assuming that today there are several cultures in the school, the opposite should be the case, i.e. differences should be considered as the driving force behind the teaching–learning process, valuing differences and enhancing heterogeneity, promoting multicultural education.

As Silva (2015, p. 4) points out, according to Banks (1985), multicultural education is seen as a set of "programmes and practices designed to help implement the academic success of ethnic and immigrant populations and/or to teach students from majority groups about the cultures and experiences of ethnic minority groups" with the aim of "reforming the school and other educational institutions so that students of different races, ethnicities and social classes experience educational equality". In this sense, the main proposals put forward to guide multicultural education concern changes in the school curriculum, the production of new teaching materials and new proposals for teaching and learning according to the characteristics of the students, their prerequisites, backgrounds and family culture.

As cultural diversity increases in the classroom, the challenges that teaching staff face are bigger, such as the language barrier, different learning styles, distinct cultural traits and the definition of effective assessment methods for all. The role of the teacher in multicultural education is therefore crucial, as it is the teacher who must be willing to change and value the different cultures in interaction.

As Gay (2010) emphasises, culturally relevant teaching involves "creating supportive learning communities where culturally diverse individuals and heritages are valued; using knowledge of ethnically diverse cultures, families, and communities to guide curriculum development, classroom climate, teaching strategies, and relationships with students; challenge racial and cultural stereotypes, prejudice, racism, and other forms of intolerance, injustice, and oppression; be agents of change for social justice and academic equity; mediate power imbalances in classrooms based on race, culture, ethnicity, and class; and accept cultural responsiveness as endemic to educational effectiveness in all areas of learning for students of all ethnic groups" (Smith 2020, p. 28).

Therefore, teachers should consider a few pillars of multicultural educational development in their teaching practice. The first will be integration, which is related to the use that teachers make of information from different cultures and groups to enrich the basic and formal contents of the curriculum. In this way, teachers help students to understand how knowledge is created and how it is influenced by racial, ethnic and social class positions of individuals and groups. Another pillar of multicultural educational development is the reduction of prejudice, as this can help students to develop more democratic attitudes and values towards their peers and can be associated with the implementation of techniques and methods in the classroom that promote the positive academic performance of students belonging to different racial, ethnic and social class groups. Finally, the third pillar involves strengthening the school culture and social structure through school organisation, teacher training and formal and informal education agents.

In this sense, and although in article 2 of the Basic Law of the Education System (1986) it is clear that 'the state cannot give itself the right to programme education and culture according to any philosophical, aesthetic, political, ideological or religious guidelines', it is also mentioned (in its Article 3) that 'it is up to the state to ensure the right to difference, through respect for personalities and individual projects of existence, as well as the consideration and valorisation of different knowledge and cultures', which meets the requirements of a multicultural education, which is based on three pillars: personal development, which emphasises the need to view cultural diversity as a means of personal, social, cultural and curricular enrichment and to commit to promoting equal opportunities for all students and to be convinced of the importance of their role in this task; curricular training, which states that the resources, knowledge and experiences that students bring to school should be valued and integrated into activities, in an attitude of constant respect for the languages and cultures of minorities; and organisational training, which stresses the importance of promoting cooperative work between teachers with a view to planning, implementing and evaluating multicultural activities (Cardoso 1994).

The Role of Geography in Multicultural Education

Geography is a pivotal discipline between the natural sciences and the social sciences that seeks to answer the questions that man poses about the physical and human environment, places students in contact with different socio-cultural realities that help them understand how spaces are interrelated.

To this extent, current geographical content can 'make' critical decision-makers students', by providing them with a global perspective and awareness of international and multicultural problems, while at the same time give them the necessary tools to understand them.

Wellens et al. (2006) argue that geography is well-placed to play a pivotal role in social transformation, lending itself to multicultural education. In the same way, Bigelow et al. (1994) outlined key characteristics for education in their vision for social justice and transformation, including that it must be: grounded in the lives of pupils; critical and linked to real-world problems; multicultural and anti-racist; participatory and experiential; culturally sensitive; academically rigorous; and concerned with issues beyond the classroom walls. These are all key characteristics of geography education (Catling 2005; Catling and Willy 2020; Dolan 2020; Martin 2008; Pike 2016, 2011; Roberts 2003, 2014). Moreover, 'geographical thinking', developed through concepts such as space, place, scale, movement and human–environmental interaction, enables pupils to develop critical thinking skills, to analyse and form opinions about real-world problems and to understand and challenge 'how the world works' (Machon and Walkington 2000; Young and Lambert 2014). Geographical perspectives encourage a deeper concept of how phenomena are interrelated, locating private, local actions, in a wider, public, global scale (e.g. climate change is a global process but concurrently plays out locally), thus enabling pupils to identify power relations and to envisage alternatives (Machon 1998; Young and Lambert 2014).

Table 1 The values, concepts and skills shared by multicultural education and geography

Values and attitudes	Concepts	Skills
Social justice A sense of place A sense of community Empathy Respect	Interdependence Sustainability Change Power relations Cultural diversity	Critical thinking Decision making Reflection Investigation Enquiry

Source Adapted from Machon and Walkington (2000)

Considering the above, it is clear in Table 1 that geography and multicultural education share many common concepts, skills and goals. We highlight values and attitudes, and critical thinking and decision-making (Table 1).

The geography professor should use methods to teach students to be: critical; to ask questions of systems, sources and society; to be inquisitive and investigative; to recognise and value similarities and differences between peoples; to explore power relations between groups; and to see things from multiple perspectives in order to foster in students respect for everyone's rights to equality, to lead the teaching of the subject towards a less ethnocentric approach, promoting citizenship education.

Methodology

This study is grounded in a literature review, with the objective of systematising the importance of multicultural education in today's school environment. It also examines the role of teachers in assessing the impact of multicultural education on the success of the teaching–learning process.

In this investigation, we implemented a sample questionnaire to 49 students of Fontes Pereira de Melo Primary and Secondary School. This questionnaire, implemented in Google Forms, is divided into two parts: the first one is related with the identification of the respondent, as age, gender, year of schooling; the second part we ask their opinion on how they evaluate the teaching–learning process at school, taking into account the growth in the number of international students in the school and their interaction with Portuguese students inside and outside the classroom.

The analysis of the results, made through Excel tool, allowed us to determine the importance of multicultural education in the success of all students, both Portuguese and from other countries.

To complement the study, we used a qualitative methodology based on semi-structured interviews to three of the five school geography teachers, to understand their teaching practices in this new school context.

Results

Sociocultural Framework of the Fontes Pereira de Melo School Group

According to the principles established by this general framework of the Portuguese education system, each school grouping establishes its own internal regulations. This document, which governs the entire school community, aims to guarantee the rules of coexistence that ensure the fulfilment of the objectives of the educational project, the harmony of interpersonal relationships and social integration, the physical, intellectual and civic development of students, the preservation of their safety and the school heritage and that of the other members of the educational community.

In its internal regulations, the Fontes Pereira de Melo School Group (Agrupamento de Escolas Fontes Pereira de Melo 2019) ensures, in its Article 65, that each and every student must 'be treated with respect and correction by any member of the Educational Community, and cannot, under any circumstances, be discriminated on the basis of ethnic origin, health, sex, sexual orientation, age, gender identity, economic condition, cultural or social convictions or political, ideological, philosophical or religious convictions', adding, in Article 91, that it is essential to 'respect the personal dignity and cultural differences of students, valuing different knowledge and cultures, preventing processes of exclusion and discrimination'.

In addition, the group's educational project identifies as its objectives 'to promote values that favour dialogue through difference and tolerance' and 'to provide students with contact with diversified cultural and artistic manifestations', thus assuming a commitment to inclusive education as a process that aims to respond to the diversity of its students, through the participation of all in learning and in the life of the school community.

In the 2023–2024 school year, the FPM primary and secondary school, the headquarters school of the group with the same name has one hundred and twenty-three international students (Fig. 1) from 21 different countries, of which eighty-one come from Brazil (66%) and ten from Angola, which allows us to say that 74% of these school's students come from CPLP countries.

Multicultural Education at EBSFPM

(a) Geography Teachers

Trying to know the perception of geography teachers regarding the process of multicultural education at school and about their role, in particular, in a school that is currently a multicultural community, we conducted interviews with three of the school's six geography teachers. Two of these teachers are female and one male, aged between 27 and 58 years old. This wide age range is naturally reflected in the

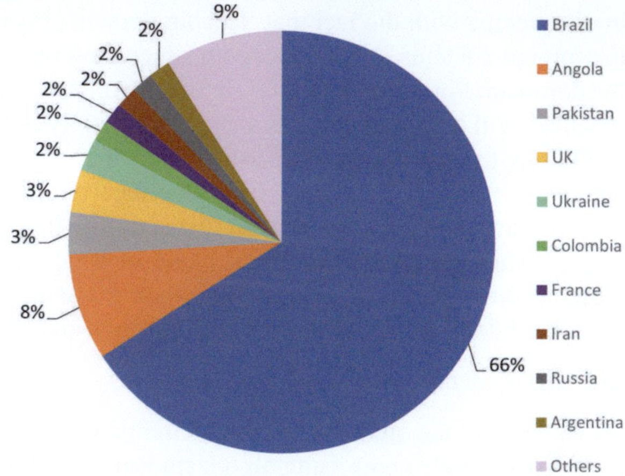

Fig. 1 Origin of EBSFPM foreign students, in the 2023/24 academic year. *Source* Own elaboration

gap in the number of career years that varies between 4 and 34 school years. These teachers teach all school grades, from the 7th to the 12th grades as well as adult education and training (EFA) courses.

All of these teachers have international students, but none have had any training on the subject and, when asked how they integrate these students into the school community, the answers were: 'without formal support', 'depending on their nationality, the integration process can be more or less easy' and 'with some difficulty, because in addition to the language barrier, there is some segregation on the part of their peers'. Curiously, and despite these findings, two teachers said they found interacting with these students challenging, both as a person and as a teacher. But the third teacher emphasised that it is indifferent to him.

To overcome the language barriers, pointed out by foreign students as the main difficulty, the teachers choose to speak English (eventually French or Spanish) and one of them says that he allows the use of the translator during classes.

Regarding the approach to the theme of multiculturalism in the geography classroom, the teachers say they approach it 'in very different ways'. One integrates the theme into the curriculum programme, in the 8th grade themes, under the sub-theme 'Migrations and Multiculturalism'. Another teacher emphasised that he adopts 'a humanist approach, where the cultural and social diversities of students of different nationalities are exposed in the classroom'. However, we would like to point out that all the teachers say that they contribute to the integration of foreign students by 'adapting the content and the language used' and 'including foreign students in different working groups, allowing them to talk about their culture and traditions, among other things' and they do so because they agree that 'the teacher's attitude interferes with the integration of these students'. One of the teachers even emphasises that 'the way we introduce, interact with and welcome these students into the class can facilitate their integration'.

When we look at school success, only two of the teachers disagree with the statement that cultural diversity can lead to school failure, but all three are unanimous in disagreeing with the fact that 'teaching practices can lead to a loss of identity for the foreign students in the classroom, since they represent a minority in relation to the dominant Portuguese culture', in other words, these teachers recognise that their practices will be decisive in enhancing the identity of their students.

Outside the classroom, when asked 'Do you feel that the school carries out enough activities to integrate the students?', we received one affirmative answer and two negative answers, pointing out that 'the activities that are carried out are not enough and are often limited to the initial phase of welcoming the students.' In addition, these teachers also believe that even when extracurricular activities are organised, the school does not consider the cultural habits and traditions of foreign students.

(b) The students

To understand the students' vision of multicultural education at school, we chose to set up a sample comprising all the students in the three secondary school classes where we teach, with a total size of 57 students. To do this, we drew up a questionnaire to which 49 students responded.

As we can see from Table 2, the sample consisted mostly of females (61.2%) and the 16 years of age dominates (40.8%). We associated these with 17 and 18-year-old students (30.6% and 22.4%, respectively), reducing the sample to only one 15-year-old student (2%) and only two twenty-year-olds (4%).

It is not surprising to note that most of the students surveyed are Portuguese (84%), followed by Brazilian and Angolan students (4% each). In these classes we found one student of India and another from Russia, with two of dual-nationality—one Portuguese Brazilian and the other Portuguese Belarusian (Fig. 2). In fact, the total number of international students in this sample is only 8 (16%).

When asked about how they feel in their class and at school (Fig. 3), even for Portuguese students, it is possible to identify a feeling of greater integration at school

Table 2 Profile of respondents

Students	Frequency	%
Gender		
Female	30	61.2
Male	19	38.8
Age		
15	1	2
16	20	40.8
17	15	30.6
18	11	22.4
19	0	0
20	2	4
Total	49	100

Source Own elaboration

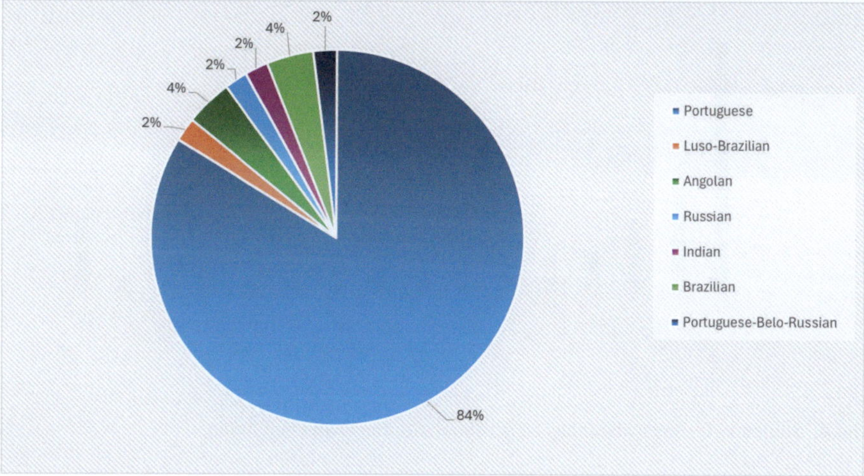

Fig. 2 Nationality of respondents. *Source* Own elaboration

level (37%) than at class level (27%), highlighting the fact that 33% of students report that school and class are indifferent to them.

It should also be noted that four of these students feel 'little integrated' or even 'not integrated at all'.

Considering the feelings most present in these students' daily lives when they are in school, as we can see by analysing the graph in Fig. 4, most of the students (31%) were happy, but they were also bored (15%), sad (9%) and angry (5%), or even tired and indifferent. In other words, we can see that most of the students don't experience positive feelings when they are at school.

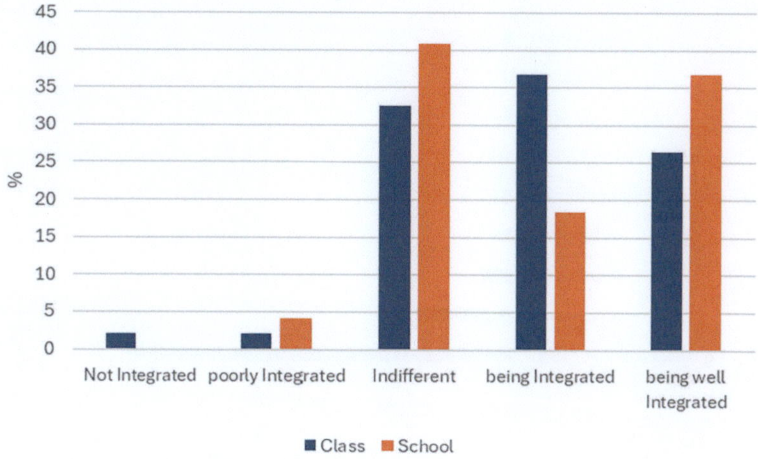

Fig. 3 Students' sense of integration into the class and school. *Source* Own Elaboration

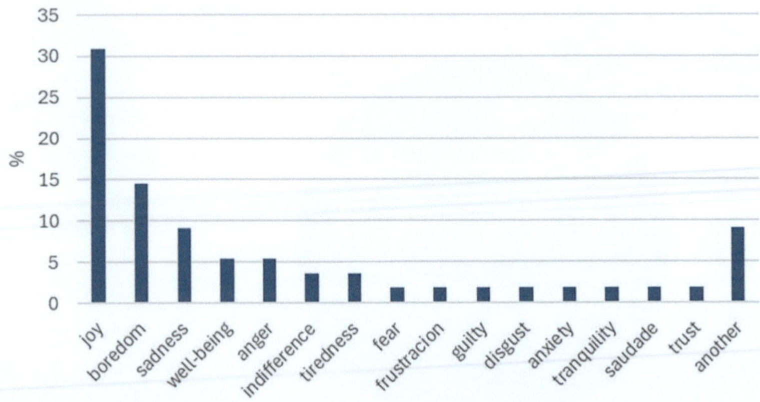

Fig. 4 Students' feelings when they're at school. *Source* Own elaboration

Trying to understand these feelings, that is, to know the greatest difficulties faced by students of other countries in their daily lives at school, we can verify by analysing the graph in Fig. 5, that language/communication is the main difficulty (63%), to which we can associate integration (16%), as linguistic limits hinder the relationship with the school community. This is followed by xenophobia (5%), the difficulty of 'adaptation', 'culture shock' and the 'different educational system' (4%). In addition, when only these students were directly questioned, they also mentioned 'shame', lack of 'conviviality' and even the simple fact of 'being foreigners' as the biggest barriers to their own integration into school.

These students pointed out that their biggest problems are essentially due to 'the fact that they have no family in Portugal and, regarding school, the fact that they came without knowing anyone'. Some added: 'when I arrived in Portugal, I had

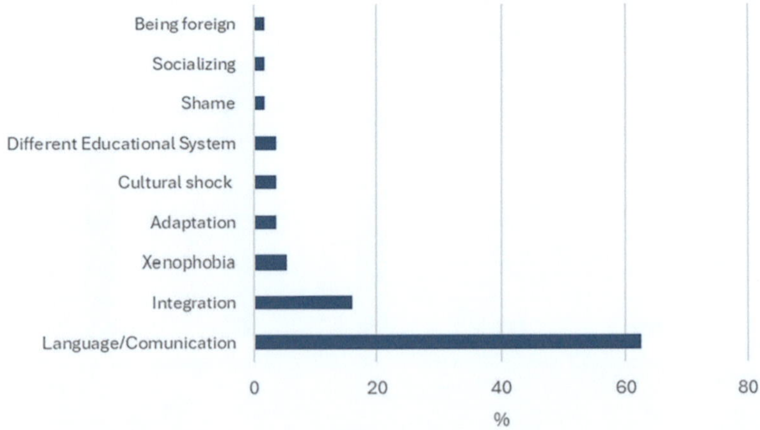

Fig. 5 The greater difficulty of foreign students in attending this school. *Source* Own elaboration

difficulty understanding (…) because it's a different language'; or said 'my difficulty was adapting to the school and the country, both are very different from mine'; or mention the fact that 'I had to do it all over again, new friends, language, etc.'

Aware of the role of the school in the integration of these students, we asked about the activities developed in the school context and, although most of these students (47%) reported that the school develops activities that meet their cultural habits (Fig. 6), 12% answered 'sometimes' and 37% answered that these are 'clearly insufficient'.

When we asked the students about which elements of the cultural identity of each community they would like to see disseminated and valued in curricular or extracurricular activities (Fig. 7), the students highlighted Gastronomy (40%), followed by Traditions and Customs (17%) and the History of the country (15%). Music was also a factor chosen by the students, surpassing Language and Religion (13%).

Fig. 6 The school should integrate the cultural habits and traditions of foreign students into extracurricular activities. *Source* Own elaboration

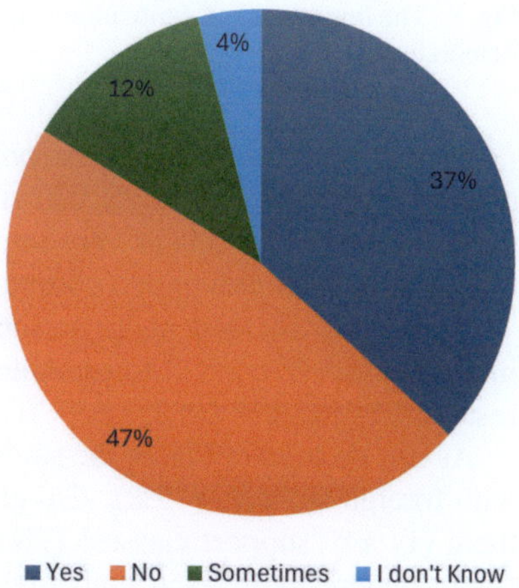

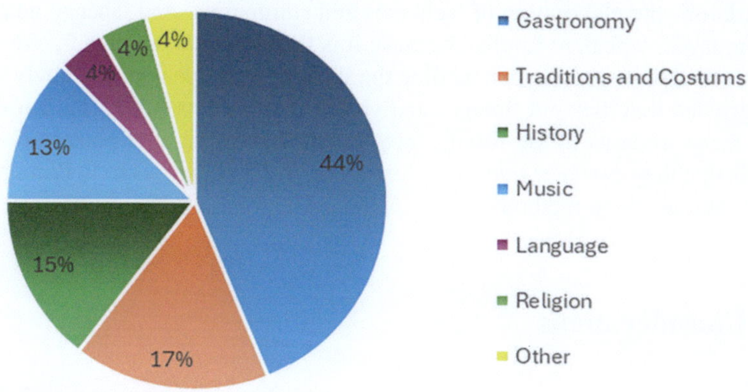

Fig. 7 Elements of cultural identity to emphasise in the school context. *Source* Own elaboration

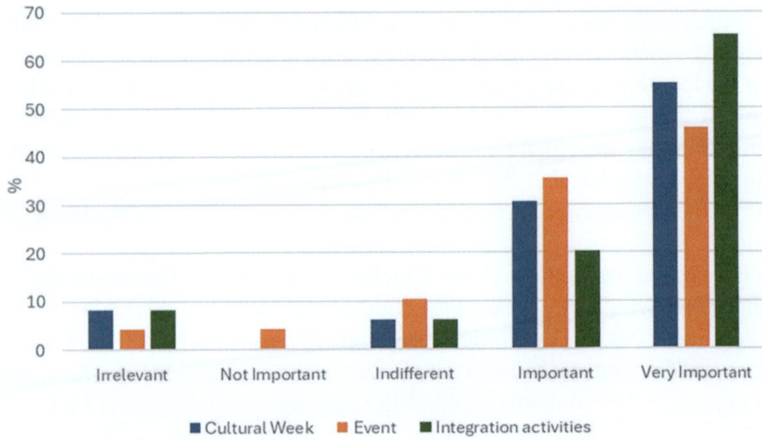

Fig. 8 The importance of multicultural activities to be developed in a school context. *Source* Own Elaboration

Facing the possibility of carrying out multicultural activities in the school context, we could also verify that students highlight any and all integration activities for foreign students as very important (32%), with multicultural events (22%) highlighting the holding of a cultural week (27%) (Fig. 8).

Finally, we ask students to present proposals for work/strategies to be developed in the classroom context to promote a multicultural education that allows positive results.

As can be seen in Fig. 9, the students' proposals highlight: 'talking individually' with foreign students (15%), also giving them 'extra attention' (9%), to 'integrate' them (11%); 'support classes' (7%), 'help in the Portuguese language' (7%) and 'more understanding and empathy' (5%). It was also suggested to 'increase the participation of foreign students' (28%), the 'permission of the mobile phone and the translator', the elaboration of 'schemes and summaries' and 'sheets' adapted to their languages, 'speak in English because it is the universal language', 'frequently question whether they are understanding the subject' and 'speak more slowly'. It was also suggested that they put 'images and videos in their language', 'put a translator next to these students in the room', 'create and make available presentations and materials that they can later translate', 'explain how teaching works in Portugal' and the 'different teaching methods'.

Final Considerations

We live in a hyper-connected and global world that poses enormous challenges to education as schools encompass a range of students with increasingly diverse cultural backgrounds. In this sense, it is fundamental to ensure an education system

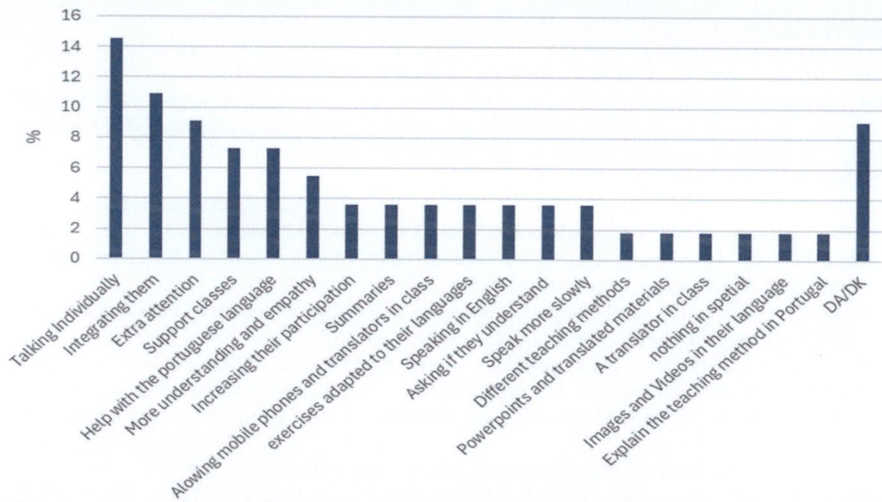

Fig. 9 Teacher proposals to minimise difficulties of foreign students

that welcomes and integrates students of different nationalities and cultural habits in a learning community with equal opportunities for success should be promoted.

This study leaves no doubt. Multicultural education is decisive for today's Portuguese schools. And Fontes Pereira de Melo Primary and Secondary School is a good example not only of the cultural range of the students who are part of the school community but also for the awareness that teachers demonstrate of the need for didactic and pedagogical innovation instruments that allow to improve the integration of all students.

As we could see, this school's geography teachers consider that their action is decisive in the process of foreign students' integration and, by listing the syllabus in which multiculturalism can be explored, they reinforce the statement of Wellens et al. (2006) that geography is well-placed to play a pivotal role in social transformation, thus lending itself to multicultural education. These teachers also say that they adapt the content, and the language used in the classroom to these students and insert them into different working groups, allowing them to talk about their cultural elements and traditions, among others.

These facts are further reinforced by the students by stressing the need for cultural interaction activities within the school and not only in the classroom, but activities also where their cultural habits and traditions are highlighted to enhance their identity and minimise the socio-cultural impact they experience. Throughout all the questions of the survey carried out, we were able to verify the value attributed by all students (not just foreigners) to the cultural integration of foreign students through an express action of the various educational agents in the entire school co-text and not only within the classroom.

References

Agrupamento de Escolas Fontes Pereira de Melo (2019) Regulamento Interno, pp 1–104. Available: Final—RI AEFPM—retificado julho2019.pdf (aefontespmelo.com), accessed [28/12/2023]
Banks J (1985) Multicultural education. In: The international encyclopedia of education, vol 6. Pergamon Press, Oxford, pp 3440–3442
Bigelow B, Christensen L, Karp S, Miner B, Peterson B (1994) Creating classrooms for equity and social justice: rethinking our classrooms. Rethinking school, Milwaukee
Canclini N (2004) Diferentes, Desiguales y Desconectados. Mapas de la interculturalidade. Gedisa, Barcelona, Espanha
Cardoso C (1994) Pedagogias diferenciadas para a educação multicultural. Pensar a avaliação, melhorar a aprendizagem. Direção Geral da Educação, pp 1–5
Catling S (2005) Seeking younger children's voices in geographical education research. Int Res Geogr Environ Educ 14(4):297–304. https://doi.org/10.1080/10382040508668362
Catling S, Willy T (2020) Understanding and teaching primary geography. Sage, London. ID: 5018867
Cavalcanti L (2012) O ensino de Geografia na escola. Papirus, Campinas, S.P
Diário da Républica Eletrónico (2024) Lei de Bases do Sistema Educativo. Lei n.º 46/86—Diário da República n.º 237/1986, Série I de 1986–10–14
Dolan A (2020) Powerful primary geography: a toolkit for 21st century learning. Routledge, New York. https://doi.org/10.4324/9781315397542
Gay G (2010) Culturally responsive teaching: theory, research and practice. Teachers College Press, Columbia University, New York
Machon P, Walkington H (2000) Citizenship: the role of geography? In: Kent A (ed) Reflective practice in geography teaching. SAGE, London, pp 53–105
Machon P (1998) Citizenship and geography. Educ Teach Geogr 23(3):115–117
Martin F (2008) Ethnogeography: towards liberatory geography education. Children's Geographies 6(4):437–450. https://doi.org/10.1080/14733280802338130
Pike S (2011) If you went out it would stick: Irish children's learning in their local environments. Int Res Geogr Environ Educ 20(2):139–159. https://doi.org/10.1080/10382046.2011.564787
Pike S (2016) Learning primary geography: ideas and inspirations from classrooms. Routledge, London. https://doi.org/10.4324/9781315684222
Roberts M (2003) Geography through enquiry: approaches to teaching and learning in the secondary school. The Geographical Association, Sheffield
Roberts M (2014) Powerful knowledge and geographical education. Curric J 25(2):187–209. https://doi.org/10.1080/09585176.2014.894481
Silva M (2015) Educação Multicultural: atualidade e pertinência. Revista De Ciências Sociais, Humanas e Engenharias 1(1):41–51
Smith C (2020) Estudantes Internacionais e as suas experiências académicas: Satisfação do aluno, desafios de sucesso dos alunos e práticas de ensino promissoras. Universidade de Windsor, pp 271–287
Viana B, Silva G (2020) Multiculturalismo no ensino de Geografia: Possibilidades através do conceito de lugar. Geografia: Publicações Avulsas. Universidade Federal do Piauí, Teresina V.2, 2:209–226
Wellens J, Berardi A, Chalkley B, Chambers B, Healey R, Monk J, Vender J (2006) Teaching geography for social transformation. J Geogr High Educ 30(1):117–131
Young and Lambert (2014) Knowledge and the future school: curriculum and social justice. Bloomsbury, London

“Teachers for What?”—Inertia and Emergency in Basic Education in Portugal

Elsa Pacheco, Laura Soares, and Paulo Lemos

Abstract The Portuguese education system has been pursuing the goal of UNICEF World Declaration on Education for All—*the fulfilment of basic learning needs.* However, there is still resistance to the change in curricula and school organisation. The perpetuation of educational evaluation centred on school results disguises the training paths fundamental to true human empowerment—*being, the essential.* In line with these ideas, some questions emerge: why is there so much resistance to changing the essentials? Do times change, but not wishes? What are the changes in Portuguese education? How can we modify schools in the present context? We revisited Smith's (1789) and Marshall's (1879) human condition for the progress and wealth of nations, looked at Schultz's Theory of Human Capital (1963, 1971), and noted Gusdorf's (Professores Para Quê? Para uma pedagogia da pedagogia, 1963) answers to the question *Teachers for what* and recalled the Modern School Movement. For Higher Education, Recommendation 4/2022 calls for the participation of young people, identifying several problems of the pedagogical model, including the fact that it is unidirectional. We hope to find some answers through a cross-analysis between (1) desires—based on the principles of education defined in legal documents; (2) supply—approached from the point of view of the curriculum and training of geography teachers; (3) demand—based on the dimension of students who choose geography.

Keywords Education · Geography · Curriculum innovation · Teaching method

E. Pacheco (✉)
Faculty of Arts and Humanities of University of Porto, Transdisciplinary Research Center for Culture, Space and Memory (CITCEM), University of Porto, Porto, Portugal
e-mail: elsap@letras.up.pt

L. Soares
Faculty of Arts and Humanities of University of Porto, Geography and Spatial Planning Research Centre (CEGOT), University of Porto, Porto, Portugal
e-mail: lauras@letras.up.pt

P. Lemos
Geography and Spatial Planning Research Centre (CEGOT), Porto, Portugal
e-mail: paulomclemos@outlook.com

L. Soares et al. (eds.), *Compromised Geography*, Key Challenges in Geography,
https://doi.org/10.1007/978-3-032-05201-8_4

Introduction: Topics on the Essence and Essentials of Teaching Geography

Geography plays a crucial role in human, social and economic development and is essential for empowering citizens with the skills and resilience needed to face the challenges of the twenty-first century. As stated in the *International Charter on Geographical Education* (IGU/CGE 2016, pp 4–5) "(...) studying geography helps people to understand and appreciate how places and landscapes are formed, how people and environments interact, the consequences that arise from our everyday spatial decisions, and Earth's diverse and interconnected mosaic of cultures and societies."

In this context, geography teachers are vital in helping students understand the complex interactions between humans and the environment. Given the dynamic nature of geographical changes, it is essential for teachers to stay updated and ensure that learning remains meaningful (Ausubel 1968), both in terms of understanding the challenges at different scales and fostering the development of citizens who recognise the increasing significance of solidarity and sustainability (Martins et al. 2017; Monteiro et al. 2017).

There is no doubt that geography is fundamental to the education of any citizen. As a science, geography finds its roots in the efforts to discover, position and understand spatial distributions and interactions, challenges that remain valid today. However, in academia, the growing complexity of geographical spaces and technological advances have strengthened scientific specialisation with the emergence of large-scale, layered approaches. In the same way, basic schools have pursued what is regarded as fundamental, elementary and essential to learn in geography. In this context, the long and complex transposition of knowledge between university and school follows the evolution of scientific knowledge and social and political issues, creating difficulties in designing a curriculum that is in line with the principles (desires) of modern education (Lemos et al. 2023).

Today, just as in the past with other technological tools, many of our young people are looking towards artificial intelligence (AI) for answers to their questions and concerns, and there is nothing wrong with that. The problem is that the education system and the curricula (in particular) have not allowed students to work independently! If we fail to develop our students' critical and creative abilities, the positive contributions that AI can and should make to education could be completely skewed. As Luckin et al. (2016, p. 34), point out,

> AIEd has the tools and techniques to conduct the fine-grained analysis that allows us to track each learner's development of skills and capabilities as they interact and learn over time. This tracking of individual learners can then be collated and interpreted as required to provide knowledge about progress at the school, district, and country level.

The rapid advance of technology has increased the potential for connection between teachers and students, both inside and outside the classroom. However, the time teachers spend outside of their working hours (i.e. communicating online) does not count for their workload! So, when students consider the content taught to be useless, irrelevant or without substance and the teacher, in turn, takes the direction of what is necessary to learn but does not see his work recognised, we can ask *School for what* or, remembering Gusdorf (1963), *Teachers for what?* These questions become even more critical when we think about what is intended, when we reflect on the role of teachers in their relationship with the formation of individuals and how this impacts economic and social development. Some answers can be retraced in Smith's (1789) vision of the human condition, Marshall's (1879) approach to the progress and wealth of nations or Schultz's (1960) theory of human capital.

Although seen from the perspective of the division of labour and productivity, in Smith's view (op. cit., 1789) education is crucial to the development of individuals' abilities, in which empathy and social interaction play an important role in understanding interdependence on multiple scales resulting from the impact of human actions. Similarly, based on the appropriate combination of physical and human capital, Marshall (op. cit., 1879) attributes education as the central engine of progress, as it improves the productive skills and innovative capacity of individuals. In line with these ideas, Schultz (op. cit., 1960) considers that investment in education produces human capital and is, therefore, one of the most effective ways of increasing the productivity and income of people.

In this context, as Gusdorf asserts, teachers are essential for raising citizens who contribute to collective well-being, not limiting learning to theoretical knowledge, and promoting economic and social progress. The three authors answer Gusdorf's question, albeit in earlier periods: teachers are fundamental to forming ethically and socially aware citizens, capable of contributing to a just society (Smith); they are responsible for driving economic and social progress, developing skills and competencies relevant to the life of nations (Marshall); they are essential for transforming education into human capital, promoting innovation and adaptability (Schultz).

If, in essence, it has been considered for centuries that education should transcend theory, valuing the integral development of the human being, then, in a world whose changes are taking place ever more rapidly, geography is essential in this formation. In essence, it not only provides a holistic view of the world, integrating cultural, natural and social dimensions but also stimulates territorial, environmental and multicultural awareness, promoting a deep understanding of the relationship between human beings and the environment.

As Kaercher states (1996, p. 113, our translation), "If our students can use geography as a useful tool for reading the world, we will be helping to build not only a school, but also a society that is more critical and indignant towards all human misery."

Emergence and Inertia of Change

As we previously mentioned, the growing complexity of geographical spaces and technological advances have evolved and consolidated academic specialisation, particularly with the large-scale approaches that seek to position themselves in (apparently contradictory) layers that cross between the global and the local. In the same way, basic and secondary schools have pursued what at each moment in history has been considered fundamental, elementary, and what is essential to learn in geography. However, because education systems are highly complex—hence the enormous difficulty in making fundamental changes—the desired transformation cannot follow the vertigo of social, economic and technological evolution.

In Portugal, the legal principles for basic and secondary education, set out in the Basic Law of the Education System (Law 46/86, with the most recent changes), already pointed to goals focused on the integral formation of the individual, namely through the promotion of essential and specific skills for the critical interpretation of the world and active participation in society. These principles echo in the teacher training curriculum—mediators of knowledge and promoters of meaningful learning—which must be permanently attentive to changes in the demand for geographical knowledge, i.e. students' learning processes. Similarly, and it may seem innovative, the *Profile of Students Leaving Compulsory Schooling* (Martins et al. 2017) also assigns great importance to the humanistic dimension and the formation of citizens prepared for the unpredictability of the world. In fact, this profile

> (…) points to a school education in which students of this global generation build and consolidate a scientific and artistic culture with a humanistic foundation. To this end, they mobilise values and skills that enable them to intervene in the lives and history of individuals and societies, make free and informed decisions on natural, social and ethical issues, and have the capacity for active, conscious and responsible civic participation (op. cit., p.10)

But over the last two centuries, we have seen the consolidation of the principle that education is fundamental to the development of nations. Smith and Marshall's approaches to the human condition support the idea that people's qualities and abilities are a form of capital that education should promote. To achieve the concept of human capital, taken from the perspective of economic advantage, was a small step. The impetus was given by the Universal Declaration of Human Rights (UN 1948)—pursuing principles of equality and justice in post-war reconstruction—which states, among other principles, that everyone has the right to education. Over a decade later, in the 1960s, the confrontation between the capitalist perspective of school functions and the universal and humanist concept of education gave rise to various reflections and movements. In his book *Teachers for What*, Gusdorf begins the second chapter by defending the importance of dialogue in the classroom—a practice he claims was ignored by pedagogical theorists. To pedagogues, teaching is reduced to a monologue, which leads Gusdorf to suggest that the teacher could be replaced by a book or a radio station… (perhaps like today!). In line with these ideas, the Modern School Movement, with the initial aspiration of developing a new model of teacher and creating a differentiated pedagogy (Lico 2023) sees the school as a free space where people learn to live.

Despite successive appeals to the risks of ongoing evolution, in other words, appeals to the emergence of change, the massification of education has become prevalent in most educational systems. As Dubet (2003) points out, "Among the direct consequences of school massification, we must consider all the mechanisms of internal differentiation that structure the system. The school offer is not homogeneous and does not always produce the same performance; it does not always have the same effectiveness. The first differentiation mechanism is the development of pathways built much more according to performance criteria rather than on the genuine choices or preferences of the students." (Dubet 2003, p. 35–36, our translation). Additionally, other factors contribute to this differentiation, such as the disparities between public and private schools and the socio-economic contexts in which they operate. Thus, it appears that the emphasis on the economic valuation of knowledge, as described in Schultz's "Investment in Human Capital" (1973), has ultimately prevented the adoption of alternative school models.

Changes in education systems, driven by technical and technological advances, have been advocated since the late nineteenth century. Evidence of this can be observed in illustrations that have reached the current public through the dissemination provided by the internet. In 1899, for instance, Jean-Marc Côté imagined what a classroom might look like in the year 2000 (Fig. 1). In his mind, the teacher stands to the side of students while they do not face him directly. Notably, the teacher's knowledge is supported by textbooks processed by a machine, from which wires transmit educational content to each student via headphones (Asimov et al. 1986).

Although there are substantial differences, we cannot forget Aldous Huxley in his Brave New World—published in 1932—and the way *hypnopaedia* was used to teach ethics to children to "manipulate the mind, create distinct conditioned reflexes in the individual and teach them to conform to their destiny and develop prejudice between the castes" (Martins Filho 2003, p. 104).

Fig. 1 Classroom 2000 envisioned by Jean-Marc Côté in 1899. Extracted from https://publicdomainreview.org/collection/a-19th-century-vision-of-the-year-2000

Nearly twenty years later, inspired by the dizzying technological evolution triggered by the launch of the Soviet satellite Sputnik in 1957, the cartoonist Arthur Radebaugh illustrated, a year later, what could be the result of the concerns in the USA about the need to reformulate the school curriculum with a strong investment in science education. From Novak (2012), we recover this image (Fig. 2), whose description explains that:

> Tomorrow's schools will be more crowded; teachers will be correspondingly fewer. Plans for a push-button school have already been proposed (...) Teaching would be by means of sound movies and mechanical tabulating machines. Pupils would record attendance and answer questions by pushing buttons. Special machines would be "geared" for each individual student so he could advance as rapidly as his abilities warranted. Progress records, also kept by machine, would be periodically reviewed by skilled teachers, and personal help would be available when necessary

Here, the teacher ceases to exist physically, and communication with the pupils takes place on individual electronic media with buttons, giving rise to what Radebaugh calls "Push-button Education"—a scenario that does not yet fully exist!

In 1990, the *World Declaration on Education for All* assumed as a principle the Satisfaction of Basic Learning Needs, indicating, with the concept of "basic needs," the necessary review of the purposes and adequacy of school education, which was one of the fundamental seeds for the *Essential Learnings* first defined, in Portugal, in 2017. Meanwhile, in the mid-twenties of the twenty-first century, new challenges are being set to education systems, revealing the consolidation of Côté and Radebaugh's visions whose evolution has resulted in the illustration, made through AI, that we present in Fig. 3.

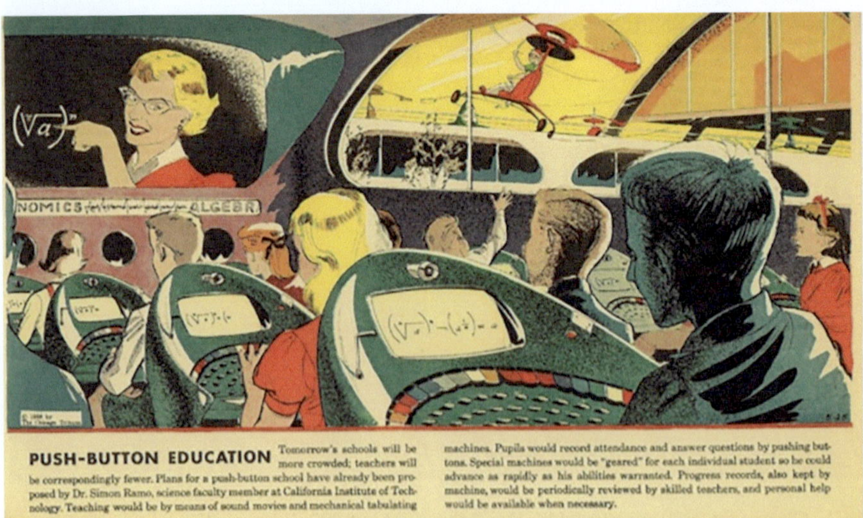

Fig. 2 "Push-button Education," by Arthur Radebaugh (1958). Extracted from Novak (2012, p. 5)

Fig. 3 Classroom in 2100 based on Côté and Radebaugh's trends. *Source* Response to the prompt "Based on the visions of Côté (1899) and Radebaugh (1958) make an illustration of a classroom in the year 2100" (Copilot, 3/5/2025)

The evolution will move towards expanding bots to support human actions, whose importance, represented by large robots, is decisive for the necessary individualisation of teaching. However, the teacher remains at the forefront for students, with an aura identical to that of the electronic devices used by students viewing different content. Teachers will eventually play a fundamental role, especially in their approach to human beings, i.e. their predisposition to learn, depending on the social and emotional context determining their motivation.

In other words, it is a highly complex picture that goes beyond what we consider essential to learn! So, suppose we already have an idea of what we want. In that case, the path suggests that we reflect more carefully on improving the alignment of essential content, with the inevitable focus on the student's individuality. However, before addressing this issue, it is necessary to resolve the fragility of the arguments that lead students to choose geography by their own choice, arguments that mostly do not recognise its meaning and/or usefulness in life.

The Desires

The school that everyone wants is the one that can prepare students for the challenges of the future, using innovative methodologies adapted to contemporary needs, a future in which geography plays a crucial role in addressing fundamental issues such as territorial diversity, sustainability and globalisation—essential topics for the development of skills that will enable citizens to face increasingly complex problems.

In recent years, school curricula have been constantly updated to include objectives, goals and essential learning outcomes (ELOs) that seek to incorporate the above-mentioned issues. Interdisciplinarity and the use of technology are emphasised, enabling students to acquire a holistic and integrated view of the world. However, the achievement of these objectives faces significant challenges, especially due to the rapid change of established methodologies and working habits.

Innovation and generational change require teachers to adapt their teaching practices appropriately and flexibly. However, many teacher trainers and educators find it challenging to keep up with the rapid pace of recent changes. The way young people learn today differs from the past, and there is still much to understand about their learning processes. In other words, the training of our teachers should be based on methodologies that engage students and on the relevance of geography in the contemporary world so that, whatever the content to be taught, they can captivate students with what makes sense and is useful. This is not always an easy task, particularly because it is necessary to take advantage of the spatial knowledge and skills that students already possess so that they can be transposed and readapted to the volatility of territorial evolution.

The foundations for change were laid when the Education System Base Law was published in 1986. It states: "in access to education and its practice, all Portuguese citizens have the right to respect and for the principle of freedom to learn and teach, with tolerance for possible choices." This principle of school autonomy underwent several developments during the 1990s. However, only thirty years later, the meaning of Autonomy and Curricular Flexibility was defined (Order No. 5908/2017, of 5 July). Ten years later, and almost forty years after the Base Law, despite the efforts made, we are still a long way from implementing the change needed to respond to the emergency. The *Profile of Students Leaving Compulsory Education* (Order No. 6478/2017, 26 July) defines the general principles of what society (or politics) wants citizens to be when they leave compulsory education. However, it also contains some contradictions with the principles of individual development and freedom of citizens.

That is how the desire to satisfy basic learning needs is pursued. However, since the twentieth century, changes in the Portuguese education system have resulted in the consolidation of massification: the extension of compulsory schooling and the democratisation of access to education are important milestones, but they also bring significant challenges (Queiroz 2024). Particularly when we consider goals such as the fundamental individualisation of teaching and inclusion, which seem contradictory to mass education. In other words, technological advances, which should promote more student-centred and differentiated teaching and learning, have brought many advantages and just as many problems: the potential to transform the learning experience, making it more personalised, collaborative, interactive and innovative, is counterbalanced by obstacles that often lie in difficulty teachers have in applying the knowledge they have acquired throughout their training, in resistance to change or simply because there is still a great deal of inequality in access to technological resources (Bocconi et al. 2022).

There seems to be no doubt that the ideal school should be a place where teaching is dynamic and interactive, respectful of difference and diversity, using advanced technologies and interdisciplinary approaches in such a way that students feel attracted to the spaces and times of teaching and learning. The school we all want is the one that prepares citizens to be critical, aware and capable of acting in a world that is only globalised because everyone lives and interacts with each other—a goal that history has shown to be challenging to achieve, as it requires a joint effort between educators, teacher trainers and educational institutions and, above all, strong political influence and involvement.

The Offer

Suppose it is obvious that the desire expressed in the legal principles for education should be reflected both in teacher training curricula and in the teaching practice of geography. Is there agreement and coordination to implement the necessary change simultaneously in initial teacher training and EBS schools?

The EBS teachers, as part of the education system, are trained for over 20 years in four main areas: general educational formation, which is transversal to all subject areas (including psychology, ethics and curriculum); scientific formation, which aims to improve and deepen knowledge related to EBS contents; didactic formation, materialised in the experiential use of teaching–learning methodologies; and professional practice, which take place in the real context of Cooperating Schools. The government determines the weighting of each of these areas in university study plans, always considering the needs of the moment.

More recently, with the shortage of teachers in basic and secondary education, it was decided to increase the number of hours of the Introduction to Professional Practice internship component at the expense of training in the scientific area and general education. The only sector that maintains the previous ECTS load is that of specific teaching methods, but this is a misleading stabilisation since the Decree Law requires the addition of aspects of inclusion and technology to these curricular units! In any case, adjustments were previously made to the distribution of credits between areas of initial teacher training. However, there was always difficulty in coordinating academic training with what is necessary (because we would have to know how young people learn, and we do not!) to be the "teacher of today and tomorrow," particularly with the banalisation of geographical knowledge and the inertia that exists in many basic and secondary schools.

> In academia, regarding teachers in Portuguese universities, where the emerging change was supposed to take place, the following can be read in *Diário da República* (Official Gazette) No. 124, of 2022 (our translation): The pedagogical model still dominant in higher education institutions can be characterised as unidirectional, (...) with the teacher exposing the knowledge to students; with little participation, (...) students do not frequently intervene and rarely take a stance of sharing knowledge; With a heavy workload; Perpetuating high levels of anxiety because it is a model based on competitiveness, where the final objective

of the entire learning process is to pass exams; It does not value the use of digital tools to support teaching for sharing content and asynchronous moments

This description highlights the urgent need to reform the pedagogical model in higher education institutions, promoting a more participatory approach focusing less on competition.

Returning to EBS, although geography is a compulsory discipline in the early years of schooling, its popularity at more advanced levels (such as secondary and university education) may be affected by fundamental aspects such as the association that candidates make with memorising maps and concepts widely disseminated in the media, games, etc. Similarly, their difficulty in transposing geographical content into interpreting their daily lives and the world or establishing a direct link with the labour market can discourage them from pursuing geography themes in greater depth. Indeed, in 1980, a survey was conducted among 137 secondary school students in Lisbon (Melo 1984), and forty years later, we repeated the question to 142 students in Porto. It is interesting to note that the main reasons remain the same and that more than half of the students choose geography in secondary school because they are not interested in other subjects and/or because it is an easy subject that allows them to obtain high grades—the latter reason rose to first place in 2020, increasing in percentage terms (Fig. 4).

Given this scenario, we must question which geography everyone must learn in basic and secondary education. How can and should geography contribute to essential learning? What is essential to teach and learn in compulsory geography? What is the usefulness of geography in the life of any citizen?

Some of these questions were asked to our first-year students of the geography undergraduate in 2023 (via Google Forms), but in an implicit way, namely, why they chose to attend this course and what geography means to them. The responses are now more positive, considering that over 95% of respondents chose geography as

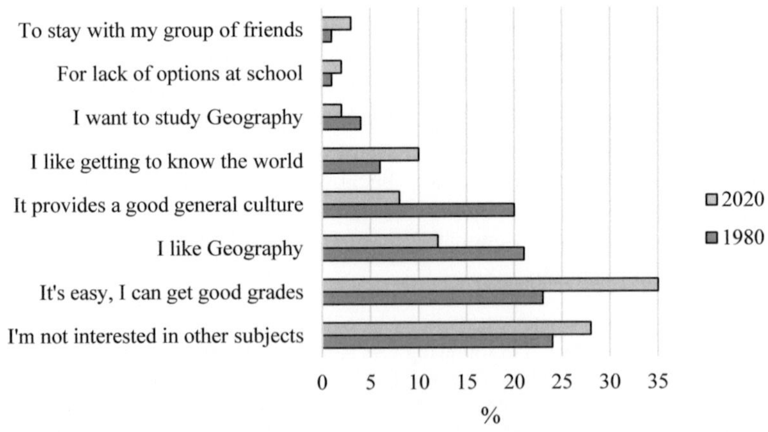

Fig. 4 Answer to the question "Why did you choose Geography?" in 1980 and 2020. Adapted from Melo (1984) and own survey (2020)

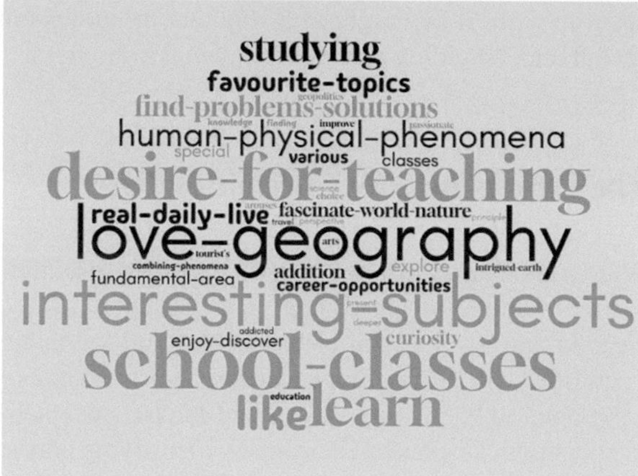

Fig. 5 Word cloud that expresses the answer to the question, “Why did you choose to attend the Geography undergraduate course?” Own survey

their first choice. Regarding the reasons that led them to select the course (Fig. 5), we can highlight the fact that they like geography because of its interesting subjects, namely the interconnected human-physical phenomena that they consider to be part of their real daily life, their fascination for world nature and their interest in finding solutions to world problems. Also noteworthy is the desire to be a teacher, a career of opportunities today. One of the factors that influenced their choice was often the geography teacher they had in their basic and secondary school classes.

Regarding the second question (what does geography mean to you?), several answers highlight the unique nature of geography as a science that studies, in an interconnected way, the physical and human environment and the relationships between them to understand the world around us. However, we would like to highlight those that we find most interesting:

(a) *“For me, geography is a motivation. Knowing that I am going to study it motivates me to learn more about the world.”*
(b) *“For me, geography means finding social or environmental problems and discovering the best solutions to them. It’s everything that surrounds us.”*
(c) *“Geography is the subject that addresses all current issues and allows us to discuss them. It is one of the most comprehensive and enlightening subjects as we learn about what is happening in our country and the world and the reasons behind these events.”*
(d) *“Geography is the most comprehensive and essential science because, without geography, we cannot understand the world we live in, whether socially or environmentally. It is the key to the organisation of society.”*

Geography plays a fundamental role in shaping informed and critical citizens, helping them to understand the world we live in, the interactions between humans and the environment, and global challenges such, among others, climate change and its associated risks, natural resources, migration, development contrasts, geostrategy. In other words, in basic and secondary education (with subjects for all citizens), geography must be taught in a way that promotes critical and networked thinking, spatial understanding and environmental awareness.

The Emergency of Overcoming Inertia

Within the framework of what we consider essential in geography, let us focus on a brief example of the content and concepts at the beginning of the study cycles (second and third cycles of basic level and secondary education). Its organisation is questionable: for example, does it not make more sense for an introduction to geography to begin with sketches of landscapes, their description and identification in geographical space? Is starting by identifying and locating the geometric elements of the Earth on a map not too abstract for ten or eleven-year-old students? Even in the fifth year (the first in which they have contact with geography), memorisation is developed through meaning. However, if they do not even have cognitive development to understand the "geometric elements of the terrestrial sphere," it will be more difficult to mobilise what they have memorised, as proposed in the strategies of Essential Learnings (DGE 2023). Is this the best way to start?

The difficulty in finding precise meanings that allow us to break with inertia is evidence that we live in a state of "educational emergency" because we cannot escape a system where alternative realities seem to coexist! In this regard, we evoke here one of Rob Gonsalves' paintings (1996) which, in a very appropriate way, depicts a "flight plan" (Fig. 6), in the style of magical realism, highlighting that its realisation always depends on the person point of view.

In Sarah Thomson's fictionalised version (2011), the writer directs Gonsalves illustrations to younger audiences to comment on this image appropriately: "Imagine a day… when the edge of the map is only the beginning of what we can explore." In other words, we admit that this perspective may be at the centre of the problem because it concerns essential knowledge and realities for our young people. However, in the eyes of their elders (teachers and those responsible for modernising the curriculum), they remain obscured!

Indeed, it seems clear that the contents do not make sense, meaning that students retain little or nothing about the subject at the beginning of each cycle and even with repetition. They arrive at university and still do not have much of the knowledge that we define as basic/essential. In other words, even though we have a curriculum structure in place for decades, we have done a lot… but there is still much more to do (Queiroz 2024). If we define Essential Learning that does not translate into skills or abilities, is it because they are not helpful and, therefore, more compulsory than essential? We know what we want for a long time. At least as long as we have been living in an educational emergency. That is, every time there is a transformation (largely technical and/or technological) responsible for more or less severe ruptures between generations.

Fig. 6 "Flight Plan," de Rob Gonsalves (1996). Source: https://www.robgonsalves.live/ (accessed in 2024/02/01)

At the same time, we have reached a situation where access to education is (or should be) universal. However, the solution took the easy path, resulting in the consolidation of school "massification," in which the definition of a student profile, albeit supported by a strongly humanistic perspective, was not accompanied by the necessary fundamental articulation with the curricula, which were heavily influenced by past guidelines. That undermines the autonomy and curricular flexibility that enable students to develop independently (Pintassilgo and Mogarro 2003).

Additionally, and no less important, the change in educational practices is not worth it. On the contrary, they seem to create more work! If essential learnings are in textbooks and countless resources available to students, teachers, as a fundamental element of didactic transposition, should be recognised for the success of their student's education, a success that we continue to evaluate and reward through grades and rankings.

To ensure that education in Portugal truly meets basic learning needs, a continuous effort of innovation and adaptation is needed, based much more on the reform of pedagogical models than on technological integration. Only then can we guarantee an education that empowers students for the challenges of the future (CNE 2024).

Final Remarks

Teacher training and basic/secondary education in Portugal face significant challenges, including in geography. As mentioned by Pezzini and Szymanski (2018, p. 1, our translation), many students "(…) attend classes as an obligation, without, however, participating in basic activities, remain apathetic in the face of any initiative by teachers, miss classes without justification, fail tests and do not seek to reschedule them, and ultimately frustrate teachers by failing to fully achieve their objectives." To solve these problems, those responsible for the education system must understand what is fundamental to learning in geography and, above all, that they can position themselves from the student's point of view. We must urgently develop strategies that change the answers to the questions: What is the purpose of geography we learn in basic education? How is geography useful in the life of any citizen? How can geography contribute to meeting the basic needs of education, as defined by the World Declaration on Education for All (1990)? In a way consistent with what we want.

We already know that geography is fundamental to educating citizens who needs to be prepared for increasingly unpredictable challenges, contributing to human, social and economic development and thus helping us to understand the world we live in, the interactions between human beings and the environment, and the global challenges we face. However, as responsible for these processes, we must further explore the effects of our actions so that geography is seen, particularly by young people, as a fundamental area of education for life.

If it were not one of the fundamental problems that urgently needs to be solved—excessive class sizes in most schools—we could develop the necessary approach to the needs of each student, treating them as explorers and discoverers in finding ways to learn and make plans for their discoveries. Once again, this has already been proven, for example, at Escola da Ponte, where students have been allowed to pursue a path of freedom and personal development imagined and/or desired since 1976. At the same time, teachers and schools should be rewarded for their ability to bring about significant change.

While these fundamental actions fail to materialise, we can desire and make changes here and there. However, significant resistance will remain to adapt (or, better yet, breaking with) the essential elements of the education system. So, it is urgent to clearly define the usefulness of geography in the lives of citizens and how it can contribute to their daily lives.

References

Asimov I, Côté JM (1986) Futuredays: a nineteenth-century vision of the year 2000. H. Holt

Ausubel D (1968) Facilitating meaningful verbal learning in the classroom. Arith Teacher 15(2):126–132. https://doi.org/10.5951/AT.15.2.0126

Bocconi S, Chioccariello A, Kampylis P, Dagienė V, Wastiau P, Engelhardt K, Stupurienė G et al (2022) Reviewing computational thinking in compulsory education: state of play and practices from computing education. ISBN 9789276472087

CNE (2024) Estado da Educação 2023. Ano: 1ªEdição: dezembro 2024. ISSN Digital: 2976–0267

Lico MD (2023) Participação democrática dos alunos na escola do século XXI, em Portugal. Contributos do Movimento da Escola Moderna. Tesis Doctoral. Faculdad de Educación. Departamento de Teoría e História de la Educación. Universidad de Salamanca. https://doi.org/10.14201/gredos.153807

DGE (2023) Aprendizagens Essenciais. Direcção-Geral da Educação. https://www.dge.mec.pt/aprendizagens-essenciais-0.

Dubet F (2003) A escola e a exclusão. Cad Pesqui 119:29–45. https://doi.org/10.1590/S0100-15742003000200002

Gusdorf G (1963–1968) Professores Para Quê? Para uma pedagogia da pedagogia. Livraria Morais Editora, Lisboa. https://archive.org/details/georges-gusdorf-professores-para-que

IGU/CGE (2016) International charter on geographical education. IGU/CGE, Beijing, China. https://www.igu-cge.org/publications/charters/

Kaercher NA (1996) A geografia é nosso dia-a-dia. Boletim Gaúcho de Geografia 21(1):109–116. http://seer.ufrgs.br/bgg/article/view/38639/26361

Lemos P, Soares L, Pacheco E (2023) Transposição didática-linhas metodológicas gerais & conceito (s). In Didática da Geografia para uma Cidadania Territorial: Uma bússola para um mundo em profunda transformação: Atas do X Congresso Ibérico de Didática da Geografia, pp 386–408. ISBN 978-989-53324-1-0

Luckin R, Holmes W, Griffiths M, Forcier LB (2016) Intelligence unleashed. An argument for AI in education. Pearson, London. ISBN: 9780992424886

Marshall A (1879) Principles of economics. https://archive.org/details/principlesecono00marsgoog/page/n8/mode/2up?view=theater

Martins GD, Gomes CA, Brocardo J, Pedroso JV, Camilo JL, Silva LM, Rodrigues SM et al (2017) Perfil dos alunos à saída da escolaridade obrigatória. Ministério da Educação/Direção Geral da Educação (DGE). ISBN 978-972-742-416-0

Martins Filho AJ (2003) Entre o visível e o invisível: reflexões acerca de um admirável mundo novo. ORG & DEMO 4:97–115. https://doi.org/10.36311/1519-0110.2003.v4n1.433

Melo M (1984) 12º ano, que Geografia? Finisterra, 19(38):242–244. https://doi.org/10.18055/Finisterra2083

Monteiro R, Ucha L, Alvarez T, Milagre C, Neves M, Silva, Macedo E et al (2017) Estratégia Nacional da Educação para a Cidadania. https://www.dge.mec.pt/estrategia-nacional-de-educacao-para-cidadania

Novak M (2012) Before the Jetsons, Arthur Radebaugh illustrated the future. Smithsonian Magazine. https://www.smithsonianmag.com/science-nature/before-the-jetsons-arthur-radebaugh-illustrated-the-future-122729342

Pezzini C, Szymanski M (2018) Por onde andará o desejo de aprender dos alunos? Dia a dia Educação. Secretaria de Educação e Esporte. Governo do Paraná. Disponível em: http://www.diaadiaeducacao.pr.gov.br/portals/pde/arquivos/853-4.pdf

Pintassilgo J, Mogarro MJ (2003) A ideia de Escola para todos no pensamento pedagógico português. In: Fernandes R, Pintassilgo J (Orgs) A Modernização Pedagógica e a Escola para Todos na Europa do Sul no Século XX. Grupo SPICAE, pp 51–71

Queiroz R (2024) As baixas qualificações, a massificação e a educação integral na educação. Reflexão de uma perspetiva sistémica 50 anos depois da revolução. Revista Portuguesa de Investigação Educacional (28):1–14. https://doi.org/10.34632/investigacaoeducacional.2024.16252

Schultz TW (1960) Capital formation by education. J Polit Econ 68(6):571–583. https://doi.org/10.1086/258393. JSTOR 1829945. S2CID 154572248

Smith A (1789) An inquiry into the nature and causes of the wealth of nations. https://era.ed.ac.uk/handle/1842/1455

Thomson SL (2011) Imagine a day. Simon & Schuster Children's Pub. ISBN-13978-1442436787

Education and Territorial Inequalities: The Portuguese Case

Hélder Santos, Catarina Maia, Márcio Ferreira, Marcelo Torres, and Teresa Sá Marques

Abstract The aim of this research is to contribute to a territorial understanding of inequalities in education, both in terms of access and of attainment. These socio-spatial inequalities appear in the distribution of educational resources and in school, family and local contexts, creating a territorial mosaic of inequalities. The focus of the analysis is on the period when the country was under severe pressure due to the COVID-19 pandemic (2020–2022). To achieve this goal, a multidimensional methodological proposal has been developed at the municipal level, using context-based information on schools, families and places. It also includes indicators for analysing resilience and existing territorial vulnerabilities. In terms of findings, we highlight the fact that great progress has been made in Portugal in recent decades in terms of access to education and combating school failure, and that we have identified spatial inequalities or injustices, providing useful information for reflection on territorial policies aimed at strengthening educational quality, equity and resilience.

Keywords Education geography · Educational resilience · Spatial justice · Territories vulnerabilities · Place-based policies

---

H. Santos (✉) · T. S. Marques
Faculty of Arts and Humanities of University of Porto, Department of Geography, Centre of Studies in Geography and Spatial Planning (CEGOT), University of Porto, Porto, Portugal
e-mail: hcsantos@letras.up.pt

T. S. Marques
e-mail: tmarques@letras.up.pt

C. Maia · M. Ferreira · M. Torres
Centre of Studies in Geography and Spatial Planning (CEGOT), Faculty of Arts and Humanities of University of Porto, Porto, Portugal
e-mail: catarina.maia2@gmail.com

M. Ferreira
e-mail: marcioferreira16@hotmail.com

M. Torres
e-mail: marcelojbt9@gmail.com

L. Soares et al. (eds.), *Compromised Geography*, Key Challenges in Geography,
https://doi.org/10.1007/978-3-032-05201-8_5

Introduction

Education is central to the development of any society or country. Portugal has made significant progress over the last 50 years[1] in raising the educational attainment of the population and reducing early school leaving and school failure. Despite these changes, which can be observed at the national level, there are still very significant differences between local places. These are socio-spatial inequalities or injustices that integrate different factors and contexts, articulated in a dynamic with cumulative effects.

In terms of objectives, this research chooses to focus on inequalities related to school, family and local contexts, linking them to educational outcomes and highlighting levels of resilience and vulnerability of territories, while recognising the importance of marginalisation processes based on individual inequalities (gender, age, specific needs, cultural or ethnic/racial origin, sexual orientation, among others).

Currently, public education policy aims to promote equity in access to resources, in the quality of services provided, and in educational outcomes and skills (CNE 2021). In Portugal, there is no territorialised research by municipality, with only limited references to the NUT III scale (CNE 2021). There is therefore a research gap that hampers the development of policies with a territorial focus. This research addresses this gap by using information at the school cluster level and official statistical data available at the county level, thus moving towards a multidimensional territorial approach.

During the period when the COVID-19 pandemic was at its peak (2020–2022), social and territorial inequalities in education were intense and significant. The analysis[2] carried out focuses precisely on this period and aims to capture the behaviour of a set of variables related to the education system during the exposure to this event. Equity in education is built by promoting inclusion and equity, with territorial contexts being essential factors in building resilience and addressing vulnerability. Thus, this research addresses a key challenge in building a Committed Geography Agenda, as education is a central issue in territorial development. In the field of education, territorial policies are fundamental to the promotion of equity and justice in education, as they make it possible to design responses tailored to the concrete needs of populations, to combat structural inequalities and to create the conditions for success at school, regardless of socio-economic background. In addition, these policies promote territorial cohesion by reducing asymmetries between more and less favoured regions and by strengthening the resilience and innovation of territories.

Regarding the structure of this chapter, Section 2 provides a theoretical reflection. It focuses first on inequalities of opportunity and socio-spatial injustices in education, followed by a reflection on the need to strengthen resilience and reduce territorial

---

[1] This study was developed under the project "50 years of April 25th: the geographies of 50 social, economic and political dynamics" (2023.10912.25ABR), funded by the Foundation for Science and Technology (FCT).

[2] The territorial analyses developed in this chapter are part of a broader project entitled: "Portugal in transformation: a mosaic of opportunities or a mirror of territorial inequalities?".

vulnerability in the event of shocks or stressful trends. In Sect. 3, a methodology is presented that aims to identify different resilience and vulnerability factors with territorial expression, using a multidimensional approach supported by a multiple correspondence analysis, followed by a cluster analysis with the aim of identifying a territorial typology. Section 4 presents the results of this research. It begins by identifying the main advances made in Portugal over the last 50 years. It then presents profiles/groups of variables (characterising the diversity of educational supply and demand, school outcomes and school, family and local contexts). Finally, a territorial typology systematises the levels of resilience and vulnerability. Section 5 concludes, raises questions and suggests avenues for further research.

Conceptual Approach

Academic literature has debated the worsening of inequalities and defended basic principles of socio-spatial justice (Heynen et al. 2018), protesting against economic growth that is mainly for the benefit of a few social groups (Etherington and Jones 2009; Florida and Melander 2016; Truelove 1993). In recent years, territorial disparities and inequalities have also become increasingly visible in several countries (Hobson 2006; Soja 1996) with poor access to mainstream services, socio-territorial segregation and the concentration of poverty (Madanipour and Weck 2015; Malheiros et al. 2024; Marques et al. 2020).

The main political agendas—namely the Sustainable Development Agenda (United Nations 2015) and the Territorial Agenda 2030 (European Union 2020)—emphasise the need to counteract social inequalities and reduce territorial disparities, to promote territorial cohesion and ensure a future for all territories. Some authors (Barca et al. 2012; Florida 2021; Rodríguez-Pose et al. 2023; Setianto and Gamal 2021) argue that reduced or unequal access to public services (education, health and public transport) in peripheral areas is a trigger for social inequalities and surges of discontent. In fact, the processes that generate social and territorial inequalities (Cravinho 2020; Costa 2020; Harvey 2015; Soja 2013) are based on different forms of marginalisation (gender, special needs, cultural or ethnic/racial origin, age, sexual orientation, social class, etc.) and different degrees of resilience (exposure, susceptibility and vulnerability) to crises, shocks or structural transitions (Adger and Kelly 2012; Cretney 2014; Fainstein 2015; Moura 2024).

Education is a universal right, enshrined in the Universal Declaration of Human Rights (United Nations 1948) and one of the pillars of social progress. Universal access to quality education is essential for building more just, equitable and empowered societies (Cerna et al. 2021; OECD 2012). The reduction of social and territorial inequalities depends on the ability of an education system to guarantee equal opportunities for all citizens. Education therefore has a transformative role for individuals and society and can contribute to promoting social and territorial cohesion and resilience, as well as encouraging greater civic participation in social change (Cerna et al. 2021; Dijkstra et al. 2017; OECD 2010; World Bank 2012).

Education: Individual and Socio-Economic Development

According to Amartya Sen (1997), development promotes freedom of choice and opportunities for people to achieve the quality of life and well-being they seek or desire. There are two approaches to the study of the role of education in development: human capital accumulation and human capability development (Leoni 2023). Sen (1997) argues that the two are related, and emphasises the role of the skills that individuals acquire—"The former concentrates on the agency of human beings—through skill and knowledge as well as effort—in augmenting production possibilities. The latter focuses on the ability of human beings to lead lives they have reason to value and to enhance the substantive choices they have" (Sen 1997, p. 1959).

Education and lifelong learning have an important transformative role to play. At an individual level, they shape people's knowledge, perspectives, values and behaviour (Benavot 2017; Wals and Benavot 2017). In other words, education triggers the mechanisms of the "social lift," leading individuals to see life differently and to transform their own lives (human capability vision). Although the image of the "social lift" has been criticised for representing a vision of education that reproduces a concept of a vertically hierarchical society (Chu 2022), a large body of literature recognises the central role of education in promoting equal opportunities, thus acting as a socio-economic "lift" and a driver of individual quality of life (Aiyar and Ebeke 2020; Andlib, Sadiq and Scicchitano 2025; De Pablos Escobar and Izquierdo 2016).

At a societal level, education helps to build the concepts, skills and tools that promote resilience and social prosperity (Benavot 2017; Leoni 2023; Wals and Benavot 2017). Capacity building is essential to bring about social change, such as reducing gender inequality, improving the quality of public debate, or increasing people's well-being and freedom (Sen 1997). In addition, education can play a prominent and cross-cutting role in achieving the goals set out in the Sustainable Development Goals (SDG) (Benavot 2017; Wals and Benavot 2017).

However, education can also be a source of individual and social stagnation, especially if it does not reach everyone with the quality required to fulfil its transformative role (Kennedy 2013). Hence, the need to promote educational equity and justice in the wake of socio-economic inequalities.

Promoting equity in education means ensuring that all learners have access to high quality education, including access to resources (diverse equipment, alternative education routes and a variety of courses), ensuring equity in the quality of services provided (by teachers and schools) and equity in school outcomes and competences (CNE 2021). Equity is promoted by creating more justice in education, i.e. by creating the necessary conditions so that the educational performance of students is free from the constraints imposed by family and local contexts. In addition, equity also implies strengthening inclusion, i.e. ensuring that everyone reaches a minimum level of competence, considering the social, cultural and territorial diversity of students (CNE 2021). To this end, as the OECD (2021) argues, it is important to promote both horizontal equity strategies, i.e. providing equal access to resources for all students in the education system, and vertical equity strategies, i.e. providing positive discrimination strategies according to specific situations, in favour of the most disadvantaged (localities, communities, schools, families and individuals).

The scale of analysis is not irrelevant when analysing quality and inequalities in education. The quality of education can also vary considerably at neighbourhood level. In other words, there are factors that determine how local school quality varies (Zheng and Graham 2022). Therefore, the study of the territorial mosaic of education, i.e. the spatial justice/injustice (Soja 2009) resulting from the different forms of how the education system is rooted in space (Butler and Sinclair 2020; McCluskey et al. 2025; Tieken and Auldridge-Reveles 2019)—the geography of education—is an important contribution to the debate on public policies for socio-economic development. In particular, because it is important to discuss the process of building territorial education policies as a way to enable local communities to escape the vicious circle of socio-economic (under)development traps (Barca 2019; Zheng and Graham 2022).

Although location is an essential dimension that shapes the geography of opportunities, as well as the experiences of students, families and interests in and around schools, and although research on the spatial context of education exists, there is a need to strengthen research on the concept of place in the geography of education. There is a significant lack of research that uses the scale of place and spatial methods to explore the nature of, and possible solutions to, educational inequality (Butler and Sinclair 2020).

Education: Resilience and Territorial Vulnerabilities

Today, education systems have to respond to a variety of crises and overlapping risk or stress factors, including pandemics and economic and financial shocks, technological transitions, conflict and insecurity, natural disasters and environmental degradation. In this context of global uncertainty, education policies have increasingly adopted the concept of resilience to help education systems respond and recover (GPE 2024; UNESCO 2021; UNHCR 2017; UNICEF 2021; USAID 2020b). Thus, in the field of education, the aim is to promote the capacity to respond to opportunities and improve educational outcomes through equitable and appropriate learning for all students.

The factors that build resilience depend not only on individuals, but also on their schools, families and other social contexts (Morote et al. 2020). Most studies on resilience in education focus on individuals (gender, age, special needs, socio-emotional dimension), especially students (Ang et al. 2021; CNE 2021) and teachers (McGaughey et al. 2021), and on social contexts (Cordero and Mateos-Romero 2021).

Broadly speaking, resilience is the ability of a system, whether it is an area, a school community, a school, a family or a person (pupils and teachers), to recover, maintain its integrity and thrive after a shock or during a period of stress (Borazon and Chuang 2023; Folke et al. 2002; Meerow et al. 2016; Keck and Saklapolrak 2013). It is therefore essential to enable all those involved in the system to resist, adapt and recover when faced with possible shocks that shake or could shake its structures (Ahmed Shafi et al. 2020; Jacobson et al. 2019; Silva and Silva 2022). It is also essential to enable the system to respond to chronic stressors, i.e. long-term trends that can weaken a system's potential and ultimately increase the vulnerability of systems and the people involved (Ahmed Shafi et al. 2020; OECD 2014).

In education, resilience is a dynamic, multifaceted response to context. (Borazon and Chuang 2023; GPE 2024).

It is a relational and multisystemic phenomenon, built from the resources available at different levels of the education system and from the interactions between its subsystems: students with each other and with teachers; schools with families; local contexts with regional and national systems (Ahmed Shafi et al. 2023Casas et al. 2021; Masten 2021). In this way, resilience develops through the interaction between individuals, groups and territories.

The education system is complex, with a wide range of players and dynamics at different levels: students, teachers, schools, families, and social and local contexts; but also regional, national and international contexts. These interactions are subject to social, economic and political pressures that can compromise their proper functioning (Borazon and Chuang 2023). It is therefore essential to understand the characteristics and dynamics of the system, especially those that contribute to its development in the light of possible disruptions, in order to make it more capable, agile and efficient (Sriskandarajah et al. 2010). The resilience concept suggests that responses to crises or shocks can be absorptive, adaptive or transformative (GPE 2024; Shah 2019; USAID 2020a) and can be developed on a *continuum* or simultaneously. Transformative responses are the most ambitious and sustainable because they focus on long-term change. Adverse responses that absorb or adapt to the shock perpetuate unequal structures, structural vulnerabilities, rather than transform them (Ebersohn 2020).

When a system lacks or loses its resilience capacities, it becomes more susceptible and, therefore, more vulnerable (Berkes 2007; Folke et al. 2002; Gallopin 2006). Vulnerability thus represents the predisposition of systems to suffer damage when exposed to shocks and a weak capacity to respond to such stress situations (OECD 2014).

In a vulnerable context, it is a priority to reduce susceptibility, either by addressing key weaknesses or by strengthening adaptive capacities. Identifying the territorial vulnerabilities of the education system, whether among students, teachers, schools, families or social contexts, is an essential step towards designing strategies that promote equity and cohesion.

In conclusion, it is essential to combat the territorial vulnerabilities that exist in the education system, and to do this we need to understand the weaknesses of the territory to develop and articulate strategies that ensure a more integrated, inclusive and equitable development of the education system (GPE 2024; OECD 2021). The different contexts "should be able to contribute to building capacities to bring to light what will work in its system, without neglecting minorities" (Borazon and Chuang 2023, p. 14) or the most vulnerable.

Methodological Approach

The challenges of the SDG emphasise the need to promote equal opportunities, particularly in education, for all people and in all places (United Nations 2015). In order to respond to global challenges and multiple shocks or stresses, it is not enough to promote strategies that help the education system to absorb and adapt to large-scale extreme adversities (unpredictable or continuous). We need to think about strategies to transform the system, because the existing structures are clearly unequal, so the capacity to respond is clearly different (Ebersohn 2020).

The education system in Portugal has shown remarkable momentum (CNE 2020a, 2024; PLANAPP 2023) in recent decades. However, a closer look reveals that individual, family, school, social and local vulnerabilities vary widely and persist. It is therefore important to tackle inequality in a comprehensive and systematic way, by promoting smooth and structural changes to increase equity throughout the education system (CNE 2020b, 2021). This involves strengthening inclusion and social and territorial equity, taking into account the different subsystems or contexts (individual, family, school, community and local).

The aim of this research is to analyse inequalities and, more specifically, to identify the vulnerabilities that exist in the different contexts of the education system, through an exploratory approach, still incomplete, that focuses on potentially interconnected territorial factors.

Given the objectives of this research, we must devise a methodological framework suited to the proposed concept, based on a multidimensional territorial approach. The analysis begins with a broad characterisation of the major changes that have taken place in recent decades (generally between 1981 and 2021), reflecting the increase in school enrolment and the decrease in school failure. However, inequalities in educational outcomes persist as equal opportunities in access to education and success at school are not yet guaranteed (Cravinho 2021).

As inequalities exist throughout the territory, at the level of municipalities and schools, our multidimensional analysis will aim to identify territorial inequalities and vulnerabilities in education in Portugal. This analysis will be limited by the information available and accessible in the national statistical system, which we agree is insufficient for the intended purpose. Despite these limitations, the analysis will provide useful information for the development of territorial policies, taking into account different scales of analysis.

Considering the conceptual framework described and the limitations resulting from the statistical information available, the methodological approach will be organised according to six analytical dimensions (Table 1): educational and training supply, educational and training demand, student performance, school contexts, family contexts and local social contexts.

Educational and Training Supply

The first dimension is education and training supply. This takes into account the diversity of schools in each municipality. The places (municipalities) that are best equipped

Table 1 Analytical dimensions and indicators

Analytical dimensions	Indicators
Educational and training supply	1. Pre-school
	2. High school
	3. Art schools
	4. Professional schools
	5. Tertiary education
	6. Technological specialisation schools (CET)
Educational and training demand	7. Students in primary and secondary education (%)
	8. Students attraction (%, national attractiveness)
	9. Higher education students in Erasmus (%, international attractiveness)
	10. Change of pupils in primary and secondary education last decade (%)
	11. Population in training courses (%, 18–65 years)
Students and scholar success	12. Students placed at university in their 1st choice in the 1st phase (%)
	13. Student success direct pathways (%)
	14. Retention or drop-out (%)
	15. Participation in dual certification courses (%)
School contexts	16. Schools with low averages in basic education
	17. Schools with high retention in primary schools
	18. Schools in Ciência Viva Network projects
	19. Number of students per computer
	20. Ageing index of teachers
Family contexts	21. Schools with more parents with tertiary education (%)
	22. Schools with low students not benefiting from "social action" (%)
	23. Schools with more parents with low levels of education (%)
	24. Schools with more students benefiting from "social action" (%)
Local contexts	25. TEIP Schools (Educational Territories of Priority Intervention)
	26. Population with tertiary education (%)
	26. Population with lower education (%)
	28. High reported incomes (Income Tax (1st to 6th): 6th bracket) (%)
	29. Low reported incomes (Income Tax (1st to 6th): 1st and 2nd bracket) (%)

are those that offer a wider range of choices to children and young people, i.e. they promote more opportunities for young people to choose and build the pathways that best suit their needs and ambitions. This analytical dimension looks at the territorial equity of educational services, as the geographical variability of educational offer is partly responsible for the differences in access to and quality of educational services provided.

The aim of this dimension is to explore the diversity of offers at the local level, from pre-school, secondary, arts, vocational and technological education to higher education, allowing us to identify the municipalities with more or less developed levels of educational and training offer, assuming that there are very different levels of polarisation in the country.

Educational and Training Demand

The second dimension relates to demand, assuming that places with higher demand offer more diverse and enriched learning environments. This means that they are attractive places at a national (supramunicipal) and international level and have a very significant and dynamic local student base. On the one hand, the large number and social and cultural diversity of the students promotes individual and collective development. On the other hand, the increased demand in certain places stimulates the social and economic development of these geographical contexts.

This dimension considers the local attractiveness for national and international students, the enrolment rate in primary and secondary education (of the total population aged 5–19) and the change in the number of students over the last decade (2010/2020).

Students and Scholar Success

This third dimension focuses on the students, with the aim of assessing levels of inclusion and student behaviour in relation to the learning process.

Student resilience is influenced by academic self-efficacy, which means that educational success enables students to navigate new and uncertain environments. A young population that invests heavily in upgrading its skills helps to reduce its vulnerability to future shocks and risks, the uncertainties arising from globalisation processes and the possible acceleration of socio-technical transitions. A relatively educated and trained population is potentially better able to respond to uncertainties and has a set of skills that make it easier to resist, adapt and transform. Conversely, the lowest achievers increase their vulnerability now and in the future. Comparing the best and the worst performers makes it possible to measure the level of inclusion (CNE 2021, p. 48).

For this dimension, we sought to identify success factors, including two indicators, one relating to students placed in first-choice higher education institution and course in the first phase, and the other relating to students with direct routes to student success. On the other hand, we have included factors that may put students in a more vulnerable situation now and in the future, such as retention and drop-out rates in basic education, and students who have not completed grade 12 and are enrolled in dual-certificate secondary education courses. Together, these four indicators help to measure inclusion.

School Contexts

School contexts in which academic failure, high levels of retention and early school leaving are prevalent show higher levels of vulnerability. These schools do not guarantee the right to equal opportunities for access and success in school and are highly vulnerable to any unexpected adversity.

In addition, the physical conditions of the schools vary greatly, with some having dilapidated buildings and no sports facilities, while others have excellent conditions and levels of comfort. The quality of digital and technological infrastructure also varies widely. This weakens the resilience of some and increases the susceptibility of others. Some schools have structured specificities to respond to adverse situations by developing alternative training offers in order to promote academic success and the resilience of the school community.

This dimension includes a series of indicators for each school, e.g. for the most vulnerable schools, because they have the lowest average scores in national exams or the highest retention rates in the first cycle, and for those that are trying to strengthen their resilience, some of which invest in the *Ciência Viva* network projects, and others (*TEIP* schools—Educational Territories of Priority Intervention) develop an alternative educational offer to address adverse contexts (local and family). It was decided to include an indicator relating to digital conditions—the number of students per computer. We also considered that the high level of ageing of the teaching staff in some school contexts may increase the vulnerability of some schools, as in some subject areas it is difficult to replace them (on a temporary or permanent basis).

Family Contexts

The factors that build resilience or increase vulnerability depend not only on the students and their school contexts, but only on their family contexts. It is well documented in national and international studies (e.g. CNE 2021, 2024; European Commission/EACEA/Eurydice 2023) that students from more disadvantaged family backgrounds tend to have lower attainment, higher retention rates and a higher risk of dropping out. Families play a crucial role in the individual development of students. Factors such as economic resources, level of qualifications and literacies (languages, maths, digital skills, etc.) of parents in particular, and of the family nucleus in general, have an impact that is felt in multiple dimensions of students' academic (and social) performance. There are many dimensions of inequality resulting from different family contexts that affect the education and future aspirations of children and young people: working conditions for students in the family home, access to goods, school and cultural facilities and services, the ability of families to accompany and support students in the educational process and to prioritise their children's education, among others.

In addition, students from families with lower incomes and qualifications are more exposed to vulnerabilities and have more limited choices, which can condition their academic pathways and hinder their expectations of individual development. The relationship between the socio-economic status of families and student performance allows us to assess how the education system contributes to social equity (CNE 2021).

In this respect, we felt it was necessary to include two types of family contexts: on the one hand, family situations in which the parents have a relatively low level of education (only basic education) and situations in which the child-students benefit from "social action"; on the other hand, family situations in which the parents have a higher education and their child-students have no access to "social action."

Local Contexts

Schools located in areas of economic and social deprivation, characterised by unemployment, poverty and social exclusion, where violence, indiscipline, drop-out and failure are more prevalent, are local contexts that present very demanding challenges for the whole school community. Schools located in more or less peripheral contexts face specific adversities and try to respond to specific challenges. For this reason, schools located in areas with a high number of children and young people at risk of social exclusion benefit from an educational policy aimed at improving inclusion and educational attainment, improving the quality of learning and combating early school leaving (*TEIP* schools). It is therefore important to bring the scale of place analysis and the analytical power of the concept of spatial inequity into the debate on educational equity.

In order to characterise the social contexts of the different municipalities in a cross-cutting way, we found it appropriate to include two indicators that would point to more vulnerable social contexts (population with low income and population with a low level of education) and two others that would indicate less vulnerable social contexts (population with high income and population with a high level of education). *TEIP* schools were also included in this dimension.

For each of these six dimensions, the most appropriate indicators have been defined to measure their representativeness and to assess their spatial distribution, considering the statistical information available at municipal level. Despite the severe limitations of the statistical information available to support the conceptual ambition of this territorial approach, we believe that, even with these limitations, the selected indicators provide a solid and diverse information base for achieving the intended objectives. Therefore, a database of 29 indicators has been structured with information for all municipalities in mainland Portugal for the period of the COVID-19 pandemic (2020–2022 data[3]). All the indicators were mapped to test the quality of the information collected and to understand the spatial distribution of the information presented.

The following steps were used to identify groups or standard profiles: First: standardisation of variables or indicators (0–100), followed by the classification of the values of each spatialised variable (per municipality) into five classes, using the natural break method (very low, low, medium, high and very high values); Second:

[3] The Portuguese government declared the end of the state of alert on October 1, 2022—the last of the official phases of the response to the pandemic. Decree-Law No. 66-A/2022 of September 30, 2022.

a multiple correspondence analysis was applied to all active variables, to eliminate random and insignificant variances; Third: given the large number of territorial units (municipalities), we performed a mixed classification—Ascending Hierarchical Classification and Moving Averages Classification—using Euclidean distance as a measure of similarity and Ward's method as a clustering criterion. We selected the number of classes for analysis by analysing the dendrogram (based on the most significant particles); Four: the description of the retained classes was carried out using the statistical significance test (V-test) and the classes related to the values of the observed variables. Five: the results were mapped, constructing a territorial typology.

Geography of Education: A Matrix of Territorial Inequalities

A Late Educational Process, Strongly in Line with European Trends

The process of literacy and schooling in Portugal was clearly late compared to the European context. As João Cravinho points out, "the democratisation of education made significant progress, based on the extension of compulsory education to Year 12 and the intensification of the fight against school failure" (Cravinho 2020, p. 540).

Although there are still significant differences between Portugal and the European Union average, Portugal has made progress in its education system (European Commission 2024; OECD 2024). One of the most striking advances has been in pre-school education, where the participation rate of 96.3% is higher than the European average of 93.1%. This indicator reflects the impact of public policies and investments that promote access to education from early childhood and recognise its importance in developing children's cognitive and social skills. Portugal also stands out as a positive performer in vocational training. The proportion of graduates with work-based learning experience is 74.1% (EU average of 64.5%). This result confirms the effectiveness of dual training models, which combine theoretical training with professional practice and respond to the demands of an ever-changing labour market. However, structural weaknesses continue to exist. In mathematics, the rate of students not attaining an adequate level of proficiency remains high and slightly above the European average, while in reading, Portugal has a lower rate (23.1% compared to 26.2% in the EU), although this is still worrying. These figures reveal difficulties in the consolidation of essential basic skills, often in combination with socio-economic and educational inequalities. Participation in lifelong learning programmes is another area that needs further attention. With a rate of 33.4% (the EU average is 39.5%), Portugal still has difficulties in promoting the continuous training of its adult population. This is particularly the case for those with lower levels of qualification. This challenge is essential to ensure that the workforce can adapt to the demands of a global economy that is increasingly driven by innovation and digitalisation.

Public policy, in particular the late commitment to a public and universal education system, is largely responsible for Portugal's backwardness. Compulsory primary education was not introduced until 1956 and was extended to 6 years in 1964.[4] After the revolution of 25 April 1974, the right of access to education was accelerated and widespread. When the Basic Law came into force in 1986, compulsory education was extended to 9 years, and in the mid-twentieth century to 12 years.[5]

However, there is a lack of consistency between the policy intent set out in the legislation and its actual application and coverage of all territories, and its effective application and effectiveness for the entire target population. The indicators are clear: Portugal has made significant progress in education in recent decades, starting with the most basic aspects of literacy:

- In 1960, i.e. four years after the law establishing compulsory primary education came into force, 33.1% of the resident population had no schooling at all;
- This figure stood at 18.6% of the resident population after the democratic revolution of April 25, 1974, according to the 1981 census;
- In 2021, this figure fell to 3.1%. This refers to an older population that did not complete their planned schooling or did not take advantage of the school catch-up programmes for the adult population.

Another achievement of recent decades has been the reduction of early school leaving:

- In 1992, 56.2% of young men and 44.2% of young women had not completed nine years of compulsory education;
- By 2023, with 12 years of compulsory education in force, this had fallen to 9.8% of young men and 6.1% of young women.

It should be noted that gender inequality is now more favourable to women, which can be seen a further change in the Portuguese educational process, in this case towards the promotion of gender equality.

The very positive development of the resident population with tertiary education is also worth mentioning. The numbers are quite significant: in 1960, the population with tertiary education was 0.6% of the resident population; after the 25 April Revolution, the 1981 census identified 1.6% of the resident population. However, the most striking increase in this growth trend has occurred since 2011, probably as a result of the introduction of compulsory 12-year education. Thus, the resident population with tertiary education- increased to 11.8% of the population in 2011 and to 17.2% in 2021.

We could list many other indicators of these positive changes, but the central aim of this paper is to show that despite the positive evolution, local variability remains, resulting in a territorially unequal mosaic. Indeed, the mapping of this indicator of

---

[4] Decree-Law No. 4964 of 31 December 1956 and Decree-Law No. 45810 of 9 July 1964.

[5] Law No. 85/2009 of 27 August.

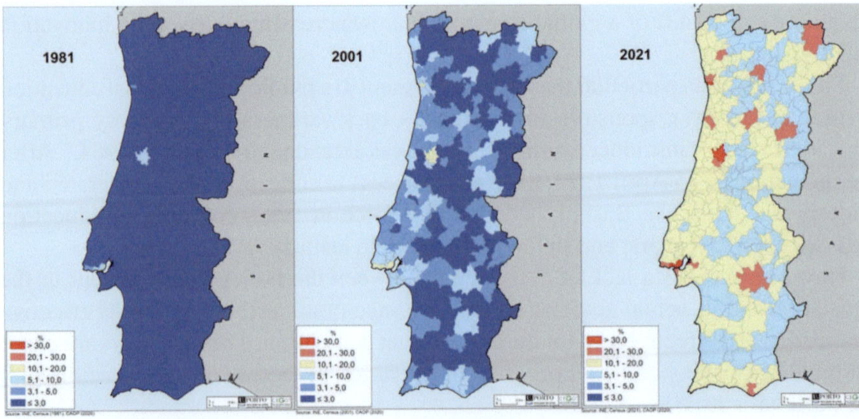

Fig. 1 Share of resident population with tertiary education in the total population (%), in 1981, 2001 and 2021, by municipality

the population with tertiary education by municipality is a good example of the local disparities that make up Portugal's territorial mosaic (Fig. 1).

While the dynamics of the resident population with tertiary education is positive for the national total, a territorial presentation of the data reveals various local disparities, resulting in an uneven territorial mosaic. The processes of territorial differentiation have two characteristics: first, change comes earlier in some places than in others, or may not come at all; and second, the intensity of change varies from place to place.

This example highlights the need to explore the territorial mosaic of education, using the scale of place and multidimensional spatial analysis methods.

Territorial Profiles in Education in Portugal

Does the education system today guarantee equity in all places? The answer is to develop geography of education that highlights the spatial inequalities and inequities that result from the local context of the education system.

The analysis of territorial contexts was structured around six dimensions:

- Two dimensions reflect the geographical distribution of resources (diversity of facilities) and their users (the population attracted by the resources), and the third dimension analyses the educational outcomes achieved.
- The other dimensions focus on three subsystems: school context, family context and local context.

The use of the multivariate methodology described above has identified five profiles that characterise the education system in mainland Portugal. The analysis in

Table 2 characterises each of the five profiles identified, considering the six dimensions described. The profiles reflect the diversity of educational problems, showing inequalities and specific vulnerabilities, but also strengths.

The five profiles were mapped, resulting in a territorial typology (Fig. 2), to illustrate the approach developed and to analyse the spatiality of each of the profiles.

The analysis of the main indicators characterising the five profiles (Table 2) shows a large gap between Profile A and Profile E. Profile A is in the best position, as there are a large number of variables that strengthen the education system, even though there may be some that weaken it. Profile E, on the other hand, is in the worst position, as a large number of variables weaken the education system. Profiles B, C and D have specific characteristics with different levels of vulnerability.

Synthesis: Profile A

- High supply of resources combined with a diversity of education and training facilities
- High local, regional and international attractiveness of the education system
- High academic achievement, clearly reflected in direct pathways to student success and low retention and dropping out

Table 2 Profiles in the education system (five classes)

Analytical dimensions	Indicators	Profile A	Profile B	Profile C	Profile D	Profile E
Educational and Training Supply	1. Pre-school	Moderate				
	2. High school	Very high	Very high	Very high		Very low
	3. Art schools	High				Very low
	4. Professional schools	Very high				Very low
	5. Tertiary education	Very high	High	Moderate		Very low
	6. Technological specialisation schools (CET)	Very high				Very low
Educational and Training Demand	7. Students in primary and secondary education (%)	Very high		Low	High	Very low
	8. Students attraction (%, national attractiveness)	Very high				Very low
	9. Higher education students in Erasmus (%, international attractiveness)	Very high		Very low		Very low
	10. Change of pupils in primary and secondary education last decade (%)			Very high	Very low	Very low
	11. Population in training courses (%, 18-65 years)	High	Moderate	Low		Very low
Students and Scholar Sucess	12. Students placed at university in their 1st choice in the 1st phase (%)					Very low
	13. Student success direct pathways (%)	Very high	Low	High	Low	Very low
	14. Retention or drop-out (%)	Low				
	15. Participation in dual certification courses (%)	Very high		Low		Very low
School contexts	16. Schools with low averages in basic education	High		Very low		Very high
	17. Schools with high retention in primary schools	High		Moderate	Low	Very low
	18. Schools in Ciência Viva Network projects	Very high	Very high	Moderate		Very low
	19. Number of students per computer	High	High			Very low
	20. Ageing index of teachers	Very high	High			Very low
Family contexts	21. Schools with more parents with tertiary education (%)	Very high			Very low	Very low
	22. Scools with low students not benefiting from "social action" (%)	Very high	High	Moderate	Low	Very low
	23. Schools with more parents with low levels of education (%)	Very low		Low	Very high	Very high
	24. Schools with more students benefiting from "social action" (%)	Very low	Low	Moderate	High	Very high
Local contexts	25. TEIP Schools (Educational Territories of Priority Intervention)	High		Very low		
	26. Population with tertiary education (%)	Very high		High	Moderate	Very low
	26. Population with lower education (%)	Very low	Low	Low	High	Very high
	28. High reported incomes (Income Tax (1st to 6th): 6th bracket) (%)	Very high	High	Moderate	Low	Very low
	29. Low reported incomes (Income Tax (1st to 6th): 1st and 2nd bracket) (%)	Very low		Low	High	Very high

indicators that greatly strengthen the education system
indicators that strengthen the education system
indicators that maintain the education system
indicators that weaken the education system
indicators that greatly weaken the education system

Note: the terms "very low", "low", "moderate", "high", "very high" indicate the classes of representative values for each variable.

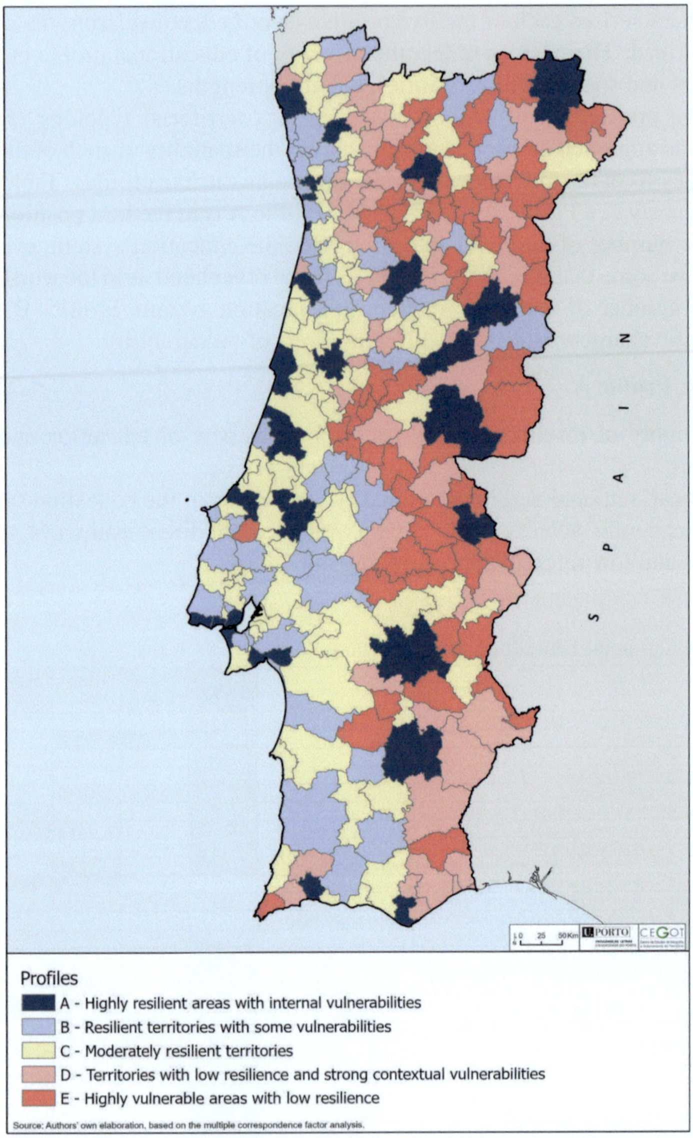

Fig. 2 Territorial profiles in education

- Strong participation in dual training models and lifelong learning programmes
- Social inequalities between schools, reflected in differing educational outcomes
- Contrasting family backgrounds, with high qualifications and incomes in some schools and very low in others
- Local contexts indicate social structures with high levels of education and average incomes, but also social problems and local adversity

Profile A shows a wide range of provision, enabling children and young people to make the most of their choices, whether they prefer an arts or technical education, vocational training or tertiary education. These places offer more opportunities for young people to build the pathways that best suit their needs and ambitions. In this profile, territorial equity in education is comparatively better guaranteed.

The student population adds to the human and social capital of these places and boosts their social and economic vitality.

The students show significantly more positive academic outcomes compared to the other profiles: the direct pathways to student success are very significant, and retention and drop-out rates in basic education are low. Lifelong learning is also more important in this profile, reflecting the strong willingness of the working age population to update their skills.

Schools show local dualities (within the municipality), as some have very high retention and drop-out rates (the worst in the country), with the slowest school grades. This profile includes schools with high levels of vulnerability, indicating some schools that are not contributing adequately to the basic learning of their pupils, jeopardising their future pathways and choices.

The digital infrastructure is relatively inadequate for the needs of students. The average age of the teachers is very high, which indicates pedagogical experience. However, in some contexts, the ageing of teachers can increase the vulnerability of some schools, as it is difficult to replace teachers in some subjects.

The family and local contexts are better compared to the other profiles, with parents on average more educated and with higher income levels. As noted above, economic resources and skill levels can influence multiple dimensions of academic and social performance, so these contexts play a crucial role in students' personal and social development.

However, the high number of *TEIP* schools (Educational Territories of Priority Intervention) demonstrates the existence of an alternative educational offer in the face of unfavourable family or social contexts. This once again illustrates the inequalities that exist at local level.

In terms of territorial distribution (Fig. 2), this profile is spread across the country, forming a polynuclear structure close to the matrix of the national urban system. These territories have a high level of territorial resilience due to the diversity of existing resources in terms of facilities and high attractiveness to students. There is also a strong presence of family, school and local contexts with favourable conditions in terms of qualifications and income.

However, territorial vulnerability is manifested in some school, family and local contexts, combined with unsatisfactory school results (low grades and high retention rates).

In national and local education strategies, it is essential to recognise the seriousness of the situation in some schools, which is reflected in school outcomes and is therefore a key issue for schools, but one to which national and local policies should pay greater attention. This situation calls for the establishment of effective monitoring systems and greater support and follow-up for the development of school projects.

Synthesis: Profile B

- A high number of mainstream education institutions and a high number of university students
- A reasonable number of active people in lifelong learning programmes
- School outcomes show little evidence of direct pathways to student success
- Schools show a high level of participation in *Ciência Viva* networks, but weaknesses in terms of digital infrastructure and a potential need to replace ageing teachers
- Families with reasonable income and low dependency on social support
- Families with reasonable income and low dependency on social support
- Local contexts show a low proportion of people with low levels of education, but a high proportion with high incomes

Profile B is relatively worse than Profile A, but better than Profile C. It has some peculiarities, where a small set of variables characterises the education system.

The supply of facilities is mainly in mainstream education, with high participation of young people in tertiary education and adults in lifelong learning.

Educational success is not clearly evident, and there is little evidence of "direct pathways to student success."

The school context reflects a high level of activity in *Ciência Viva* networks, indicating the existence of active school communities that are curious and open to innovation. However, schools may lack digital and technological resources and the average age of teachers is relatively high.

The family context is characterised by a low level of dependency on social support and a reasonable average income. The local context reinforces this characteristic, with good incomes despite the low level of education of the local population. They seem to be active in jobs do not require a high level of school education, but which are reasonably well paid.

In territorial terms, this profile is mainly characterised by attractive urban, peri-urban or rural contexts. The size of the school community (students, teachers and others), participation in *Ciência Viva* networks and the ability to access higher education, lead to resilience in the education system. Territorial vulnerability manifests itself mainly in schools.

Synthesis: Profile C

- A high supply of facilities for mainstream education and a reasonable tertiary education student population
- Strong student population dynamics, but low attractiveness of Erasmus students, low attractiveness of vocational training
- In terms of school results, direct routes to student success are highly represented, but dual certification courses have a low take-up rate
- Schools with reasonable participation in the *Ciência Viva* networks, but also a fair number of schools with high retention rates
- Families have low qualifications, low income and high levels of "social action" support

- Favourable local conditions, as the resident population is highly qualified and the population with high incomes is reasonable, while low incomes are less significant

Profile C is in an intermediate situation. The supply of facilities is evident in mainstream education, with many students in tertiary education, but the numbers in adult lifelong learning are scarce. The number of pupils in primary and secondary education is increasing and "direct pathways to student success" are high, while participation in dual certification courses is low.

The school context reflects reasonable activity in the *Ciência Viva* networks, in a context where retention and drop-out rates are reasonable. The family context shows some need for social support, but the declared incomes of the families are not low on average. The local context is favourable, with high levels of schooling and a reasonable proportion of the population with high incomes, and relatively few people with low levels of schooling and low incomes.

In territorial terms, this profile is identified along the coast (between Aveiro and Caldas da Rainha), on the Lisbon-Évora axis, and more dispersed throughout the country.

There is little vulnerability, demonstrating territorial resilience, supported by favourable local and family contexts and good levels of educational attainment.

Synthesis: Profile D

- A significantly high number of pupils in mainstream education, but with a downward trend
- Few direct pathways to help students succeed
- Schools show low levels of retention in primary education
- Family backgrounds with low or very low qualifications and income, with many students receiving "social action" support
- Local contexts indicate social structures with reasonable or low levels of education and predominantly low incomes

The number of pupils in primary and secondary education has fallen and "direct pathways to student success" are low, although retention in primary schools is low.

Here, the specificity of the family context highlights social vulnerability, with parents having a low level of education and families needing social support for their student children.

The local context is also somewhat unfavourable, as the resident population is predominantly poorly educated and has a low declared income.

In territorial terms, this profile shows a certain *continuum*: in the north, in an urban–rural strip between the densely north-west and the sparsely populated north-east; in the south Alentejo and Algarve, especially in the interior. These are territories with significant family and local vulnerabilities (low levels of schooling and low income are associated with a need for social support) where, despite low levels of retention in primary schools, educational attainment is not observed in the direct pathways of student success. In this profile, some schools are losing pupils, exacerbating existing vulnerabilities.

Synthesis: Profile E

- Few educational and training facilities and a very small school population and pupils attending training
- Very low participation in dual training models and lifelong learning programmes
- Very poor school outcomes, insignificant direct pathways to student success, dual certification and access to tertiary education
- Schools show low grades in basic education, but retention in primary education is low
- Schools with sufficient computers and young teachers
- Families with very low qualifications and income
- Local contexts indicate social structures with very low levels of education and average income

Facilities are scarce and not very diverse, which is reflected in the low polarisation capacity of places (Profile E). School outcomes reflect low learning capacity associated with contexts (family and local) with low levels of income and schooling.

The schools have low retention and drop-out rates, a relatively good supply of digital resources, and relatively young teachers. However, schools have pupils with low grades in basic education. The schools do not participate in the *Ciência Viva* networks.

In territorial terms, we find this profile in low-density areas, with a strong concentration in the Douro Valley and in the north of the Alentejo and border areas. Here, structural weaknesses are most evident, with a scarce supply of school facilities and a small and declining school population. As such, it is an educational territory with very little resilience. They are also vulnerable territories because the education system has major weaknesses in terms of school outcomes and in family and local contexts where low income and schooling levels are strongly felt.

In short, educational resources and the educational community interact with the contexts (local, school and family), influencing school outcomes. In some municipalities, this interplay of forces can strengthen the education system (Profile A), while in others it can make it vulnerable (Profile E). Territorial inequalities are therefore the result of a combination of factors that ultimately influence school performance.

When the factors at play are clearly favourable (positively interacting), school performance tends to be better (Profile A), leading to greater territorial resilience. Conversely, when the factors at play are unfavourable (weakening the system as a whole), school outcomes tend to be worse (Profile E), leading to greater territorial vulnerability. Between these two ends of the spectrum, there are a variety of hypotheses (different forces at work, with different intensities and directions) with different outcomes (profiles B, C and D), leading to greater or lesser vulnerability.

In the same municipal context, different games can be played at the same time, fragmenting the educational system: in some schools it creates resilient contexts, while in others it places them in situations of great vulnerability. This reality is particularly evident in Profile A territories.

Local characteristics reveal territorial inequalities: while in some municipalities the education systems are stronger, and therefore, the territories are more resilient, less susceptible to the impact of possible shocks (and therefore less vulnerable), in others the education systems have weaker various components and are therefore increasingly susceptible to shocks, resulting in high territorial vulnerability.

The analysis carried out for mainland Portugal identified five types of territories:

- Highly resilient areas with internal vulnerabilities (Profile A)
- Resilient territories with some vulnerabilities (Profile B)
- Moderately resilient territories (Profile C)
- Territories with low resilience and strong contextual vulnerabilities (Profile D)
- Highly vulnerable areas with low resilience (Profile E)

These types of territories support the need to develop territorial policies adapted to local specificities, identifying the factors that weaken or strengthen education systems, thus providing important clues for the design of specific action-oriented measures.

Conclusion

Despite the progress made in the Portuguese education system in recent decades, structural inequalities persist and call for a transformative response aimed at promoting equity and social and territorial justice.

In conceptual terms, this research raises the issue of socio-spatial justice and equal opportunities in the education system, given the need to make progress in achieving equal opportunities in terms of access to quality education and academic achievement. At the same time, the education system is under pressure from shocks (e.g. the COVID-19 pandemic) and major trends (e.g. socio-demographic or technological transitions), with different levels and profiles of resilience and vulnerability across territories.

In methodological terms, it adopted a territorialised approach based on multidimensional methodologies structured around six analytical dimensions: educational supply and demand, student performance, and school, family and local contexts. It then develops an approach that combines a perspective focused on equity in education (equity in educational outcomes and access to resources) with contextualisation at different levels (school, family and local contexts) to identify profiles/groups of variables based on their interaction. Finally, it identifies a territorial typology at the municipal level, which helps to fill some knowledge gaps.

In terms of findings, five profiles characterise the education system in mainland Portugal, highlighting educational resources, school results and the factors that strengthen or weaken school, family and local contexts. These profiles show the presence of inequalities in the different dimensions analysed, which supports the need to promote policies of educational equity and inclusion. The spatialisation of the profiles creates a territorial typology that shows the geography of educational equity. This forms a mosaic that makes it possible to systematise levels of territorial resilience and vulnerability. These findings are an important contribution to the design of more equitable public policies in the future, adapted to local specificities and problems.

The construction of geographies of education at different scales shows how the development of individuals, places and countries is at risk and contributes to the design of more transformative public policies that encourage the construction of contextualised responses to the challenges of development for all people and all places.

This study provides relevant input, although we recognise the limitations associated with the availability and scope of the data and the need to supplement this analysis with qualitative information. A number of complementary activities are underway to mitigate these limitations, e.g. we are strengthening the database to include information on individuals (especially gender, special needs, migrant pupils), on pupils' levels of knowledge and key competences (in reading, maths and science) and on the quality of services provided, and we are also preparing to conduct a survey of schools in order to deepen the qualitative component and to capture the perceptions of the different actors involved in the education system.

References

Adger WN, Kelly PM (2012) Social vulnerability and resilience. In: Living with environmental change. Routledge, pp 19–34

Ahmed Shafi A, Templeton S, Middleton T, Millican R, Vare P, Pritchard R, Hatley J (2020) Towards a dynamic interactive model of resilience (DIMoR) for education and learning contexts. Emot Behav Diffic 25(2):183–198. https://doi.org/10.1080/13632752.2020.1771923

Aiyar S, Ebeke C (2020) Inequality of opportunity, inequality of income and economic growth. World Dev 136:105115. https://doi.org/10.1016/j.worlddev.2020.105115

Andlib Z, Sadiq M, Scicchitano S (2025) Intergenerational (im)mobility in a developing economy context: Is the social elevator broken? Eurasian Econ Rev. https://doi.org/10.1007/s40822-025-00313-2

Ang WHD, Shorey S, Hoo MXY, Chew HSJ, Lau Y (2021) The role of resilience in higher education: a meta-ethnographic analysis of students' experiences. J Prof Nurs 37(6):1092–1109. https://doi.org/10.1016/j.profnurs.2021.08.010

Barca F (2019) Place-based policy and politics. Renew J Soc Democ 27(1):84–95

Barca F, McCann P, Rodríguez-Pose A (2012) The case for regional development intervention: place-based versus place-neutral approaches. J Reg Sci 52(1):134–152. https://doi.org/10.1111/j.1467-9787.2011.00756.x

Benavot A (2017) Education for people, prosperity and planet: can we meet the sustainability challenges? European J Educ 52(4):399–403. https://www.jstor.org/stable/26609386

Berkes F (2007) Understanding uncertainty and reducing vulnerability: lessons from resilience thinking. Nat Hazards 41:283–295. https://doi.org/10.1007/s11069-006-9036-7

Borazon EQ, Chuang HH (2023) Resilience in educational system: a systematic review and directions for future research. Int J Educ Dev 99:102761. https://doi.org/10.1016/j.ijedudev.2023.102761

Butler A, Sinclair KA (2020) Place matters: a critical review of place inquiry and spatial methods in education research. Rev Res Educ 44(1):64–96. https://doi.org/10.3102/0091732X20903303

Casas EV Jr, Pormon MM, Manus JJ, Lejano RP (2021) Relationality and resilience: environmental education in a time of pandemic and climate crisis. J Environ Educ 52(5):314–324. https://doi.org/10.1080/00958964.2021.1981205

Cerna, L., et al. (2021). Promoting inclusive education for diverse societies (No. 260). OECD Publishing. https://doi.org/10.1787/94ab68c6-en

Chu Z (2022) Social transformation and human education. In: People-Oriented education transformation: the great transformation of China. Palgrave Macmillan, Singapore, pp 1–14. https://doi.org/10.1007/978-981-16-6353-6_2

CNE (2020a) Estado da Educação 2019. Conselho Nacional de Educação, Lisboa. https://www.cnedu.pt/content/edicoes/estado_da_educacao/EE2019_Digital_Site.pdf

CNE (2020b) Desempenho e Equidade: uma análise comparada a partir dos estudos internacionais TIMSS e PIRLS. Conselho Nacional de Educação, Lisboa

CNE (2021) Efeitos da pandemia COVID na educação: Desigualdades e medidas de equidade. Conselho Nacional de Educação, Lisboa. https://www.cnedu.pt/content/noticias/estudos/Estudo_AssembleiaRepublica-Efeitos_da_pandemia_COVID-19.pdf

CNE (2024) Estado da Educação 2023. Conselho Nacional de Educação, Lisboa. https://www.cnedu.pt/content/EE2023/Versao_Integral/EE2023.pdf

Cordero JM, Mateos-Romero L (2021) Exploring factors related to resilience in primary education: evidence from European countries. Stud Educ Eval 70:101045. https://doi.org/10.1016/j.stueduc.2021.101045

Costa AF (2020) Desigualdades sociais e pandemia. In: Renato Miguel do Carmo, Inês Tavares, Ana Filipa Cândido (orgs). Um Olhar Sociológico sobre a Crise Covid-19 em Livro, 4–16. CIES-ISCTE/Observatório das Desigualdades, Lisboa. https://www.observatorio-das-desigualdades.com/2020/11/29/umolharsociologicosobreacovid19emlivro/

Cravinho J (2020) As desigualdades de resultados entre regiões e escolas e o direito constitucional à igualdade de oportunidades. In: CNE. Estado da Educação 2019, pp 540–556

Cretney R (2014) Resilience for whom? Emerging critical geographies of socio-ecological resilience. Geogr Compass 8(9):627–640

De Pablos Escobar L, Izquierdo MG (2016) Intergenerational educational and occupational mobility in Spain: does gender matter? Br J Sociol Educ 37(5):721–742

Dijkstra AB, Daas R, Motte PI, Ehren M (2017) Inspecting school social quality: assessing and improving school effectiveness in the social domain. J Soc Sci Educ 16(4):75–84. https://doi.org/10.4119/jsse-850

Ebersohn L (2020) Resiliência coletiva ao desafio global: transformar uma agenda coletiva de bem-estar rumo a uma educação equitativa sustentada. Práxis Educativa, Ponta Grossa 15:1–15. https://doi.org/10.5212/PraxEduc.v.16344.082

Etherington D, Jones M (2009) City-regions: new geographies of uneven development and inequality. Reg Stud 43(2):247–265. https://doi.org/10.1080/00343400801968353

European Commission/EACEA/Eurydice (2023) Structural indicators for monitoring education and training systems in Europe—2023: equity in school and higher education. Eurydice report. Publications Office of the European Union. https://doi.org/10.2797/829723

European Commission (2024) Education and training monitor 2024: Portugal. Publications Office of the European Union. https://op.europa.eu/en/publication-detail/-/publication/39776d27-ac6f-11ef-acb1-01aa75ed71a1/language-pt

European Union (2020) Territorial agenda 2030: a future for all places. https://territorialagenda.eu/

Fainstein S (2015) Resilience and justice. Int J Urban Reg Res 39(1):157–167

Florida R (2021) Discontent and its geographies. Camb J Reg Econ Soc 14:619–624. https://doi.org/10.1093/cjres/rsab014

Florida R, Mellander C (2016) The geography of inequality: difference and determinants of wage and income inequality across US metros. Reg Stud 50(1):79–92. https://doi.org/10.1080/00343404.2014.884275

Folke C, Carpenter S, Elmqvist T, Gunderson L, Holling CS, Walker B (2002) Resilience and sustainable development: building adaptive capacity in a world of transformations. Ambio 31(5):437–440. https://doi.org/10.1579/0044-7447-31.5.437

Gallopin GC (2006) Linkages between vulnerability, resilience, and adaptive capacity. Glob Environ Chang 16:293–303. https://doi.org/10.1016/j.gloenvcha.2006.02.004

GPE (2024) Evidence for system transformation: education resilience. Global Partnership for Education, Washington, DC

Harvey D (2015) The right to the city. In: LeGates RT, Stout F (eds) The city reader, 6th ed. Routledge, pp 270–278

Heynen N, Aiello D, Keegan C, Luke N (2018) The enduring struggle for social justice and the city. Ann Am Assoc Geogr 108(2):301–316. https://doi.org/10.1080/24694452.2017.1419414

Hobson K (2006) Enacting environmental justice in Singapore: performative justice and the green volunteer network. Geoforum 37(5):671–681. https://doi.org/10.1016/j.geoforum.2005.08.004

Jacobson MJ, Levin JA, Kapur M (2019) Education as a complex system: conceptual and methodological implications. Educ Res 48(2):112–119. https://doi.org/10.3102/0013189X19826958

Keck M, Sakdapolrak P (2013) What is social resilience? Lessons learned and ways forward. Erde 67:1–18

Kennedy KJ (2013) Education: social elevator or holding area? In: Hughes P (ed.) Achieving quality education for all (vol 20). Springer, pp 167–185. https://doi.org/10.1007/978-94-007-5294-8_8

Leoni S (2023) A historical review of the role of education: from human capital to human capabilities. Rev Polit Economy 37(1):227–244. https://doi.org/10.1080/09538259.2023.2245233

Madanipour A, Weck S (2015) Social exclusion and poverty in Europe: territorial patterns. Local Econ 30(7):715–720. https://doi.org/10.1177/0269094215601628

Malheiros J, Silva D, Júnior L (2024) The contemporary geographies of urban inequality: Insights from Portugal and the Lisbon Metropolitan Area. In: Urban change in the Iberian Peninsula: a 2000–2030 perspective. Springer Nature Switzerland, pp 69–85

Marques TS, Saraiva M, Ribeiro D, Amante A, Silva D, Melo P (2020) Accessibility to services of general interest in polycentric urban system planning: the case of Portugal. Eur Plan Stud 28(6):1068–1094. https://doi.org/10.1080/09654313.2019.1658718

Masten AS (2021) Multisystem resilience: pathways to an integrated framework. Res Hum Dev 18(3):153–163. https://doi.org/10.1080/15427609.2021.1958604

McCluskey MS, Narayanan M, Milne E (2025) Reform for "These" Kids: neoliberal reform, theorized target populations, and spatial injustice. Educational Policy. https://doi.org/10.1177/08959048251326829

McGaughey F, Watermeyer R, Shankar K, Suri VR, Knight C, Crick T, Hardman J, Phelan D, Chung R (2021) 'This can't be the new norm': academics' perspectives on the COVID-19 crisis for the Australian university sector. High Educ Res Dev. https://doi.org/10.1080/07294360.2021.1973384

Meerow S, Newell J, Stults M (2016) Defining urban resilience: a review. Landsc Urban Plan 147:38–49. https://doi.org/10.1016/j.landurbplan.2015.11.011

Morote R, Las Hayas C, Izco-Basurko I, Anyan F, Fullaondo A, Donisi V, Zwiefka A, Gudmundsdottir DG, Ledertoug MM, Olafsdottir AS (2020) Co-creation and regional adaptation of a resilience-based universal whole-school program in five European regions. European Educ Res J 18(5):615–636. https://doi.org/10.1177/1474904120947890

Moura RV (2024) Spatial justice in fragile peripheral areas: Analysing Italian strategies and approaches to counter territorial inequalities and spatial injustice. Seminario Internacional de Investigación en Urbanismo, (16)

OECD (2010) Improving health and social cohesion through education. Educational Research and Innovation. OECD Publishing, Paris. https://doi.org/10.1787/9789264086319-en

OECD (2012) Equity and quality in education: supporting disadvantaged students and schools. OECD Publishing. https://doi.org/10.1787/9789264130852-en

OECD (2014) Guidelines for resilience systems analysis. OECD Publishing

OECD (2021) Education policy outlook 2021: shaping responsive and resilient education in a changing world. OECD Publishing. https://doi.org/10.1787/75e40a16-en

OECD (2024) PISA 2022 technical report. OCDE Publishing. https://doi.org/10.1787/01820d6d-en

PLANAPP (2023) Relatório sobre as desigualdades—2023. Lisboa: PLANAPP. https://www.planapp.gov.pt/wp-content/uploads/2023/11/PlanAPP_Relatorio-Desigualdades_2023.pdf

Rodríguez-Pose D, Dijkstra L, Poelman H (2023) The geography of EU discontent and the regional development trap. European Commission, Directorate-General for Regional and Urban Policy

Sen A (1997) Editorial: human capital and human capability. World Dev 25(12):1959–1961. https://doi.org/10.1016/S0305-750X(97)10014-6

Setianto MAS, Gamal A (2021) Spatial justice in the distribution of public services. IOP Conf Series: Earth Environ Sci 673(1):012024. https://doi.org/10.1088/1755-1315/673/1/012024

Shah R (2019) Transforming systems in times of adversity: Education and Resilience. White paper, USAID Education in Conflict and Crisis Network. https://www.gcedclearinghouse.org/sites/default/files/resources/190448eng.pdf

Silva AS, Silva SM (2022). Resilient Schools: Indicators from Literature and Border Schools' Projects. https://doi.org/10.1590/198053148438_en

Soja E (1996) Thirdspace: journeys to Los Angeles and other real-and-imagined places. Basil Blackwell

Soja E (2009) The city and spatial justice. Justice Spatiale/spatial Justice 1(1):1–5

Soja E (2013) Seeking spatial justice (vol 16). University of Minnesota Press

Sriskandarajah N, Bawden R, Blackmore C, Tidball KG, Wals AEJ (2010) Resilience in learning systems: case studies in university education. Environ Educ Res 16(5):559–573. https://doi.org/10.1080/13504622.2010.505434

Tieken MC, Auldridge-Reveles TR (2019) Rethinking the school closure research: school closure as spatial injustice. Rev Educ Res 89(6):917–953. https://doi.org/10.3102/0034654319877151

Truelove M (1993) Measurement of spatial equity. Eviron Plann C Gov Policy 11(1):19–34. https://doi.org/10.1068/c110019

UNESCO (2021) Strengthening education management information systems for increased resilience to crises: a synthesis of case studies. UNESCO, Paris. https://unesdoc.unesco.org/notice?id=p::usmarcdef_0000378150

UNHCR (2017) Resilience and self-reliance from a protection and solutions perspective. UNHCR, Washington, DC. https://www.unhcr.org/excom/standcom/58ca4f827.pdf

UNICEF (2021) Reopening with resilience: lessons from remote learning during COVID-19. UNICEF, New York and Florence

United Nations (1948) Universal declaration of human rights (217 [III] A). Paris

United Nations (2015) Transforming our world: the 2030 agenda for sustainable development

USAID (2020a) A map of resilience capacities for the education sector: absorptive, adaptive, and transformative capacities for learners, schools, communities, and institutions. USAID, Washington, DC

USAID (2020b) Education & resilience: transforming systems in adversity. Policy brief. USAID, Washington, DC

Wals AEJ, Benavot A (2017) Can we meet the sustainability challenges? The role of education and lifelong learning. Eur J Educ 52(4):404–413. https://doi.org/10.1111/ejed.12250

World Bank (2012) World bank education strategy 2020: learning for all investing in people's knowledge and skills to promote development

Zheng A, Graham J (2022) Public education inequality and intergenerational mobility. Am Econ J Macroecon 14(3):250–282. https://doi.org/10.1257/mac.20180466

Teacher Training in Geographical Education with Artificial Intelligence Applications

Javier Álvarez-Otero and Isabel María Gómez-Trigueros

Abstract Teacher training to address global challenges, such as climate change, is fundamental to geographical education. This chapter explores how to integrate artificial intelligence (AI) into teacher training in order to strengthen education for sustainability through transformative actions in their immediate environment and to foster critical thinking. AI applications offer possibilities for managing and representing geographical data, as well as for visualizing and interpreting phenomena at local and global scales. Properly implemented, these tools facilitate the understanding of the environmental, social and economic dimensions of sustainability. However, an ethical use of AI and a critical reflection of its implications are essential. The proposed methodology has a mixed methodological design of experimental cut with pre-test and post-test (quantitative analysis) through surveys with three dimensions: sustainability, IA and critical thinking ($n = 149$), and focus group (qualitative analysis) with 45 students (30% of the sample), organized in a group of five people. An improvement in participants' self-perception was observed about their ability to integrate AI into the teaching of sustainability and critical thinking in geography. The focus groups confirmed its relevance and urged more teacher training. In conclusion, specific training in AI and sustainability is essential to prepare future educators capable of fostering critical thinking and citizen responsibility in the twenty-first century.

Keywords Geography · University education · Digital literacy · Educational technology · Critical thinking competence

J. Álvarez-Otero (✉)
Universidad a Distancia de Madrid (UDIMA), Collado Villalba, Spain
e-mail: javier.alvarez.o@udima.es

I. M. Gómez-Trigueros
Universidad de Alicante, Alicante, Spain
e-mail: isabel.gomez@gcloud.ua.es

L. Soares et al. (eds.), *Compromised Geography*, Key Challenges in Geography,
https://doi.org/10.1007/978-3-032-05201-8_6

Introduction

Forming educators to address complex global challenges is increasingly important in geographical education, especially as issues like climate change, social inequality, war and conflicts become more pressing. This chapter explores how artificial intelligence (AI) can be integrated into teacher training in geographical education to enhance sustainability education with transformative actions in the nearest environment and foster critical thinking skills amongst future teachers. AI applications offer several possibilities for geographical education, from data analysis and its management until the cartographic representation, and therefore, the visualization and its interpretation about the geographical phenomena. When effectively implemented, AI applications can provide teacher trainees tools to better understand sustainability in all these dimensions (environmental, social and economic). However, it is necessary that the use of AI in education is guided by ethical principles and critical reflection.

AI has its origins in the 1950s, when Walter Pitts and Warren McCulloch researched the development of the first mathematical models of artificial neurons and neural computation. Likewise, Alan Turing proposes the Turing test to evaluate the intelligence of machines and published an article on this possibility (Computing Machinery and Intelligence). For its part, the term 'artificial intelligence' was coined at the Dartmouth Conference (1956) and began to be recognized as an academic field with its own entity. During the following decades, AI evolved thanks to significant advances related to robotics and robotics programming (Unimate, ELIZA, Shakey, PROLOG). For example, in the 1970s, the Stanford Cart becomes one of the first mobile robots capable of navigating autonomously in obstacle environments. All these advances ended up materializing in the last years of the twentieth century and the beginning of the twenty-first century, when some domestic appliances began to integrate some AI functionalities and software applications were developed based on its principles. Therefore, it can be stated that AI is not a recent technology. However, its popularization skyrocketed with the launch of ChatGPT in November 2022. Since then, generative artificial intelligence (GenAI) applications have grown exponentially, both on the web and in their integration with existing technologies.

Faced with the challenges posed by AI in education, higher education institutions and, in particular, Faculties of Education, have a responsibility to respond to these challenges through GenAI literacy and responsible and ethical use when using it. This will address both the opportunities and challenges posed by this emerging technology. To fulfil this task, universities and research centres can focus their actions on several key areas: (a) research in educational innovation, leading advanced research on new applications of AI and GenAI in the educational field, developing technologies that personalize the teaching–learning process and, in turn, improving the educational quality of institutions and educational centres; (b) teacher training, the educational role of universities and, more specifically, the Faculties of Education being essential. These must lead the integration of AI and GenAI in education, through university programmes that train both students in training and active teachers to incorporate AI strategies in educational environments at all levels; (c) promotion of interdisciplinary collaboration, generating innovative solutions in educational processes and adaptive technology, through synergies between experts in Education and its relationships with disciplines, like Geography; (d) ethics and the responsible use of AI and AGI, an area in which universities are called upon to lead and guide the debate on their ethical implications in education, establishing regulatory frameworks and protocols that guarantee their responsible use by all agents of the educational community; and finally, (e) knowledge transfer, where the Faculties of Education, again, must lead the dissemination of research in educational innovation with AI and GenAI, sharing findings with various social sectors to promote and guide its use throughout the education system. In this context, the Education Sciences must play a transformative and critical role in the process of integration and implementation of emerging technologies such as GenAI by being responsible for analysing, understanding and improving teaching and learning processes. The incorporation of new technologies has generated various changes in recent decades, transforming both scientific research and innovation processes (UNESCO 2024a, 2024b).

The objective of this research is to carry out a diagnosis and analysis of the perception of students, future teachers, on a set of items on artificial intelligence applied as an educational technology, critical thinking as a strategy and geographical education related to climate change as a pedagogical approach.

Theoretical Framework

Artificial Intelligence and Digital Education

Education and technological innovation are intrinsically linked, since the incorporation of technologies into the educational field has enhanced, consolidated and perfected the teaching–learning processes. Currently, education is undergoing a profound transformation driven by technological advances. These transformations are reflected, amongst other aspects, in the way of establishing new ways of communicating knowledge, and, therefore, learning, which is enhanced by the use of technologies in the educational process (Ahmad et al. 2021).

AI is emerging as a transformative technology, offering new opportunities to improve teaching and learning (De Lázaro 2024). In addition, these technologies allow for the personalization of curriculum and educational content, adapting to the individual needs of students, which improves the learning experience and knowledge retention (Chen et al. 2020). Other authors (Akinwalere and Ivanov 2022; Moncayo Cueva et al. 2024) point out that the implementation of AI in education through the use of data will serve to improve educational equity and quality.

This technological evolution and its different challenges have always made possible the emergence of new didactic models that have served to establish teacher training plans and programmes, amongst other purposes. One of the most widespread models since its appearance has been the TPACK model (Technological Pedagogical Content Knowledge), designed by Mishra and Koehler (2006), which adds a technological dimension to the model previously created by Shulman (1986) for the knowledge of pedagogical content. This model is based on three areas of knowledge: pedagogical, disciplinary and technological, with the particularity that none of these three elements alone would produce a satisfactory learning outcome. For this reason, the interaction of these three elements bridges the distances between content and teachers, between teachers, between teachers and students and between students themselves, which facilitates the collaborative learning advocated by many teachers through what some have defined as an inclusive approach to ICT–TAC–TEP (Rodríguez and Gallardo 2020).

This panorama underlines the importance of educational technology training becoming an essential pillar for the continuous updating of teachers. Various authors (Latorre Iglesias et al. 2019; García-Valcárcel-Muñoz-Repiso et al. 2014) agree on the need to structure a training itinerary that contemplates the progressive use of these technologies and the development of digital skills. This itinerary includes several stages: firstly, Information and Communication Technologies (ICT) are addressed, focused on storing, organising and communicating information through digital media, facilitating interaction between the different members of the educational community. Secondly, Learning and Knowledge Technologies (LKT) enhance the didactic use of ICTs. This facilitates, amongst other aspects, a greater adaptation and personalization of learning, collaborative learning, etc., whilst promoting critical thinking about the use made of the digital tools we have at our disposal. Finally, TACs give way to Technologies for Empowerment and Participation (TEP), which seek to promote digitalization in schools and reduce the digital divide. These technologies strengthen digital skills aimed at citizen participation and addressing social problems, offering innovative solutions through the use of technology.

Therefore, the objective of training future graduates, future education professionals and future technicians to join this new reality is reaffirmed, where GenAI is a technology that allows teachers to expand their capacities through putting into practice their creativity, curiosity and ingenuity (De la Higuera and Iyear 2024), and they can also focus on the task of educating from the personalization of learning and from the approach of educational inclusion. In this way, this training will have an impact on the acquisition of Digital Skills, framed in the six areas of the DigCompEdu framework (Punie 2017), and on the acquisition of the recently published Artificial Intelligence Competence framework (Miao and Cukurova 2024). It develops through five dimensions: (1) human-centred mindset; (2) AI ethics; (3) fundamentals and applications of AI; (4) AI pedagogy; and (5) AI for professional learning. These dimensions in turn have a level of graduation and progression in the following categories: acquisition, deepening and creativity, resulting in fifteen performance indicators of these competencies. The incorporation of these skills in teacher training and education qualifications, both digital and in AI, presents a set of significant challenges. This transformation process requires that efforts should focus on integrating learning about AI and GenAI, both in the curricula of future teachers and in the training of university teachers who train them (Barra Arias et al. 2024). However, the integration of AI content into education qualifications remains limited and, in many cases, depends on particular initiatives and specific contexts (Estévez and Sánchez-Vera 2024). Based on the AI Competence Framework for teachers (Miao and Cukurova 2024), it can be said that this set of competences is in the accelerated development phase by many institutions, motivated by the digital transformation in the training and updating of pre-service teachers and in-service teachers. However, there are inequalities in its implementation, both in initial training and in the continuous updating of teachers. In teacher education qualifications, AI is not yet uniformly included. Some educational programmes have begun to incorporate specific modules on educational technology and AI tools, but their scope remains limited. Therefore, teacher education institutions should promote AI literacy amongst teachers to use it with their students, trainee teachers, and teach it ethically and responsibly.

Critical Thinking in Geographical Education for Sustainability

Geographical education plays a crucial role in promoting sustainable development by integrating physical and human concepts, enabling students to understand the interrelationships between the environment and human activity (Maude 2022). This approach to education for sustainability is essential to address global challenges and foster critical thinking, which is so necessary in the formation of citizenship, which must be committed to its closest and local environment.

In today's rapidly changing world, fostering critical thinking through geographical education plays a crucial role in promoting sustainable development (UNESCO 2017; Palacios et al. 2017). By integrating analytical skills with a comprehensive understanding of spatial relationships and environmental dynamics, students become empowered to evaluate complex issues and contribute to building more sustainable societies (Risopoulos-Pichler et al. 2020).). Geography, by its very nature, provides a unique platform for cultivating the analytical and evaluative skills necessary to comprehend and address sustainability challenges (Kerski 2008; De Lázaro-Torres et al. 2023; Nicholson et al. 2023).

The study of geography transcends the mere memorization of locations and physical features; it involves the interpretation of patterns, interactions and processes that define the human–environment relationship (De Miguel González 2024). From climate change and deforestation to urbanization and resource distribution, geographical education equips students with the ability to assess the interconnectedness of global and local sustainability issues, in the framework of global change (Leininger-Frézal et al. 2023). More importantly, it nurtures a sense of responsibility by fostering an understanding of the consequences of human actions on the environment and encouraging proactive solutions. A geographical perspective enables students to question how societies interact with their natural surroundings and how these interactions shape the world (Favier and Van der Schee 2014).

Critical thinking is the foundation of an informed and responsible approach to sustainability. In geographical education, this cognitive skill involves the ability to analyse information, recognize biases, challenge assumptions, and propose innovative solutions to environmental and global issues (Frank et al. 2024). Critical thinking encourages learners to move beyond passive acceptance of information and instead engage in evidence-based reasoning and problem-solving (Gambrill 2018). The integration of critical thinking in geographical education is particularly relevant when addressing sustainability challenges. A critically thinking student is equipped to evaluate the strengths and limitations of these approaches, consider their feasibility across different regions and advocate for solutions that align with scientific evidence and ethical considerations (Brookfield 2011; Abrami et al. 2015). Furthermore, the ability to engage with geospatial data enhances the analytical dimension of critical thinking (García de la Vega 2019). Tools such as Geographic Information Systems (GIS), remote sensing, and cartographic analysis allow students to assess trends in land use, monitor environmental changes and predict potential sustainability risks (Kerski 2024). These technological advancements not only facilitate a deeper understanding of spatial dynamics but also promote data literacy, an essential skill in modern sustainability research.

Methodology

Research Design

This research is ascribed to studies with a mixed experimental methodological design based on a pre-test and post-test questionnaire (quantitative) and focus groups (qualitative), cross-sectional (Sánchez-Gómez et al. 2018). The research has been developed during the academic years 2023–2024 and 2024–2025, in the context of learning in a Faculty of Education of a Spanish university.

The study process has been configured over four phases (Gómez-Trigueros 2024): the first phase, based on a theoretical review on AI, critical thinking and geographical knowledge related to climate change and sustainable development, the TPACK model and Geographic Information Technologies (TIG); the second phase, ad hoc design and validation of the instruments (pre-test and post-test questionnaires and open questions for focus groups) based on the collaboration of teaching staff from other national and international universities; the third phase, collecting information through validated instruments and the development of classroom practice; and, the fourth phase, the emptying of information and the elaboration of the conclusions.

The study was carried out according to the guidelines of the Declaration of Helsinki (Declaration of the World Medical Association), guaranteeing the philosophical ethical commitment and the unwavering respect for human dignity, privacy, physical and moral integrity, as well as the protection of personal data in the treatment of the survey and throughout the research. Privacy regulations were complied with, taking into account the personal data protection code (Organic Law 3/2018), on anonymity and informed consent. Likewise, the study was reviewed and

Table 1 Sociodemographic characteristics

Studies in progress	♀		♂		Total	
	n	%	n	%	n	%
Degree in early childhood education	22	100	0	0	22	14.7
Degree in primary education	55	68.7	25	31.3	80	53.6
Master's degree in secondary education	19	40.4	28	59.6	47	31.7
Total	96	64.4	53	35.6	149	100

Note n: number; ♀ women; ♂: men

approved by the Ethics Committee of the university responsible for the research (Exp. UA-2023–12-05_2).

Participants

The selection of participants was carried out through intentional or convenience sampling, according to the possibilities of access of the research team to the field of study and according to their degree of adequacy to the objectives formulated. A total of 149 students of the Bachelor's Degree in Early Childhood Education, Bachelor's Degree in Primary Education and Master's Degree in Compulsory Secondary Education (68% women and 32% men) with an age range between 19 years and more than 23 years of age agreed to participate in the research. All of them enrolled in a Spanish university. With a margin of error of 5% and a confidence level of 95%, the sample is considered significant with respect to the total population from which it comes ($N = 1022$) (Otzen and Manterola 2017) (Table 1).

Design and Validation of Instruments

The instruments of this research are, on the one hand, the pre-questionnaire, distributed prior to the implementation of the work with AI in the classroom (questionnaire 1), and the post-questionnaire (questionnaire 2) applied after the didactic intervention; and, on the other hand, the discourses and opinions of the focus groups. The pre- and post-questionnaires consist of 16 questions (P) measured on a five-point Likert scale (1, *Strongly Disagree* to 5, *Strongly Agree*) and are organized into three dimensions of study, which are as follows:

- Dimension 1. Sociodemographic characteristics (questions 1–3);
- Dimension 2. Students' self-perception of:
  - Subdimension 2_1. sustainability (questions 4–6);
  - Subdimension 2_2. AI (questions 7–11);

  - Subdimension 2–3. critical thinking (questions 12–13).

- Dimension 3. Self-perception in teaching skills to work on sustainability and critical thinking with AI in the classroom (questions 14–16).

To avoid training the sample in the responses, for questionnaire 2, the order of the questions was modified, as well as their wording (Table 2) (Gargallo-López et al. 2017); the intention has been to achieve a diachronic or longitudinal analysis, with the study of one or more phenomena over time as is the case (López-Roldán and Fachelli 2015).

To facilitate the understanding of some of the issues raised in questionnaires 1 and 2, a pilot test was carried out that allowed the instruments to be tested. An incidental sample of 15 university students was used for this purpose, which allowed the wording of the questions to be nuanced.

For the validation of the questionnaires, the opinion of four experts from Spanish and intentional universities (Andorra, Greece, Portugal) was obtained through the expert panel method (Table 2). Each expert was asked to assess the importance and operability of each question using a 5-point scale (1, *Strongly Disagree* to 5, *Strongly Agree*). They were also encouraged to provide suggestions for each question. Based on the feedback received from the experts, three questions were removed from the scale due to their similarities in content, whilst one new question was included as recommended by the experts.

Finally, questionnaires 1 and 2 were made up of 16 questions that allow us to gather the opinion and self-perception of the trainee teachers on the dimensions indicated (Table 3).

As for the focus groups, they were configured in small groups of five people, in which, in order to maintain the nature of this type of instrument, they were left free to engage in an open dialogue, based on questions that activated their participation that were configured based on the dimensions and subdimensions of the questionnaires and taking into account the objectives of the research (Table 4).

The triggering questions were validated by the panel of experts (Table 2) who evaluated the degree of agreement between the ratings given to the attributes of each axis-question (P-E) (AERA 2018). To do this, a range of 1 to 4 points was used, where

Table 2 Professional characteristics of the group of experts

Categoría profesional	Área de conocimiento	Años docencia	Código Identificador
Professor	General didactics	14	A
Professor	Didactics and school organization	12	B
Lecturer	Educational research and innovation	10	C
Lecturer	Educational research and innovation	9	D

Note A, B, C, D Expert identification codes

Table 3 Research questionnaires (questionnaire 1; questionnaire 2)

Questionnaire 1			
Dimension		Questions	
D 1		P1	Gender
		P2	Age
		P3	Studies you carry out
D 2	Subd. 2_1	P4	I know the meaning of the concept of sustainability and its relationship with geography
		P5	I believe that sustainability is key to education in the twenty-first century
		P6	I feel qualified to research the approach to education for sustainability (SDGs) from geography
	Subd. 2_2	P7	I know what Artificial Intelligence (AI) is
		P8	I have used AI in my geography training assignments
		P9	I am able to use AI for teaching and learning geography
		P10	I consider AI to be a resource for the teaching of geography in the twenty-first century
		P11	I feel qualified to do educational research on geography with AI
	Subd. 2_3	P12	I understand the meaning of competence in critical thinking
		P13	I believe that critical thinking is a key factor in education in the twenty-first century
D 3		P14	I believe that AI can help achieve the SDGs in education and research in geography
		P15	I believe that competence in critical thinking allows the achievement of the approach to education for sustainability (SDGs) from geography
		P16	I think that AI is a technological resource that enables the acquisition of critical thinking in geography
Questionnaire 2			
Dimension		*Questions*	
D 1		P1	Gender
		P2	Age
		P3	Studies you carry out
D 2	Subd. 2_1	P4	I know what sustainability is from geography

(continued)

Table 3 (continued)

Questionnaire 1			
Dimension		Questions	
		P5	I think it is necessary to educate the new generations in sustainability
		P6	After the intervention in class, I feel qualified to investigate the approach of education for sustainability (SDGs) from geography
	Subd. 2_2	P7	After using AI in the classroom, I know more tools to be able to use it in my future teaching work
		P8	After the class activity I learned other ways to use AI to learn and teach geography didactics
		P9	The intervention has enabled me to use AI in the teaching and learning of geography
		P10	After the activity, I consider that AI is a resource for teaching geography for today's citizens
		P11	After the classroom proposal, I feel qualified to do educational research on geography with AI
	Subd. 2_3	P12	I think that competence in critical thinking is key in citizenship education
		P13	After the intervention, I consider that critical thinking is a key factor in the formation of citizenship
D 3		P14	After carrying out the activity in class, I think that AI is a key resource to achieve the SDGs in education and research in geography
		P15	The intervention has shown me that competence in critical thinking allows the achievement of the approach of education for sustainability (SDGs) from geography
		P16	After the didactic proposal, I think that AI is a technological resource that helps in the acquisition of critical thinking in geography

Source Own elaboration

1 implied a low level of compliance with the criterion and 4 its excellent satisfaction. The independent assessment of the experts ($n = 4$) was composed of four specific evaluation criteria (EC), applied to the instrument as a whole: Coherence (Ce1): 'The axis-question presents a logical relationship with the measurement indicator'; Significance (Ce2): The axis-question is pertinent and has an appropriate level of significance to explain the measurement objective'; Representativeness (Ce3): 'The axis-question brings together inferential semantic evidence appropriate to the scientific literature that is the subject of this research'; Clarity (Ce4): 'The axis-question

Table 4 Focus group core questions

D		P-E
D2	Subd. 2_ 1	P-E1 Why do you think the SDGs and sustainability are key in education?
	Subd.2_2	P-E2 What is your opinion about AI and its implementation in teaching?
	Subd.2_3	P-E3 Do you think that critical thinking should be developed in the classroom?
D3		P-E4 What is your opinion about AI as a tool for achieving the SDGs from the field of geography?
		P-E5 What importance do you give to the acquisition of critical thinking in the formation of citizenship? Explain your answer

Note D Dimension; *P-E.* Question-axis (focus group)

does not incur partial readings of the indicator, the objective of this research to be measured' (Table 5).

Subsequently, the researchers, moderators of the focus groups, guided the interventions and collected the responses through the recording of the speeches. Subsequently, they were transcribed and organized.

In order to verify the reliability of the questionnaires, Cronbach's alpha coefficient (α) (Raykov and Marcoulides 2017) and McDonald's omega coefficient (ω) were calculated (Table 6). The results obtained (questionnaire 1 $\alpha = 0.893$; questionnaire 2 $\alpha = 0.901$) confirm the existence of a high and adequate internal consistency of both instruments for the proposed study (Bisquerra 2004). Similarly, Pearson's chi-square index was found with results of p-value $< 1 =$ Sig. 0.001 in both instruments (Cohen et al. 2007), indicative of the high correlation of the questions posed and illustrative of the validity of the questions and the structure of the instruments implemented. The SPSS V26 statistical program was used for reliability and correlations.

Table 5 Dimensional descriptive statistics group of experts

Ce	P-E1				P-E2				P-E3				P-E4				P-E5			
	A	B	C	D	A	B	C	D	A	B	C	D	A	B	C	D	A	B	C	D
Ce_1	4	4	4	4	4	4	3	4	4	4	4	4	4	4	4	4	4	4	4	4
Ce_2	4	4	4	4	4	4	4	4	4	4	4	4	4	4	4	4	4	4	4	4
Ce_3	4	3	4	4	3	4	3	4	4	4	4	4	4	4	4	4	4	4	4	4
Ce_4	4	3	3	3	4	4	4	4	4	4	4	4	4	3	4	4	4	4	4	4

Note Ce_1 Coherence. Ce_2 Significance. Ce_3 Representativity. Ce_4 Clarity. *P-E* Questions-axis. *A, B, C, D* experts

Table 6 Reliability analysis by variable

Questionnaire 1	α	ω	Questionnaire 2	α	ω
P4	0.851	0.866	P4	0.862	0.866
P5	0.853	0.869	P5	0.865	0.870
P6	0.788	0.789	P6	0.798	0.799
P7	0.806	0.809	P7	0.810	0.817
P8	0.813	0.821	P8	0.822	0.830
P9	0.825	0.832	P9	0.835	0.838
P10	0.809	0.810	P10	0.812	0.817
P11	0.826	0.830	P11	0.833	0.835
P12	0.822	0.833	P12	0.841	0.844
P13	0.821	0.828	P13	0.836	0.837
P14	0.823	0.826	P14	0.837	0.841
P15	0.820	0.827	P15	0.831	0.837
P16	0.809	0.812	P16	0.830	0.838

Note α Cronbach's alpha. ω Omega de McDonald

Procedure

In relation to the procedure, two questionnaires were distributed, one prior to the intervention (questionnaire 1) and the other after working with the participants on content related to geography, sustainability and critical thinking with AI (questionnaire 2) (Table 3). They were provided by email through the free Google Forms application, during the second semester of the academic year analysed. The students received questionnaires through their institutional university email and were informed of the objective of the research, as well as the confidentiality of the answers.

Regarding the intervention in the classroom, a group practice was designed, oriented to cooperative work with the aim of working on critical thinking and AI from geographical sustainability. Over the course of three sessions of 120' each, a work proposal was designed organized in the following phases (Table 7):

Subsequently, in a final phase, focus groups composed of five members were configured. Forty-five students of the total participating sample participated in the focus groups, representing 30% of the analysed sample ($n = 149$), enough to be able to saturate the discourses in the answers issued (Kuckartz 2014). For a representation of all the groups analysed, nine focus groups were configured, for discussion, for each participating study. For the processing and coding of the data, we use the ATLAS.ti software. The textual records were labelled with alphanumeric codes according to the study carried out by (F), student (E) and focus group (Gf): $F_{\cdot\text{(Infant Grade; Primary Grade; Master's Degree)}}$-$E_{\cdot\text{x(student)}}$-$Gf_{\text{.(focus group)}}$. Likewise, their discursive transcriptions (focus groups) complied with the conventions proposed by Jefferson (1984).

Table 7 AI classroom intervention

Phase	Session	Action and objective	T
Research phase	First session	The introduction to Chat GPT, with a brief presentation of the IAG, explanation of its operation and its application in text generation	25'
		The formulation of the research question to be developed related to sustainability and its relationship with geography. The aim is to formulate an open question, which encourages reflection and critical analysis of the students	25'
		The implementation and use of Chat GPT. In cooperative groups of five members; students use AI to get answers to the question posed. Each group must record the answers generated	70'
Verification phase	Second session	The information is verified. Students look for reliable sources that review the information provided by the AI application, detecting possible shortcomings or inaccuracies. To do this, the presentation of reliable sources is carried out, offering a brief guide on how to identify reliable and reliable sources online	30'
		Subsequently, a research and verification stage is implemented, where the groups verify the information obtained through Chat GPT using reliable sources and document possible shortcomings or inaccuracies	60'
		Finally, a sharing session is carried out where each group of students shares their findings with the rest of the class	30'
Reflective and observational phase	Third session	Individual reflection on the experience and its learning with annotation of their observations	30'

(continued)

Table 7 (continued)

Phase	Session	Action and objective	T
		Subsequently, the debate takes place in class; a space is opened for the exchange of opinions and experiences; the role of AI in education, ethical challenges and educational possibilities for sustainability work from geography are discussed	60'
		Finally, as a conclusion, the synthesis of the key points and outstanding lessons learned from the debate is sought	30'

Note T Timing

The call for the different focus groups was scheduled after a meeting. The procedure that was followed was: through Google Forms, a calendar was proposed for the participation of the different focus groups, discussions, created; once the day and time had been agreed upon, the students were summoned to the spaces of the centre where the research was carried out and were informed of the objectives of the study and their informed consent was requested, guaranteeing the anonymity and confidentiality of the answers, as well as their subsequent processing; then, the researcher in charge of collecting the answers launches the questions (P-E) elaborated to work on in the focus groups (Table 4); the time for the development of the discussions with a duration of no more than 45' was adjusted for all groups; Finally, the speeches of the focus groups were transcribed and their coding and categorization were carried out.

Results

Descriptive Results

The descriptive statistical results ($M = mean; SD = standard\ deviation =$) of each of the dimensions (Table 7) underline the participants' knowledge of AI prior to the intervention and contrasts with their deficient self-perception for the implementation of this resource in the work of sustainability in the geography classroom.

In dimension 2 (D2) (questions 4–13), which provides information on students' self-perception of sustainability from geography (Subd.2_1) (questions 4–6); their self-perception in the capacity for the didactic use of AI for sustainability education (Subd.2_2) (questions 7–11); and their self-perception referring to critical thinking as a key formative competence (Subd.2_3) (questions 12–13) is where the answers present the most significant differences between both questionnaires (Table 8).

It is worth highlighting question 6 (Subd. 2_1) where important differences are observed between questionnaires 1 ($M = 3.16$; $SD = 0.369$) and 2 ($M = 4.99$; $SD = 0.245$). These values indicate the weak pre-intervention training of participants on their ability to investigate sustainability in the geography classroom. Also, question 9 (Subd. 2_2) yields contrasting results between the answers to questionnaires 1 ($M = 2.09$; $SD = 0.488$) and 2 ($M = 4.23$; $SD = 0.375$), which point to the poor training perceived by the participants to be able to implement AI in the teaching of sustainability. Something similar is observed in question 13 (Subd. 2_3), where the answers to questionnaire 1 ($M = 2.16$; $SD = 0.489$) contrast with the answers to questionnaire 2 (M = 4.90; $SD = 0.223$), confirming a weak recognition of critical thinking as a key factor in the formation of citizenship before the didactic intervention.

Regarding dimension 3 (D3) (questions 14–16), the values show the negative perception of the participants, prior to the didactic classroom intervention, in relation to the value of AI as a didactic resource (P14 $M = 3.26$; $SD = 0.620$) and its ability to promote critical thinking (P15 $M = 3.02$; $SD = 0.663$) and the sustainable development of citizenship (P16 $M = 3.07$; $SD = 0.678$) (questionnaire 1). For its part, this same dimension yields positive values, which are close to the answer option 'Strongly agree' in all these questions ($M \geq 4.96$; $SD \leq 0.298$) throughout questionnaire 2.

Table 8 Descriptive results (mean = *M*; standard deviation = *SD*) questionnaire 1 and questionnaire 2

	Í Variable	Questionnaire 1		Questionnaire 2	
		M	*SD*	*M*	*SD*
Dimension 2 subdimension					
Subd. 2_1	P.-4	3.98	0.587	4.92	0.263
	P.-5	4.19	0.380	4.98	0.223
	P.-6	3.16	0.369	4.99	0.245
Subd. 2_2	P.-7	4.00	0.289	4.97	0.227
	P.-8	3.71	0.961	4.95	0.443
	P.-9	2.09	0.488	4.23	0.375
	P.-10	4.17	0.381	4.84	0.313
	P.-11	3.04	0.861	4.13	0.799
Subd. 2_3	P.-12	4.00	0.263	4.89	0.312
	P.-13	2.16	0.489	4.90	0.223
	Dimension 3				
	P.-14	3.26	0.620	4.96	0.298
	P.-15	3.02	0.663	4.98	0.287
	P.-16	3.07	0.678	4.97	0.282

Source Own elaboration

Parametric Results

To determine whether there are significant differences between the dimensions analysed and taking into account the differences that have been observed in the results of the descriptive dimensions (M and SD), the scores obtained in the variables of each phase have been compared by inferential statistics (questionnaires 1 and 2). To do this, we applied Student's t-test for related samples (Table 9), where we have also incorporated the information regarding the verification of the parametric assumptions of homoscedasticity and normality. In this regard, it should be noted that the value of the homoscedasticity of the variables yields values compatible with the application of parametric tests. Regarding normality, the Kolmogorov–Smirnov test (with Lilliefors significance correction), the significance level is higher than 0.05 ($p \geq 0.09$), adjusting to normality, so the parametric Student's t-test can be applied for related samples.

With regard to the results of the parametric assumptions, we can observe the existence of statistical significance in all the variables with respect to the differences between both phases of the study (questionnaires 1 and 2). This denotes the positive assessment regarding the didactic possibilities of the intervention carried out in the different dimensions that have been considered for the study.

However, although at the statistical level there are significant differences in all the dependent variables, dimension 2, with the subdimension 2_1, has been the area where the distance is greatest in the evaluations given by the participating students. The t-value for dimension 3 should be highlighted with a unilateral significance value of 0.006 (≤0.05) as well as a significant correlation ($p = 0.003$) between the dependent variables of this dimension.

Table 9 Results of parametric assumptions and correlational analysis (Student's t-test)

Dimension		Questionnaire	*t* de Student		Sample correlation	
			t	Sig. Uni	Correlation	Sig
2	Subd. 2_1	*1*	−2.660	0.009	0.803	0.005
		2				
	Subd. 2_2	*1*	−3.051	0.010	0.754	0.011
		2				
	Subd. 2_3	*1*	−3.029	0.010	0.812	0.001
		2				
3		*1*	−2.369	0.006	0.834	0.003
		2				

Source Own elaboration

Qualitative Results

After transcribing the comments of the focus groups, a volume of 497 textual excerpts has been obtained. In their speeches, the participants show a proactive self-perception towards AI as a resource to teach sustainability in students. In fact, 71% of these discourses (ni = 352.8) justify the importance of the inclusion of AI in citizenship education based on the training of teachers in their training. From this perspective, the stories return a naturalized contribution of the use of AI in a context ‘mediated by technologies’ (Gómez-Trigueros 2024) in the Information and Communication Society of the twenty-first century. In this sense, the speeches transmitted by the participants to the question *P-E2 What do you think about AI and its implementation in teaching?* denote its importance:

> AI, like technologies, is here to stay and is necessary in an environment like the current one where [technologies] are needed for any social, economic or communicative activity. [F.$_{G\text{-}I}$-E.$_{8}$-Cf.$_{2}$]
>
> Without it [AI] it is not possible to educate for the twenty-first century. [F.$_{G\text{-}P}$-E.$_{17}$-Cf.$_{4}$]
>
> It is incomprehensible to have citizens without skills in technologies and knowledge of AI to participate, actively and critically [in their context]. [F.$_{M}$-E.$_{42}$-Cf.$_{9}$]

To questions *P-E1 Why do you think the SDGs and sustainability are key in education?* and *P-E4 What is your opinion on AI as a tool for achieving the SDGs from the subject of geography?,* the focus groups respond, for the most part, with a clear recognition of its importance in training responsible and resilient citizens. Their comments on the absence of in-depth work on the subject of sustainability throughout their preparation as future teachers are striking:

> [Sustainability] is the cornerstone of developed and developing societies. If we want to achieve the survival of the human being, teachers cannot be oblivious to teaching this key concept. [F.$_{G\text{-}I}$-E.$_{3}$-Cf.$_{1}$]
>
> Because it is useless for us to know the characteristics of the landscape, if we do not know what to do to take care of it. [F.$_{G\text{-}P}$-E.$_{21}$-Cf.$_{5}$]
>
> Although I consider it one of the most important contents of education from geography, throughout my training we have hardly dealt with this topic and it has been superficially [in the Master’s Degree]. [F.$_{M}$-E.$_{36}$-Cf.$_{7}$]

Similarly, to the question *P-E3 Do you consider that critical thinking should be developed in the classroom?* and to question *P-E5: What importance do you give to the acquisition of critical thinking in the formation of citizenship? Explain your answer,* the participants explain the need for this competence for the twenty-first century; they give it a special interest in the fake news ecosystem and the development of AI and, as with the previous question, they manifest a lack of training for their future work in the classroom:

> [Critical thinking] must be one of the skills to be worked on in the classroom, in particular, in the face of the widespread use of AI. [F.$_{G\text{-}I}$-E.$_{15}$-Cf.$_{3}$]
>
> Training in critical thinking means training responsible citizens. I don’t think I have the capacity for it because I don’t know how to deal with fake news. [F.$_{G\text{-}P}$-E.$_{29}$-Cf.$_{6}$]

Undoubtedly, it is a complicated task, but necessary. We need [the Master's Degree] to give us tools to fight against misinformation. [$F._M$-$E._{40}$-$Cf._8$]

Conclusion

This study demonstrates that the integration of artificial intelligence (AI) into the training of future geography teachers has a significant impact on strengthening key competences for teaching sustainability and developing critical thinking. The quantitative findings, based on a comparison of pre- and post-intervention results, show marked improvements in students' perceptions in three fundamental dimensions: the understanding of sustainability from a geographical perspective, the ability to use AI as an educational tool, and the recognition of critical thinking as an essential element in forming responsible citizens. For instance, the notable difference observed in self-assessment regarding the capacity to investigate sustainability (question 6) and the use of AI in teaching (question 9) underscores the effectiveness of the implemented pedagogical approach. In addition, the qualitative results gathered from the focus groups confirm the relevance of both AI and the development of critical thinking in preparing teachers to address the challenges of the twenty-first century. The testimonies collected reveal a growing awareness amongst participants of the need to incorporate technological tools and innovative teaching strategies to combat misinformation and promote education that is aligned with the Sustainable Development Goals (SDGs). Furthermore, the study highlights the importance of teacher training that comprehensively integrates the ethical and responsible use of AI, underpinned by theoretical frameworks such as the TPACK model and up-to-date digital competences. Although improvements have been achieved through the intervention, challenges remain in uniformly integrating these contents into teacher training programmes, indicating a need for curricular review and updates that systematically address digital literacy and critical thinking.

In conclusion, this research provides empirical evidence of the transformative potential of AI in geographical education for promoting sustainability and enhancing critical thinking. It is concluded that, in order to educate responsible citizens capable of tackling current environmental, social and economic challenges, teacher training institutions must promote initiatives that integrate emerging technologies and innovative pedagogical approaches, thereby ensuring a high-quality education committed to sustainable development.

Funding or Acknowledgements This work was supported by the Knowledge Generation Project "El conocimiento geográfico sobre España, Europa y el Mundo entre los estudiantes de ESO (COGE-SO)" (PID2021-124390OB-I00), funded by MICIU/AEI /10.13039/501100011033 and the European Union (FEDER).

References

Abrami P, Bernard R, Borokhovski E, Waddington D, Wade C, Persson T (2015) Strategies for teaching students to think critically. Rev Educ Res 85:275–314. https://doi.org/10.3102/0034654314551063

AERA (2018) Estándares para pruebas educativas y psicológicas. Am Educ Res Assoc. https://doi.org/10.2307/j.ctvr43hg2

Ahmad SF, Rahmat MK, Mubarik MS, Alam MM, Hyder SI (2021) Artificial intelligence and its role in education. Sustainability 13(22), Article 22. https://doi.org/10.3390/su132212902

Akinwalere SN, Ivanov V (2022) Artificial intelligence in higher education: challenges and opportunities. Border Crossing 12(1). https://doi.org/10.33182/bc.v12i1.2015

Barra Arias E, Quemada Vives J, López Fernández D, Conde Díaz J, Badenes Olmedo C, Gordillo Méndez A (2024) Integración de la Inteligencia Artificial en la Educación Superior: Potencial, Desafíos y Oportunidades. E.T.S.I. Telecomunicación (UPM). https://oa.upm.es/83317/

Bisquerra R (2004) Metodología de la investigación educativa. La Muralla

Brookfield SD (2011) Teaching for critical thinking: tools and techniques to help students question their assumptions. John Wiley & Sons

Chen L, Chen P, Lin Z (2020) Artificial intelligence in education: a review. IEEE Access 8:75264–75278. https://doi.org/10.1109/ACCESS.2020.2988510

Cohen L, Manion L, Morrison K (2007) Research methods in education. Routledge

de Miguel González R (2024) The discipline of geography and geography education. In: Bednarz SW, Mitchell JT (eds) Handbook of geography education. Springer Nature Switzerland, pp 59–80. https://doi.org/10.1007/978-3-031-72366-7_4

De la Higuera C, Iyer J (2024) AI for teachers, an open textbook. Pressbooks. https://pressbooks.pub/aiforteachers/

De Lázaro Torres ML (2024) La utilidad de ChatGPT en la enseñanza universitaria a distancia de geoinformación para futuros docentes de enseñanza secundaria. Interdisciplinary J Didactics (1):83–104. https://doi.org/10.14198/ijd.28204

De Lázaro-Torres M-L, Puertas-Aguilar M-Á, Álvarez-Otero J (2023) Education for sustainability using cloud-based geographic information systems at University. In: Klonari A, De Lázaro y Torres ML, Kizos A (eds) Re-visioning geography: supporting the SDGs in the post-COVID era. Springer International Publishing, pp 137–152. https://doi.org/10.1007/978-3-031-40747-5_8

Estévez Cedeño B, Sánchez-Vera F (2024) Integración de la inteligencia artificial en la educación superior: Un análisis con perspectiva de género. Revista Iberoamericana de Ciencia, Tecnología y Sociedad—CTS 19(56). https://doi.org/10.52712/issn.1850-0013-557

Favier TT, van der Schee JA (2014) The effects of geography lessons with geospatial technologies on the development of high school students' relational thinking. Comput Educ 76:225–236. https://doi.org/10.1016/j.compedu.2014.04.004

Frank P, Henkel G, Lysgaard JA (2024) Between evidence and delusion—a scoping review of cognitive biases in environmental and sustainability education. Environ Educ Res 30(9):1477–1499. https://doi.org/10.1080/13504622.2024.2371507

Gambrill E (2018) Critical thinking and the process of evidence-based practice. Oxford University Press

García de la Vega A (2019) Spatial thinking ability acquisition through geospatial technologies for lifelong learning. In: de Miguel González R, Donert K, Koutsopoulos K (eds) Geospatial technologies in geography education. Springer International Publishing, pp 21–40. https://doi.org/10.1007/978-3-030-17783-6_2

García-Valcárcel-Muñoz-Repiso A, Basilotta-Gómez-Pablos V, López-García C (2014) Las TIC en el aprendizaje colaborativo en el aula de Primaria y Secundaria. Comunicar: Revista Científica de Comunicación y Educación 21(42):65–74. https://doi.org/10.3916/C42-2014-06

Gargallo-López B, Pérez-Pérez C, Verde-Peleato I, García-Félix E (2017) Estilos de aprendizaje en estudiantes universitarios y enseñanza centrada en el aprendizaje. RELIEVE—Revista Electrónica De Investigación Y Evaluación Educativa 23(2). https://doi.org/10.7203/relieve.23.2.9078

Gómez-Trigueros IM (2024) Interdisciplinary gamification with LKT: new didactic interventions in the secondary education classroom. Multidisciplinary J Educ Res 14(1):115–133. https://doi.org/10.17583/remie.10622

Jefferson G (1984) On the organization of laughter in talk about troubles. In: Atkinson JM, Heritage J (eds) Structures of social action: Studies in conversation analysis. Cambridge University Press, pp 346–369. https://doi.org/10.1017/CBO9780511665868.021

Kerski JJ (2008) The role of GIS in digital earth education. Int J Digital Earth 1(4):326–346. https://doi.org/10.1080/17538940802420879

Kerski JJ (2024) Integrating web mapping and geospatial technologies. In: Thriving in an Academic Career. Routledge. https://doi.org/10.21203/rs.3.rs-5410327/v1

Kuckartz U (2014) Qualitative text analysis: a guide to methods, practice and using software. SAGE. https://doi.org/10.4135/9781446288719

Latorre Iglesias EL, Castro Molina KP, Potes Comas ID (2019) Las TIC, las TAC y las TEP: Innovación educativa en la era conceptual. Universidad Sergio Arboleda. https://repository.usergioarboleda.edu.co/handle/11232/1219

Leininger-Frézal C, Sprenger S, Lázaro-Torres M-LD, Rodríguez Domenech MÁ, Heidari N, Pigaki M, Naudet C, Lecomte A, Gallardo M (2023) Global change challenge in the higher education curriculum on the approach of blended learning. European J Geography 14(2):1–14. https://doi.org/10.48088/ejg.c.lei.14.2.001.014

López-Roldán P, Fachelli S (2015) Metodología de la investigación social cuantitativa. Universidad Autónoma de Barcelona

Maude A (2022) Using geography's conceptual ways of thinking to teach about sustainable development. Int Res Geograph Environm Educ 32:4–19. https://doi.org/10.1080/10382046.2022.2079407

Miao F, Cukurova M (2024) AI competency framework for teachers. UNESCO: United Nations Educational, Scientific and Cultural organization. https://unesdoc.unesco.org/ark:/48223/pf0000391104

Mishra P, Koehler MJ (2006) Technological pedagogical content knowledge: a framework for teacher knowledge. Teach Coll Rec 108(6):1017–1054. https://doi.org/10.1111/j.1467-9620.2006.00684.x

Moncayo Cueva HLM, Cuesta-Chávez G, Ramírez A, Pintado RNZ (2024) Utilizing emerging technology trends and artificial intelligence in higher education. J Higher Educ Theory Practice 24(3), Article 3. https://doi.org/10.33423/jhetp.v24i3.6847

Nicholson D, Vargas V, Skelly G (2023) Enquiry-based skills education for sustainable development in a UK geography module as a catalyst for organisational change. Int J Sustain High Educ 24(8):1897–1914. https://doi.org/10.1108/ijshe-07-2022-0225

Otzen T, Manterola C (2017) Técnicas de muestreo sobre una población a estudio. Int J Morphol 35(1):227–232. https://doi.org/10.4067/s0717-95022017000100037

Palacios FA, Oberle A, Quezada XC, Ullestad M (2017) Geographic education for sustainability: developing a bi-national geographical thinking curriculum. In: Brooks C, Butt G, Fargher M (eds) The power of geographical thinking. International perspectives on geographical education. Springer, Cham. https://doi.org/10.1007/978-3-319-49986-4_8

Punie Y (ed) (2017) European framework for the digital competence of educators: DigCompEdu. Publications Office of the European Union

Raykov T, Marcoulides GA (2017) Ecuation of true criterion validity for unidimensional multicomponent measuring instruments in longitudinal studies. Struct Equ Model 24(4):599–608. https://doi.org/10.1080/10705511.2016.1172486

Risopoulos-Pichler F, Daghofer F, Steiner G (2020) Competences for solving complex problems: a cross-sectional survey on higher education for sustainability learning and transdisciplinarity. Sustainability 12(15):6016. https://doi.org/10.3390/su12156016

Rodríguez Tenjo JDP, Gallardo Pérez ÓA (2020) Perfil docente con visión inclusiva: TIC-TAC-TEP y las habilidades docentes. Encuentro Internacional de Educación en Ingeniería. https://doi.org/10.26507/ponencia.731

Sánchez-Gómez MC, Rodrigues AI, Costa AP (2018) Desde los métodos cualitativos hacia los modelos mixtos: tendencia actual de investigación en ciencias sociales. Revista Ibérica de Sistemas e Tecnologias de Informação 28:9–13. https://doi.org/10.17013/risti.28.0

Shulman LS (1986) Those who understand: knowledge growth in teaching. Educ Researcher 15(2):4–14. https://doi.org/10.3102/0013189X015002004

UNESCO (2017) Education for sustainable development goals: learning objectives. https://doi.org/10.54675/CGBA9153

UNESCO (2024a) Guía para el uso de IA generativa en educación e investigación. Organización de las Naciones Unidas para la Educación, la Ciencia y la Cultura. https://unesdoc.unesco.org/ark:/48223/pf0000389227

UNESCO (2024b) Chat GPT e IA en la educación superior. Guía de inicio rápido. Organización de las Naciones Unidas para la Educación, la Ciencia y la Cultura. https://unesdoc.unesco.org/ark:/48223/pf0000385146_spa/PDF/385146spa.pdf.multi

Challenges in Tracking Early Education: A European and Serbian Perspective on SDG Target 4.2

Stefana Matović, Suzana Lović Obradović, and Darko Vuković

Abstract Early childhood education (ECE) plays a critical role in shaping a child's cognitive, social, and emotional development, laying the foundation for future academic and professional success. This chapter explores the importance of ECE within the framework of Sustainable Development Goal (SDG) 4.2, which advocates for equitable access to quality early education. It highlights disparities across European countries, with a focus on Serbia, where regional, gender, and socioeconomic inequalities persist despite legislative reforms and increased investments in preschool infrastructure. The analysis underscores that while many European nations have made substantial progress in expanding early education, significant challenges remain—particularly in marginalized and rural communities. Countries with strong policy frameworks, such as Malta and Russia, demonstrate near-universal participation rates, while others, like North Macedonia and Bosnia and Herzegovina, face low enrollment and gender disparities. Serbia mirrors these trends, with notable regional disparities between urban and rural areas and an overwhelming gender imbalance among educators. The chapter emphasizes the need for targeted interventions to bridge educational gaps, including improving inclusivity, teacher training, and infrastructure development. It also stresses the importance of integrating innovative teaching methods and data-driven policymaking to ensure that all children, regardless of socioeconomic background, have access to high-quality early education opportunities.

Keywords Early childhood education · Pre-primary education · SDGs · Europe · Serbia

S. Matović (✉) · S. L. Obradović · D. Vuković
Geographical Institute "Jovan Cvijić" SASA, Belgrade, Serbia
e-mail: s.babovic@gi.sanu.ac.rs

S. L. Obradović
e-mail: s.lovic@gi.sanu.ac.rs

D. Vuković
e-mail: d.vukovic@gi.sanu.ac.rs

L. Soares et al. (eds.), *Compromised Geography*, Key Challenges in Geography,
https://doi.org/10.1007/978-3-032-05201-8_7

Introduction

Early childhood education is increasingly recognized as a vital determinant of a child's future academic and vocational success. This foundational period, which includes various forms of educational programs like pre-primary education, preschool education, kindergarten, and early childhood education, is a critical time for a child's development. These terms all imply organized institutions that cater to children between ages of 2, 5 to 6 years, a phase when the brain is rapidly developing (Pankajam 2017). During these years, children's cognitive, social, and emotional abilities are shaped, forming the basis for their lifelong learning and development.

The importance of early childhood education is underscored by global frameworks such as STEP, which is introduced by the World Bank (2012). This comprehensive model emphasizes the role of skill development throughout the lifecycle, from early childhood to adulthood, with the first step—"Getting Children Off to the Right Start"—highlighting the pivotal nature of early childhood in shaping human skills. Indeed, according to human capital theory, the skills acquired during pre-primary education are building blocks for future capabilities, with each stage of development enhancing what came before (Heckman 2006). This early period of development is particularly crucial for brain growth. During the ages of 0–5, the brain learns at its fastest rate, making this an ideal time for children to develop essential skills, including cognitive, social, and emotional abilities. Pre-primary education, which provides this early environment for learning, plays a key role in ensuring that children develop not only the academic skills needed for future learning but also the personal attributes that will support their long-term success in various aspects of life. Research by Hanusheck and Woessman (2010) found that pre-primary education provides the highest returns on investment, both for individuals and society. These benefits extend beyond academic success to include improved health, better social skills, and long-term economic outcomes. The role of early childhood education in shaping child's future is well-documented. Research shows that children who receive quality pre-primary education tend to fare better in subsequent stages of their educational journey. Brooks-Gunn (2003) and Karoly et al. (2005) have found that these children are better prepared for primary school, exhibit stronger cognitive and non-cognitive skills, and show higher levels of social competence. More specifically, OECD (2011) data suggest that children who attended preschool for more than a year score significantly higher on academic assessments such as the PISA reading test, compared to their peers who do not have access to preschool education. Additionally, countries that invest more in early education, like those in Northern Europe, tend to have higher cognitive scores and less inequality in educational outcomes, while nations with less investment, like those in Southern Europe, face poorer school outcomes and greater inequality (OECD 2013).

The social and emotional development fostered in preschool is equally important. Incorporating social and emotional learning into preschool programs has been shown to encourage positive attitudes toward school and enhance overall academic performance (Denham 2016; Ornaghi et al. 2017). Children begin developing vital skills such as empathy, emotional awareness, and prosocial behaviors from their early interactions with peers and caregivers. This type of emotional and social skill-building not only supports academic achievement but also promotes healthy relationships and self-regulation, which are crucial for success in later life (Allen and Kelly 2015). The cognitive benefits of preschool education are also significant. Studies have demonstrated that attending preschool for even one year can give children an average learning advantage of about three months, particularly in skills like literacy, mathematics, and language development (Li et al. 2020). These cognitive skills are reinforced by social interactions that foster language development and promote independence, preparing children for the academic challenges they will face in primary school. In addition to cognitive benefits, early childhood education is associated with improved long-term health outcomes. The early brain development that occurs during preschool years is influenced by numerous factors, including language, activities, and social experiences. These interactions help build a foundation for self-regulation, attention control, and behavior management, which are essential for academic success and overall well-being (Cetin et al. 2018; Yoshikaw et al. 2016). Ultimately, preschool education is not just about teaching academic content; it is about developing the whole child. Through innovative educational strategies, preschool programs today aim to nurture individual, social, and cultural skills while also promoting healthy lifestyles (Bohodirovna and Usmanova 2022). This holistic approach ensures that children are not only academically prepared but also equipped to thrive as active, engaged, and responsible members of society.

Ensuring equitable access to quality education is a cornerstone of sustainable development. As part of the broader Sustainable Development Goal (SDG) 4 (Quality Education), Target 4.2 aims to guarantee that by 2030, all children, especially those in vulnerable situations, have access to quality early childhood development, care, and pre-primary education. This is a crucial step in fostering lifelong learning and reducing inequality in education. However, not everyone has equal opportunities for educational success. Research highlights significant disparities based on socioeconomic factors within families (Marks 2013). These inequalities can limit children's access to quality early education and impact their long-term academic and career prospects. Parents, children, young adults, and adult learners all play a central role in education at the micro-level. They participation is reflected in statistics, policy discussions based on benchmarks and indicators, and the broader expectation of acquiring strong knowledge and skills for success in the global economy (Boeren 2016).

A positive parental attitude toward education is particularly crucial in this context (SDG 4.1 and SDG 4.2). It not only enhances their children's educational outcomes but also benefits their own future learning opportunities, such as vocational or tertiary education (SDG 4.3) and workforce integration (SDG 4.4) (Boeren 2019). Early experiences with education shape their confidence and motivation, influencing their approach when they become parents and support their own children's schooling. To assess progress toward SDG 4.2, measuring early childhood development is essential. Without understanding children's starting points in formal schooling, it is difficult to determine whether education systems are effectively closing gaps. Accurate measurement informs policy decisions, targets investments, shapes curricula, improves teaching methods, and tracks skill changes over time (Bassok and Latham 2017). Additionally, it is crucial for scaling up early childhood programs to reach a broader population (Frongillo et al. 2014). One of the key challenges in measuring early childhood development is the classification of the "developmentally on track" indicator. The SDG monitoring framework includes national, regional, thematic, and global indicators, but only the thematic and global indicators are designed for cross-country comparability. Target 4.2 has been classified as a "Tier III" indicator by the United Nations Statistical Commission (2017) due to significant gaps in data availability and the ongoing developed of standardized data collection methods.

ECE is widely recognized as a distinct stage before primary education, traditionally focusing on care and school readiness rather than formal learning. Although ECE is not mandatory in most European countries, participation rates are high, with 95.4% of children in the EU attending preschool in 2017 (European Statistical Office [EUROSTAT] 2019). This widespread participation ensures that most children enter primary school with prior learning experiences that support their cognitive, social, and emotional development. However, the management of preschool education varies across countries due to differences in governance structures. The European education system is decentralized, with pre-primary education serving as the first formal stage of learning. These programs, typically held in schools or dedicated centers, help young children—often starting at age three—adapt to structured educational environments while fostering their overall development.

The administration of preschool education differs significantly across European nations. In Germany, the Ministry of Public Affairs oversees preschool education, while local departments for the Welfare of the Young manage nursery education at the municipal level (Sergeeva 2010). In Russia, the Ministry of Education is responsible for setting policies and legal frameworks, while regional and municipal authorities manage the implementation and supervision of preschools (Lysova 2006). In contrast, France integrates preschool education into the formal school system, treating kindergartens as an extension of primary education rather than a separate entity. Schools follow the national academic calendar, and the Ministry of Education governs policies and curriculum standards (Tokarev et al. 2020). Similarly, in the Czech Republic, the Ministry of Education, Youth, and Sports oversees preschool education, ensuring consistency in funding, curriculum, and development from early childhood through secondary education (Vladimíra 2018).

Preschool teacher training programs across Europe reflect diverse pedagogical philosophies. These programs are often based on O. Dekroli's theory, which emphasizes structuring education around children's interests and developmental needs (Dychkivska 2017). The training curriculum typically includes child development, interaction with nature, and social environment studies. Additionally, Eastern European countries integrate innovative teaching methods, such as video lessons in Poland, simulations in Romania, and case studies, all aimed fostering consistency, effectiveness, and adaptability in preschool teaching (Dychkivska and Uliukaeva 2019). Western European preschool education places a strong emphasis on innovation and research-based quality improvement. Unlike traditional skill-based training, these systems focus on knowledge acquisition, creativity, and values-driven individualized learning. This approach ensures that teachers are equipped not only with pedagogical skills but also with a deeper understanding of child development, allowing them to adapt teaching strategies to meet diverse student needs (Vandenbroeck 2010).

The starting age for compulsory primary education varies by country, influencing the structure and necessity of pre-primary education. According to European Commission (2000), countries such as Belgium, Denmark, France, Italy, the Czech Republic, Hungary, and Sweden offer three years of pre-primary education, while Austria, Germany, and the Netherlands provide two to three years. In Finland, Greece, and Portugal, it lasts two years. The United Kingdom follows a different model, where primary education starts earlier, reducing the duration of pre-primary schooling. The process of transition from preschool to primary education can be examined on three levels. At the micro-level, it occurs within personal environments such as family, friends, teachers, classmates, and the school setting. The macro-level encompasses the broader educational framework, including institutions and the education system. The meso-level serves as a bridge between the micro- and macro-levels, focusing on how individuals interact across different groups and the wider educational context. This multitiered perspective helps policymakers and educators understand the complexities of early childhood education transitions. National education authorities establish policies that shape how countries manage early childhood and primary education, including teacher training standards (Tomaszewska-Pękała et al. 2017). A key distinction in teacher professional development exists between northern and southern EU countries. In the south, governments set national guidelines for teachers' lifelong learning, ensuring consistency across the country. In contrast, northern EU countries follow a decentralized model, where schools determine their own professional development programs, allowing for greater adaptability to local needs. Additionally, school management training is a priority for southern EU governments, emphasizing the need for leadership and administrative skills in preschool institutions (González-Moreira et al. 2021). According to Vuković et al. (2019, 2023a), economic, social, and ecological goals of sustainable development align well naturally and improve regional development, suggesting that a balanced approach to these goals is essential.

Legislative Frame and Programs of Preschool Education in Serbia

Research on sustainable development in Serbia has addressed various aspects, including educational reforms, economic growth, and the broader sociocultural context, reflecting the country's ongoing efforts to align with global sustainability goals. Various papers analyzed SDG4 (Matović and Lović-Obradović 2021), some of three pillars of sustainable development (Babović et al. 2016; Lović Obradović et al. 2016), or different branches of the economy development in Serbia (Gajić et al. 2024; Vuković et al. 2023b). Serbia has undergone significant preschool education reforms in recent decades. The 2010 Preschool Education Act aimed to promote child development, support families, and prepare children for further education while highlighting key principles such as accessibility, democracy, and openness. This law also standardized preschool programs, favoring half-day or full-day formats while eliminating shorter sessions. However, one limitation of the act was the lack of provisions for research collaboration in preschool education (Krnjaja and Pavlovic-Breneselovic 2013). Further development came with the 2012 Strategy for the Development of Education in Serbia until 2020, which sought to enhance children's well-being from birth to school age. However, the concept of "well-being" remained undefined, despite its importance in educational policies and practices (Breneselovic 2010). A new concept introduced in 2018 emphasized holistic learning, integrating play, practical activities, and structured education. This approach highlighted the importance of relationships with peers, adults, family, the environment, and the wider community. Rooted in sociocultural theory, childhood sociology, postmodernism, and poststructuralism, the new model views children as active, capable learners and promotes collaboration in building a community of practice (Stojanovic et al. 2018). According to Krnjaja and Pavlović-Breneselović (2017), this new framework aligns with international educational polices and is based on principles from UN Convention on the Rights of the Child, Serbian education policies, and successful global preschool education models. However, despite this alignment, challenges remain in translating these principles into practice, particularly in ensuring equal access and quality across different regions. Krnjaja and Pavlović-Breneselović (2017) examined shortcomings in the preschool system and found a gap between strategic and legislative goals and their implementation. While official documents emphasize equal opportunities and positive discrimination, in practice, preschool education remains regionally uneven, with underdeveloped areas having the lowest enrollment rates and limited access to facilities. Additionally, according to them, Serbia lacks a unified database for preschool children and the overall preschool education system.

Methodology

The primary objective of this study is to assess Target 4.2 of the Sustainable Development Goals (SDGs), specifically analyzing the state of pre-primary education across European countries. The research is based on a comprehensive review of scientific literature in this domain, as well as reports that provide insights into the progress made toward achieving SDG 4.2. Although the official SDG framework (United Nations Department of Economic and Social Affairs 2023) includes two primary indicators for this target, UNESCO (2023) has expanded its database with additional indicators to offer a more detailed perspective on the progress of various countries.

These supplementary data points form the foundation of this study. The first official indicator, 4.2.1, refers to the proportion of children aged 24–59 months who are developmentally on track in health, learning, and psychosocial well-being, disaggregated by sex. However, many European countries lack sufficient data for this indicator. The study presents findings for the seven countries that have conducted relevant measurements, covering the period from 2015 to 2023. The second indicator, 4.2.2, examines the participation rate in organized learning (one year before the official primary entry age), by sex. This indicator is analyzed in greater detail due to the availability of data for 41 European countries. The study evaluates both overall participation rates and gender-specific trends. The data span from 2015 to 2022, with the most recent available statistics for each country being considered. Notably, data for seven countries are available for 2022, while for most others, the latest records date to 2021. Exceptions include Liechtenstein (2020) and Russia (2019).

In addition to these two main indicators, the UNESCO database provides information on several other relevant indicators related to early childhood education, which are also considered in this study:

- 4.2.4: Net enrollment rate in early childhood education (both sexes)
- 4.2.4b: Net enrollment rate in early childhood educational development programs (both sexes)
- 4.2.4c: Net enrollment rate in pre-primary education (both sexes)
- 4.2.5: Number of years of free pre-primary education guaranteed in legal frameworks and number of years of compulsory pre-primary education guaranteed in legal frameworks.

For Indicator 4.2.4, data are available for 39 countries from 2015 to 2023. The most recent data for most countries are from 2023, while eight countries have data for 2024. Exceptions include Greece (2021), Liechtenstein (2018), Malta (2019), and Russia (2021). Indicator 4.2.4b is examined for 26 countries over the period from 2015 to 2021, with Greece and Russia having their latest available data from 2019. Indicator 4.2.4c covers 40 countries from 2015 to 2022, with six countries providing data for 2022, the majority reporting figures for 2021, and Russia's latest available data being from 2019. Regarding Indicator 4.2.5, data on the number of years of free pre-primary education guaranteed in legal frameworks are analyzed for 26 countries over the period from 2015 to 2024. Additionally, the number of years of compulsory pre-primary education guaranteed in legal frameworks is examined for 18 countries within the same timeframe.

This study aims to provide a comprehensive and comparative analysis of pre-primary education across Europe, highlighting disparities, trends, and policy implementations. By leveraging these diverse datasets, the research seeks to offer valuable insights into the progress and challenges of achieving universal access to quality early childhood education as outlined in the SDG framework. United Nations Statistical Commission (2017) structured methodological framework for revising the Early Childhood Development Index (ECDI). The revision process is being carried out in collaboration with an expert advisory panel, which comprises academic scholars and researchers specializing in early childhood development measurement and tool development, as well as technical experts with expertise in validity and reliability assessments and cognitive testing for instrument validation. The methodological approach emphasizes rigorous testing and evaluation to ensure the accuracy, reliability, and applicability of the revised index. Most studies show that early between-school tracking increases socioeconomic disparities in academic achievement compared to comprehensive or within-school tracking systems (Blossfeld et al. 2016; Burger 2016; Heisig and Solga 2015). Also, many use comparative method for analyzing SDG4 achievement (Matović et al. 2024).

The methodology for progress tracking follows Eurostat's approach, using compound annual growth rate (CAGR) to assess changes over time (Babović 2024). This method is particularly relevant for objectives without predefined quantitative targets for 2030. The EU evaluates progress over both long-term (15 years) and short-term (5 years) periods. In Serbia, since specific numerical targets are not established, progress is measured against a baseline year, examining both the trend direction and intensity of change. Reports use symbols to represent whether a goal is being approached, stagnated, or deviated from. If conflicting trends are observed within sub-indicators, the overall trend is classified as inconsistent. For Serbia, additional parameters for which data are available have been analyzed to provide a more comprehensive overview of the state of pre-primary education across different parts of the country. The data refer to the year 2022 and are presented both at the national level and by regions, highlighting differences observed in certain indicators (SORS 2022). The results are presented for the following six key indicators:

- The number of children in pre-primary education, categorized by age and gender;
- The number of live births eligible for enrollment in Preschool Preparatory Program (PPP);
- The number of children enrolled in preschool education institutions, classified by age;
- The rate of inclusion of children in mandatory PPP by gender (%);
- The number of preschool education institutions; and
- The number of employees in preschool education, categorized by type of job position and gender.

These indicators allow for a detailed examination of the preschool education system, offering insights into disparities and trends across different regions of Serbia. Preschool education in Serbia has a tradition of over 150 years, during which various changes have shaped educational programs, perceptions of childhood, and approaches to early learning. Throughout this long development, shifts have been influenced by national circumstances and dominant pedagogical philosophies, ultimately leading to a modern view of children as active, competent participants in their own learning and development (Stojanović and Bogavac 2016). The PPP was implemented in the 2007/08 school year to boost primary school enrollment, reduce dropout rates, and enhance social inclusion and educational outcomes nationwide (Pešikan and Ivić 2016). However, its introduction faced two key challenges: incomplete child participation and insufficient monitoring and analysis of its implementation.

Results

Sustainable Development Goal (SDG) 4.2 focuses on ensuring that all children have access to quality early childhood development, care, and pre-primary education, enabling them to be well-prepared for primary schooling. This goal is measured through two key indicators. However, significant data gap exists for the first indicator, which assesses whether children 24 to 59 months old are developmentally on track in terms of health, learning, and psychosocial well-being. Many European countries lack reliable data for this metric, making it difficult to assess overall progress comprehensively. Currently, data for 4.2.1 indicator are available for only seven European countries between 2015 and 2023, all of which are classified as developing countries (UNICEF 2023). These countries can be categorized into three distinct groups based on their results. First group is high developmental readiness with almost all children on track. Serbia (97%) and Bosnia and Herzegovina (96%) have the highest percentages of children who are developmentally on track. The gender differences in these countries are minimal. In Serbia, nearly all girls (99%) fall within this category, with a 3% lower rate for boys. The same trend is observed in Bosnia and Herzegovina, where 98% of girls are developmentally on track, while boys lag behind by the same 3% margin. Second group is moderate developmental readiness with approximately 90% children on track. Montenegro (90%), Ukraine (89%), and Belarus (87%) form the second group. Gender differences vary across these nations. In Ukraine, there is no significant distinction between boys and girls, while in Montenegro, the gap is minor, with girls outperforming boys by just 1%. However, Belarus exhibits a more pronounced gender gap, with 89% of girls meeting developmental milestones, compared to 5% fewer boys. Third group is lower developmental readiness with less than 85% of children on track. The Republic of Moldova (84%) and North Macedonia (82%) fall into this category. While Moldova shows only a minor gender disparity (1% more girls than boys), North Macedonia presents the most significant gap among all measured countries. In this country, 89% of girls are developmentally on track, but the rate for boys is 13% lower, highlighting a concerning gender imbalance.

Given the limited data availability for the first sub-indicator, most analyses have instead focused on the second sub-indicator—participation in organized learning one year before primary school, which is the main subject of this study. An analysis of participation rates over an eight-year period (2015–2022) reveals that many European countries are on track to meet this indicator, though there are notable exceptions (Fig. 1). The lowest participation rates are observed in Bosnia and Herzegovina (B&H), where less than a third of children attend the pre-primary education, and North Macedonia, where participation remains below 50%. Countries such as

Ukraine and Montenegro show moderate participation, with approximately three-quarters of children involved in organized learning before primary school. Romania, Bulgaria, and Slovakia have rates around 85%, while Albania, Malta, and Russia report higher participation levels, reaching 90%. However, it is important to note that Russia's last available data are from 2019, whereas for most other countries, 2022 data were used. The majority of the 41 European countries studied have near-universal pre-primary participation rates, close to 100%.

When analyzed by sexes (Fig. 2), gender disparities emerge in some countries. In six out of 41, participation rates differ noticeably between boys and girls. The largest gender gap is seen in Albania and Belarus, where approximately 5% more boys than girls participate in pre-primary education. San Marino follows closely, with a 4% higher participation rate for boys. In Andorra and Ukraine, the difference is 2%, again favoring boys. Luxemburg stands out as the only country where more girls (3% higher) are enrolled than boys.

To further refine the assessment of early childhood education, the United Nations has suggested additional indicators beyond the core SDG 4.2 framework. One such indicator is 4.2.4, and it refers to net enrollment rate of early childhood education. Figure 3 illustrates the variation in net enrollment rates across European countries. The results reveal significant disparities. The lowest enrollment rates (40% or lower) are seen in Greece, Ireland, Romania, North Macedonia, and particularly Bosnia and Herzegovina, where only 13% of children are enrolled in early childhood education. In contrast, countries that have recognized the critical importance of early childhood education—such as Belarus, France, Gibraltar, Italy, the Netherland, Malta (latest data from 2017), and Poland—report rates exceeding 90%. Other countries generally fall into 10% incremental groups, ranging from 50 to 90%. Serbia, Croatia, Austria, Hungary, Montenegro, and United Kingdom are in group from 50 to 60%. Better results are in Estonia, Germany, Lithuania, Moldova, Russia (2019 data), Spain, and Ukraine with 60–70%. The biggest group with 70–80%includes Bulgaria, Denmark, Iceland, Latvia, Liechtenstein (2016 data), Norway, San Marino, Slovakia, and Switzerland. The higher values of 80–90% are found in Andorra, Czech, Finland, Slovenia, and Sweden.

The most significant gender difference appears in Liechtenstein (Fig. 4), where boys' net enrollment rate is 6% higher than girls. Belarus follows with a 4% higher enrollment rate for boys. In Gibraltar (4%), Netherlands (3%), and Andorra (2%), slightly more girls are enrolled. Italy and San Marino have a reverse trend, with more boys enrolled than girls.

A more specific measure, the net enrollment rate for early childhood educational development programs, is available for fewer countries and lacks continuous data across the observed period (Fig. 5). However, notable trends include the highest enrollment rate (around 70%) in Belarus, Gibraltar, Slovenia, and Sweden. The lowest enrollment (less than 20%) was in Austria, Greece, Hungary, North Macedonia, Moldova, Romania, and United Kingdom, which is seven out of 26 measured countries. Gender differences for this indicator are negligible.

Net enrollment rate of pre-primary education was also measured as one of indicators (Fig. 6). While most countries report high values for this parameter, some

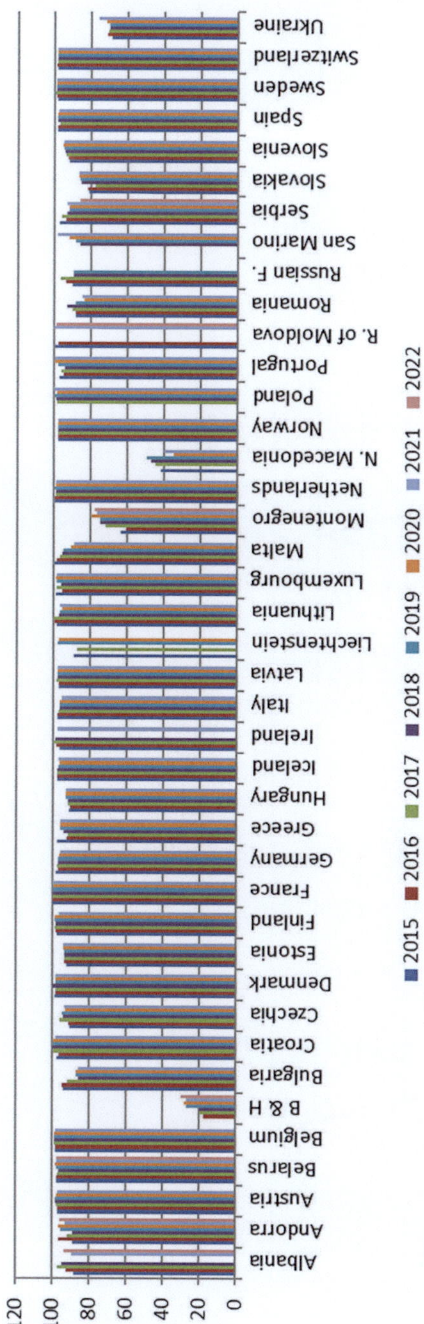

Fig. 1 Adjusted net enrollment rate, one year before the official primary entry age, both sexes (%)—Indicator 4.2.2

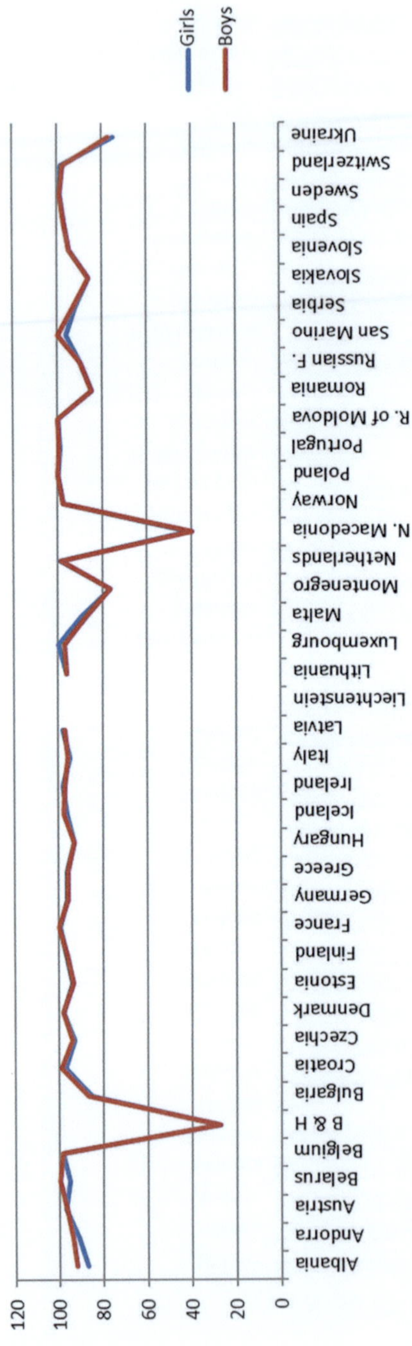

Fig. 2 Adjusted net enrollment rate, one year before the official primary entry age, by sexes (%)

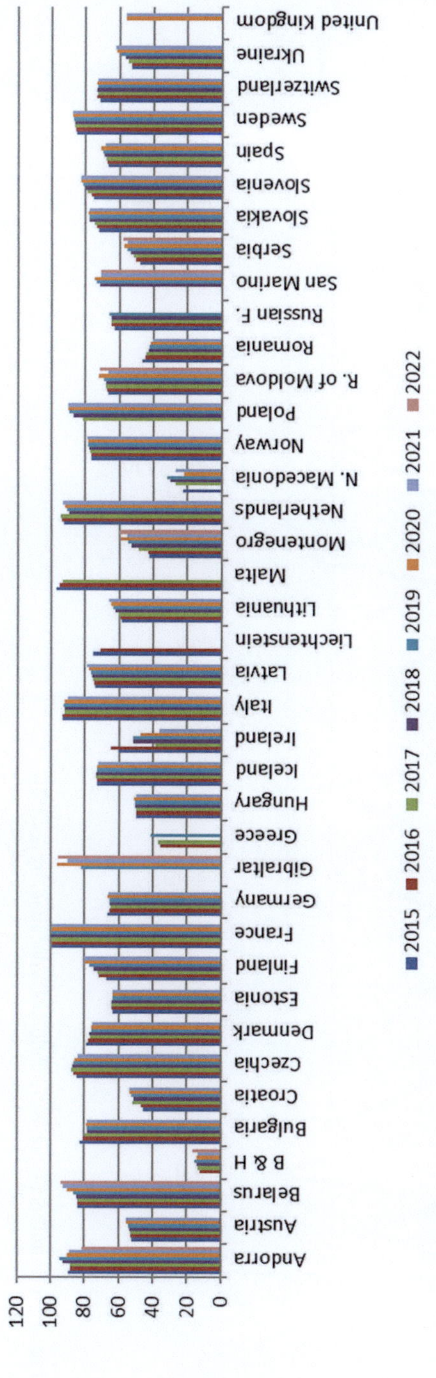

Fig. 3 Net enrollment rate, early childhood education, both sexes (%)—Indicator 4.2.4

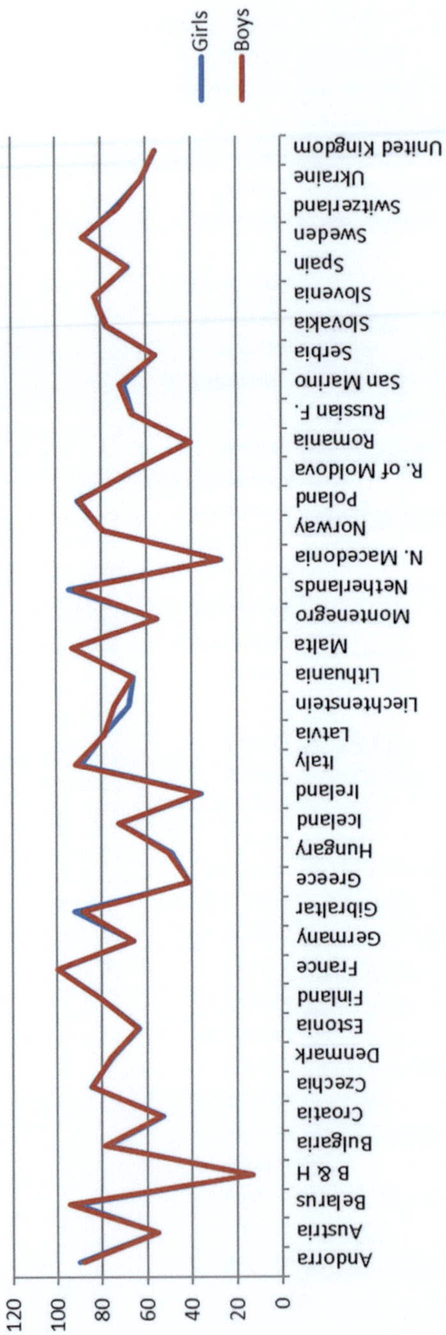

Fig. 4 Net enrollment rate, early childhood education, by sex (%)

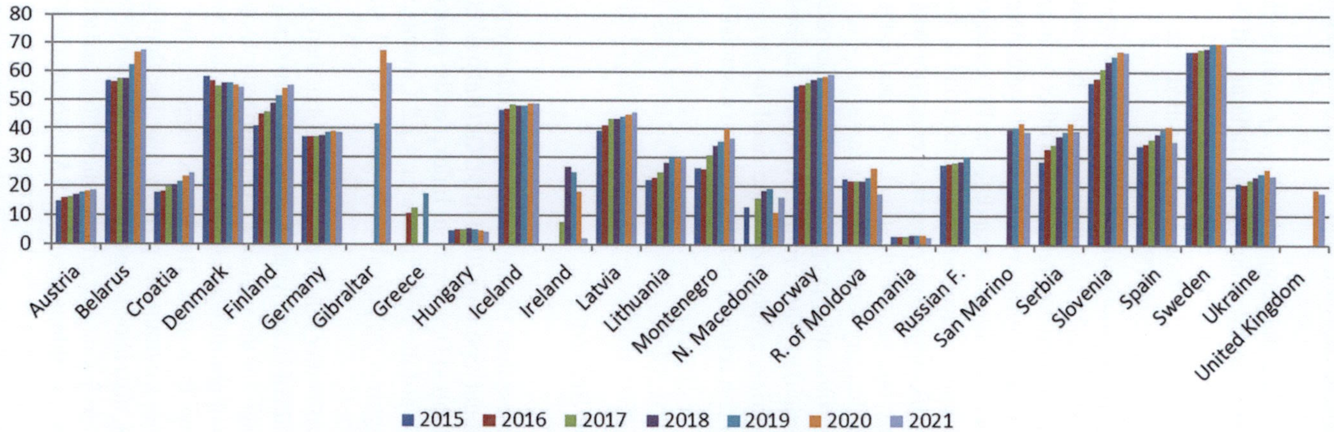

Fig. 5 Net enrollment rate, early childhood educational development programs, both sexes (%)—Indicator 4.2.4b

exceptions indicate room for improvement, such as Bulgaria, Croatia, Liechtenstein, Montenegro, Romania, Serbia, Slovakia, and Switzerland, and the lowest was in North Macedonia (34%).

Gender disparities appear in certain countries (Fig. 7), with Liechtenstein showing the largest gap (9% more boys enrolled than girls). Belarus and the UK show a 4% advantage for boys, while Albania (3%), Hungary (2%), and Italy (2%) follow. In contrast, the Netherlands reports 3% more girls enrolled than boys (4.2.4c: Net enrollment rate, pre-primary, both sexes (%)).

The number of years of pre-primary education guaranteed by law varies across countries and that is Indicator 4.2.5. This indicator can be assessed as free or as compulsory. This was not presented through years, as other indicators, because there is no change through time. Free pre-primary education is guaranteed in 26 countries, and compulsory pre-primary education is mandated in 18 countries. Figure 8 shows that the duration of pre-primary education ranges from one to four years. Also, in Fig. 9, the number of years of compulsory preschool education was presented. While most countries require only one year, Bulgaria, Latvia, and Luxembourg mandate two years, and France and Hungary require three years.

The analysis highlights substantial progress in ECE participation across Europe but also reveals gaps, particularly in certain developing nations. Gender disparities remain in some countries, emphasizing the need for targeted interventions. Expanding data collection and refining policies will be essential to achieving SDG 4.2 comprehensively.

Additional Results for Serbia

Changes greater than 0 but less than 1% are considered moderate progress toward the goal. According to the national report for 2023 of SDGs (Babović 2024), the value for Indicator 4.2.1, Early Childhood Development Index, for the period 2014–2019 was 0.4 with trend progress toward the goal. Indicator 4.2.2 Coverage of children with preparatory preschool programs for period 2018–2022 was 0.1 with trend progress toward the goal.

Serbia was compared with other European countries, but there are also some other national data that can be added in analysis. The analysis first examined the number of children by age and gender for the year 2022, categorized into specific age groups. The largest group consists of children aged 5.5 to 6.5 years (one year before starting school), with 7% more boys than girls. The second largest group includes children aged 4 to 5 years, where boys outnumber girls by 6%. The third group, comprising children aged 3 to 4 years, also shows a gender disparity, with 5% more boys than girls. Next in size is the 2- to 3-year-old group, where boys exceed girls by 6%. The fifth largest group consists of children aged 5 to 5.5 years, again with 6% more boys than girls. Slightly smaller in number is the group of children aged 6 months to 2 years, with boys outnumbering girls by 4%. The least numerous age group is 6.5 to 7.5 years, yet even here, boys slightly outnumber girls by 4%.

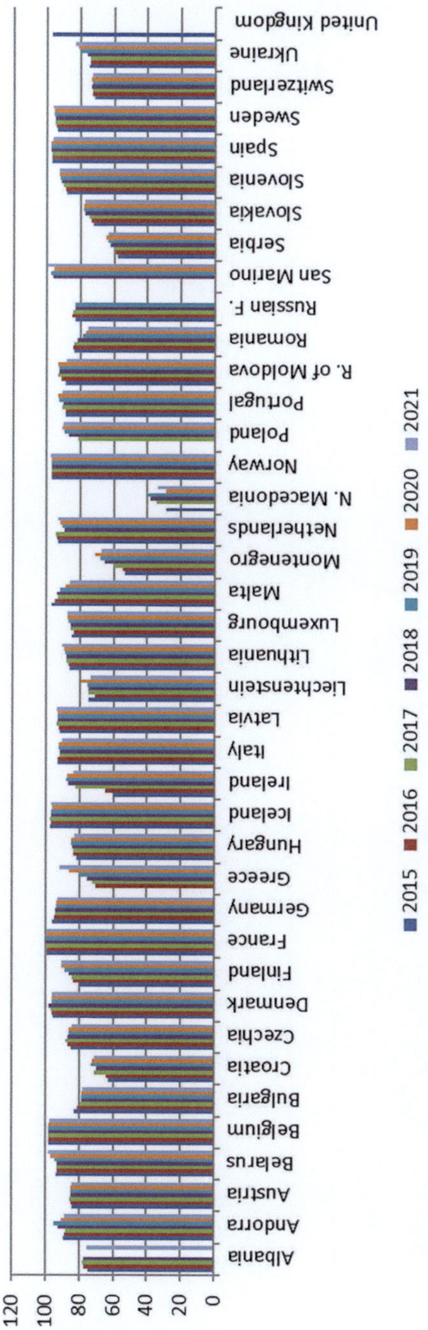

Fig. 6 Net enrollment rate of pre-primary education, both sexes (%)—Indicator 4.2.4c

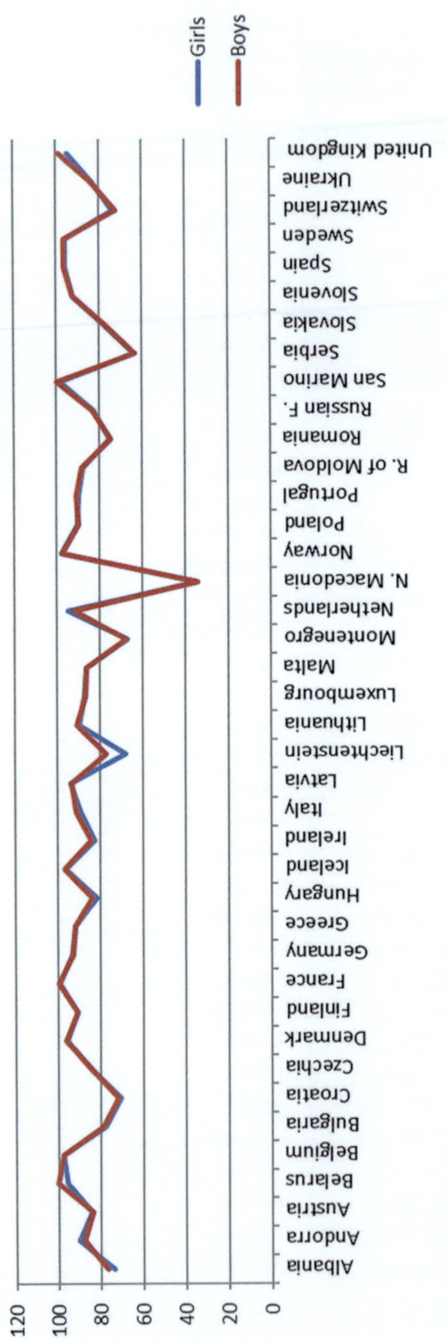

Fig. 7 Net enrollment rate, pre-primary, by sex (%)

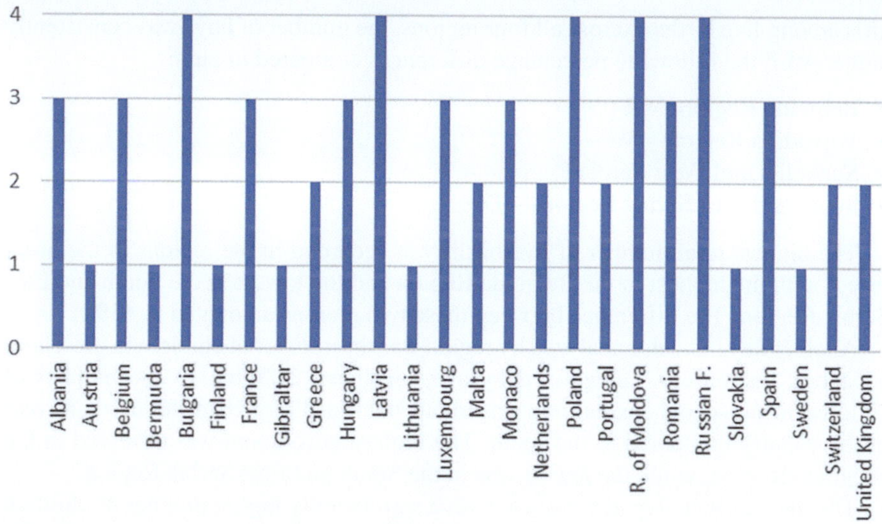

Fig. 8 Number of years of free pre-primary education guaranteed in legal frameworks—Indicator 4.2.5

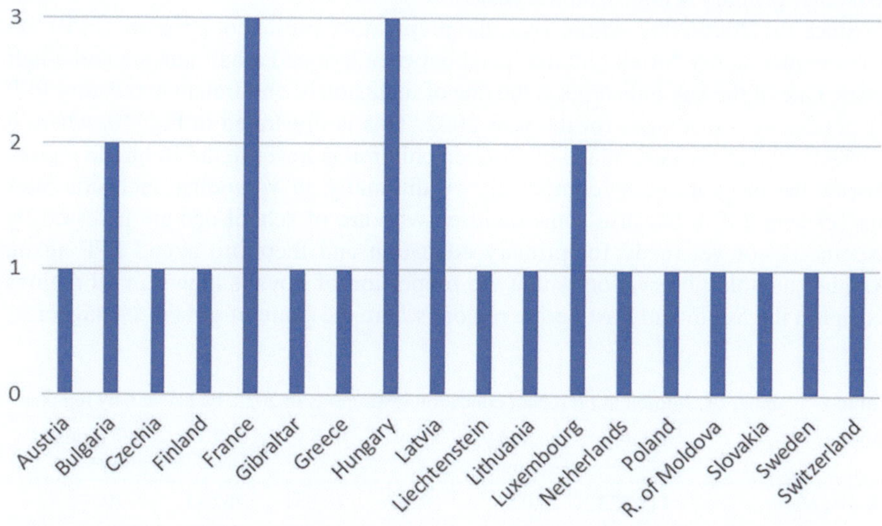

Fig. 9 Number of years of compulsory pre-primary education guaranteed in legal frameworks—Indicator 4.2.5

Following these results, the question arises as to whether the overall number of boys is generally higher than that of girls. To explore this, the number of live births for enrollment in the PPP was analyzed by region and for Serbia as a whole. The results indicate that at the national level, there were approximately 7% more boys than girls among live births. Across all four regions, the number of boys was consistently higher, with the following percentage differences compared to girls:

- Belgrade Region—9%;
- Vojvodina Region—5%;
- Šumadija and West Serbia Region—6%; and
- South and East Serbia Region—10%.

The highest total number of live births was recorded in the regions in the order listed, with the highest in the Belgrade Region and the lowest in the South and East Serbia Region. The difference between these two regions amounted to 43%.

Another key indicator analyzed by region was the number of children in preschool education institutions, categorized by age, as shown in Table 1. The number of children in preschool institutions was about one-third higher in the northern part of the country compared to the south. The highest enrollment was observed in the Belgrade Region, while the lowest was in the South and East Serbia Region.

The data include five age groups, with a significantly higher number of children aged 5 to 7 years compared to other groups. This is expected, as this period precedes primary school enrollment, which is mandatory in Serbia. Additionally, there are mixed-age groups that include children aged 3 to 7 years, among whom a portion will enter primary school within a year.

Since the 2006/2007 school year, the preparatory preschool program (PPP) has become mandatory for all children aged between five and a half and six and a half years. One of the key indicators is the rate of inclusion of children in mandatory PPP by gender in percentages for the year 2022. This is illustrated in Fig. 10, where it is observed that the percentage of enrolled children is lower in the Belgrade region, despite the program being compulsory. Additionally, in Vojvodina, inclusion rates may exceed 100% because some children who are of school age are assessed by experts as not yet ready for primary education and therefore attend PPP again. Another notable observation is that the proportion of boys is higher in all regions except in the South and East Serbia region, where the share of girls is 2% higher.

Table 1 Number of children in preschool education institutions in 2022, by age and by regions

Serbia	Total	up to 3 y	3–4 y	4–5 y	5–7 y	3–7 y, mixed groups
Serbia North	112,533	20,035	18,086	20,031	46,611	7770
Belgrade region	57,195	12,096	9943	10,020	22,125	3011
Vojvodina region	55,338	7939	8143	10,011	24,486	4759
Serbia South	76,771	11,914	10,030	11,163	37,276	6388
Šumadija and West Serbia region	45,260	7205	5701	6116	20,922	5316
South and East Serbia region	31,511	4709	4329	5047	16,354	1072

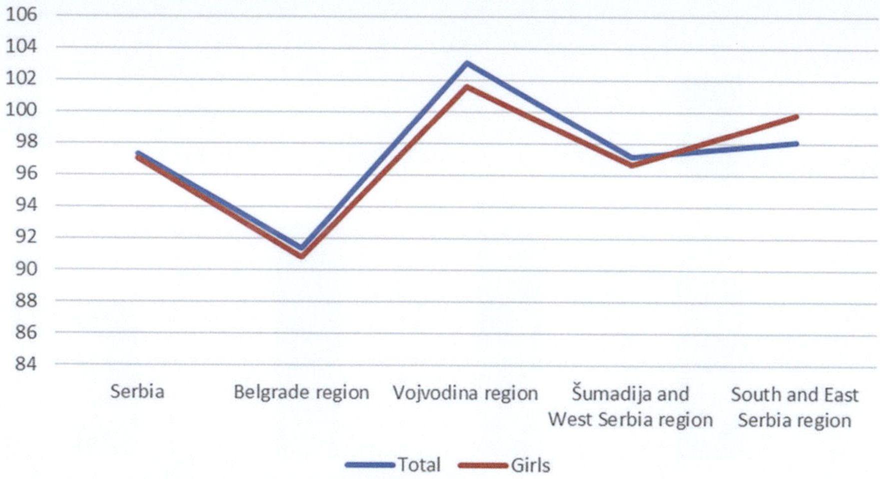

Fig. 10 Rate of inclusion of children in mandatory PPP, by sex (%)

Furthermore, it is possible to analyze the number of preschool education institutions. Data indicate that in 2016, there was a 5% increase in the number of these institutions in Serbia compared to the previous year. Regionally, this percentage remained unchanged in the South and East Serbia region, while in Vojvodina, the increase was only 2%. However, more significant growth was recorded in the Šumadija and West Serbia region (8%) and the Belgrade region (12%).

The analysis also included the number of employees in preschool education and care by type/job title and gender. This profession in Serbia is predominantly performed by women, with men accounting for only 1% of the workforce both at the national level and across regions. The only exception is the Šumadija and West Serbia region, where men make up 2% of employees. The total figures by region are presented in Fig. 11, showing that the number of employees in the Belgrade region is nearly 60% higher than in the South and East Serbia region.

Discussion

Sustainable Development Goal (SDG) 4.2 emphasizes equitable access to quality ECE and development, ensuring children are well prepared for primary school. This analysis highlights both progress and challenges in European countries, with a specific focus on Serbia. The discussion below interprets the findings, contextualizes disparities, and suggests policy implications. A major challenge in evaluating SDG 4.2 is the lack of comprehensive data, particularly for Indicator 4.2.1, which assesses whether children aged 24 to 59 months are developmentally on track in terms of health, learning, and psychosocial well-being. The availability of this indicator is limited to seven European developing countries, making it difficult to assess progress

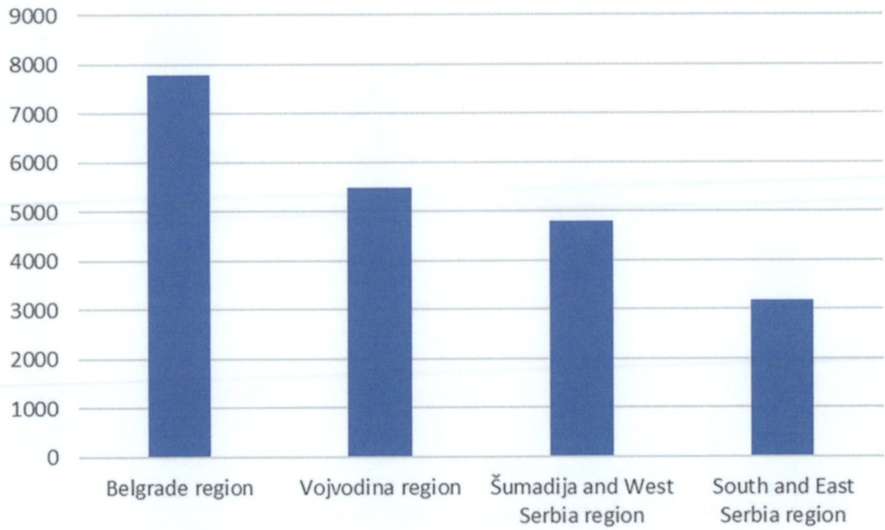

Fig. 11 Number of employees in preschool education and care by type/job title and gender

across the continent. The data gaps in high-income European nations raise concerns regarding whether developmental monitoring was systematically conducted. Without a standardized approach to collecting and reporting early childhood development data, comparisons between countries remain difficult. The reliance on another official indicator (participation rate in organized learning one year before the official primary entry age—4.2.2) and alternative indicators, such as net enrollment rates in ECE (4.2.4) and number of years of free/compulsory pre-primary education guaranteed in legal frameworks (4.2.5), provides more complete picture of children's developmental progress.

The results reveal notable differences in the proportion of children who are developmentally on track across European countries. While some countries demonstrate high levels of developmental readiness among nearly all children, others show moderate or lower readiness levels. In countries with moderate readiness, such as Belarus, gender disparities become more apparent, with boys lagging behind girls by approximately 5%. This trend is even more pronounced in countries with lower developmental readiness, such as North Macedonia, where a significant gender gap is evident—89% of girls are developmentally on track compared to only 76% of boys. These findings suggest that gender disparities emerge in specific countries, possibly due to social and cultural factors affecting early childhood experiences. The report highlights key challenges in delivering preschool and general education in Moldova under a decentralized governance system, including regulatory issues and weak implementation. Major concerns include financial barriers preventing rural children from attending preschool, societal resistance to enrolling children with special needs, overcrowding in urban preschools despite infrastructure investments, and regulatory obstacles to expanding private preschool options in cities (Hoftijzer et al. 2018).

Financial constraints disproportionately affect families in remote areas, where access to affordable ECE is already limited. On the other side, rapid pace of urbanization has outstripped efforts, leading to long waiting lists and diminished quality of care and education due to high child-to-teacher ratios. Preschool education in North Macedonia has very low coverage, with significantly fewer children attending compared to European averages, especially among marginalized communities (Sharlamamov and Mitevska Petrusheva 2023). Addressing these disparities requires targeted interventions that go beyond expanding physical infrastructure and address social and cultural barriers too. Key priorities for improvement include enhancing content quality, expanding access, improving inclusivity, and strengthening human resource capacities (Petroska-Beshka et al. 2021).

The analysis of organized learning one year before primary school reveals significant disparities across European countries. While many nations demonstrate high participation rates, notable gaps persist, reflecting a complex interplay of economic, infrastructural, and sociocultural factors. The average preschool participation rate for the 40 analyzed European countries is 94%. Fourteen countries fall below this average, which can be divided into two groups: those close to the average (Albania, Andorra, the Czech Republic, Hungary, Russia, and Serbia) and those significantly below the average, such as Bosnia and Herzegovina, North Macedonia, Montenegro, and Ukraine. Bosnia and Herzegovina and North Macedonia continue to face considerable challenges in expanding access to early education. The preschool curricula across all ten cantons in Bosnia and Herzegovina lack a central database, making them difficult to access, with some available only in paper form; reports from ministries and organizations are similarly fragmented. Efforts aim to consolidate these resources to assess the overall quality of preschool education, which currently favors children from urban areas over those from rural regions (Camovic and Hodzic 2017; Camovic and Becirovic-Karabegovic 2022). In contrast, nations like Ukraine and Montenegro exhibit moderate participation levels, reflecting steady progress toward broader access. However, ongoing issues, such as geopolitical instability in Ukraine, may impede further advancements and highlight the vulnerability of early education systems to external disruptions. High participation rates in countries such as Romania, Bulgaria, and Slovakia indicate significant progress toward achieving near-universal enrollment. These nations have made substantial investments in early childhood education, yet they still fall short of the highest-performing European countries, signaling the need for sustained policy focus and resource allocation. In rural parts of Romania, socioeconomic exclusion leads to unequal opportunities, with limited access to early childhood services for disadvantaged families, inconsistent and inaccessible data, and a need for stronger data-informed policymaking and greater focus on early development from the prenatal stage to pre-primary school entry (Adams 2019). At the upper end of the spectrum, countries like Malta, Albania, and Russia approach universal participation, supported by strong policy frameworks and considerable public investment in early childhood education. These nations exemplify the potential outcomes of targeted interventions and comprehensive support systems. Nevertheless, gender disparities remain evident in some contexts. In Albania and Belarus, boys are slightly more likely to be enrolled in early education programs, whereas Luxembourg presents an inverse trend, with higher participation rates among girls. These imbalances could be consequence of parental attitudes and societal biases regarding gender roles, but also it could be result of different number of live births by sex. While significant progress has been made across big part of Europe, persistent disparities in participation rates underscore the need for targeted policy interventions. Addressing financial barriers, expanding infrastructure, and promoting inclusive practices are essential to ensuring equitable access to early education for all children, regardless of their socioeconomic background or gender.

The net enrollment rate in ECE (4.2.4) reveals significant disparities in access across different European countries. At one end of the spectrum, countries like Bosnia and Herzegovina, North Macedonia, Romania, Greece, and Ireland face notably low enrollment rates. In the case of Bosnia and Herzegovina, the particularly low participation highlights severe barriers that demand urgent policy action, such as addressing financial constraints, improving infrastructure, and increasing awareness of the importance of early education. On the other hand, nations like Belarus, France, Gibraltar, Italy, the Netherlands, and Poland demonstrate high enrollment rates, reflecting a strong recognition of early education as a vital foundation for lifelong learning. The average net enrollment rate of ECE for the 38 observed European countries is 71%. Differences between countries based on this indicator are substantial, with 18 countries falling below the mentioned average. Significantly below-average countries include Austria, Bosnia and Herzegovina, Croatia, Greece, Hungary, Ireland, Montenegro, North Macedonia, Romania, Serbia, and the United Kingdom. Cornelissen et al. (2018) found that while expanding subsidized universal ECE in Lower Saxony increased overall participation, children from low socioeconomic and immigrant backgrounds, who would benefit most, continued to participate less. Their study suggests that expanding universal access alone is insufficient to significantly boost enrollment among disadvantaged groups. These countries should serve as a model of good practice. In Northern Italy, municipal Scuola dell'Infanzia programs, notably in Reggio Emilia and Pistoia, offer universal ECE with favorable child-to-teacher ratios, supporting exploratory, play-based learning through a holistic curriculum (Leseman and Slot 2020). It should be emphasized that these countries also face some challenges. Despite heavily subsidized and means-tested ECE in France, significant inequalities persist, with a 65-percentage-point attendance gap between the highest and lowest income groups in 2014, and only 19% of families below the poverty line accessing Early Childhood Development and Care in 2018, potentially worsening disparities in child development and academic success (Carbuccia 2024). Although gender disparities in enrollment are relatively minor, they are still noticeable in certain countries. In Liechtenstein and Belarus, boys are slightly more likely to attend early education programs, whereas in the Netherlands, a marginally higher enrollment of girls is observed.

For sub-indicator 4.2.4b, data are available for 25 European countries. The average net enrollment rate of early childhood educational development programs is 37%, with half of the countries falling below this average. In Romania, Ireland, and Hungary, the percentages are exceptionally low, ranging from 2 to 4%. The reasons for this are socioeconomic, but also politics. Several European countries, such as

Sweden, Finland, England, Scotland, and France, transitioned from early-tracking to comprehensive education systems between the 1950s and 1980s, while some Central and Eastern European countries advanced between-school tracking after the fall of communism, offering insights into the distributional consequences of educational tracking (Van de Werfhorst 2019).

The net enrollment rate of pre-primary education (4.2.4c) was analyzed for 39 European countries, with an average value of 88%. Eighteen countries fall below this average, with the most significant deviations observed in Montenegro, North Macedonia, and Serbia. Gender differences of a few percentage points are present in several countries, with the largest gaps recorded in Belarus and the United Kingdom (4%) and the highest in Liechtenstein (9%). Regional aggregates reveal significant differences in access to pre-primary education, with Eastern Europe excluded due to the survey covering less than 50% of the population aged one year before primary education (Zaw et al. 2021).

Across Europe, the duration and nature of pre-primary education vary significantly, reflecting differing national priorities, funding mechanisms, and views on early childhood development. While 26 countries guarantee free access to pre-primary education, only 18 mandate compulsory attendance, highlighting varying perspectives on the state's role in early education. Most countries require just one year of compulsory pre-primary education, but Bulgaria, Latvia, and Luxembourg extend this to two years, whereas France and Hungary mandate three years, emphasizing a stronger institutional focus on early learning. These differences reveal deeper cultural, political, and economic factors shaping educational policies, with longer compulsory durations often reflecting efforts to reduce educational disparities and promote social equity from an early age. The Ministry of Education expanded its proposal by funding infrastructure improvements and mandating three years of preschool, while the final policy, influenced by parental pressure in Sofia, granted municipalities authority over fee exemptions, encouraged universal coverage, and allocated funds to reimburse parents forced into private kindergartens due to a shortage of public spaces (Volen and De Laat 2021). Analysis of a global database on pre-primary education policies shows that only the combination of free and compulsory pre-primary education is significantly associated with higher primary gross graduation rates, while free education alone has no notable effect (Earle et al. 2018).

Serbia's Progress of Target 4.2

The case of Serbia presents a nuanced alignment with broader European trends in pre-primary education while revealing persistent regional and demographic disparities. A notable gender imbalance exists across preschool age groups, with boys consistently outnumbering girls by 7%, a reflection of national birth patterns rather than educational access issues. Regionally, the Belgrade area leads with the highest preschool enrollment, whereas South and East Serbia continue to face significant challenges, including fewer preschool institutions and lower enrollment rates. These two regions represent contrasting poles of development, differing not only in economic growth but also in their ability to attract both international and internal migrants (Lukić et al 2022, 2023), a dynamic that underpins the observed disparities. Despite making the Preparatory Preschool Program (PPP) compulsory in 2006/2007, enrollment remains uneven; paradoxically, even Belgrade, despite its overall high participation rates, experiences gaps in attendance. Meanwhile, Vojvodina reports enrollment rates surpassing 100%, likely due to the inclusion of school-aged children attending preschool for developmental support or parental preference. Additionally, Serbia's preschool education workforce is overwhelmingly female (99%), pointing to a broader gender imbalance in the sector. This highlights the need for strategies that encourage male participation, promoting gender diversity and offering children a wider range of role models during their formative years.

Despite legislative and strategic efforts to improve preschool education, significant challenges remain in the practical implementation of these policies. For instance, while the 2010 Preschool Education Act introduced standardized programs and promoted key principles such as accessibility and democracy, research collaboration in preschool education was not prioritized (Krnjaja and Pavlović-Breneselović 2013). Additionally, although the 2012 reforms aimed to enhance children's well-being, the concept of "well-being" itself was left undefined, which has created a gap in the practical application of these strategies (Breneselović 2010). This highlights that despite legislative progress, regional and demographic disparities in access to preschool education remain persistent, particularly in the southern and eastern parts of Serbia, where preschool facilities are limited. The introduction of a new holistic learning approach in 2018, emphasizing social interaction and integrated play, has the potential to improve educational quality. However, the application of these principles remains challenging in practice. This points to the need for further efforts in terms of resource allocation, teacher training, and the development of a centralized data system that would enable more effective monitoring and evaluation of these policies across regions, especially in underdeveloped areas.

Conclusion

As nations continue to recognize the importance of investing in the earliest stages of education, the benefits of such investments will undoubtedly resonate across generations, contributing to stronger, more equitable societies worldwide. Addressing these challenges requires a commitment to improving data collection and analysis while ensuring that ECE remains a priority in national and international policy frameworks. By investing in high-quality early education and strengthening parental engagement, societies can work toward a future where all children—regardless of their background—have the opportunity to thrive in school and beyond.

While significant progress has been made across many European countries in improving access to quality early childhood education, notable disparities persist, particularly between regions, genders, and socioeconomic groups. Countries like

Belarus, North Macedonia, and Bosnia and Herzegovina face substantial barriers to expanding preschool enrollment, with financial constraints, limited infrastructure, and social resistance hindering progress. On the other hand, nations such as France, Malta, and Italy demonstrate strong policies and high participation rates, reflecting the positive impact of comprehensive and inclusive early education systems. Despite the strides made, gender imbalances, insufficient resources in rural areas, and gaps in the implementation of policy reforms continue to be challenges for many nations. Serbia, while aligning with broader European trends, still faces persistent regional disparities and uneven policy implementation, highlighting the need for more targeted efforts to address these challenges and ensure equitable access to ECE for all children.

Acknowledgements This research was funded by the Ministry of Science, Technological Development and Innovation of the Republic of Serbia, contract number 451-03-136/2025-03/ 200172.

References

Adams LR (2019) Early childhood development. SABER Country Report. World Bank

Allen LR, Kelly BB (2015) Transforming the workforce for children birth through age 8: a unifying foundation. National Academies Press (US). https://www.ncbi.nlm.nih.gov/books/NBK310550/

Babović M (2024) Report on progress in achieving the Sustainable Development Goals by 2030 in the Republic of Serbia [In Serbian]. Republički zavod za statistiku Srbije

Babović S, Lović Obradović S, Prigunova I (2016) Depopulation of villages in Southeastern Serbia as hindrance to economic development. J Geogr Inst Cvijic 66(1):61–74. https://doi.org/10.2298/IJGI1601061B

Bassok D, Latham S (2017) Kids today: the rise in children's academic skills at kindergarten entry. Educ Res 46:7–20

Blossfeld HP, Buchholz S, Skopek J, Triventi M (2016) Models of secondary education and social inequality: an international comparison. Edward Elgar, Cheltenham, UK

Boeren E (2016) Lifelong learning participation in a changing policy context: an interdisciplinary theory. Palgrave-Macmillan, London

Boern E (2019) Understanding sustainable development goal (SDG) 4 on "quality education" from micro, meso and macro perspectives. Int Rev Educ 65:277–294. https://doi.org/10.1007/s11159-019-09772-7

Bohodirovna SG, Usmanova H (2022) Model of innovative strategy on the basis of variative approach to preschool education. In: Conference zone, pp 197–199

Breneselović D (2010) Child welfare in the program versus welfare programs [In Serbian]

Brooks-Gunn J (2003) Do you believe in magic? What we can expect from early childhood intervention programme. Soc Policy Rep 17(1):3–14

Burger K (2016) Intergenerational transmission of education in europe: do more comprehensive education systems reduce social gradients in student achievement? Rese Soc Stratification Mobility 44:54–67

Camović D, Hodžić L (2017) An analysis of preschool education in Bosnia and Herzegovina: Fairness and equal opportunities for all children. J Contemporary Educ Stud/Sodobna Pedagogika 68(3)

Camovic D, Becirovic-Karabegovic J (2022) Educational quality of early childhood education in Bosnia and Herzegovina. European J Educ Res 11(4):1923–1936

Carbuccia L (2024) Demand-side determinants of the socioeconomic gap in ECEC enrollment in France. Doctoral dissertation, Laboratoire Interdisciplinaire d'Evaluation des Politiques Publiques; Caisse Nationale d'Allocations Familiales
Cetin ÖŞ, Gulhan M, Katranci M (2018) A study on the effect of pre-school education on early literacy skills. Int Online J Educ Sci 10(5)
Cornelissen T, Dustmann C, Raute A, Schönberg U (2018) Who benefits from universal child care? Estimating marginal returns to early child care attendance. J Polit Econ 126(6):2356–2409
Denham SA (2016) Assessment of SEL in educational contexts. In: Durlak JA, Domitrovich CE, Weissberg RP, Gullotta TP (eds) Handbook of social and emotional learning: research and practice. Guilford Press, pp 285–300
Dychkivska I, Uliukaeva I (2019) Comparative analysis of future preschool teacher training in foreign education systems. Sustainable development under the conditions of European integration, Part I, 79
Dychkivska IM (2017) Preparing future preschool teachers for innovative pedagogical activity: theory and methodology: monograph. O. Zen, Rivne [In Ukrainian]
Earle A, Milovantseva N, Heymann J (2018) Is free pre-primary education associated with increased primary school completion? A global study. Int J Child Care Educ Policy 12:1–19
European Commission (2000) European report on the quality of school education. Sixteen quality indicators. Report based on the work of the working committee on quality indicators. Directorate-General for Education and Culture 2000
European Statistical Office [EUROSTAT] (2019) Share of children between the age of four and the starting age of compulsory education participating in early childhood (pre-primary) education. https://bit.ly/2v0TUsW. Accessed 20 April 2024
Frongillo EA, Tofail F, Hamadani J, Warren A, Mehrin S (2014) Measures and childhood development. Ann N Y Acad Sci 1306:68–88
Gajić T, Vukolić D, Bugarčić J, Đoković F, Spasojević A, Knežević S, Đorđević Boljanović J, Glišić S, Matović S, Dávid LD (2024) The adoption of artificial intelligence in Serbian hospitality: a potential path to sustainable practice. Sustainability 16(8):3172. https://doi.org/10.3390/su16083172
González-Moreira A, Ferreira C, Vidal J (2021) Comparative analysis of the transition from early childhood education to primary education: factors affecting continuity between stages. European J Educ Res 10(1):441–454. https://doi.org/10.12973/eu-jer.10.1.441
Hanuschek E, Woessmann L (2010) Education and economic growth. In: Brewer DJ, McEwan PJ (ed) Economics of Education. Elsevier, pp 60–67
Heckman J (2006) The economics of investing in children (Policy Briefing no. 1). UCD Grary Institute, Dublin
Heisig JP, Solga H (2015) Secondary education systems and the general skills of less- and intermediate-educated adults: a comparison of 18 countries. Sociol Educ 88(3):202–225
Hoftijzer M, Cretu V, Dillinger W, Casap L (2018) Moldova preschool and general education: transitioning to a decentralized service delivery model. Selected Issues. World Bank
Karoly LA, Kilburn MR, Cannon SJ (2005) Early childhood interventions: proven results, future promise. RAND Corporation, Santa Monica, CA
Krnjaja Z, Pavlovic-Breneselovic D (2013) Where is the quality—quality building policy in pre-school education, Book 1 [In Serbian]. Institut za pedagogiju i andragogiju, Beograd
Krnjaja Ž, Pavlović-Breneselović D (2017) The justness of preschool education in Serbia: Ensuring equal opportunities. J Contemporary Educ Stud/Sodobna Pedagogika, 68(3)
Leseman PPM, Slot PL (2020) Universal versus targeted approaches to prevent early education gaps. The Netherlands as case in point. Z Erziehungswiss 23:485–507. https://doi.org/10.1007/s11618-020-00948-8
Li W, Duncan GJ, Magnuson K, Schindler SH, Yoshikawa H, Leak J (2020) Timing in early childhood education: how cognitive and achievement program impacts vary by starting age, program duration, and time since the end of the program. Ed. WorkingPaper, pp 20–201. Retrieved from Annenberg Institute at Brown University: https://doi.org/10.26300/5tvg-nt21

Lović Obradović S, Babović S, Shpak N (2016) Serbia and Russia on the demographic map of Europe two decades after the fall of communism. TRAMES: J Humanities Soc Sci 20(1):59–73
Lukić V, Lović Obradović S, Ćorović R (2022) COVID-19 and internal migration in Serbia—geographical perspective. J Geographical Institute "Jovan Cvijić" SASA 72(2):191–205
Lukić V, Lović Obradović S, Stojilković Gnjatović J (2023) Did the Covid-19 pandemic change internal rural migration patterns in Serbia? Erdkunde 77(3):233–249
Lysova EB (2006) New tendencies in teachers' training on the west. Pedagog J 3:12–15
Marks GN (2013) Reproduction of economic inequalities: are the figures for the United States and United Kingdom too high? In: Birkelund GE (ed) Class and stratification analysis (Comparative Social Research), vol 30. Emerald Group, Bingley,341–363
Matović S, Demirović Bajrami D, Gajić T, Lakić M, Martinov D, Šuput S, Radosavac A, Đervida R (2024) Beyond numbers: challenges in measuring SDG4 targets—Serbia's perspective. Sustainability 16(22):10006. https://doi.org/10.3390/su162210006
Matović S, Lović Obradović S (2021) Assessing socio-economic vulnerability aiming for sustainable development in Serbia. Int J Sustain Dev World Ecol 29(1):27–38. https://doi.org/10.1080/13504509.2021.1907629
OECD (2013) PISA 2012 assessment and analytical framework: mathematics, reading, science, problem solving and financial literacy. OECD Publishing https://doi.org/10.1787/9789264190511-en
Organisation for Economic Co-operation and Development (OECD) (2011). PISA in Focus 1: does participation in pre-primary education translate into better learning outcomes at school? OECD, Paris
Ornaghi V, Brazzelli E, Grazzani I, Agliati A, Lucarelli M (2017) Does training toddlers in emotion knowledge lead to changes in their prosocial and aggressive behavior toward peers at nursery? Early Educ Dev 28(4):396–414. https://doi.org/10.1080/10409289.2016.1238674
Pankajam G (2017) Pre-Primary education: philosophy and practice. Concept Publishing Company
Pešikan A, Ivić I (2016) The sources of inequity in the education system of Serbia and how to combat them. Center Educ Policy Stud J 6(2):101–124
Petroska-Beshka V, Mickovska G, Ajruali A, Conteva Z, Mehmedi Lj (2021) Report for realized activities [In Macedonian]. Pestalizzi Children Foundation, Skopje. Available at: http: izvestaj za realizirani aktivnosti_2021.pdf
Sergeeva NB (2010) The impressions on the modern German preschool education. Pedagog J 3:22–27
Sharlamamov K, Mitevska Petrusheva K (2023) The challenges in education system in North Macedonia. Int J Soc Sci Human Res 6(3):1398–1503
SORS (Statistical Office of the Republic of Serbia) (2022) Preschool upbringing and education for 2022. https://data.stat.gov.rs/?caller=1101&languageCode=en-US. Accessed on 15 Nov 2023
Stojanovic A, Bogavac D (2016) Methodological strategies for the development of Children's Potential in a Modern Preschool Institution [In Serbian]. Uciteljski fakultet, Beograd
Stojanovic A, Kovacevic Z, Bogavac D (2018) From obsolete to contemporary preschool education (experiences in Serbia). Early Years 1–15. https://doi.org/10.1080/09575146.2018.1444584
Tokarev MA, Kosicyna IB, Batalov IV (2020) Comparative analysis of preschool education systems in Russia, France, and Germany. Proceeding of the International Science and Technology Conference "FarEastCon 2019" October 2019. Russian Federation, Far Eastern Federal University, Vladivostok, pp 491–498
Tomaszewska-Pękała H, Marchlik P, Wrona A (2017) RESL publication 6 finding inspiring practices on how to prevent ESL and school disengagement. Lessons from the educational trajectories of youth at risk from nine EU countries. University of Warsaw. https://cutt.ly/Cjk57fY
UNESCO (2023) Institute for Statistics Technical Cooperation Group. SDG4 Indicators Dashboard. https://sdg4-data.uis.unesco.org/. Accessed 15 Nov 2023
UNICEF (the United Nations International Children's Emergency Fund) (2023) Early childhood development index 2030. https://data.unicef.org/resources/early-childhood-development-index-2030-ecdi2030/. Accessed on 15 Nov 2023

United Nations Department of Economic and Social Affairs (2023) Sustainable development. https://sdgs.un.org/goals/goal4. Accessed 15 Nov 2023
United Nations Statistical Commission (2017) Work Plans for Tier III Indicators. Background paper prepared for March 2017 meeting of the United Nations Statistical Commission. Retrieved from https://unstats.un.org/unsd/statcom/48th-session/documents/BG-2017-3a-Tier-III-Work-Plans-E.pdf
Van de Werfhorst HG (2019) Early tracking and social inequality in educational attainment: educational reforms in 21 European Countries. Am J Educ 126(1):65–99. https://doi.org/10.1086/705500
Vandenbroeck M (2010) The social and political construction of early childhood education. Br J Educ Stud 58(2):139–153
Vladimíra P (2018) European school net. Czech Republic. Country report on ICT in education. http://www.eun.org
Volen E, De Laat J (2021) Building evidence for pre-school policy change in Bulgaria. Front Public Health 9:594029
Vuković D, Shams R, Vujko A (2019) Special issue "contemporary trends in sustainable development: a double-ECO (edged) sword. Sustainability ISSN 2071-1050
Vukovic DB, Spitsin V, Akerman E, Gammerschmidt I (2023a) Triple bottom line, sustainability, and regional development. J Eurasian Stud 1–12. https://doi.org/10.1177/18793665231215802
Vuković DB, Maiti M, Petrović MD (2023b) Tourism employment and economic growth: dynamic panel threshold analysis. Mathematics 11(5):1112. https://doi.org/10.3390/math11051112
World Bank (2012) Leading with ideas: skills for growth and equity in Thailand. The World Bank, Bangkok, Thailand
Yoshikaw H, Weiland C, Brooks-Gunn J (2016) Does preschool matter? Future Child 26(2):21–35. https://doi.org/10.1353/foc.2016.0010
Zaw HT, Mizunoya S, Yu X (2021) An equity analysis of pre-primary education in the developing world. Int J Educ Res 109, Article 101806

Geospatial Technologies and Artificial Intelligence

Artificial Intelligence in Compromised Geographies: Redefining Power and Territorial Challenges

Simona Epasto

Abstract Artificial intelligence (AI) is revolutionising geopolitics and territorial dynamics, reshaping global power relations in profound and multifaceted ways. This chapter explores AI's dual role: as a tool for enhanced territorial control and surveillance and as a source of new vulnerabilities and instabilities in the geopolitical landscape. Through the analysis of critical case studies, including the Israel-Hamas conflict, the Russia-Ukraine war, and the technological rivalry between USA and China, the chapter reveals how AI-driven technologies, such as drones and predictive analytics, are transforming conflict dynamics, deepening territorial inequalities, and exacerbating geopolitical tensions. AI's integration into geopolitical strategies has accelerated the evolution of warfare, challenged traditional power structures, and introduced risks of unforeseen escalations. At the same time, the global competition for AI supremacy exemplifies how technological advancements redefine international alliances and influence global governance. To address these challenges, the chapter advocates for the integration of science diplomacy as a framework to promote international collaboration and ethical governance of AI technologies. By fostering interdisciplinary dialogue and cooperative approaches, this framework seeks to mitigate risks whilst maximising AI's potential to address global challenges. Balancing AI's benefits and risks, the chapter emphasises the need for governance mechanisms that ensure AI contributes to global stability, equitable development, and sustainable peace.

Keywords Artificial intelligence (AI) · Geopolitical conflicts · Technological sovereignty · Science diplomacy · Digital geopolitics

---

S. Epasto (✉)
University of Macerata, Department of Political Science, Communication and International Relation, Macerata, Italy
e-mail: simona.epasto@unimc.it

L. Soares et al. (eds.), *Compromised Geography*, Key Challenges in Geography,
https://doi.org/10.1007/978-3-032-05201-8_8

Introduction

Artificial intelligence (AI) is emerging as a transformative force in the contemporary geopolitical landscape, reshaping power dynamics and territorial configurations on a global scale. In a world marked by compromised geographies, AI functions both as a tool for strengthening territorial control and surveillance and as a source of new vulnerabilities and instabilities in international relations. The concept of 'algorithmic nations', as proposed by Calzada (2018), highlights the transcendence of traditional borders and the rise of new forms of sovereignty, where citizenship and power intertwine with digital logics.

The ethical and regulatory implications of this transformation are central. Many scholars emphasise the urgency of a regulatory framework that not only mitigates AI-related risks but also promotes fairness and accountability in the adoption of emerging technologies (Iphofen and Kritikos 2021). Trust in global systems, already undermined by complex political dynamics, is further weakened using AI as a manipulative tool, with direct consequences for decision-making processes (Pauwels 2018; Lizza 2024). These tensions are amplified in the context of technological competition between USA and China, a phenomenon that Aresu (2024) interprets as a new 'technological cold war', where AI becomes both a symbol of power and a crucial element of global rivalry.

Through the analysis of case studies, this chapter explores three central research questions:

1. How does AI influence geopolitical conflicts and the global distribution of power? The Israel-Hamas conflict and the Russia-Ukraine war provide examples of how AI is transforming territorial control and accelerating the pace of warfare, revealing new risks of unintended escalations and exacerbating territorial inequalities.
2. What are the implications of US-China technological competition for global governance? This rivalry not only reshapes international alliances but also poses significant challenges to traditional governance mechanisms, creating opportunities for new forms of geopolitical fragmentation.
3. How can scientific diplomacy tools mitigate risks and maximise the benefits of AI? Scientific diplomacy is proposed as an integrated approach to fostering interdisciplinary dialogues and international collaborations, developing governance mechanisms that balance ethics, technology, and geopolitics.

These questions fit into a broader framework of 'compromised geographies', where interactions between technology, the environment, and society redefine not only urban dynamics but also geographical analysis paradigms. Cities, increasingly influenced by technologies such as Geo-AI and big data, emerge as epicentres of innovation but also as sites of heightened vulnerability. Such transformations necessitate a critical rethinking of geographical methodologies to address emerging challenges.

Ultimately, AI represents both a challenge and an opportunity for the evolving geopolitical landscape. Its capacity to reshape territorial dynamics and influence power relations calls for a comprehensive, multidisciplinary, and ethically guided approach. Through collaborative and inclusive governance, it is possible to harness AI's potential to promote stability and sustainable development, ensuring that this technology contributes to global cohesion rather than exacerbating divisions and instability.

Theoretical Framework

'Compromised geographies' represent a key concept for analysing how the intersection of technological innovation, critical resources, and geopolitical rivalries is redefining territorial dynamics and power relations. This theoretical framework aims to provide an interdisciplinary foundation for understanding these ongoing transformations, with particular attention to their economic, social, and environmental implications.

A crucial aspect concerns the link between critical materials and technological innovation. As highlighted by Diemer et al. (2022), critical and conflict materials are not only the foundation of contemporary technological infrastructures but also the core of deep territorial inequalities. The extraction of resources such as cobalt and lithium is concentrated in regions of the Global South, often marked by political instability and economic exploitation, whilst the value generated from these resources is predominantly captured by advanced technological hubs in the Global North. This phenomenon, described as a 'resource-technology fracture', underscores the need for a critical and spatially aware approach to addressing the consequences of technological development on a global scale.

At the same time, the concept of sovereignty is undergoing significant transformation under the influence of digitalisation. Martin et al. (2023) analyse how the introduction of digital technologies, such as blockchain and big data, is reshaping power dynamics in humanitarian contexts and beyond. In these spaces, sovereignty is no longer merely an exercise of territorial control but becomes an experimental ground for new forms of authority and governance. However, these transformations raise pressing questions about transparency, decision-making autonomy, and power relations between states, international organisations, and private actors, revealing deep tensions in the management of digital sovereignty.

Geoeconomics offers another analytical tool for understanding how digital technologies are strategically used in state rivalries. Csurgai (2018) highlights how twenty-first-century economic competition is increasingly shifting towards the systematic use of non-market instruments, including economic intelligence and the manipulation of global value chains. This approach emphasises how states leverage technologies to consolidate control over strategic markets and protect key sectors of their economies. The interaction between geopolitics, geoeconomics, and geotechnology is central to understanding the ongoing global transformations. Khanna (2012) describes this interplay as an 'Iron Triangle' shaping power distribution in the twenty-first century. This perspective is essential for analysing compromised geographies, where emerging technologies redefine not only territorial boundaries but also economic and security dynamics.

When integrated, these theoretical perspectives outline a complex framework in which technology, resources, and geopolitics intertwine to reshape territorial relations on a global scale. The approach of this chapter seeks to build a critical and interdisciplinary understanding of these dynamics, questioning their implications for the future of global geographies. In this context, the concept of 'compromised geography' serves as an analytical lens to explore how technological innovations can both amplify existing inequalities and open new possibilities for cooperation and sustainable development.

AI and Geopolitical Theory

AI is not merely an enabling technology; it is a pivotal element in understanding the geopolitical transformations of our time. Its impact extends beyond the technical domain to become a crucial factor in the reorganisation of global power, influencing regional balances, economic inequalities, and strategic stability.

The territorial dynamics of technological innovation highlight how AI does not distribute benefits equitably. Pinheiro et al. (2022) analyse the 'shadows' of the geography of innovation, where regions with greater technological and infrastructural capacity continue to accumulate advantages, leaving marginal areas trapped in cycles of low productivity. This divide underscores the need to address the systemic inequalities that AI adoption risks exacerbating, further deepening existing fractures between the Global North and South.

At the same time, technological fragmentation emerges as one of the most significant challenges of the twenty-first century. Larsen (2022) emphasises how the rise of digital sovereignty is reshaping the geopolitical landscape, with USA, the European Union (EU), and China developing increasingly autonomous technological ecosystems. This trend reflects not only differences in governance models—where the EU leans towards a rights- and privacy-based approach and China towards state control—but also a strategic competition for dominance over digital supply chains. This fragmentation, manifesting in the decoupling of digital infrastructures and conflicts over semiconductor access, directly challenges the idea of a globally integrated technological ecosystem.

On the front of strategic stability, Nadibaidze and Miotto (2023) highlight how AI integration into military operations is reshaping the foundations of global deterrence. Recent technologies, such as autonomous systems and AI-driven surveillance and drones, introduce elements of uncertainty and risks of unintended escalation. This evolution is not merely technical but deeply social and political: the way global powers perceive their rivals' technological capabilities can exacerbate cycles of mutual distrust, further destabilising an already fragile international balance.

Finally, Taneja and Zakaria (2023) describe the ongoing competition between USA and China as a veritable 'digital cold war', with AI at its core. This rivalry extends beyond technological innovation, evolving into a struggle for control over critical infrastructure, data resources, and global narratives. In this context, Western democracies face the challenge of coordinating their efforts to counter alternative models of technological governance whilst upholding principles of openness and international cooperation.

This analysis reflects a complex landscape in which AI is not only an accelerator of change but also a mirror of contemporary geopolitical tensions and opportunities. Understanding these dynamics is essential for developing strategies that integrate technological innovation and global stability, fostering a more equitable and sustainable future.

Methodology

This chapter adopts an integrated methodological approach to explore the impact of AI on geopolitical and territorial dynamics. The methodology is structured around four main components: methodological objectives, case study selection, sources and tools, and study limitations. Additionally, a geopolitical methodology is integrated to expand the analysis by considering spatial and strategic variables essential for studying power relations in the AI era.

Methodological Objectives

The objective of this chapter is to critically analyse how AI is influencing territorial control and power relations on a global scale. The methodology combines qualitative and quantitative analysis, with a focus on the critical interpretation of official documents, scientific literature, and representative case studies. This analysis aims to highlight both the opportunities and risks associated with AI integration into geopolitical strategies.

This approach is complemented by a geopolitical perspective, which emphasises three fundamental aspects of analysis:

1. Definition of spaces and actors: Identification of areas of interest, influence, and action, along with key actors involved (states, technology corporations, international organisations).
2. Assessment factors: Examination of economic, technological, and security-related variables associated with AI.
3. Interconnections between actors and factors: Analysis of the intersection between technology and geopolitical dynamics, with a particular focus on emerging power relations.

Case Study Selection

The selected case studies meet criteria of geopolitical relevance and significant AI applications:

- Israel-Hamas conflict: Analysis of AI technologies in surveillance (e.g. drones and predictive analytics) and their impact on territorial control and regional stability.
- Russia-Ukraine conflict: Examination of AI deployment in hybrid warfare, demonstrating how AI has transformed military strategies and reshaped global power balances.
- US-China technological competition: Assessment of the geopolitical impact of the AI race, focussing on global trade implications and strategic alliances.

Following the geopolitical methodology, the definition of spaces of interest considers not only the states involved but also the distribution of technological power in patchwork-like spheres of influence rather than homogeneous blocs.

The selection of case studies in this chapter—the Israel-Hamas conflict and the Russia-Ukraine war—is based on three main factors: geopolitical significance, AI impact in military operations, and availability of reliable data.

Geopolitical significance and impact on global dynamics: Both conflicts represent critical junctures in international relations, with ramifications extending beyond the directly involved regions. The Russia-Ukraine war has redefined security balances in Europe and the relations between NATO, the EU, and Russia, whilst the Israel-Hamas conflict is embedded in a broader context of regional instability with potential global repercussions.

Significant AI use in conflict theatres: These conflicts exemplify AI deployment in military and surveillance operations. In Ukraine, AI is employed in reconnaissance systems, automated targeting, and predictive troop movement analysis, whilst in the Israel-Hamas conflict, it is used for advanced surveillance, autonomous drones, and digital disinformation management. Both contexts provide valuable insights into AI's impact on modern warfare.

Availability of data and analytical sources: Compared to other ongoing conflicts, such as the civil war in Sudan or the Yemen crisis, these selected conflicts offer greater transparency in data availability, thanks to institutional reports, think tank analyses, and international organisations' assessments (e.g. SIPRI, NATO, Brookings Institution). This enables a more detailed and rigorous examination of AI's impact on warfare and geopolitical dynamics.

The integration of these criteria has allowed for the selection of two highly representative contexts illustrating AI's emerging role in contemporary geopolitical conflicts. However, the methodological approach remains open to future comparative analyses with other scenarios where AI is reshaping military and security governance strategies.

Sources and Analytical Tools

The analysis is based on a combination of primary and secondary sources, with a mixed-methods approach integrating geopolitical tools with qualitative and quantitative analysis techniques.

Primary Sources: Political speeches and strategic documents (e.g. government policy papers and reports from international institutions such as the UN and the EU) and internal reports from major technology firms involved in AI development.

Secondary Sources: Academic studies and scientific literature, Think tank and advisory firm reports (e.g. Lazard Geopolitical Advisory 2023; Brookings Institution 2021).

Analytical Tools:

1. Qualitative analysis: Critical discourse analysis is applied, focussing on political narratives regarding AI governance.
2. Geopolitical impact analysis: Based on geopolitical methodologies, interactions between actors and factors are examined, particularly regarding shifts in technological power distribution.
3. Probabilistic modelling: To analyse AI's impact on geopolitical conflicts and technological competition, we employ scenario-based simulations integrated with probabilistic modelling. Specifically, Bayesian probabilistic modelling has been applied to estimate escalation risks in the Israel-Hamas and Russia-Ukraine conflicts, as well as technological fragmentation in the US-China competition. Additionally, qualitative scenario analysis identifies possible AI evolution trajectories based on historical trends, government reports, and industry studies. This combined approach enables an exploration of AI's strategic implications through both numerical projections and structured geopolitical assessments.

Integration of Geopolitical Methodology

The geopolitical methodology provides a crucial framework for evaluating AI's impact on global dynamics. This approach emphasises the spatial dimension, power hierarchies amongst state and non-state actors, and the interconnections between political, economic, and technological variables.

According to the principles of new geopolitics, AI's impact assessment must consider:

The concept of 'spaces and flows': Power is no longer exclusively territorial but is distributed through digital infrastructures and data networks. AI transcends physical borders, redefining spheres of influence through information control.

The principle of differentiated power attenuation: An actor's ability to influence a region depends on technological resource availability and digital power projection capacity, rather than mere geographical proximity.

The use of cross-impact analysis: This methodological technique examines the relationships between AI, security, economy, and global governance, providing more flexible scenarios compared to rigid models like game theory.

Integrating these elements enables a dynamic, multi-layered assessment of AI as an emerging geopolitical force.

Study Limitations

As with any research on geopolitical transformations induced by an emerging technology, there are methodological constraints:

- AI dynamism and geopolitical uncertainty: The rapid evolution of AI makes long-term predictions challenging.
- Data access and transparency: Many strategic aspects of AI remain classified as military or corporate secrets, limiting empirical analysis.
- Subjectivity in interpretations: Geopolitical analysis requires a qualitative assessment of global actors' interests, introducing an interpretative component in conclusions.

Methodological Contribution

The integration of qualitative analysis, geopolitical frameworks, and quantitative models provides a robust analytical framework for understanding AI's impact on international relations.

This methodology offers a spatial perspective on global transformations induced by AI; a dynamic and interconnected analysis of forces at play, moving beyond deterministic approaches; and an interdisciplinary vision, linking geopolitics, technology studies, and international relations theory.

Methodological Note on Case Studies

The initial probabilities used in our probabilistic modelling were estimated based on multiple sources, including historical conflict dynamics and military escalation data (e.g. NATO studies, SIPRI reports); economic and technological trends related to AI investments (sources: OECD, World Economic Forum); expert analyses on geopolitical scenarios, derived from think tank reports and academic institutions.

For example, military escalation data were obtained from SIPRI's global military expenditure database and NATO deterrence strategy studies. AI investment trends were extracted from the OECD's (2023) annual reports on technological innovation and McKinsey Global Institute (2023) projections on AI sector spending in China, USA, and Europe.

These considerations underscore that, whilst probabilistic modelling is a useful tool for analysing AI's impact on geopolitical relations, it must be complemented with qualitative analyses and alternative scenarios to provide a comprehensive evaluation of future trajectories.

Case Studies

As previously highlighted, AI is a driver of geopolitical transformation, with implications ranging from the redefinition of territorial control to the emergence of new strategic vulnerabilities. This section examines three emblematic cases—the Israel-Hamas conflict, the Russia-Ukraine conflict, and the technological competition between USA and China—to illustrate how AI is reshaping the global geopolitical landscape.

To quantify the potential impact of AI on geopolitical dynamics, we have also applied a probabilistic model to estimate the likelihood of key events under different AI adoption scenarios. The analysis focuses on three major geopolitical contexts: the Israel-Hamas conflict, the Russia-Ukraine conflict, and the USA-China technological competition. By comparing initial probabilities with updated ones influenced by AI, we provide a quantitative assessment of AI's role in conflict escalation and global technological fragmentation.

Israel-Hamas Conflict

The Israel-Hamas conflict serves as a case study for observing AI deployment in high-density urban environments, where technology is integrated into surveillance, logistics, and disinformation processes. Graham (2008) highlights how robotic and surveillance technologies, such as drones and predictive analytics systems, are designed to operate in complex urban settings, optimising territorial control and minimising operational errors. These technologies enable real-time intelligence acquisition, enhancing both tactical and strategic response capabilities.

At the same time, a new dimension of the conflict emerges with the use of AI-generated imagery as a tool for disinformation. Through deepfakes and manipulated content, conflict narratives are distorted to influence global public opinion and undermine the credibility of the involved parties.

This phenomenon raises new ethical and regulatory challenges, underscoring the growing difficulty in distinguishing between reality and digital manipulation.

In an increasingly complex battlefield, Raza et al. (2023) analyse the operational use of autonomous drones and AI-assisted targeting systems. Whilst these tools enhance precision, they also raise questions regarding compliance with international humanitarian law and the impact of technological militarization on civilian populations.

The integration of AI in the Israel-Hamas conflict is transforming war strategies and security management. The deployment of autonomous systems in drones and predictive surveillance is reducing response times whilst increasing the risk of unintended escalation. To estimate the potential impact of these technologies on conflict stability, we applied a Bayesian probabilistic model. Currently, the probability of significant military escalation is 35%. If AI is extensively utilised for predictive surveillance and automated strikes, this probability rises to 49.6%. These findings suggest that, although AI improves threat prevention capabilities, its large-scale use may heighten the risk of strategic miscalculations and uncontrollable automated decisions, increasing the likelihood of accidental escalations.

The implications of AI in the Israel-Hamas conflict align with Graham's (2008) arguments in *Robowar Dreams*, which describe how autonomous technologies are reshaping urban warfare. Our probabilistic model supports this trend, showing that AI deployment in surveillance and autonomous strikes could raise the escalation risk from 35 to 49.6%.

Russia-Ukraine Conflict

The Russia-Ukraine conflict is one of the most significant examples of AI's strategic impact on modern warfare. The technological disparities between the regions involved have amplified inequalities in innovation capacity and military operations management (Pinheiro et al. 2022). This asymmetry underscores AI's critical role in reinforcing the competitive advantages of technologically advanced powers.

AI has undeniably accelerated the conflict's dynamics by introducing advanced technologies for monitoring, information management, and battlefield coordination (Marr 2024). These developments have altered traditional power paradigms, shifting the focus from numerical superiority to technological dominance.

In this context, Ruppert (2024) introduces the concept of 'hyperwar', a conflict characterised by unprecedented operational speed enabled by AI-driven military decision-making systems. This warfare model emphasises real-time prediction and response, transforming the battlefield into a highly technological and dynamic space. However, such technological advancements raise concerns about strategic stability and the risks of unintended escalation.

The introduction of AI in military systems has reshaped modern warfare, introducing new strategic uncertainties. To assess AI's potential role in escalating the conflict over the next decade, we again applied Bayesian probabilistic modelling. The initial probability of global escalation due to AI is 30%. If AI-related errors in military systems (e.g. autonomous drones malfunctioning) become more frequent, the escalation probability increases to 48%. This projection highlights how AI's integration into warfare could amplify geopolitical risks, necessitating greater regulatory and strategic oversight.

AI is becoming a key component of hybrid warfare, transforming battlefields through data-driven predictive analytics. The probabilistic results indicate that AI's integration into surveillance and attack systems could raise the escalation risk to 49.6% in the Israel-Hamas conflict and 48% in the Russia-Ukraine war. This scenario underscores the urgent need to develop international regulations to prevent the uncontrolled deployment of autonomous systems in warfare, avoiding unintended escalations caused by rapid algorithmic decision-making (United Nations Office for Disarmament Affairs (UNODA), 2023).

Marr (2024) highlights how AI in contemporary conflicts is redefining deterrence strategies. Our model confirms this perspective: the increasing use of AI in advanced defence systems raises escalation risks from 30 to 48%, suggesting that warfare automation may have destabilising effects.

USA-China Technological Competition

The technological competition between USA and China is the dominant paradigm of contemporary global rivalry. Diemer et al. (2022) explore the role of critical materials in constructing digital infrastructures. The extraction and control of key resources, such as rare earth elements and lithium, not only influence technological production but also serve as geopolitical tools that reinforce global North–South inequalities.

Miailhe (2018) introduces the concept of 'digital empires', describing how USA and China are building global technological ecosystems based on proprietary digital infrastructures. This competition extends beyond innovation, influencing global supply chains and reshaping strategic alliances.

The idea of global infrastructural competition, or 'competitive connectivity', proposed by Khanna (2016), is essential for understanding how geotechnology is reshaping power relations. Competition is no longer just between states but between digital networks, supply chains, and technological hubs, determining economic and political control on a global scale. This model helps interpret new forms of sovereignty and influence within the context of technological fragmentation.

Finally, Schindler et al. (2024) provide a systemic perspective, analysing how USA and China compete for dominance in digital, production, and financial networks. This competition, driven by network logic rather than territorial blocs, makes AI a crucial battleground for twenty-first-century economic and strategic supremacy. The implications for third-party countries, often caught in a balancing act between these technological giants, highlight the growing complexity of the new world order.

China is striving to reduce its reliance on the United States in semiconductor production, investing massively in AI and digital infrastructure. To estimate China's likelihood of achieving technological independence by 2030, we applied Bayesian probabilistic modelling. The initial probability of China's technological independence is 40%. If China continues investing over $200 billion annually in AI, this probability rises to 51.4%. This projection is based on Bayesian modelling applied to China's AI investment growth data, with an annual increase rate of 16.5% according to Trivium China's *China AI Strategy* report (2023). However, if the U.S.A significantly restricts the export of advanced semiconductors, China's technological independence probability drops to 38.2%, indicating continued reliance on Western technology (Carnegie Endowment for International Peace, 2022).

These findings suggest that, despite U.S.A bans on advanced chip exports, China could still develop an autonomous supply chain in the coming years.

China's AI dominance extends beyond hardware and semiconductors to open-source software. According to Khanna (2019), companies like DeepSeek demonstrate that open-source fine-tuning of large language models (LLMs) can compete with proprietary American models, reducing dependence on centralised infrastructure and posing a direct challenge to Western technological supremacy. This reinforces the hypothesis of global technological fragmentation and China's increasing digital autonomy, with both economic and geopolitical implications.

This case study analysis demonstrates how AI has become a central force in global geopolitical dynamics. By examining conflicts and technological competitions, the chapter highlights both the opportunities and risks associated with AI's integration into conflict and international cooperation settings.

Geopolitical Implications

AI has become a cornerstone of global geopolitical competition, transforming not only economic and technological models but also international relations, security, and notions of sovereignty. The geopolitical implications of AI can be structured into three main dimensions: the need for shared ethical regulations, technological fragmentation, and the crucial role of international cooperation.

As previously mentioned, one of the most pressing issues concerns the urgency of establishing global ethical guidelines for the design, implementation, and use of AI. Iphofen and Kritikos (2021) highlight how the absence of a shared regulatory framework risks exacerbating global inequalities and intensifying conflicts. The proposal of an 'ethics by design' approach aims to integrate ethical principles from the preliminary stages of technological development, promoting transparency, accountability, and inclusivity. However, the implementation of such principles is hindered by the global race for innovation, where any slowdown is perceived as a strategic loss. This creates a tension between the need for regulation and the desire to maintain international competitiveness, further complicating AI governance.

At the same time, technological fragmentation represents one of the most significant challenges to the geopolitical landscape. As Larsen (2022) points out, the emergence of divergent approaches to digital governance is accelerating 'technological decoupling', with profound consequences for international relations. USA,

China, and the EU are increasingly adopting independent strategies to consolidate their digital infrastructures and reduce technological dependence on rivals. China, for instance, is developing a highly centralised and state-controlled technological ecosystem, whilst the EU focuses on data protection and ethical regulation. USA, on the other hand, adopts a more flexible and innovation-driven approach, prioritising technological leadership over regulatory control. This fragmentation not only disrupts the global market but also creates tensions between states, complicating international cooperation and amplifying the risks of conflict. In recent years, contrary to its initial policies aimed at containing China, Europe has progressively reopened its doors to Chinese investments, particularly in electrification and digital infrastructure (Khanna 2019). This reflects the logic of Eurasia's 'fluid multipolarity', where no single actor holds complete dominance, but rather a dynamic interaction takes place amongst economic, political, and technological powers. China's integration into European markets underscores how trade and industrial networks tend to be more resilient than short-term geopolitical tensions.

In this context, international cooperation emerges as both a crucial opportunity and a complex challenge. Lazard (2023) underscores that despite its divisive potential, AI could serve as a catalyst for inclusive global governance—provided that geopolitical rivalries can be overcome. The competition between USA and China is creating bottlenecks in AI's value chain, linked to critical resources such as computational power, human talent, and data access. These factors not only constrain global technological development but also hinder efforts to address shared challenges, such as international security and sustainability. To foster effective cooperation, Lazard identifies three categories of policies: promotion (e.g. incentives for research and development), protection (e.g. export controls), and ethical principles (to guide responsible innovation). However, these initiatives require significant political commitment and a shared willingness to bridge ideological and strategic divides.

As already emphasised, the global context makes it clear that AI is not merely a technological tool but a geopolitical lever capable of reshaping international relations and global stability. The lack of a common vision and growing fragmentation risk turning AI into a divisive force rather than an opportunity for cooperation. Establishing a shared regulatory framework and committing to inclusive governance are therefore essential to fully harness AI's potential and ensure a more equitable and stable global future.

Political and Strategic Implications and Discussion of Results

The probabilistic analysis conducted highlights how AI is redefining power dynamics and geopolitical strategies on a global scale. The adoption of AI in conflicts and technological competition is amplifying existing tensions, accelerating technological fragmentation, and reducing diplomatic manoeuvring space in international relations.

One of the most significant findings of this study concerns the risk of military escalation linked to AI use in conflicts. In the cases of Israel-Hamas and Russia-Ukraine, the automation of surveillance and attack systems has increased the speed of military response, reducing the time available for strategic and diplomatic decision-making. Probabilistic modelling indicates that AI integration in military systems can raise the risk of escalation by up to 49.6% in the Israel-Hamas conflict and 48% in the Russia-Ukraine context. This acceleration of warfare dynamics raises questions about the need for new international regulations to limit the indiscriminate use of AI in warfare scenarios, preventing unintended escalations.

At the same time, the competition between USA and China for technological supremacy is driving a fragmentation of the global digital ecosystem. The probability that China will achieve technological independence in semiconductors is estimated at 51.4% if annual AI investments exceed $200 billion. This scenario confirms the trend of 'technological decoupling', where rival blocs develop separate digital ecosystems, impacting not only global supply chains but also economic security and international innovation capacity. This competition is also reflected in key technology control strategies, where AI has become a tool of geopolitical power. AI is emerging as a true instrument of power in the geopolitical landscape. States are implementing strategic policies to control access to advanced AI technologies, using them as instruments of influence and coercion. A key example is the United States' decision to restrict exports of advanced semiconductors to China—measures aimed at slowing Beijing's ability to develop independent AI models. These restrictions, imposed in 2022 and reinforced in 2023, demonstrate how control over digital infrastructures has become a central element of geopolitical competition.

The EU, by contrast, is adopting a regulatory approach with the AI Act, which sets guidelines for AI use in critical sectors, limiting the adoption of potentially dangerous technologies in military and surveillance domains (European Commission 2023). This legislative act reflects the EU's intention to position itself as a leader in AI ethical governance, in contrast to the more aggressive approaches of the US and China. However, for European regulations to have a global impact, closer coordination with other regulatory frameworks and greater convergence in AI governance strategies will be necessary. The UN, OECD, and G7 are attempting to develop shared principles to balance innovation, security, and fundamental rights protection, but the lack of a binding multilateral framework makes true regulatory harmonisation difficult. The EU could leverage its regulatory soft power to influence emerging regulations, but the divergence between US and Chinese approaches remains a significant obstacle to establishing a global AI governance framework.

These examples illustrate how AI is not only a driver of innovation but also a strategic battleground where states compete for control over the most advanced technologies (ISPI 2023; Il Sole 24 Ore 2023). One of the key challenges is that the absence of a shared regulatory framework risks exacerbating global inequalities and conflicts. If AI becomes a strategic power instrument without proper regulation, the gap between technologically advanced actors and countries with limited digital capabilities is likely to widen, fuelling new economic and political tensions.

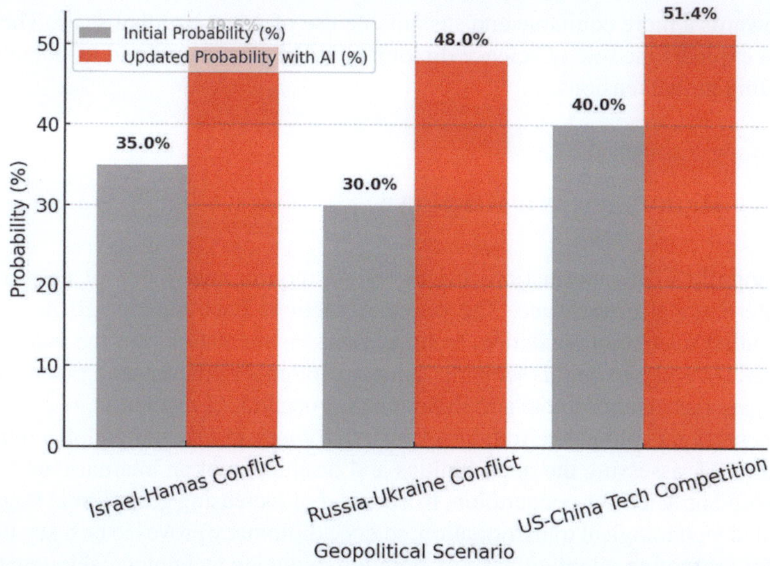

Fig. 1 Geopolitical risk trends with AI adoption in Israel-Hamas and Russia-Ukraine conflicts and US-China technologies competition. *Source* Own elaboration based on Bayesian probabilistic modelling, with initial probabilities estimated from geopolitical scenarios and updated considering the impact of AI on conflict escalation and global technological fragmentation

To summarise AI's impact on geopolitical dynamics, the following chart illustrates the evolution of conflict escalation probability and technological fragmentation based on increasing AI integration. The data reveal a significant rise in escalation risk in the Israel-Hamas and Russia-Ukraine conflicts, as well as a growing likelihood that China will achieve technological independence (Fig. 1).

In this context, scientific diplomacy, which will be discussed in the next section, emerges as an opportunity to mitigate risks and foster international cooperation (The Economist, 2019). The implementation of multilateral agreements on AI governance, as suggested by the literature on digital trust and ethical regulation, could be a crucial step in preventing a new technological arms race. AI is not merely an enabling technology; it is a geopolitical lever capable of redefining global balances and governance models. Its political and strategic implications require a coordinated response amongst state and non-state actors to ensure that technological acceleration does not exacerbate instability and fragmentation but instead promotes security and sustainable development.

In this regard, scientific diplomacy stands out as a promising avenue to ease tensions and encourage a collaborative approach to AI regulation. Initiatives such as the TRUST project (2023) demonstrate that international cooperation can be built on the sharing of technological knowledge, fostering digital trust and reducing asymmetries between the Global North and South. The construction of shared regulatory frameworks, combined with science diplomacy strategies, could represent a first step towards a more equitable and sustainable use of AI on a global scale. The next section explores the role of science diplomacy in AI governance and its potential in mitigating global tensions.

Science Diplomacy as a Response

The geopolitical transformations driven by AI require new forms of cooperation and international governance. The previous sections have highlighted how AI is contributing to conflict escalation, technological fragmentation, and the redefinition of digital sovereignty. In this scenario, science diplomacy emerges as a potential tool to mitigate risks and promote a multilateral approach to AI regulation.

This section explores the role of science diplomacy in managing global tensions related to AI, assessing the opportunities and limitations of an international framework based on scientific cooperation. In a context of increasing geopolitical fragmentation and technological transformation, science diplomacy proves to be a key instrument for promoting international collaboration, reducing technological inequalities, and fostering relationships based on digital trust. This concept, central to the TRUST project (Digital TuRn in EUrope: Strengthening Relational Reliance through Technology) led by Professor Francesca Spigarelli at the University of Macerata, demonstrates how advanced technologies such as blockchain and AI can be used not only to improve transparency and traceability but also to address the digital divide between advanced and less developed regions (TRUST Project 2023).[1]

Inclusive Innovation Policies

Science diplomacy can play a role in promoting innovation policies that reduce inequalities between regions, addressing technological divides that exacerbate economic and social disparities. Pinheiro et al. (2022) highlight how peripheral regions are often excluded from innovation processes, leading to the concentration of benefits in a few advanced areas. The TRUST project illustrates how a collaborative approach based on the Quadruple Helix model (involving governments, industries,

---

[1] TRUST "digital TuRn in EUrope: Strengthening relational reliance through Technology" is a Horizon 2020—MSCA (Marie Skłodowska-Curie Actions)—RISE (Research and Innovation Staff Exchange) program aimed at understanding the role of trust in the implementation of blockchain technology and suggesting actual means of development. The TRUST project promotes an interdisciplinary research program, involving academic and non-academic institutions from different backgrounds (legal, economic, engineering), belonging to EU, as well as Israel and China, to understand the mutual influence between trust and digital technologies. This project has received funding from the European Union's Horizon 2020 Research and Innovation Programme under the Marie Sklodowska-Curie grant agreement n. 101007820. This document reflects only the author's view. The Research Executive Agency is not responsible for any use that may be made of the information it contains.

academia, and citizens) can strengthen territorial cohesion and enhance resilience in more vulnerable regions (Carayannis and Campbell, 2012; Kuzior and Kuzior, 2020; Epasto, 2025).

Global Alliances in the AI Era

Miailhe (2018) emphasises that building global alliances is essential to harmonising regulations and mitigating AI-related risks. Science diplomacy can facilitate this process by providing platforms for dialogue to bridge political differences and promote mutual trust between nations. In particular, the TRUST project demonstrates how digital trust can serve as a catalyst for inclusive development, using technologies like blockchain to ensure decentralised and transparent governance. This approach is crucial for balancing asymmetries between the Global North and South and creating a common foundation for international cooperation Stockholm International Peace Research Institute (SIPRI) (2023).

Transatlantic Partnership and Regional Cooperation

Cooperation between Europe and USA is a key element in addressing AI challenges and technological fragmentation. Can and Kaplan (2020) propose a transatlantic alliance that leverages science diplomacy to strengthen ties between both sides of the Atlantic, fostering the sharing of technological resources and the definition of common standards. At the same time, the TRUST project exemplifies how collaborations between regions with different technological capabilities can promote inclusion and reduce digital inequalities, demonstrating that science diplomacy can operate at both global and regional levels.

Digital Trust as a Lever for Science Diplomacy

Digital trust is not just a technical concept but a strategic element for restructuring geopolitical interactions and promoting sustainable development. The TRUST project shows how advanced technologies can be employed to build bridges between regions, strengthen territorial cohesion, and contribute to long-term sustainability. This approach, integrated with the principles of science diplomacy, highlights the potential of science as a soft power tool to address urgent global challenges.

Science diplomacy, reinforced by the concept of digital trust, offers a unique opportunity to address geopolitical fragmentation and foster international cooperation. In a world where AI and other advanced technologies are reshaping global dynamics, integrating science into diplomatic decision-making represents a strategic response to ensure equity, transparency, and sustainability.

Furthermore, science diplomacy could provide concrete tools to tackle challenges emerging from the case studies analysed. In the Israel-Hamas conflict, an international framework regulating AI use in armed conflicts could reduce the risk of escalation resulting from the automation of drones and surveillance systems. In the Russia-Ukraine context, diplomatic initiatives for sharing data on hybrid threats and AI-driven disinformation could limit the strategic use of technology to manipulate public opinion.

Finally, in the USA-China technological competition, science diplomacy could be leveraged to negotiate agreements on the transfer of critical technologies and establish shared AI standards. The absence of a multilateral framework is contributing to global technological fragmentation, increasing the risks of economic and cybersecurity conflicts.

The adoption of AI in geopolitical dynamics raises urgent questions about the need for international coordination. Science diplomacy offers an opportunity to reduce technological asymmetries and create shared rules, but its success will depend on the political will of global actors to collaborate on security, development, and innovation issues.

The analysis conducted suggests that science diplomacy could be a strategic tool to address AI-related risks, but it alone is not sufficient. It will be essential to combine diplomatic initiatives with effective regulatory policies and adaptive governance mechanisms capable of responding to the rapid pace of technological change.

Conclusion

This chapter has explored the geopolitical and territorial implications of the interaction between critical resources, technological innovation, and global transformations, using the concepts of *Technology, Resources, and Geography* (Diemer et al. 2022) and *Robowar Dreams* (Graham 2008) as analytical frameworks. The analysis clearly reveals how critical resources such as cobalt and lithium not only constitute the infrastructural foundation of emerging technologies but also serve as a battleground for geopolitical competition, exacerbating regional inequalities and global tensions. Their extraction in the Global South and the concentration of added value in the Global North underscore a structural divide between extractive and technological economies (Diemer et al. 2022).

Advanced technologies, particularly AI, are at the heart of a paradigm shift that is redefining territorial control and global power dynamics. As discussed, AI is not only a lever for enhancing surveillance and governance capabilities but also a catalyst for new vulnerabilities and geopolitical fragmentations (Larsen 2022). The competition between USA and China (Schindler et al. 2024) highlights how dominance in digital, productive, and financial infrastructure networks has become the new battleground for global hegemony. This dynamic is further intensified by China's growing effort to reduce its dependence on Western semiconductors, with investments that, if maintained above $200 billion annually, could lead to near-total technological independence by 2030 (estimated probability: 51.4%). The global governance of AI will also have to address the increasing competition over data control and hardware infrastructures. Strategic dependence on advanced semiconductors is already shaping new geopolitical alliances, whilst generative AI raises concerns about technological sovereignty and access to computational resources. In this context, regulation must not only mitigate the risks of military AI applications but also ensure that access to these technologies does not become a new instrument of global inequality. This trajectory heightens the risk of an irreversible fragmentation of the global technological ecosystem, with direct consequences not only for economic security but also for the governance of emerging technologies. Unlike the first Cold War, the current phase is characterised by a networked competition model, where control over strategic nodes in global networks defines geopolitical power.

A striking example is the increasing urbanisation of conflicts, as described by Graham (2008), where cities in the Global South emerge as epicentres of political tensions and violence. The integration of automated systems and military technologies in urban contexts poses significant challenges to global stability and security, exacerbating territorial inequalities and complicating governance dynamics. Furthermore, the regulatory implications of these transformations are critical, as highlighted in *The Geopolitics of AI* (Larsen 2022), which emphasises the urgency of an ethical and multilateral approach to the regulation of emerging technologies.

Looking ahead, it is essential to acknowledge that AI and other advanced technologies will continue to reshape power relations and the global political geography. However, these transformations must be accompanied by an inclusive and interdisciplinary governance framework. Global AI governance must address the challenge of balancing innovation and security by developing a regulatory framework that limits the risk of uncontrolled technological escalation. An international AI treaty, like the Nuclear Non-Proliferation Treaty, could serve as an initial step towards regulating the development and use of AI in security and conflict contexts, preventing the emergence of automated warfare and destabilising cyber warfare scenarios. In this context, scientific diplomacy emerges as a crucial tool for fostering international cooperation and mitigating the risks associated with emerging technologies. In an increasingly fragmented scenario, scientific diplomacy presents an opportunity to counteract the risks of technological competition by promoting knowledge-sharing and the development of global standards. Initiatives such as the TRUST project demonstrate that building digital trust can help reduce inequalities in AI access and foster more inclusive governance. Adopting approaches that balance technological innovation, sustainability, and social justice is a priority for ensuring more equitable global development.

In conclusion, this work calls for a critical reflection on the role of resources and technologies in the future geopolitical dynamics. As suggested by Schindler et al. (2024), the competition for dominance in global networks could amplify inequalities but also offers opportunities to rethink alliances and governance models. Only through an ethical and collaborative approach will it be possible to transform current challenges into opportunities for a more stable and inclusive global order. Finally, AI governance will need to address the growing risk of a technological oligarchy, where a few private companies—predominantly American and Chinese—hold a monopoly over critical AI infrastructures. This power imbalance raises pressing questions about the need for antitrust regulations and policies to ensure a more equitable distribution of access to advanced technologies.

References

Aresu A (2024) Geopolitica dell'intelligenza artificiale. Feltrinelli, Milano

Brookings Institution (2021) The geopolitics of AI and the rise of digital sovereignty. Brookings Reports. https://www.brookings.edu

Calzada I (2018) Algorithmic nations. Reg Stud Reg Sci 5(1):267–289. https://doi.org/10.1080/21681376.2018.1507754

Can M, Kaplan H (2020) Transatlantic partnership on artificial intelligence: realities, perceptions and future implications. Global Affairs 6(4–5):537–557. https://doi.org/10.1080/23340460.2020.1854049

Carayannis E, Campbell D (2012) Triple Helix, Quadruple Helix and Quintuple Helix and how do knowledge, innovation and the environment relate to each other? Int J Soc Ecol Sustain Dev 7(1):41–69. https://doi.org/10.4018/jsesd.2010010105

Carnegie Endowment for International Peace (2022) The AI cold war: US-China competition and the future of technological leadership. Carnegie Publications. https://carnegieendowment.org

Csurgai G (2018) The increasing importance of geoeconomics. Geopolitics 23(1):1–9. https://doi.org/10.1080/14650045.2017.1359547

Diemer A, Iammarino S, Perkins R, Gros A (2022) Technology, resources and geography in a paradigm shift: the case of critical and conflict materials in ICTs. Regional Studies, 1–13. https://doi.org/10.1080/00343404.2022.2077326

Epasto S (2025) Blockchain e intelligenza artificiale per la governance delle smart cities. Modelli inclusivi, sfide etiche e implicazioni geografiche. In fase di pubblicazione

European Commission (2023) Regulatory framework on artificial intelligence (AI). European Commission Publications. https://digital-strategy.ec.europa.eu/en/policies/regulatory-framework-ai

Graham S (2008) Robowar dreams: US military technophilia and global south urbanisation. City 12(1):25–49. https://doi.org/10.1080/13604810801933511

Il Sole 24 Ore (2023) Chip, nuova stretta degli Stati Uniti contro la Cina. https://www.ilsole24ore.com/art/chip-nuova-stretta-stati-uniti-contro-cina-AGn2U3VB

International Strategic Policy Institute (ISPI) (2023) Chip war: un nuovo capitolo nella competizione tecnologica USA-Cina. ISPI Report. https://www.ispionline.it/it/pubblicazione/chip-war-un-nuovo-capitolo-197675

Iphofen R, Kritikos M (2021) Regulating AI. Contemporary Soc Sci 16(2):170–184. https://doi.org/10.1080/21582041.2018.1563803

Iqbal S, Rizvi SWA, Haider MH, Raza S (2023) Artificial intelligence in security and defense: explore the integration of AI in military strategies, security policies, and its implications for global power dynamics. Int J Human Soc 3(4):341–353

Khanna P (2012) Hybrid reality: thriving in the emerging human-technology civilization. Ted Books

Khanna P (2016) Connectography: mapping the future of global civilization. Random House

Khanna P (2019) The future is Asian: commerce, conflict and culture in the 21st century. Simon & Schuster

Kuzior A, Kuzior P (2020) The Quadruple Helix model as a smart city design principle. Virtual Economics 6(3):39–57. https://doi.org/10.34021/ve.2020.03.01(2)
Larsen BC (2022) The geopolitics of AI and the rise of digital sovereignty. Brookings. https://www.brookings.edu/articles/the-geopolitics-of-ai-and-the-rise-of-digital-sovereignty/
Lazard Geopolitical Advisory (2023) The geopolitics of artificial intelligence. https://www.lazard.com/research-insights/the-geopolitics-of-artificial-intelligence/
Lizza G (2024) Il tarlo. La manipolazione della mente nella nuova geopolitica. UTET, Torino
Marr B (2024) The geopolitics of AI. Forbes. Sep 18, 2024. https://www.forbes.com/sites/bernardmarr/2024/09/18/the-geopolitics-of-ai/
Martin A, Sharma G, de Souza SD, Taylor L, van Eerd B, McDonald SM, Marelli M, Cheesman M, Scheel S, Dijstelbloem H (2023) Digitisation and sovereignty in humanitarian space: technologies, territories and tensions. Geopolitics 28(3):1362–1397. https://doi.org/10.1080/14650045.2022.2047468
McKinsey Global Institute (2023) Artificial intelligence and the future of work: global economic trends. McKinsey & Company. https://www.mckinsey.com
Miailhe N (2018) The return of empires. Politique étrangère. https://www.ifri.org/en/articles-politique-etrangere/geopolitics-artificial-intelligence-return-empires
Nadibaidze A, Miotto N (2023) The impact of AI on strategic stability is what states make of it: comparing us and Russian discourses. J Peace Nuclear Disarmament 6(1):47–67. https://doi.org/10.1080/25751654.2023.2205552
Organisation for Economic Co-operation and Development (OECD) (2023) AI policy observatory: global AI investment trends. OECD Publishing. https://www.oecd.org/digital/ai/
Pauwels E (2018) The new geopolitics of AI. World Economic Forum. https://www.weforum.org/stories/2018/10/artificial-intelligence-ai-new-geopolitics-un/
Pinheiro FL, et al (2022) Geography of innovation. Regional Studies 1–16. https://doi.org/10.1080/00343404.2022.2106362
Raza S, Iqbal S, Abbas S, Hasnain M (2023) Artificial intelligence in security and defense: explore the integration of AI in military strategies, security policies, and its implications for global power dynamics. Int J Hum Soc 3(4):341–353
Ruppert L (2024) Geopolitics of technological futures. Geopolitics 29(2):581–606. https://doi.org/10.1080/14650045.2023.2174431
Schindler S, Alami I, DiCarlo J, Jepson N, Rolf S, Bayırbağ MK, Zhao Y et al (2024) The second cold war: US-China competition for centrality in infrastructure, digital, production, and finance networks. Geopolitics 29(4):1083–1120. https://doi.org/10.1080/14650045.2023.2253432
Stockholm International Peace Research Institute (SIPRI) (2023) The future of AI in warfare: Risks and regulations. SIPRI Publications. https://www.sipri.org
Taneja H, Zakaria F (2023) L'intelligenza artificiale e la nuova guerra fredda digitale. Harvard Business Review. https://www.hbritalia.it/homepage/2023/09/22/news/lintelligenza-artificiale-e-la-nuova-guerra-fredda-digitale-15653/
The Economist (2019) Artificial intelligence is changing every aspect of war. https://www.economist.com/science-and-technology/2019/09/07/artificial-intelligence-is-changing-every-aspect-of-war
Trivium China (2023) China AI Strategy Report 2023. Trivium China Research. https://www.triviumchina.com
TRUST Project (2023) Digital turn in Europe: strengthening relational reliance through technology. https://trustrise.eu/
United Nations Office for Disarmament Affairs (UNODA) (2023) Lethal autonomous weapons systems (LAWS): status and negotiations. UNODA Reports. https://www.un.org/disarmament

The Impact of Artificial Intelligence on Urban Transformation with a Prominent Role in Smart Cities

Ivana Vyslúžilová

Abstract This chapter outlines the revolutionary process of boundless new technologies with a focus on artificial intelligence (AI) in the most progressive smart cities in Europe. Technological change and the impact of digitalization affect not only today but have played a significant role in the past decades by creating a space for innovations and development in the field of data science. Not long ago we observed an acceleration of higher use of digital technologies, not only through statistical methods of machine learning and deep learning algorithms but also advancements in deep learning artificial intelligence applications. In addition, advantages from speech recognition, image classification, computer vision, and natural language processing have in recent times created a significant phenomenon in which we are more connected than ever before. Different routes to the digital transformation taking advantage of smart infrastructure led to the reduction of boundaries in urban areas. The principal approach of the utilization of smart urban solutions can lead to several interests, together with urban sustainability that can be seen through reduced pollution, more efficient energy, water, and waste management. One of the defining components of innovations of today is the use of a large amount of data. Therefore, the government is applying these data to make more correct decisions and to support more operations that are more efficient. However, we should not exclude the observation that data collection has some real concerns in the context of information security and privacy. In general, we can conclude that it will be crucial to find the balance between the risks and opportunities of AI services that can affect the future of technological trends in smart cities.

Keywords Artificial intelligence (AI) · Smart city · Smart infrastructure · Urban development · Ethical risks

---

I. Vyslúžilová (✉)
Comenius University, Department of Human Geography and Demography, Bratislava, Slovakia
e-mail: ivana.vysluzilova@gmail.com

L. Soares et al. (eds.), *Compromised Geography*, Key Challenges in Geography,
https://doi.org/10.1007/978-3-032-05201-8_9

Introduction

Adapting to emerging trends, artificial intelligence (AI) is one of the fastest-expanding and most discussed technologies globally. Particularly, urban smart technologies support a set of technologies that permit computers to implement various ultramodern functions. This category relates mainly to computer simulation and problem-solving through image and visual AI tools, AI-powered machine translation, and AI recommendation engines. In addition, it is important to investigate the rising demand for proposing a solution mechanism to implement various sources of information about human preferences on the environmental dimension of sustainability toward spatial urban forms.

Green buildings and using renewable energy sources explicitly demonstrate the significance of sustainable cities in transforming our cities into smart cities of the future. Already, Europe presents a continent-wide system of interconnected cities with a supranational level of government where incorporating the latest innovations and development can contribute to protecting biodiversity with a focus on the most effective solutions to combat climate change. In response, AI tools are necessary to provide more accurate predictions of the risks of climate change via machine learning algorithms. We may live in a global world, but even the most minor change in local access to advanced technology can significantly impact urban forms in surrounding areas. Predicting and managing the effects of climate change will help achieve better policy planning for urban or rural settlements. In this light, being digitally and environmentally connected in smart cities is the next step of the globalization process.

From a societal perspective, technological development and the subsequent introduction of coordination mechanisms to achieve a sustainable environment concentrate mainly on their core socio-economic activities from the city centers to their surroundings to satisfy the citizens' needs. This process is followed by accelerating smart city trends transforming urban living. In relation to this, building eco-friendly alternatives involves various initiatives within urban areas. In general, building a complex system of urban spatial units creates an argument with the key determinants of technological growth. Regardless, current development models have progressed, and it is becoming evident that the difference between cities in more technologically developed environments and less connected areas is that they do not fully leverage this cutting-edge technology. Digital innovation hubs are becoming global axes connecting other promising cities. As a result, a crucial research question arises: What unique conditions are essential for cities to grow and develop into smarter and more sustainable environments in the future?

Our collective actions influence the emphasis on citizens' environmental responsibility. However, in terms of sustainable objectives, a growing population presents significant challenges for today's world. Thus, delivering smarter and more sustainable technologies to undeveloped digital settlements is important to provide its ability to counterbalance the difference. Indeed, the rapid dynamics of cities represented by the changes in the European population in the past and future predictions are significant. Sustainable smart cities require a strong strategy to strengthen Europe's position in global technology standard-setting. The current state of urbanization testifies to the scope of technological opportunities that are mainly in line with the size of the population over time. The latest UN (2024) revision reveals that the world's population is expected to grow significantly, with an estimated peak of around 10.3 billion people in the mid-2080s. Conversely, the global population is projected to decline gradually to 10.2 billion people by the end of the century. In the case of Europe, the terminology in this prospect presents more developed regions comprising all countries and areas (UN 2024). At this point, it is interesting to pay attention to the different approaches to the terminology of urban units.

The purpose of this chapter is to investigate technological responses to today's challenges, including urbanization, dynamics of digitalization, and climate change. This chapter outlines the impact of AI on citizens' lives in European smart cities. Empirical research presents some general observations that may be used to incorporate new ideas in developing innovative actions included in urban development policy. This chapter ends with a discussion of the potential risks of using AI tools in the field of ethics. In this light, the focus should be directed toward understanding the limits of AI technology and its further impacts.

The Expansion of Boundaries in Urban Units

Over the past few decades, researchers have dedicated considerable effort to establishing the basic concepts of Urban Geography. Overall, research on spatial units has grown in recent decades to emphasize the size and distribution of cities through extended analysis using data collection techniques. Especially nowadays, we are witnessing an acceleration of the processes of creating urban structures in the world, and the explanation of terms such as city, urban agglomeration, and urban region is gaining importance.

The city is a long-researched concept within the terminology. Glaeser (2011, p. 7) describes that the very cities that now constitute urban agglomerations on a global scale were "engines of innovation" in the time of Plato and Socrates. According to the classical definition in the field of urban geography in France, a city is defined as "an agglomeration of agglomerated populations characterized by a numerical population minimum and a form of economic and social organization" (Beaujeu-Garnier 1997, p. 45). Different approaches to investigation can shape the definition of a city. In France, Dumont (2012, p. 12) provides a detailed examination of the principled definition of cities based on legislative, historical, and administrative frameworks. From a legislative perspective, the law has long ignored the term city. "In fact, the term city in colloquial speech represents an enterprise." The naming of the town, which appeared in a ministerial document on December 19, 1990, was restrictive in defining the town's boundaries. Historically, the beginning of urban development is attributed to port towns. From an administrative point of view, the city develops according to the functions and distribution of areas (Dumont 2012). However, there is general agreement that defining the cities' boundaries in a wider geographical area is difficult. There is a growing trend toward establishing a single definition of cities for more compatible comparisons of these units across space. The OECD and the European Commission defined a new definition of cities and their commuting zones in 2011 (Dijkstra and Poelman 2012).

The concept of "future cities" is often discussed, which addresses the issues of tomorrow's cities. These future cities support their substantial economic and political position among other cities. They strongly support technological development and education. They also seek to mitigate the effects of climate, geopolitical, and depopulation changes (Moir et al. 2014, p. 8). Urban agglomerations are synonymous with extensive areas where the basic principles of sustainable development can be developed not only in their central core but also in their agglomerated territory.

In today's sense, agglomeration results from the acceleration of urbanization caused by a significant increase in urban population. In more detail, due to the inertial forces of administrative traditions, most large cities and medium-sized towns have widely transcended their initial boundaries, so we can no longer speak only of cities but agglomerations. The representation in Fig. 1, for example, directly reveals the issue when we focus only on large metropolitan areas, such as Paris and Rome. It is clearly demonstrated that the difference in population density between these cities depends mainly on the boundary definition. However, the actual population distribution in cities plays a secondary role (OECD 2012). In general, there is a diversity of opinions when considering administration boundaries. This inconsistency may raise an important question concerning the inclusion or non-inclusion of specific spaces located outside the core of the agglomeration to its whole. Several studies have been conducted on urban agglomerations in Europe and other urbanized countries worldwide. Unfortunately, the terminology used for these spatial settlement units is inconsistent. In French literature, the terms aires urbaines (more similar to a metropolitan area), unités urbaines (administrative term), or agglomérations urbaines (common term) are used. In German literature, terms such as die Agglomerationsräume, der Ballungsraum (more similar to a conurbation), or die Stadtregion (urban region) are used. The term der Ballungsraum has been used by some authors in their studies (Egner 1950; Boustedt 1960).

Functional urban regions (FURs) are a frequently used term within the settlement literature. Several authors have undertaken detailed research on these units (e.g., Doxiadis 1968; Hall and Hay 1980; Berry 1970; Bezák 2014). Defining the internal structure supports the precise delimitation of functional urban regions. A functional urban region consists of an urban core composed of one or more cities and a hinterland. Bezák (2014, p. 123) states that these units represent "spatially continuous territories that are internally coherent and externally relatively closed concerning the daily movement of inhabitants for work, education, services, recreation, and social contacts." Moreover, many countries have revised the definitions of travel-to-work areas, including European projects like ESPON, and whether EU Local Administrative Units (LAU level 2, formerly NUTS5) data could be equivalent to functional regions based on journey-to-work flows (Williams et al. 2012).

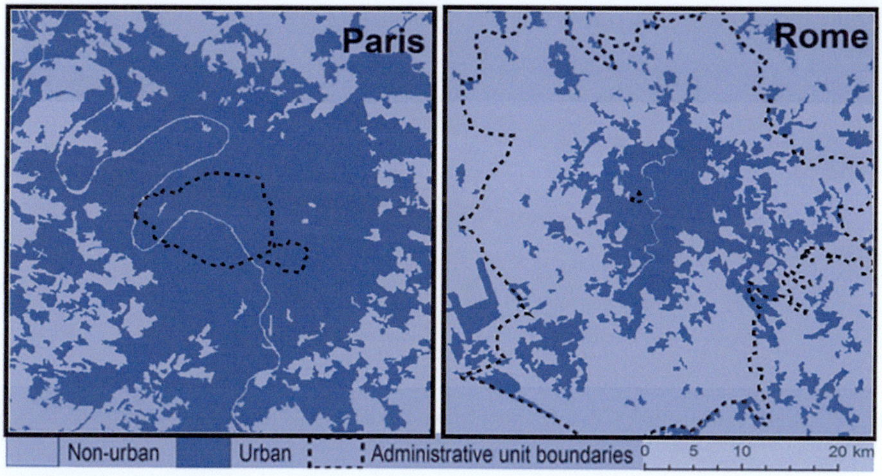

Fig. 1 Urban and non-urban population density: Paris and Rome (OECD 2012)

According to European Commission (2019), functional urban areas (FUAs) may be superimposed onto NUTS level 3 regions to identify metropolitan and non-metropolitan regions, defined as a NUTS level 3 region or a group of NUTS level 3 regions with the criteria that at least 50% of the population resides within a functional urban area with more than 250,000 inhabitants. In fact, functional urban areas usually consist of a relatively densely populated city center and a less densely populated suburban area. The methodology for the typology of cities, towns, suburbs, and rural areas, with relatively small (1 km^2) and uniform grid cells, means the basic concept for the degree of urbanization. This approach examines larger local administrative units, resulting in more accurate data that can be aggregated to generate national statistics. Reliable and comparable datasets for various territorial typologies can only be produced based on coherent building blocks. Location is the foundation for collecting, processing, storing, analyzing, and aggregating data. Often, the framework established by a specific geographic feature, such as a national border or proximity to a coastline, is the only common element shared among different datasets (European Commission 2019).

The Statistical Atlas is an interactive tool that allows users to view and explore statistical maps. In the context of the digital society proposed from the Eurostat regional yearbook 2024, there were 63 NUTS level 2 regions in the EU where the share of households with access to the internet was at least 95%. Internet access takes various forms, such as satellite internet, which provides connectivity to remote areas where other technological options might be limited (European Commission 2024b). Although FUAs representation differs exceedingly across the EU.

As illustrated in Fig. 2, the increase in the range of FUAs is visible in Stuttgart and Frankfurt am Main, where the population of the commuting zone was 3.0 and 2.5 times excessively higher than the city. On January 1, 2020, Paris represented

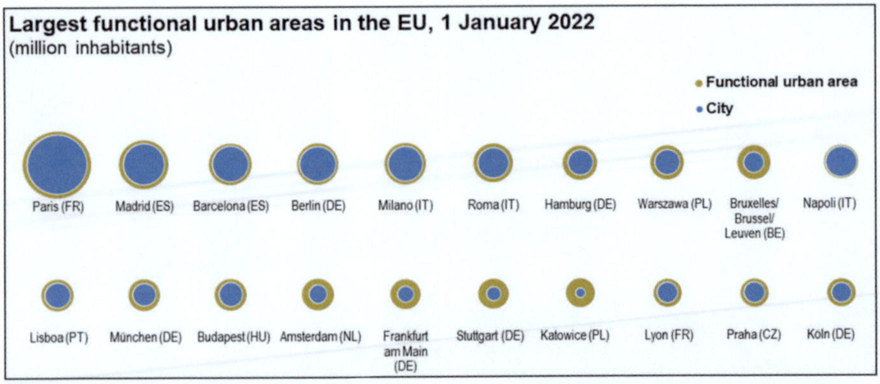

Fig. 2 Largest FUAs in the EU by city's population, 1 January 2022 (European Commission 2024c)

the largest FUAs in the EU with 13.1 million inhabitants. This value comprised just under 10.3 million people living in the city and almost 2.9 million in the surrounding commuting zone. Spain belongs to one of the most populous countries in the EU, with Madrid and Barcelona the country's largest FUAs (data for January 1, 2022). It is common that many Iberian cities feature compact city centers, with clear boundaries separating them from the sparsely populated surrounding areas. The EU's eight largest FUAs reported more inhabitants living in their city centers than in their surrounding commuting zones. By contrast, in seven of the EU's twenty largest FUAs, the commuting zone's population exceeded the city's population. The largest share of commuters was in the FUA of Katowice (European Commission 2024c).

Generally, the delimitation of FUAs can be defined using commuting flows, demographics, or administrative boundaries. However, with the arrival of advanced analytical tools and AI, the presented process can be significantly improved. The AI method establishes dynamic boundaries based on daily/weekly data, utilizing an adaptive algorithm to update FUA boundaries in response to changing mobility patterns.

From a geopolitical view, Lacoste (1993) claims that urban agglomeration units are at the forefront of the world's great cities, competing with the most powerful metropolises, attracting capital, being a place of exchange and contact, and, in short, being part of a capital that has a global reach.

The definition of settlement units is not uniform, as individual authors have tried to define their definitions. Defining the base of how differently a city and the surrounding areas can be seen is a closer step to the smart cities, which are merging technologies outcomes to provide a range of benefits for their residents regarding higher sustainability. However, there is no agreed definition of a smart city. Angelidou (2015) has outlined that the concept of a smart city is associated with utilizing human, collective, and technological capital in urban agglomerations. Scientific knowledge grows by producing numerous scientific publications related to smart cities (e.g., Giffinger et al. 2007, 2010; Kominos 2011; Albino et al. 2015, McClellan et al. 2018, Balderacchi and Piacentini 2024). With global urbanization progressively, cities are turning to smart cities. The next chapter presents the revolution of modern technological systems incorporated into smart cities on a global scale.

Built Infrastructure in Smart Cities

A smart city integrates technology into its urban infrastructure. Overall, smart infrastructure as a cyber-physical system provides for an integral management of all elements that it comprises. Enormous amounts of data must be analyzed using various technological tools to provide sustainability, efficiency, and safety objectives in smart cities. Today, we attempt to access more data than we can even elaborate. In this line of thinking, a challenge can only be the ability to understand how to work with such smart systems and how these data could be used more efficiently. Alpaydin (2016) indicates that information repositories available online today contain massive amounts of digital text and are now so enormous that they cannot be processed manually. Indeed, the rapid development of AI tools can help this infrastructure collect data, process information, and take appropriate action without human intervention.

Ferguson (2017) states that behind the data lies technology, which includes algorithms, network analysis, data mining, machine learning, and various computer technologies that are continuously being refined and improved. Machine learning algorithms, which artificial intelligence models drive, can analyze large volumes of data in unprecedented ways. However, these algorithms and models learn by examining historical data. As a result, their ability to learn is limited to identifying correlations within past data (Ferguson 2017).

It is clear that, for future needs, this enormous historical data should be stored somewhere. Alpaydin (2016) identifies two parallel trends in computing. On the one side is building general-purpose computers that can be programmed for different tasks and purposes, such as those used in data centers. On the other side is to build specialized computing devices for particular tasks, packaged together with specialized input and output. He further claims that well-known parametric estimation, hypothesis testing, and regression methods are now insufficient and mostly univariate for statistical purposes. Similarly, we may agree with Alpaydin's (2016) statement that making such systems smart is a meaningful research direction.

Implementation of Smart Systems for Smart Mobility

The approach, which supports synergies between different activities and public services, concentrates attention on strategic city management. This concept is not only about using smart technologies but also about helping cities' overall economic growth.

Vuchic (2010) claims that cooperation between competent civic leadership, transportation, and city planning experts authorizes cities to develop policies that define the goals for the city, where the development of an efficient transformation system is an element for achieving a livable and sustainable city. Issues arising from a growing population and increasing car ownership are leading to inefficient transportation use and significant traffic congestion. In addition, chronic traffic congestion harms the health of the population. It follows the fact that cities are becoming continuously more unlivable. Creating a green infrastructure of the city with vegetation is one of the solutions and creates the precondition for stability and balance of flora and fauna in the urban environment. Sustainable urban mobility aims to develop urban transport systems that are resilient, smart, safe, and emission-free. The arrival of new technologies and AI solutions is significantly shaping the landscape of urban development, particularly in the quality of life of the city's inhabitants. It confirms the fact, that green infrastructure should be in symbiosis with smart mobility.

Current trends in most developed cities are to build green lines with vegetation near the public transport corridors. In addition, in this area, the noise is less because of the soil in the ground. Consequently, simple improvements can also have a positive impact on cities. It is common for not every smart city to have a metro rail system. Vuchic (2010) pointed out that the metro rail system is a rational solution for essential transportation networks in large cities because it is the highest-performance transit mode that positively impacts urban development and livability. Nevertheless, their investments in costs are very high, so several medium-sized cities have alternatively found Light Rail Transit, with the main characteristic being that it is pedestrian-friendly (Vuchic 2010).

In this context, smart mobility management creates a new city's traffic system. Indeed, we can observe local government initiatives that aim to support walking and cycling in cities rather than individual car transportation to avoid traffic congestion—changes in land-use planning result in converting motor vehicle lanes into bicycle lanes. Such use of routes differs in each country, but the aim remains the same. Also, with the current digital development of several apps, there are opportunities to take advantage of shared mobility. This system combines mobility services such as long-distance and suburban trains, buses, e-scooters, and bike or car sharing. All tendencies are moving forward to enhance sustainable urban areas.

A challenging goal will be to electrify the majority of vehicles. Examples where AI is significantly transforming how we are used to seeing the field of mobility. For example, AI-powered traffic management systems help to analyze real-time data from various means of transportation, including cars, buses, and trains. Intelligent solutions can represent a new kind of mobility through the concept of driverless electric shuttle vehicles with integrated services. For a future of connected automated driving, vehicles must be able to communicate seamlessly with each other and their surroundings. Likewise, digital key management enables a digital car key or phone-as-a-key, revolutionizing vehicle access for private consumers and mobility app use cases. In addition, updated navigation systems can find the fastest route with charging stations. In particular, there is a strong potential for a cloud-based solution that sends warnings to all wrong-lane drivers and all vulnerable road users within a few seconds.

From the perspective of environmental protection, engineers focus mainly on developing sustainable materials and smart manufacturing processes to prevent environmental damage. For example, at the University of Warwick, they are working on a hydrogen-powered car fueled by sewage.

AI tools align data analysis, and digital platforms support the effective planning of the urban mobility system in Smart cities. Smart cities are emerging due to Internet of Things (IoT) connectivity. Smart objects can be helpful in several areas.

Network of Smart City Lights

Smart lighting is a public light system integrating IoT technologies to lower energy consumption. Public lighting is a necessary condition for a smart city. There is a relationship between urban green space and smart public lighting. With the help of modern lighting technologies, night lighting is more plant-friendly and fulfills citizens' purpose. Public lighting offers an energy infrastructure that can synergize with other modern technologies in the city. Simultaneous sensors for traffic telematics, Wi-fi signal transmitters, or a charging station for electric vehicles can be connected to one public lighting column. Overall, public lighting also has broader social benefits that are reflected in city life. In public spaces, it is possible to install a smart lighting system, the basis of which is a software platform that allows efficient control of public lighting according to immediate requirements using web applications. Conversely, restrictions on public lighting can lead to increased crime or traffic accidents. Although in Europe, the network of lightning coverage is widely present in large and medium-sized urban agglomerations, small urban agglomerations or rural areas are less represented. If the municipality's budget allows it, it can undergo modernization by implementing smart technologies in its territory. For example, in implementing energy and non-energy-related services on public lighting are leading European cities Paris and London (Energy Cities 2023).

Figure 3 illustrates the lights on human activity shining in NASA's image of northwestern Europe at night. It clearly demonstrates that likewise, in both European global cities, London and Paris, there is an illuminated line, including partially a corridor of the urbanized area known as a European megalopolis. In particular, the concept of smart cities with their modern public services can be a pleasant place to live and do business.

City Design in the Context of Artificial Intelligence

With the boom of digital technologies and tools, urban designers and planners have more public data sources than ever. Planning compact and accessible smart cities must include sophisticated information technologies that permit instantaneous communication among each other. One of the most important attributes is a connection to

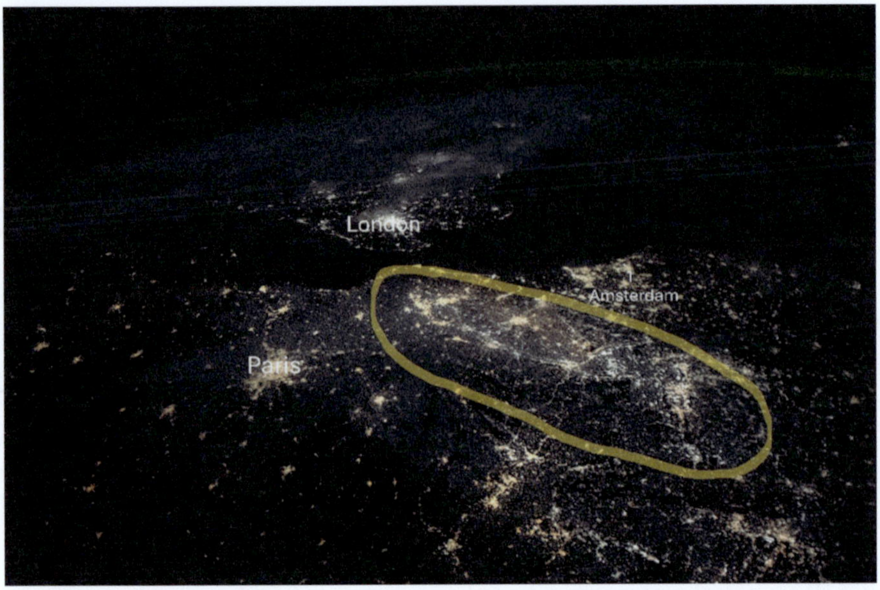

Fig. 3 Northwestern Europe at night (NASA 2024)

IoT devices. It follows big data processing and using the necessary land-use planning tools. Generative design combines digital computation and human creativity. Within the design process, the human professional vision is critical to prioritize the needs and perceptions of citizens. It is typical for architecture in smart cities to go beyond functionality and aesthetics and bring to the forefront the overall quality of life for citizens. Therefore, the value of aesthetics is more or less questionable. As traditional architectural creations decline, the architecture of various cultures is being lost. Barnett (2011) highlighted that traditional city design, including shared expectations for future buildings, has been recovering in historic European cities—for example, the rebuilding of the central parts of Berlin with modernist buildings constrained by traditional guidelines. Also, the area around St. Paul's Cathedral in London, where a modernist plan was implemented just after World War II, has been replaced with new buildings that traditionally hold the street walls. However, it is important also to consider the meaning of sustainability. Barnett (2011) emphasizes that designing within the constraints of natural systems is essential for sustainability, both to mitigate climate change and to adapt to its new challenges. Linking traditional and modern architecture within historical centers should not be chaotic.

We can say that sometimes it is better to place modern architecture outside of a core city, where there is more space for new creations and where art can be in harmony. Like Ludwig Mies van der Rohe said, the famous dictum, "Less is more." The urban growth outside of central cities and economic, social, and cultural development in other zones in urban agglomerations, with the protection of their historical cores, gives more opportunities to support this hypothesis (Vyslúžilová 2016).

Barnett (2011) concludes that understanding cities as systems and creating systematic ways of producing new cities will require a fusion of sophisticated mathematics and an accurate understanding of the systemic behavior that constantly characterizes cities.

Taking advantage of AI can lead to designing user-centric experiences. Automatically generating textual descriptions of images is a key challenge in AI that integrates computer vision and natural language processing.

Figure 4 shows an example of using the OpenAI tool with the specialized deep learning model DALL-E, which generates complex digital images based on textual descriptions. We can prompt the deep learning model to create an imaginative picture of AI in a smart city overlooking the horizon, and the program can generate a utopian architectural vision of a future city.

Currently, the creation of deep learning models is expanding, and among the pioneers like Dall-E or Midjourney can be found other AI tools (e.g., DeepAI, Canva,

Fig. 4 AI vision of smart city via deep learning model DALL-E (Vyslúžilová)

Crayion AI, SoulGen, Leonardo AI). With AI tools in city design, it is possible to extend imagination for more sustainable urban agglomerations (e.g., sophisticated and greener urban infrastructure, biophilic air-purification towers, or complex units with energy-efficient technological systems).

Artificial Intelligence Technologies in Smart Environment

As technology advances, AI and IoT tools are essential in shaping the future of sustainable smart cities. Sustainable urban development is currently the subject of several scientific conferences on urban development. The World Urban Forum, organized by the United Nations Human Settlements Programme, is the foremost global event dedicated to sustainable urban development and will host its fourteenth session (WUF 2025). Urban strategies expert Lawrence (2009, p. 5) supports sustainability in cities, and according to him, sustainability and sustainable development are "how the natural systems of the world function, how they influence and are themselves influenced by individual and aggregate human behavior." He perceives the lack of understanding regarding the causes and long-term effects of the intersection between natural phenomena and human behavior as a barrier to sustainability.

Arimah et al. (2009) highlighted a key goal of the Millennium Development Goals: to significantly reduce biodiversity loss and reverse the decline of environmental resources. In some cases, during the process of urbanization, the loss of biodiversity can occur. Nevertheless, Chou (2010) argues that in dense city centers, biodiversity remains a remote and unrealistic privilege as it stands in the way of city development, occupies expensive real estate, and is unprofitable for economic considerations. Despite this, there are some possibilities to benefit from rooftop gardens by increasing city biodiversity. At the same time, they can reduce heat transfer to the interior of buildings (Chou 2010).

Optimalization of Smart Living with IoT Devices

The latest generation of smart buildings is fostering a climate-centric design. IoT revolutionizes the building sector through intelligent solutions. IoT devices enable automation by responding to changes without human input. Smart buildings are constructed with advanced technological systems that automate various internal processes, including heating, lighting, security, and other building systems.

In addition, investments in energy-saving technologies that reduce consumption, minimize environmental impact, and lower energy costs. For example, heat pumps represent an alternative to air-conditioners by moving heat rather than generating it. Also, advanced machine learning thermostats are produced to make micro-adjustments to temperature settings while optimizing energy efficiency.

Sørensena et al. (2017) analyzed a pilot project of an energy-balanced office building within the research center on zero-emission buildings in Norway. They have argued that predicted energy production from the solar cells was calculated using the command-line interface, considering simulations of meteorological data. The positive results of analyzing the energy performance according to the different energy use compared with the predictions confirm that building a net zero-energy office is possible (Sørensena et al. 2017).

Domotics or home automation relates to all the technologies and automated systems used to control and manage a home's equipment. Smart home technology is transforming modern living where people can benefit from innovative solutions and be guided by intelligent assistants like Amazon Alexa, Google Assistant, and Apple's Siri.

Several companies with forward-thinking approaches to smart living have introduced smart home solutions, highlighting the effectiveness of AI-driven features. For example, it can be the solution for households that enable automatic food identification utilizing artificial intelligence technology in smart refrigerators with additional storage recommendations—the touchscreen countertop with integrated 3D sensors. Likewise, with the virtual touchscreen, people can use a personal assistant to create smart shelves in their closets. Among several smart home solutions, there are also voice-controlled heating systems that can make everyday life easier. Integrating AI and IoT home systems creates a productive sphere by using less energy, reducing carbon dioxide emissions, and helping fight climate change.

Climate Change Predictions *via* Machine Learning Algorithms

It is crucial for future generations to implement interventions to develop technologies that address climate change. Intelligent operations centers should be equipped with the latest disaster management and response technology. Zhuhadar et al. (2017) emphasized that selective data integration into the Intelligent operations centers makes efficient problem-solving solutions readily available to city authorities. One of the primary advantages of Intelligent operations centers is their ability to optimize planned and unplanned operations using a holistic reporting and monitoring approach (Zhuhadar et al. 2017). Real-time monitoring of urban activities directly related to social life can be achieved through surveillance cameras and IoT sensor networks, providing timely alerts about potential incidents. According to Jiang et al. (2018), utilizing machine learning algorithms to mine monitoring information can significantly reduce energy consumption in data centers.

Huntingford et al. (2019) highlight the current trend of Earth monitoring, mainly by satellites collecting climate-related data, which requires advanced algorithms to characterize any overall trends. In this line of thinking, satellite mapping information is extremely valuable in investigating the effects of climate change.

The report of ESA and World Bank (2013) discusses that these mapping can have several uses, such as monitoring controlled urban sprawl and associated fragmentation of communities through informal settlements, loss of green urban spaces, degradation of landscapes, or supporting crisis operations.

Furthermore, Di Nunno et al. (2023) examined a regional study based on a novel hybrid artificial intelligence algorithm to provide short-term forecasts in the United Kingdom. According to them, the hybrid machine learning deep learning model is less affected by performance reductions as the forecasting horizon increases compared to simpler models (Di Nunno et al. 2023). Thus, the diffusion of machine learning and artificial intelligence algorithms offers support for the future load situation (Jiang et al. 2018).

According to Tupin (2025), Synthetic Aperture Radars have revolutionized earth observation by enabling the capture of high-resolution images. She concludes that these radars can contain remarkable granularity despite their effectiveness, causing a lower interpretation. On the contrary, the MERLIN method differs from the spectral bands captured by optical sensors (Tupin, 2025). New strategies with broad, innovative solutions provide a critical approach to estimating the impacts of climate change.

Other major global issues related to natural resource depletion, such as deforestation, soil degradation, desertification, and biodiversity loss, also reveal alarming trends. As objectives in Industry 5.0, the European Commission (2021) has identified three key elements: resilience, sustainability, and human-centricity. Another initiative, the United Nations Global Pulse, provides real-time monitoring and predictions for development and aid programs. This UN initiative seeks to promote innovations to help achieve the Sustainable Development Goals, respond effectively to crises, and tackle emerging global challenges (UN Global Pulse, 2025). Logar et al. (2020) argue that using AI assistants to augment human analysts is a sensible approach for transitioning from fully manual operational pipelines to more productive ones.

In particular, a collaborative satellite image analysis tool, PulseSatellite, can be used, among others, to reduce the time needed to load satellite imagery into machine learning models. Daily, millions of images are captured from space by a steadily increasing number of satellites. Logar et al. (2020) describe that the possibilities of using the system represent settlement mapping, damage assessment, monitoring population displacement, damage to transportation networks, fire detection associated with human rights violations, and identifying the direct impact of earthquakes, volcanoes, cyclones, and landslides. Thus, we can observe that high levels of accuracy are possible with machine learning methods.

European Smart Cities Toward an Efficient and Sustainable Future

Europe has a broad potential to affect important key players that can help by supporting progress on climate. In addition, they can influence the masses to change their daily habits to get closer to a genuinely waste-free process and achieve a more sustainable-oriented population. Therefore, there is a need for support in introducing technological innovations for a broader spectrum of knowledge to achieve a more sustainable-oriented population. The degree of digital knowledge is rising more in urbanized environments. Each year, European cities are getting smarter with digital data processing and using this data for other purposes in different areas of interest. However, it is very complex to figure out what smart cities create toward upcoming world challenges and how to define their boundaries, as new technologies are emerging significantly fast.

The Future of Geospatial Technology in Smart Cities

In a recent work, Othengrafen et al. (2025) identify that the notion of a smart city has become an umbrella concept that divides into multiple approaches to digitizing services. Some leading European cities, such as Zurich, Oslo, Geneve, Lausanne, London, Paris, Amsterdam, Helsinki, Madrid, and Lisbon, are most successfully implementing smart technologies. Furthermore, in recent years, European cities have become more connected regarding digitalization and sustainable development through several public initiatives or legal declarations. For example, according to the EU decision, the European Commission (2024a) set up the European Digital Infrastructure Consortium for Networked Local Digital Twins toward the CitiVERSE. This document confirms the role of AI as a key driver in developing and introducing Local Digital Twins by helping cities operate more efficiently. As technology progresses toward a more sophisticated society, this highlights the power of open data initiatives. Pellegrin et al. (2021) introduce a city project for Pilsen using tools for policymaking, building on earlier research (H2020) related to traffic modeling and utilizing open geodata (OpenTransportNet, PoliVisu). These models will create Digital Twins to develop interaction tools among traffic, air quality, and noise pollution paradigms. As a result, the city will be able to simulate different urban development scenarios (Pellegrin et al. 2021). In more detail, Othengrafen et al. (2025) describe that AI-based planning processes include tools that identify specific topics in the planning process, perform AI-based keywording, group similar comments, anonymize personal data, and send evaluation results to the involved groups. Overall, implementing a digital twin solution can significantly enhance urban projects and investments.

From the assumption of technological complexity, the impact of AI is frequently perceived as ultramodern technology. There is a significant opportunity for this innovative technology, combined with advancing geospatial tools for urban planning, to offer substantial potential to improve planning outcomes. This idea aligns with Radecky's (2023) findings, which show that a specific group of planners and architects from Vienna participated in piloting the digital building permission process using building information model (BIM), AI, and augmented reality (AR), delivering proof that the system can achieve significant improvements in real-life building inspection processes. Planners and investors can submit their building plans and upload their 3D BIM digitally through the city's servers. The system also includes AI-based verification routines and AR-based visualizations to enhance citizen engagement (Radecky 2023). This approach differentiated and put to the forefront the implementation of new technologies in comparison with previous processes.

Wang and Yorke-Smith (2025) even go one step further and promote that successfully leveraging AI for urban planning requires developing technologies that are not only proficient in handling data but also understanding the contextual distinctions that impact human behavior, urban interactions, and the associated social-technical implications. AI improves the efficiency and accuracy of analyzing satellite imagery, LiDAR, drone data, and GPS information, leading to better urban development planning in smart cities. Multiple platforms are available for geospatial data analysis using AI technologies, e.g., QGIS with Python AI integrations, Esri ArcGIS with AI/ML plugins, Sentinel Hub with deep learning frameworks, Carto, or Google Earth Engine. Geospatial tools powered by AI can predict various phenomena, including traffic congestion, flood zones, crop yields, and the spread of diseases. In data investigation, it is necessary to store enormous amounts of data more efficiently. Dritsas and Trigka (2025) emphasize that Cloud computing has revolutionized geospatial data processing by providing scalable, on-demand resources capable of handling massive datasets without local infrastructure.

Smart Cities as Key Drivers in Technology and Sustainability

AI could also assist in monitoring critical infrastructure and advancing sustainability. A different approach presents Smart Neighborhoods. Manville et al. (2014) describe Smart Neighborhoods as fully developed infrastructure projects that can be implemented on greenfield sites or through retrofitting existing developments and can improve a city's capabilities and promote economic growth by establishing the city as a leader in technology and sustainability. Examples include the London suburb of Hackbridge, Hafencity in Hamburg, Nordhavn in Copenhagen, Stockholm Royal Seaport, Oulu Arctic City, Lyon Smart Community, and Aspern in Vienna (Manville et al. 2014).

Giffinger et al. (2010) note that six elements are important in identifying a smart city: economy, people, governance, mobility, environment, and living. These dimensions of research are standard in several methodologies. For example, Manville et al. (2014) state that participants and government officials regard the open data projects currently under deployment as providing better outcomes for Smart Governance and Smart Economy than traditional methods. They have outlined that London,

Amsterdam, Helsinki, Barcelona, and Florence are notable cities developing ICT-enabled citizen participation platforms. In addition, their methodological analysis is based on the alignment between the objectives and characteristics of each city's portfolio of projects and the relevant objectives for Europe 2020. Overall, they identified the six most successful cities: Amsterdam, Barcelona, Copenhagen, Helsinki, Manchester, and Vienna (Manville et al. 2014).

Furthermore, Berrone and Ricart (2024) from IESE present a Cities in Motion Index every year. The last results show that beyond the top 10—London (1st), Paris (3rd), Berlin (5th), Oslo (7th), and Amsterdam (8th), Europe rated more than half of the cities in the top 50 smart cities worldwide. They used several indicators to assess the quality of cities in the technology sector, including the proportion of smart uses of ICT technologies, broadband availability, and the percentage of households with computers or mobile phones. Regarding urban planning, they considered not only housing policy, sanitation services, and health infrastructure but also artificial intelligence projects that promote urban growth and enhance overall well-being.

Specific European smart cities' rankings are enlarger with other international surveys (Forbes 2023; Earth.Org 2024; ProptechOS 2024). European smart cities' rankings differ according to the use of indicators. Therefore, the results of delimitations in smart cities are difficult to compare. The difficulties of identifying correct indicators also lie in the inconsistency of views about which set of indicators is relevant. One analysis can place much more emphasis on one specific dimension, so another dimension is less included. For example, when some specific environmental factors are missing, the final analysis cannot be complex and comparable to the other analysis. When considering smart cities as key drivers of digitalization, it is essential to avoid technology obsolescence and follow the latest trends. In London, lamp posts or bollards can serve multiple functions, including public electric vehicle charging points (Fig. 5) and benches that act as Wi-Fi hotspots.

The role of citizens is an essential part of the transformation of cities. Citizens are empowered with innovative AI tools that support digital transformation and help cities become smarter and more resilient. A more sustainable-oriented population toward science and technology emphasizes outcomes or the overall impact on other cities in the technological transition toward sustainable stabilization.

According to Conti and Simone (2019), the environmental performances observed in the various Green City Index, besides verifying the degree of environmental sustainability, allow analyzing urban phenomena closely linked to environmental sustainability, for example, wealth, governance, or the degree of social inclusion. The results presented in several reports confirm strong correlation between the size of urban centers and the efficiency of their environmental performance (Conti and Simone 2019). In recent times, Vilnius has become the European Green Capital 2025 due to its strong commitment to becoming carbon neutral by 2030. The city's tech-oriented approach to citizen participation is innovative, as is its method of monitoring the well-being of its citizens through data-driven monitoring (Labbe 2023). Monitoring citizen data can investigate correlations in large amounts of raw data and lead to further decisions.

Fig. 5 Public smart charging solution in London (Manthey 2023)

Information Security and Privacy in Smart Cities

As AI increasingly influences different areas of human life, AI systems can learn and adapt their methods for each area of research separately. Prior to delving into such an evolving method, it is helpful to elucidate the potential of AI as a tool to improve citizens' security in smart cities. If the city is equipped with the latest AI technology with IoT devices, it can provide a more complex overview of the urban pattern that citizens create daily. These patterns are recorded through robust surveillance systems and stored as significant gathered data.

The advantages of AI-powered video analytics in urban areas include sending alerts about crimes and warnings regarding accidents, as well as facilitating immediate assistance to crime victims. In the field of measuring AI systems, elementary ethics provide further needs in the development of data security. It is important to emphasize the segregation of sensitive data from less critical components in the frame of citizens' data. In particular, as noted by Nissenbaum (2010), a normative conception of privacy incorporates a presumption that privacy is valuable, worthwhile, and deserving of protection, contrary to neutral conceptions. She provides as an example a normative account of privacy based on individual preferences or interests that might claim that people have a right to control information about themselves or have a right to impose restrictions on access to information about themselves because this best serves their interests or because this accords with most people's preferences. Defining the scope of privacy protection through the public/private dichotomy is a common approach in both policy and legal practice (Nissenbaum 2010). The extent to which modern technologies contribute to citizen security varies among European countries. AI-supported surveillance systems can help reduce crime and enhance safety, creating a more secure environment for residents in smart cities.

Intelligent Video Analytics in Monaco's Smart City

Monaco is a smart city where citizens' security is at the forefront, among other areas. Its powerful security system results in low crime rates and makes it one of the safest places to live. In 2023, a report from Enact (2023) revealed that Monaco's crime score was 2.58%, based on its Global Organized Crime Index, an innovative tool for measuring criminality. As digitalization advances, crime increasingly occurs in the digital sphere. In digital security, utilizing modern technological advancements more systematically is crucial. The European Union (2013) emphasizes that cybersecurity incidents are rising in frequency and complexity without knowing no borders. In Monaco, the public security system is implemented in various locations, including public spaces and coastal areas (Fig. 6), to ensure the protection of citizens.

Borak (2024) highlights that the Monaco Principality police force already uses over 1000 CCTV cameras to improve public security in the city, but these cameras are currently not connected to facial recognition software. In fact, facial recognition technology is an AI form that can authenticate and identify individuals, such as finding someone within a group, image, database, or specific location. Borak (2024) argues that the national parliament is going further. The Principality of Monaco is set to debate its first law on public video surveillance systems with remote biometric identification, and the draft bill, number 1087, has already been submitted (Borak 2024).

Fig. 6 Intelligent video surveillance system in Monaco's smart city (Vyslúžilová)

Citizen engagement should be part of the response to privacy and ethical concerns. Smart cities produce enormous video data analytics, making data collection a potentially double-edged sword for their residents. In addition, these data can be accompanied by algorithmic bias. Addressing these issues is crucial for government authorities to prioritize the secure storage of sensitive information while respecting citizens' privacy rights. Proactively selecting new data sources or enhancing algorithms can improve future risks.

Overall, expectations for AI solutions in urban areas are high, and their impacts on societal change are unpredictable. However, the future direction of AI development will undoubtedly shape our future.

Conclusion

AI can significantly change our society, especially from the perspective of living and working. It is evident that our world has become more digital than ever before. The theoretical crossing line is becoming increasingly blurred between the physical and digital worlds. In the urban context, smart cities represent a concept of a sovereign zone where advanced technologies in smart infrastructure foster a sustainable and efficient future. In the age of globalization, cutting-edge technologies are blurring the boundaries within smart cities. Many European cities are at the forefront of technological and sustainable development initiatives reshaping urban life. The integration of AI and IoT into smart infrastructure reduces the need for human intervention by utilizing AI tools. Creating the cities of the future will necessitate the support and implementation of advanced solutions, such as smart vehicles, smart lighting, and net zero-energy buildings. Effective management of big data will be essential for implementing smart city projects in government. Therefore, a set of standards for the transparent use of AI is essential across various sectors. A framework with a significant impact should be established to address and find solutions to moral and ethical issues in an intelligent manner. But first, it is essential to understand current AI routing to maintain a sustainable and secure future.

References

Albino V, Berardi U, Dangelico RM (2015) Smart cities: definitions, dimensions, performance, and initiatives. J Urban Technol 22(1):3–21. https://doi.org/10.1080/10630732.2014.942092

Alpaydin E (2016) Machine learning: the new AI. MIT Press, Cambridge

Angelidou M (2015) Smart cities: a conjuncture of four forces. Cities 47:95–106. https://doi.org/10.1016/j.cities.2015.05.004

Arimah BC, Jensen I, Mutizwa-Mangiza ND, Yemeru EA (2009) Planning sustainable cities: global report on human settlement 2009. UN Habitat, London

Balderacchi C, Piacentini A (2024) The path towards sustainable development in Europe: SDGs, EU institutions, and smart cities. J Eur Integr 46(8):1275–1281. https://doi.org/10.1080/07036337.2024.2423463
Barnett J (2011) City design: modernist, traditional, green and systems perspectives. Routledge, London and New York
Beaujeu-Garnier J (1997) Géographie urbaine. Armand Colin, Paris
Berrone P, Ricart JE (2024) IESE cities in motion index 2024. IESE. Business School University of Navarra. https://www.iese.edu/media/research/pdfs/ST-0649-E.pdf. Accessed 11 Feb 2025
Berry BJL (1970) The changing internal structure of the daily urban systems. In: The great lakes megalopolis. A comparative study of growth trend. Doxiadis Associates, Athens, pp 171–230
Bezák A (2014) Some preliminary considerations concerning the internal structure of functional urban regions in Slovakia. Acta Geogr Univ Comenianae 58(2):123–130
Borak M (2024) Law on public facial recognition surveillance submitted in Monaco. https://www.biometricupdate.com/202401/law-on-public-facial-recognition-surveillance-submitted-in-monaco. Accessed 12 Feb 2025
Boustedt O (1960) Die Stadtregionen in der Bundesrepublik Deutschland. In: Forschungs und Sitzungsplanung der Akademie für Raumforschung und Landesplanung, Band 14, Bremen
Chou LM (2010) Biodiversity in sustainable cities. In: Ooi GL, Yuen B (eds) World cities: achieving liveability and vibrancy. World Scientific Publishing, Singapure, pp 131–154
Conti ME, Simone C (2019) Smart cities: are they really sustainable? In: Ravina Ripoll R, Tobar Pesantes LB, Núnez Barriopedro E (eds) Happiness management and creativity in the XXI century: intangible capitals as a source of innovation, competitiveness and sustainable development. Granada, Comares, pp 5–21
Di Nunno F, de Marinis G, Granata F (2023) Short-term forecasts of streamflow in the UK based on a novel hybrid artificial intelligence algorithm. Sci Rep 13:7036. https://doi.org/10.1038/s41598-023-34316-3
Dijkstra L, Poelman H (2012) Cities in Europe: the new OECD-EC definition. Regional focus a series of short papers on regional research and indicators produced by the directorate-general for regional and urban policy RF 01/2012. 2012. https://ec.europa.eu/regional_policy/sources/focus/2012_01_city.pdf. Accessed 09 Feb 2025
Doxiadis CA (1968) An introduction in the science of human settlements. Hutchinson, London
Dritsas E, Trigka M (2025) Remote sensing and geospatial analysis in the big data era: a survey. Remote Sensing 17(3):550. https://doi.org/10.3390/rs17030550
Dumont GF (2012) La France en villes. Sedes/Cned, Paris
Earth.Org (2024) Top 7 smart cities in the world in 2024. https://earth.org/top-7-smart-cities-in-the-world/. Accessed 11 Feb 2025
Egner E (1950) Wirtschaftliche Raumordnung in der industriellen welt. Dorn, Bremen
Enact (2023) Monaco: global organized crime index. https://ocindex.net/2023/country/monaco. Accessed 12 Feb 2025
Energy Cities (2023) Handbook smart city an other ICT technology in EPC. Smart EPC. https://energy-cities.eu/wp-content/uploads/2023/03/Smart-EPC_Handbook-on-Smart-City-and-other-ICT-technology-in-EPC.pdf. Accessed 09 Feb 2025
ESA and World Bank (2013) Earth observation for sustainable development partnership report, June 2013. European Space Agency and World Bank. https://eo4society.esa.int/wp-content/uploads/2021/11/eoworld1_red.pdf. Accessed 07 Feb 2025
European Commission (2019) Eurostat, methodological manual on territorial typologies—2018 edition. European Union, pp 1–132. https://doi.org/10.2785/930137
European Commission (2021) Industry 5.0—towards a sustainable, human-centric and resilient European industry. Directorate-General for Research and Innovation, Brussels. https://doi.org/10.2777/308407
European Commission (2024a) Commission implementing decision (EU) 2024/459 of 1 February 2024 on setting up the European digital infrastructure consortium for networked local digital

twins towards the CitiVERSE (LDT CitiVERSE EDIC). Official J Eur Union. http://data.europa.eu/eli/dec_impl/2024/459/oj. Accessed 10 Feb 2025

European Commission (2024b) Eurostat, Eurostat regional yearbook—2024 edition. Publications Office of the European Union. https://doi.org/10.2785/174221. Accessed 8 May 2025

European Commission (2024c) Eurostat, Urban—rural Europe-demographic developments in cities. European Union. https://ec.europa.eu/eurostat/statistics-explained/index.php?title=Urban-rural_Europe_-_demographic_developments_in_cities. Accessed 6 May 2025

European Union (2013) EU cybersecurity plan to protect open internet and online freedom and opportunity. European Union External action. https://ec.europa.eu/commission/presscorner/detail/en/ip_13_94. Accessed 12 Feb 2025

Ferguson AG (2017) The rise of big data policing. New York University Press, New York

Forbes (2023) Here are the 20 smartest cities in the world. https://www.forbesmiddleeast.com/industry/economy/here-are-the-20-smartest-cities-in-the-world. Accessed 11 Feb 2025

Giffinger R, Fertner C, Kramar H, Kalasek R, Milanović N, Meijers E (2007) Smart cities—ranking of European medium-sized cities. Centre of Regional Studies, Vienna UT. https://www.smart-cities.eu/download/smart_cities_final_report.pdf. Accessed 10 Feb 2025

Giffinger R, Haindlmaier G, Kramar H (2010) The role of rankings in growing city competition. Urban Res Pract 3(3):299–312. https://doi.org/10.1080/17535069.2010.524420

Glaeser E (2011) Triumph of the city: how our greatest invention makes us richer, smarter, greener, healthier, and happier. The Penguin Press, New York

Hall P, Hay D (1980) Growth centres in the European urban system. Heinemann, London

Huntingford H, Jeffers ES, Bonsall MB, Christensen HM, Lees T, Yang H (2019) Environ Res Lett 14:124007. https://doi.org/10.1088/1748-9326/ab4e55

Jiang C, Fan T, Wan J (2018) An edge computing platform for intelligent operational monitoring in internet data centers. 7:16–28. https://doi.org/10.1007/978-981-32-9987-0_2

Kominos N (2011) Intelligent cities: variable geometries of spatial intelligence. Intell Build Int 3:177–188. https://doi.org/10.1080/17508975.2011.579339

Labbe AF (2023) Jury report EGCA 2025, Phrenos. https://environment.ec.europa.eu/topics/urban-environment/european-green-capital-award/winning-cities/vilnius-2025_en. Accessed 11 Feb 2025

Lacoste I (1993) Dictionnaire de géopolitique. Flammarion, Paris

Lawrence G (2009) Drivers of change: building our urban future. In: Urban world. UN Habitat, pp 5–6

Logar T, Bullock J, Nemni E, Bromley L, Quinn JA, Luengo-Oroz M (2020) PulseSatellite: a tool using human-AI feedback loops for satellite image analysis in humanitarian contexts. Proc AAAI Conf Artif Intell 34(09):28–29. https://doi.org/10.1609/aaai.v34i09.7101

Manthey N (2023) Infrastructure: ubitricity enables public smart charging at UK. Lampposts, Ubitricity. https://www.electrive.com/2023/02/15/ubitricity-enables-public-smart-charging-at-uk-lampposts/. Accessed 11 Feb 2025

Manville C, Cochrane G, Cave J, Millard J, Pederson JK, Thaarup RK, Kotterink B (2014) Mapping smart cities in the EU. Directorate General for Internal Policies, European Parliament. http://www.europarl.europa.eu/RegData/etudes/etudes/join/2014/507480/IPOL-ITRE_ET(2014)507480_EN.pdf. Accessed 10 Feb 2025

McClellan S, Jimenez JA, Koutitas G (2018) Smart cities: applications, technologies, standards, and driving factors. Springer, San Marcos

Moir E, Moonen T, Clark G (2014) What are future cities? Origins, meanings 147 and uses. Government Office for Science

NASA (2024) Europe at night. NASA Earth Observatory. https://earthobservatory.nasa.gov/images/152693/europe-at-night. Accessed 04 Feb 2025

Nissenbaum H (2010) Privacy in context: technology, policy and the integrity of social life. Stanford University Press, Stanford

OECD (2012) Redefining "urban": a new way to measure metropolitan areas. OECD Publishing, Paris. https://doi.org/10.1787/9789264174108-en

Othengrafen F, Sievers L, Reinecke E (2025) From vision to reality: the use of artificial intelligence in different urban planning phases. Urban Plann 10:168–185. https://doi.org/10.17645/up.8576
Pellegrin J, Colnot L, Delponte L (2021) Research for REGI committee: artificial intelligence and urban development. European Union. https://www.europarl.europa.eu/RegData/etudes/STUD/2021/690882/IPOL_STU(2021)690882_EN.pdf. Accessed 5 May 2025
ProptechOS (2024) The cities best prepared for a smart city future in 2024. https://proptechos.com/smart-city-index-2024/. Accessed 11 Feb 2025
Radecky AV (2023) Piloting the BRISE-Vienna results—Journal N° 3. Urban Innovative Actions. https://www.uia-initiative.eu/en/news/piloting-brisevienna-results-journal-ndeg-3. Accessed 07 May 2025
Sørensena ÅL, Mysena M, Andresenb I, Hårklauc B, Lunde A (2017) Zero emission office building in Bergen: experiences from first year of operation. In: 11th Nordic symposium on building physics, Trondheim, Energy Procedia, vol 132. pp 580–585. https://doi.org/10.1016/j.egypro.2017.09.747
Tupin F (2025) "Denoising" radar satellite images with artificial intelligence. AI Action Summit. https://www.ip-paris.fr/en/news/denoising-radar-satellite-images-artificial-intelligence. Accessed 09 Feb 2025
UN (2024) United nations department of economic and social affairs, population division. In: World population prospects 2024: summary of results. https://www.un.org/development/desa/pd/world-population-prospects-2024. Accessed 02 Feb 2025
UN Global Pulse (2025) Our common Agenda—report of the secretary-general. https://www.unglobalpulse.org/un-global-pulse/. Accessed 09 Feb 2025
Vuchic VR (2010) Transportation for liveable cities: problems, obstacles and successful solutions. In: Ooi GL, Yuen B (eds) World cities: achieving liveability and vibrancy. World Scientific Publishing, Singapure, pp 105–130
Vyslúžilová I (2016) Cultural changes and impact of globalization in European urban agglomerations. Eur J Geogr 7(1):36–44. https://eurogeojournal.eu/index.php/egj/article/view/331. Accessed 05 Feb 2025
Wang T, Yorke-Smith N (2025) AI for and in urban planning. Urban Plann 10:5–10. https://doi.org/10.17645/up.i388
Weihling HG (1978) Dorfpolitik: Fachwissenschaftliche Analysen und didaktische Hilfen. Opladen: Leske Verlag—Budrich
Williams AM, Foord J, Mooney J (2012) Human mobility in functional urban regions: understanding the diversity of mobilities. Int Rev Sociol 22(2):191–209. https://doi.org/10.1080/03906701.2012.696961
WUF (2025) UN-Habitat invites countries and cities to host WUF14: are you ready to lead? United Nations Human Settlements Programme. https://wuf.unhabitat.org/news/un-habitat-invites-countries-and-cities-host-wuf14-are-you-ready-lead. Accessed 09 Feb 2025
Zhuhadar L, Thrasher E, Marklin S, Ordóñez de Pablos P (2017) The next wave of innovation—review of smart cities intelligent operation systems. Comput Hum Behav 66:273–281. https://doi.org/10.1016/j.chb.2016.09.030

Compromised Spaces of Health: Geospatial Exploration of Cardiovascular Mortality Patterns

Suzana Lović Obradović, Stefana Matović, Gorica Stanojević, and Nina Ćurčić

Abstract Serbia continues to face a significant public health challenge, with cardiovascular mortality (CVM) as the leading cause of death. The spatial distribution of CVM is shaped by a complex interaction of demographic, socio-economic, healthcare, and environmental factors, reflecting deep regional disparities. This study employs a geospatial approach to analyze municipal-level (LAU 1) CVM patterns in 2023, using Geographically Weighted Regression (GWR) modeling to assess localized relationships between CVM and its key determinants. The results reveal distinct spatial variations in the influence of demographic structure, socio-economic conditions, and healthcare accessibility on CVM across Serbia's 168 municipalities. Through cartographic visualization, the study identifies critical factors driving mortality rates at a local level, providing understanding of compromised health spaces. By highlighting spatial inequalities, this research offers essential insights for policymakers, enabling the development of targeted, place-based interventions to mitigate CVM risks and promote health equity. Addressing these localized vulnerabilities is crucial for designing effective public health strategies that acknowledge Serbia's diverse geographic and socio-economic landscape. Ultimately, this study underscores the importance of spatial analysis in public health research, laying the foundation for more data-driven, regionally adaptive responses to the burden of cardiovascular diseases.

Keywords Cardiovascular mortality · Spatial analysis · Health policy · Serbia

---

S. L. Obradović (✉) · S. Matović · G. Stanojević · N. Ćurčić
Geographical Institute "Jovan Cvijić" SASA, Belgrade, Serbia
e-mail: s.lovic@gi.sanu.ac.rs

S. Matović
e-mail: s.babovic@gi.sanu.ac.rs

G. Stanojević
e-mail: g.stanojevic@gi.sanu.ac.rs

N. Ćurčić
e-mail: n.curcic@gi.sanu.ac.rs

L. Soares et al. (eds.), *Compromised Geography*, Key Challenges in Geography,
https://doi.org/10.1007/978-3-032-05201-8_10

Introduction

Over the course of human history, there has been a notable evolution where chronic diseases have increasingly supplanted infectious diseases as the primary contributors to morbidity and mortality. This epidemiological transition is largely due to advancements in sanitation, hygiene, and medical care, alongside demographic changes and lifestyle changes. As societies have developed, the incidence of chronic conditions such as cardiovascular and respiratory diseases, cancer, and diabetes has increased. This rising prevalence imposes substantial demands on global public health infrastructures, necessitating comprehensive strategies to manage and mitigate the impact of these long-term health issues.

Cardiovascular mortality (CVM) represents a significant global and local problem, impacting population worldwide. Globally, cardiovascular diseases (CVDs) account for nearly one-third of all global deaths annually (World Health Organization n.d.), making them the leading cause of mortality worldwide even though it is preventable. Factors such as sedentary lifestyles, unhealthy diets, tobacco use, and increasing rates of obesity contribute to this alarming trend.

Disparities are evident not only between countries but also within them. At a local level, CVM presents complex issues influenced by physical-geographic conditions, socio-economic disparities, access to healthcare, and cultural norms. Certain regions and populations may have a disproportionate burden due to limited resources, inadequate preventive measures, and disparities in healthcare infrastructure.

To assess these differences, there is a growing global trend toward employing spatial approaches in CVM research. Spatial analyses play a crucial role in understanding CVM patterns by examining geographical variations and identifying high-risk areas. These analytical methods enable researchers to map disease hotspots, explore environmental influences, and assess healthcare access disparities. By integrating spatial data with epidemiological models, researchers can uncover spatially dependent risk factors and develop targeted interventions to mitigate cardiovascular health disparities among populations.

This paper aims to explore spatial variation, specifically non-stationarity, in the factors associated with CVM at a subnational (local) level and to predict their potential impact on future CVM trends. The primary objective is to identify compromised areas that require targeted interventions to reduce CVM rates. The findings are intended to inform local public health policies that support targeted approaches for each area, moving beyond national strategies. Addressing CVM effectively necessitates comprehensive strategies encompassing public health initiatives, lifestyle education, equitable healthcare access, and early detection and management of risk factors. Collaborative efforts across global and local scales are crucial to mitigate the burden of CVDs and enhance health outcomes.

Background and Literature Review

According to the World Heart Federation (2023), over 500 million people worldwide still grapple with the impact of cardiovascular diseases, resulting in approximately 20.5 million fatalities in 2021. Furthermore, this staggering figure represents nearly one-third of all global deaths, marking a significant rise from the previously estimated 12.1 million deaths attributed to CVDs. Eurostat data (n.d.) indicated that in 2020, among European countries, Bulgaria led with a rate of 1119.7 per 100,000 population, followed closely by Romania at 918.2, while Serbia occupied the third position with a rate of 810.3 per 100,000 population. Cardiovascular diseases have been the leading cause of death in Serbia for decades with an average of approximately 50,000 deaths annually (Lović Obradović et al. 2024).

CVM is affected by numerous risk factors, both modifiable and non-modifiable. Key modifiable risk factors include hypertension, hyperlipidemia, diabetes, smoking, obesity, physical inactivity, and an unhealthy diet. Non-modifiable risk factors comprise age, gender, and genetic predisposition. Moreover, socio-economic determinants like income, education, and access to healthcare services play a significant role in shaping cardiovascular health outcomes.

Age is a key determinant of health outcomes, with advancing age often correlating with increased susceptibility to various health conditions and mortality. An increased proportion of the elderly population within the total population, along with an unfavorable ratio between the elderly and the younger population, adversely affects CVM. CVD's mortality rates exhibit a notable upward trend increasing the age of the population (Fadah et al. 2022). Lakatta (2002) states that advancing age itself emerges as the paramount risk factor for CVDs, considering a striking statistic in the United States: over 80% of all cardiovascular-related deaths manifest among individuals aged 65 years and older.

Single-member households, as a proxy for social isolation and feelings of loneliness, are often associated with increased vulnerability to adverse health outcomes and diminished overall well-being. Lee and Singh's (2021) findings indicate that adults, who reside alone, are prone to experiencing diminished health and shorter lifespans. Specifically, their study reveals that individuals encountering solitary life exhibit significantly elevated relative risks of both all-cause mortality and mortality due to CVDs in the United States when compared to individuals living with others. Batista-Zamora and Natera-Rivas (2025) argue that women aged 75 and older who live alone face dual vulnerability, as solitary living can negatively impact both physical and mental health. Jensen et al. (2019) utilizing a thirty-year series of data from The Copenhagen Male Study, revealed that living alone emerged as a standalone risk factor for CVM. Living alone, commonly associated with feelings of loneliness, social isolation, lack of emotional support, and potentially poorer physical and mental health outcomes due to reduced social interaction and support systems, may potentially have adverse effects on cardiovascular health outcomes.

Significant life transitions, such as the loss of a spouse through widowhood or divorce, can lead to deterioration in living situations. Individuals who transition from cohabitation to solitary living due to divorce or widowhood may be particularly susceptible to health-related difficulties. Consequently, many studies have confirmed the negative impact of individuals who have experienced divorce or the loss of a partner on CVM (Manfredini et al. 2017; Singh Dhindsa et al. 2020). John Molloy and his colleagues (2009) investigated the correlation between marital status and CVM while considering health behaviors. Their findings indicate that unmarried individuals, including widows and divorcees, irrespective of gender, exhibit a significantly greater inclination toward smoking, excessive alcohol consumption, and heightened psychological distress, compared to married individuals. These factors are indeed primary risk factors for CVDs and mortality.

The influence of income, as a determinant of the socio-economic status of the population, on cardiovascular health is quantifiable and profound. Most of the fatalities attributed to CVDs transpire within the realms of low to middle-income countries (among which is Serbia), encompassing more than three-quarters of the total occurrences (World Health Organization's report 2021). On the other hand, high-income countries have recorded a noticeable trend of declining CVM, although with certain fluctuations in recent years. Lopez and Adair (2019) conducted an evaluation of recent trends in CVDs mortality within high-income populations. Their findings indicate a substantial deceleration in the decline of CVM rates across high-income countries. Furthermore, individuals with lower income bear a significant burden of CVDs, facing heightened event rates and poorer outcomes compared to their higher income counterparts (Wang et al. 2024).

Economic stability can significantly influence an individual's health by enabling access to adequate healthcare, quality nutrition, and other basic necessities. Conversely, economic instability, particularly among the unemployed, may lead to deteriorating health. Several studies have confirmed a positive correlation between unemployment and CVM (Kawachi et al. 2023; Sayed et al. 2022). However, this pattern does not hold in countries with predominantly high living standards, as demonstrated in a study conducted by Dadgar and Norström (2020). The authors suggest that unemployment has a limited capacity to affect trends in CVD mortality.

The level of education attained by individuals has a significant impact on public health, as higher educational attainment is often associated with greater health literacy, access to healthcare resources, adoption of healthier behaviors, and ultimately, improved health outcomes. Extensive scholarly research consistently demonstrates a strong inverse correlation between levels of educational attainment and CVM. In a study conducted by Baptista and Queiroz (2022), the analysis revealed a significant correlation: nations boasting higher income and educational attainment levels demonstrated lower instances of mortality related to coronary heart disease. A significant association between lower educational attainment and heightened risk of CVM was also uncovered, with individuals possessing less than a high school education demonstrating the highest risk (Khan et al. 2023).

Primary health care, facilitated by physicians, plays a crucial role in CVDs management by providing preventive care, early detection, and comprehensive treatment of cardiovascular risk factors. Through regular examination and patient education, primary care physicians contribute significantly to reducing the CVM. To ensure this, it is necessary to provide an adequate number of doctors in accordance with the number of patients affected by CVDs. Basu and colleagues (2019) conducted a study on the changes in the number of primary care physicians across US counties from 2005 to 2015, as well as the associations between these changes and population mortality rates. Their findings revealed a noteworthy correlation: for every increase of 10 primary care physicians per 100,000 populations, there was a corresponding decrease in cardiovascular mortality by 30.4 deaths per million.

Although the aforementioned studies suggested a correlation between associated factors and CVM, they fail to unveil any localized or geographical variations among them. To gain a better understanding of the spatial distribution of CVM rates, a spatially explicit approach is essential for examining the factors that contribute to variations in these rates, as their effects may vary depending on geographical location. This kind of research became increasingly favored as geographical information systems evolved and the accessibility of georeferencing data expanded (Jung et al. 2019). Geographically Weighted Regression (GWR) (Brunsdon et al. 1996) is a valuable analytical tool utilized in detecting spatially varying relationships between phenomena within a given geographic area. It is applicable in analyzing phenomena where the spatial dimension significantly influences the interaction among variables (Lović Obradović and Vojković 2021).

GWR finds extensive application in health studies due to its ability to reveal spatial variability in health outcomes and associated determinants. It is widely used in studies investigating spatial variations in the influence of specific factors on CVM. Lim et al. (2014) confirmed the association between PM10 concentration and CVM, while also revealing spatial variations in the strength of this association across different zones within the Seoul metropolitan area. Although most of Serbia's territory meets national air quality regulations, the entire country surpasses the stricter safety standards set by the WHO (Stanojević et al. 2019; 2023; 2024). Exploring the spatial association between social deprivation and CVM at the neighborhood level in Harris County, Texas, Ford and Highfield (2016) highlighted significant spatial heterogeneity in the relationship between these two variables across the study area. The literature review revealed that GWR presents clear advantages in capturing localized variations among associated variables, facilitating the creation of customized strategies for specific geographic regions. By addressing spatial heterogeneity, GWR enhances comprehension of how relationships between variables vary across different locations. This enables policymakers and practitioners to identify compromised areas and local-specific factors influencing outcomes and design targeted interventions aimed at reducing CVM.

On a global level, but also in Serbia, CVDs and CVM are primarily the focus of medical research. The medical research on CVD risk factors, which is the focus of this study, was conducted by Jović (2016) as part of her doctoral dissertation. The results indicated that CVDs are influenced not only by demographic factors (gender, age, marital status), socio-economic characteristics (education level, employment/work status), and lifestyle, but also by the characteristics of the municipalities (population density, level of local government development, unemployment, etc.) where the population resides. This highlights the importance of the neighborhood environment and underscores the need to incorporate spatial components in CVDs research. Excluding the extensive body of literature on CVM research from a medical perspective, previous studies in Serbia have concentrated on identifying risk factors associated with CVM. For instance, a link between air pollution and CVM has been established (Todorović et al. 2019). The influence of risk factors, such as smoking (Marinković 2017), alcohol consumption (Jakovljević et al. 2004), or biometeorological factors (Arsenović 2014), was also the focus of scientific research. Stanojević et al. (2014a, b) established a link between extreme temperature events and human health in Serbia. Furthermore, their study found that temperature has the most significant impact on CVM in Belgrade, suggesting that the city is a strong candidate for implementing a Heat Health Warning System (Stanojević et al 2014). Research evaluating cardiovascular health (CVH) in relation to demographic and socio-economic factors, as well as smoking, physical activity, diet, BMI, blood pressure, cholesterol, and glucose levels, found a low prevalence of optimal CVH in Serbia (Janković et al. 2019). Radovanović et al. (2023) emphasize the importance of promoting CVH, believing that even minimal improvements can lead to a decrease in CVM.

These studies, conducted at a local or state level, neglect to account for spatial disparities in the researched phenomena, potentially leading to findings that overlook hidden variations across different geographic areas. Consequently, no progress has been made in comprehending how geographical location affects the factors associated with CVM. To bridge this gap, additional research is necessary. In this paper, geospatial approach, more specifically GWR, was leveraged to capture spatial variations between associated factors and CVM.

Methodology

Study Area

The analysis of varying influence of associated factors on CVM in space was conducted within the geographical scope of Serbia. Municipalities, representing the lowest administrative division within the European Union's Nomenclature of Territorial Units for Statistics hierarchy (Local Administrative Units level 1), were selected for study. A total of 168 units were analyzed using data sourced from the Statistical Office of the Republic of Serbia (SORS 2024), with the latest available data from 2023 utilized. Due to variations in population sizes, relative data was utilized for the analysis. Municipalities located in the Autonomous Province of Kosovo and Metohija were not included in the study due to its status under United Nations administration as outlined in Council Resolution No. 1244/99 (United Nations 1999). Furthermore, the most recent accessible data from the SORS dates back to 1998.

Data

The dependent variable in this study is the CVM rate, i.e., the number of deaths from CVDs per 1000 inhabitants of the municipality. The dataset is based on CVDs as the leading factor contributing to mortality coded according to the guidelines of the International Classification of Diseases, tenth revision (ICD-10), which is implemented by the Institute of Public Health of Serbia "Dr Milan Jovanović Batut" (2020) (in collaboration with the World Health Organization). At the municipal level, CVM rates vary widely, ranging from 3.2 to 24.4 per 1000 inhabitants.

Eight explanatory variables were selected based on established correlations with CVM in scientific studies and the data available to the authors at the time of writing. These selected variables can be conditionally categorized into four groups: demographic, psychosocial, socio-economic, and health.

Age, alongside gender, is a fundamental demographic category and serves as a starting point in demographic research. Two variables have been selected: the aging index, which represents the ratio of the elderly population (aged 60 and over) to the young population (aged 0–19), and the proportion of the population aged 65 and over per 1000 inhabitants of the municipality. As indicated in the literature review section, previous research has indicated an increase in CVM rates with advancing age, suggesting that areas with a higher proportion of elderly residents relative to younger or total population may experience higher CVM rates.

The proportion of single-member households among total households and the proportion of divorced and widowed individuals aged 15 and over among the total population aged 15 and over, two variables from the psychosocial factors group, have been identified as significant influencers of CVM. Living alone can be linked to increased feelings of loneliness, social isolation, and reduced social support, which may negatively impact health. The loss of a partner, whether through divorce or death, can represent a significant source of psychological stress. A positive correlation between these two variables is also expected.

Income, employment, and education are indicators of socio-economic status and represent significant predictors of CVM. This study utilized the average gross salary per municipality, the number of unemployed individuals per 1000 inhabitants of the municipality, and the population without schooling or with incomplete primary education per 1000 inhabitants. It is expected that lower average salaries, higher unemployment rates, and lower educational attainment will have a negative impact on CVM.

Finally, the analysis also included the population per doctor ratio as an indicator of healthcare accessibility. A higher population per doctor ratio may be associated with increased CVM rates due to potential barriers in accessing timely medical care, preventive interventions, and management of cardiovascular risk factors. Consequently, a negative correlation between these two variables is expected.

Analytical Strategy

The initial step of the study involved using the Global Moran Index (Anselin 1995) to assess spatial autocorrelation. Based on the obtained results, the null hypothesis, which posits that CVM rates at the municipal level are randomly distributed in space, was either accepted or rejected. The outputs of the analysis include Moran's I index, p-value (probability), and z-score (standard deviation). The p-value and z-score serve as measures of statistical significance, while Moran's I values range from − 1 to + 1. A statistically significant ($p < 0.05$) Moran's I value closer to + 1 indicates stronger spatial clustering of CVM, whereas values closer to − 1 suggest greater dispersion of CVM in space, meaning each municipality has different levels of CVM compared to its neighbors. A Moran's I index equal to zero indicates spatial randomness of CVM, meaning no identifiable spatial pattern is present.

To enhance comprehension of the quantified relationships between a dependent variable and a set of explanatory variables at the state level, this study initially employed Ordinary Least Squares (OLS) linear regression. This model operates under the assumption that the relationships between the dependent and explanatory variables remain consistent across space. It facilitates estimating the coefficient values between variables, which remain consistent in direction and magnitude throughout the studied area. This implies that all selected variables exert uniform influence on CVM across the entirety of Serbia's territory. Significant outputs of this analysis include: R^2 (providing an indication of how well the model fits the data), Akaike Information Criterion ([AIC] indicates the model's performance in explaining the relationship between the dependent and explanatory variable(s)), and the Variance Inflation Factor ([VIF] assesses the degree of multicollinearity or the level of interdependence, among the variables in a model).

After the obtained results, the spatial autocorrelation method (Global Moran I) was applied again for the residuals of the OLS model. The importance of this step in the analysis is reflected in the detection of potential spatial grouping of the residuals (the difference between the observed and the predicted value), which would indicate that this model was not evaluated as the best for assessing the links between associated factors and CVM, which would justify using the GWR model.

Following the previous steps, this study implemented GWR to assess and quantify spatial variations in the relationships between selected explanatory variables (identical to those employed in the OLS model) and CVM rates across each municipality. The basic function of the GWR model is described in Eq. 1 (Brunsdon et al. 1996).

$$Y_i = \beta_0(_i) + \beta_1(_i)X_{1i} + \beta_2(_i)X_{2i} \cdots + \beta_n(_i)Xn_i + \varepsilon_i \tag{1}$$

where Y_i denotes CVM rate in municipality, with a center determined by coordinates, $\beta_0(_i)$ is the local intercept for municipality $_i$, and $\beta_n(_i)$ represents the local coefficient for predictor n for municipality $_i$. By integrating location data into the GWR model, it becomes feasible to establish a unique regression equation for each municipality, enabling the calculation of regression coefficients. These coefficients signify the impact of specific explanatory variables on CVM within each locality (the larger coefficient, the stronger relationship), while the sign indicates the nature of correlation (positive or negative).

R^2 is another significant component of the obtained results, pointing to goodness-of-fit for each local regression model. Ranges from 0 to 1; values approaching 1 indicate superior model performance (larger proportion of the variability in the dependent variable is explained by the explanatory variables).

Predicted values represent estimated CVM rates for each municipality based on selected explanatory variables and their spatial variations. These values depict the expected CVM rates derived from the GWR model, accounting for local spatial variations and effects that may vary across municipalities in Serbia. The significance of this parameter lies in its ability to assess changes in CVM rates across space and identify municipalities with high or low values compared to modeled predicted values. As this measure diminishes, the alignment between the GWR model and the observed data becomes more precise. In our analysis, we focus on observations characterized by standardized residuals exceeding 2 in the positive direction, indicative of model underprediction, or falling below -2 in the negative direction, signifying model overprediction. These criteria identify instances where the model deviates significantly from observed values, providing insights into areas of potential model inadequacy or misfit.

Ultimately, a statistical determination was reached utilizing a significance threshold of $p < 0.05$, accompanied by a 95% confidence interval, for both OLS and GWR. The regression coefficients, R^2, and predicted values are also visually represented on maps to provide a clearer understanding of their spatial dimension. All analyses were performed in ArcGIS Pro (Version 3.1).

Results and Discussion

Global Moran

The Global Moran's I analysis reveals statistically significant spatial autocorrelation for all variables, as indicated by *p*-values below 0.05. This suggests that the distribution of variables across the study area is not random, exhibiting clear patterns of clustering or dispersion, thereby leading to the rejection of the null hypothesis. The *Z*-score values demonstrate that, with the exception of the Population per doctor variable, which is dispersed, all other variables show a tendency to cluster. The greatest degree of grouping is observed in the divorced and widowed variable.

OLS

The results of the OLS model indicate that all eight explanatory variables are statistically significant at the 0.05 confidence level. The R^2 value reveals that 55% of the variation in CVM is accounted for by the selected explanatory variables. Among these variables, "Population per Doctor" has the smallest effect on CVM, while "Unemployment" exhibits the most significant impact on CVM. VIF values were assessed with a threshold of 7.5. The aging index exhibited a VIF value higher than 7.5, indicating potential multicollinearity issues. Considering that, the aging index was excluded from further analysis to avoid multicollinearity and improve model reliability.

Global Moran's I analysis of the residuals from the OLS model revealed that they are not randomly distributed ($p < 0.05$) but exhibit spatial clustering across the study area. This suggests that the OLS model is not the best-fitting model for identifying spatial variations in the relationship between the explanatory variables and CVM rate at the municipality level in Serbia. The remainder of the paper discusses the results of the GWR model, the application of which is justified by these findings.

GWR

The performance of the GWR model, compared to the OLS model, is notably improved. The selected variables account for a greater degree of variation in the CVM rate at the municipal level, as evidenced by higher R^2 values and lower AICc values. The R^2 values are not uniform across the study area. Results indicate that the model with the highest explanatory capacity is located in the northeastern regions of Serbia, where it accounts for 75–81% of the variability in the CVM rate through the selected independent variables. Conversely, the explanatory power of the model diminishes progressively toward the southwestern and southeastern regions of the country, ranging between 62 and 65%. Areas with lower R^2 values suggest that the model could be improved by incorporating more explanatory variables (Fig. 1).

Spatial heterogeneity in the relationships between selected variables and CVM rates is displayed on 2D maps (Fig. 2). These maps clearly show that the relationships are not constant but vary spatially.

The spatial variation in the relationship between CVM and the proportion of the population aged 65 and over across different locations in Serbia is evident. The estimated coefficients are positive, suggesting that a higher proportion of older population corresponds to higher CVM values. The range of coefficients from 0.72 to 5.38 signifies spatial variation in the strength of this relationship, with the highest positive coefficient observed in western Serbia and the lowest in the northeastern part. This spatial variation suggests localized factors influencing the relationship. These findings are consistent with those of Sidney et al. (2019), who reported that the increase in the U.S. population of adults aged 65 years and older paralleled an increase in the

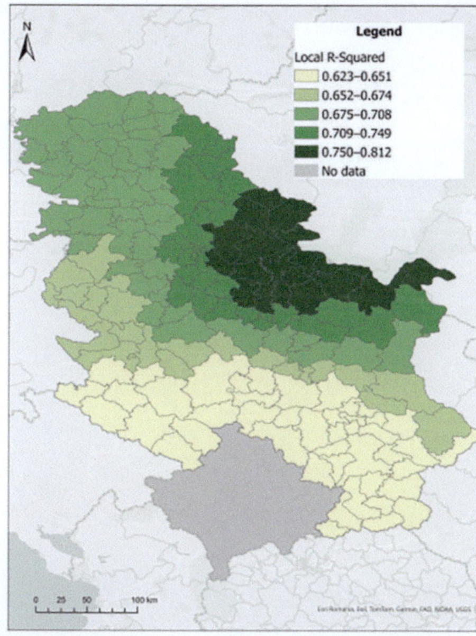

Fig. 1 Spatial variation of the model's explanatory power

number of deaths attributed to CVDs, with the majority occurring within this age group. The predictor seniors proved to have the most significant influence on the CVM rate, with the highest estimated coefficient value. Given that, at the national level, with the exception of a few oases, there has been an increase in the proportion of those over 65 years old in the total population over the years (Lović Obradović et al. 2016; Stojilković Gnjatović 2023), it can be expected that this factor will play the most significant role in the variation of CVM. Demographic trends, particularly the aging population, are expected to have a substantial impact on CVM rates.

The link between single-member households and the CVM rate demonstrates a varying trend, shifting from negative to positive. A coefficient with a negative sign, indicating an inversely proportional relationship between these two variables, was identified in municipalities in the central and western part of the country. Low values of the index indicate weak connection strength. Nevertheless, an increase in the share of single-member households in the total number of households can lead to a reduction in the CVM rate, and vice versa. This result is completely surprising considering that research has demonstrated that being single heightens the probability of encountering social isolation and sensations of loneliness, both of which have emerged as contributory factors to overall mortality risk, particularly impacting cardiovascular health (Abell and Steptoe 2019). A study involving just under half a million participants documented similar results. Specifically, Gan et al. (2021) aimed to determine whether living alone might increase the likelihood of cardiovascular events and found a significant association: living alone appears to heighten the risk of cardiovascular outcomes.

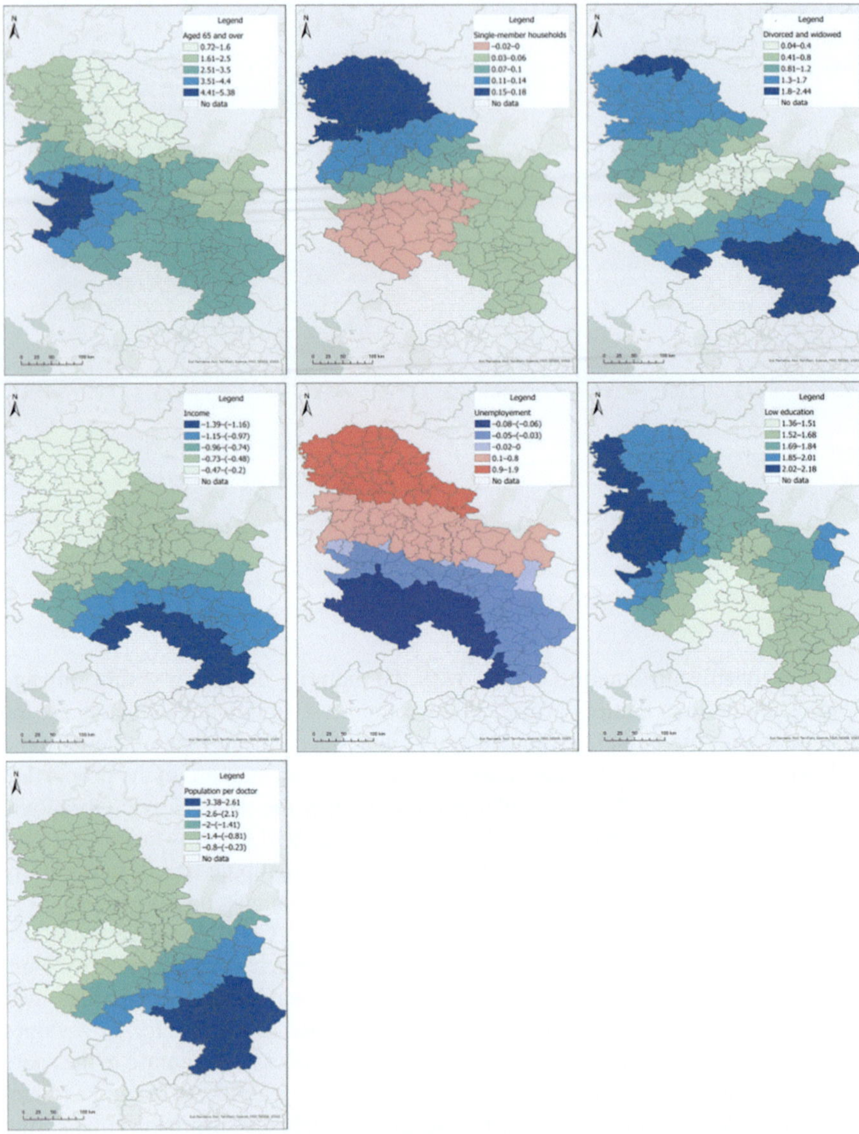

Fig. 2 Spatial variations in the GWR model's relationship between explanatory variables and CVM

Most municipalities exhibit a proportional relationship, as indicated by positive coefficient values. The most significant impact is observed in the northern municipalities of the country, where the highest recorded estimated coefficient values are clustered. This suggests that solitary living has a more pronounced impact on the population in this part of Serbia, leading to a significant increase in deaths from

CVDs compared to others. An observation worth noting is the prominence of single-person households in this part of Serbia in contrast to their prevalence in central Serbia (Bobić 2015).

A positive association between CVM and the proportion of divorced individuals and widows has been identified. The obtained results are consistent with the findings of Jović (2016), who found that the risk of CVDs of the population in Serbia is higher (30.9%) for participants living without a partner (unmarried/divorced/widowed) compared to those living with a partner. The regression coefficients are positive, suggesting that an increase in the proportion of divorced individuals and widows leads to a rise in CVM rates. These effects are least pronounced in the central part of the country, gradually increasing toward the north and south in a nearly symmetrical manner. Particularly high impacts were observed in municipalities in the south and southeast, while municipalities in the southwest and north-northeast were identified as clusters with isolated impacts. Indeed, in the northeastern and east-southeastern parts of the country, the proportion of divorced and widowed individuals in the total population is among the highest. On average, one-fifth of the population falls into this category (SORS 2023a).

Across Serbia, the estimated coefficients illustrating the impact of income on CVM demonstrate spatial variability, ranging from − 1.39 in the south and south-west to − 0.2 in the west and northwest. This spatial variation underscores the nuanced nature of the income–CVM relationship, highlighting differing degrees of influence across regions. A negative sign indicates their inversely proportional relationship. Municipalities in the south and southeast of our country, where the highest negative regression coefficients are found, also have the lowest average incomes. This part of southeastern Serbia is one of the most economically disadvantaged areas in the country, and high unemployment rates are prevalent (Matović and Lović 2022). The region has long been marked by underdevelopment (Miletić 2022) and has recently experienced substantial depopulation (Milošević et al. 2010, 2011; Panić et al. 2022). Therefore, the unfavorable economic status of the population makes these areas highly compromised zones for CVM.

The relationship between the variable of Unemployed and the CVM rate exhibits a sign-changing pattern since the regression coefficients vary within the range of − 0.08 to 1.9. Positive coefficient values in the northern, as well as parts of the western, central, and eastern regions of the country, indicate a direct relationship between the share of employees in the total population and the CVM rate. This means that in the municipalities within these areas, an increase in the share of the unemployed will lead to an increase in the CVM rate. Radević et al. (2018) examined the influence of socio-economic inequalities on non-communicable diseases in Serbia, emphasizing notable health disparities linked with unemployment. Their findings highlighted that unemployed individuals generally reported poorer health statuses compared to their employed counterparts. Additionally, the study noted that the unemployed utilized health services less frequently, including visits to general practitioners or occupational medicine specialists, and had limited access to hospital treatment. Conversely, in other parts of the country, negative values of the regression coefficient suggest an inversely proportional relationship between these two variables. Although the negative values are small, they imply that a certain decrease in the share of the unemployed may lead to an increase in the CVM rate.

The outcomes derived from the GWR analysis present compelling evidence of a significant positive correlation between CVM and low education across diverse spatial units within Serbia. The estimated coefficients, ranging from 1.36 to 2.18, delineate spatially variant magnitudes of this positive relationship. Our study corroborates the findings of Baptista and Queiroz (2022), as well as Khan et al. (2023), both of whom also observed a significant positive association between educational attainment and cardiovascular mortality. The concentration of higher positive coefficients in the peripheral regions of western and northwestern Serbia can be noted, indicative of the strongest association between education attainment and CVM in these locales. Serbia is making concerted efforts to transform its education system by addressing existing challenges, implementing reforms, and improving access to quality learning opportunities (Matović et al 2024). The research by Anđelković-Stoilković (2016) showed that border and small municipalities have lower educational attainment than the average of Serbia. Since in most cases the highest values of the estimated coefficients are detected in small border municipalities with a high share of the population with primary education in this population cause doubts whether these border municipalities can count on education as a development resource that can lead to a decrease in the CVM. Conversely, within the central region of Serbia, coefficients exhibit comparatively lower positive values, implying somewhat weaker linkage between education and CVM.

The variable "Population per doctor" demonstrated a negative association with the CVM rate across the entire study area. A well-organized healthcare system is crucial in combating CVM, as it ensures timely access to preventive care, effective management of risk factors, and comprehensive treatment options. Equally important is the role of the individual. The lowest negative coefficient values were observed in western Serbia, suggesting that municipalities in this region would experience the lowest increase in CVM rate if the number of doctors decreases. Conversely, changes in the number of doctors would have the highest impact on the CVM rate in southeastern Serbia. The medical center of this part of Serbia is the City of Niš, which serves patients from the entire region. An event from 2023, relevant to this research, is the elimination of the waiting list for patients awaiting cardiovascular surgery. At the start of the year, 149 patients were on the waiting list. The doctors at Niš Cardiosurgery performed surgeries voluntarily for 12 consecutive weekends. It is important to note that the mortality rate, including emergency cases, remained unchanged, with no patient deaths during these operations (Tanjug 2023). Consequently, it can be expected that the highest negative regression coefficient values will decrease in the future.

The predicted values of the CVM for each municipality, considering selected independent variables and local spatial variations, range from 3.5 to 16.6. The lowest predicted values are associated with large urban centers with over 100,000 inhabitants (SORS 2023b), such as the regional centers Novi Sad, Niš, and most municipalities within the capital city of Serbia, Belgrade (Lukić et al. 2023). Additionally, municipalities in southwestern Serbia (Novi Pazar, Sjenica, and Tutin) and southern Serbia (Bujanovac and Preševo) belong to this group. These municipalities notably have a higher proportion of young population compared to the national average (Penjišević 2012; Penjišević and Nikolić 2011). Higher predicted values are found in the eastern and southeastern parts of Serbia. Babović et al. (2016) highlighted that this specific area has a significant elderly population.

Differences between predicted and estimated values of CVM provide insight into how well the GWR model forecasts actual CVM values based on available independent variables, indicating the model's accuracy and reliability. Small differences suggest that variations in CVM can largely be explained by the selected associated factors. This is evident in a large number of municipalities (133, or 79.2%). The smallest differences, just 0.1, were observed in the municipalities of Bujanovac (south of Serbia) and Vršac (northeastern Serbia). This finding is intriguing given the different population sizes, age structures, CVM rates, etc., in these two municipalities. In 16 municipalities (9.5%), marked with light red, CVM rates are underestimated or overestimated (19, or 11.3%). These larger discrepancies indicate the need to incorporate additional variables in these municipalities to enhance the explanation of local variations in CVM rates (Fig. 3).

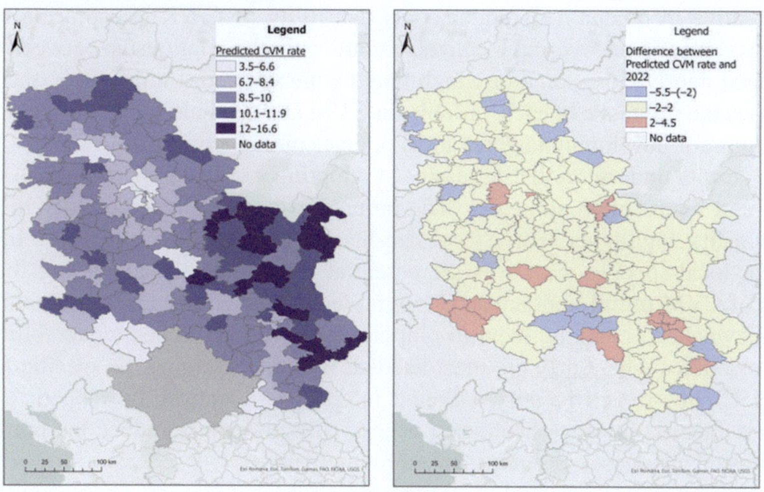

Fig. 3 Predicted CVM rate (left) and difference between predicted CVM rate and 2022 (right)

Conclusion

CVMs are becoming an increasingly prevalent issue worldwide, posing a significant threat to human health. In order to improve population health, it is crucial to identify compromised areas where this issue is more pronounced compared to others, as well as the impact of associated factors. From a geographical standpoint, integrating spatial determinants into CVM research would greatly enhance insights into both spatial patterns and the underlying causes shaping these patterns.

The findings of this study provide crucial insights into the spatial dynamics of CVM in Serbia. By employing the GWR model, significant spatial variations in the relationships between CVM and associated demographic, socio-economic, and health-related factors across 168 municipalities were identified. It was found that there is a positive relationship between variables: aged 65 and over, divorced and widowed and education on one side and CVM on another. This suggests that an increase in the population with these demographic characteristics will contribute to a rise in CVM. In contrast, income and population per doctor exhibit a negative relationship with CVM, indicating an inverse proportionality between these independent variables and the dependent variable. The findings align with previous studies, with the exception of two factors—unemployment and single-member households—where the direction of impact varies across municipalities, being positive in some and negative in others. This underscores the utility of GWR in capturing spatial heterogeneity and uncovering localized variations in the relationship between CVM and social deprivation, some of which may be unexpected. Furthermore, the study quantifies the influence of each variable on CVM at the municipal level, providing a foundation for developing targeted local policies aimed at mitigating CVM.

The results underscore the necessity of adopting localized approaches to health policy and intervention strategies. The pronounced spatial heterogeneity in CVM determinants highlights that a one-size-fits-all strategy is inadequate. Instead, tailored, local-specific initiatives are essential for effectively addressing the root causes of CVM in different areas. Cartographic visualizations offer a valuable tool for policymakers, enabling them to pinpoint areas of heightened vulnerability and direct resources where they are most needed. By understanding the specific factors driving CVM in each locality, targeted interventions can be developed to mitigate risk factors, improve healthcare access, and ultimately reduce the burden of cardiovascular diseases in compromised municipalities.

However, there were several sources of limitations in this study. The original intention to include a broader array of associated factors was constrained by data limitations. Specifically, climatological variables, such as the influence of temperature fluctuations on CVM, were considered. Although data on mean annual temperatures were accessible to the researchers, their minimal variation across extreme values led to their exclusion from the analysis to avoid potential bias in the findings.

Given that this research was conducted at the municipal level, there is a possibility that disparities between urban and rural areas may be masked, considering distinct variations in population age structure, health behaviors, access to healthcare services, etc. Future research should continue to explore these spatial patterns and refine intervention approaches encompassing a broader array of associated factors to attain a more nuanced understanding of CVM.

Acknowledgements This research was funded by the Ministry of Science, Technological Development, and Innovation of the Republic of Serbia, contract numbers 451-03-136/2025-03/200172.

References

Abell JG, Steptoe A (2019) Living alone and mortality: more complicated than it seems. Eur Heart J Qual Care Clin Outcomes 5(3):187–188

Anđelković-Stoilković M (2016) Can border municipalities count on education as a resource development? J Geogr Inst "Jovan Cvijić" SASA 66(3):433–448

Anselin L (1995) Local indicators of spatial association—LISA. Geogr Anal 27(2):93–115

Arsenović D (2014) Uticaj temperature vazduha na sezonalnost mortaliteta u Novom Sadu (doktorska disertacija). Prirodno-matematički fakultet Univerziteta u Novom Sadu

Babović S, Lović Obradović S, Prigunova I (2016) Depopulation of villages in southeastern Serbia as hindrance to economic development. J Geogr Inst "Jovan Cvijić" SASA 66(1):61–74

Baptista EA, Queiroz BL (2022) Spatial analysis of cardiovascular mortality and associated factors around the world. BMC Public Health 22:1556

Basu S, Berkowitz SA, Phillips RL, Bitton A, Landon BE, Phillips RS (2019) Association of primary care physician supply with population mortality in the United States, 2005–2015. JAMA Intern Med 179(4):506–514

Batista-Zamora AE, Natera-Rivas JJ (2025) Female senior one-person households in Malaga municipality (Spain). Temporal evolution and intraurban spatial distribution. J Geogr Inst "Jovan Cvijic" SASA 75(1):153–165

Bobić M (2015) The households of Serbia at the dawn of third millennium—socio-demographical analysis. Sociologija XLVI(4):349–372

Brunsdon C, Fotheringham AS, Charlton ME (1996) Geographically weighted regression: a method for exploring spatial nonstationarity. Geogr Anal 28(4):281–298

Dadgar I, Norström T (2020) Is there a link between cardiovascular mortality and economic fluctuations? Scand J Public Health 48(7):770–780

Eurostat (n.d.) Causes of death—diseases of the circulatory system, residents, 2020

Fadah K, Hechanova A, Mukherjee D (2022) Epidemiology, pathophysiology, and management of coronary artery disease in the elderly. Int J Angiol 31(4):244–250

Ford MM, Highfield LD (2016) Exploring the spatial association between social deprivation and cardiovascular disease mortality at the neighborhood level. PLoS ONE 11(1):e0146085

Gan T, Yang J, Jiang L, Gao Y (2021) Living alone and cardiovascular outcomes: a meta-analysis of 11 cohort studies. Psychol Health Med 1–13

Institute of Public Health of Serbia Dr Milan Jovanović Batut, World Health Organization (2020) Međunarodna statistička klasifikacija bolesti i srodnih zdravstvenih problema, Deseta revizija [International statistical classification of diseases and related health problems, Tenth revision]

Janković J, Davidović M, Bjegović-Mikanović V, Janković S (2019) Status of cardiovascular health in the republic of Serbia: results from the national health survey. PLoS ONEe 14(3):e0214505

Jakovljević B, Stojanov V, Paunović K, Belojević G, Milić N (2004) Alcohol consumption and mortality in Serbia: twenty-year follow-up study. Croat Med J 45(6):764–768

Jensen MT, Marott JL, Holtermann A, Gyntelberg F (2019) Living alone is associated with all-cause and cardiovascular mortality: 32 years of follow-up in the Copenhagen Male Study. Eur Heart J Qual Care Clin Outcomes 5(3):208–217
John Molloy J, Stamatakis E, Randall G, Hamer M (2009) Marital status, gender and cardiovascular mortality: behavioural, psychological distress and metabolic explanations. Soc Sci Med 69(2):223–228
Jović DP (2016) Višestepena analiza faktora povezanih sa obolevanjem od hroničnih nezaraznih bolesti [Multilevel analysis of factors associated with noncommunicable diseases morbidity]. Doctoral Dissertation
Jung M, Ko W, Cho Y (2019) Spatial variations in fertility of South Korea: a geographically weighted regression approach. Int J Geo-Inform 8:262
Kawachi I, Kyriopoulos I, Vandoros S (2023) Economic uncertainty and cardiovascular disease mortality. Health Econ 32(7):1550–1560
Khan N, Javed Z, Acquah I, Hagan K, Khan M, Valero-Elizondo J, Chang R, Javed U, Taha MB, Blaha MJ, Virani SS, Sharma G, Blankstein R, Gulati M, Mossialos E, Hyder AA, Achirica MC, Nasir K (2023) Low educational attainment is associated with higher all-cause and cardiovascular mortality in the United States adult population. BMC Public Health 23(1):900
Lakatta EG (2002) Age-associated cardiovascular changes in health: impact on cardiovascular disease in older persons. Heart Fail Rev 7(1):29–49
Lee H, Singh GK (2021) Social isolation and all-cause and heart disease mortality among working-age adults in the United States: the 1998–2014 NHIS-NDI record linkage study. Health Equity 5(1):750–761
Lim YR, Bae HJ, Lim YH, Yu S, Kim GB, Cho YS (2014) Spatial analysis of PM10 and cardiovascular mortality in the Seoul metropolitan area. Environ Health Toxicol 29:e2014005
Lopez AD, Adair T (2019) Is the long-term decline in cardiovascular-disease mortality in high-income countries over? Evidence from national vital statistics. Int J Epidemiol 48(6):1815–1823
Lović Obradović S, Vojković G (2021) The impact of spatial-demographic disparities on fertility variations in Serbian municipalities—an example of hidden content research. Hidden Geogr 379–399
Lović Obradović S, Matović S, Shpak N (2016) Serbia and Russia on the demographic map of Europe two decades after the fall of communism. Trames 20(70/65):1, 59–73
Lović Obradović S, Matović S, Stanojević G, Ćurčić N (2024) Leveraging geospatial technologies to capture spatial variation between associated factors and cardiovascular mortality. In: EUROGEO conference "Compromised geography: spreading a new world", pp 58–59
Lukić V, Lović Obradović S, Stojilković GJ (2023) Did the COVID-19 pandemic change internal rural migration patterns in Serbia? Erdkunde 77(3):233–249
Manfredini R, De Giorgi A, Tiseo R, Boari B, Cappadona R, Salmi R, Gallerani M, Signani F, Manfredini F, Mikhailidis DP, Fabbian F (2017) Marital status, cardiovascular diseases, and cardiovascular risk factors: a review of the evidence. J Womens Health 26(6):624–632
Marinković I (2017) Smoking as the main factor of preventable mortality in Serbia. Stanovništvo 55(1):87–106
Matović S, Lović OS (2022) Assessing socio-economic vulnerability aiming for sustainable development in Serbia. International J Sustain Dev World Ecology 19(1):27–38
Matović S, Demirović Bajrami D, Gajić T, Lakić M, Martinov D, Šuput S, Radosavac A, Đervida R (2024) Beyond numbers: challenges in measuring SDG4 targets—Serbia's perspective. Sustainability 16(22):10006
Miletić R (2022) Regional specialization in Serbia during the period 2001–2015. J Geogr Institute "Jovan Cvijić" SASA 72(1):67–83
Milošević M, Milivojević M, Ćalić J (2010) Spontaneously abandoned settlements in Serbia—part 1. J Geogr Inst Jovan Cvijić SASA 60(2):39–57
Milošević M, Milivojević M, Ćalić J (2011) Spontaneously abandoned settlements in Serbia—part 2. J Geogr Inst Jovan Cvijić SASA 61(2):26–35

Panić M, Drobnjaković M, Stanojević G, Kokotović Kanazir V, Doljak, D (2022) Nighttime lights—innovative approach for identification of temporal and spatial changes in population distribution. J Geogr Inst "Jovan Cvijić" SASA 72(1):51–66

Penjišević I (2012) The directions of regional development of Raška Municipality. J Geogr Inst Jovan Cvijić SASA 62(2):33–48

Penjišević I, Nikolić MM (2011) Changes in population dynamics of Raška region. J Geogr Inst Jovan Cvijić SASA 61(3):81–94

Radević S, Radovanović S, Djonović N, Simić Vukomanović I, Mihailović N, Janićijević K, Kocić S. (2018) Socioeconomic inequalities and non-communicable diseases in Serbia: national health survey. Vojnosanit Pregl 75(9): 926–934

Radovanovic M, Jankovic J, Mandic-Rajcevic S, Dumic I, Hanna RD, Nordstrom CW (2023) Ideal cardiovascular health and risk of cardiovascular events or mortality: a systematic review and meta-analysis of prospective studies. J Clin Med 12(13):4417

Sayed A, Aldosoky W, Khalid N, Diab M, Elsaiey A, Dar T, Abohashem S (2022) Impact of unemployment on cardiovascular mortality in united states: a nationwide county-level analysis 2010–2019. Eur Heart J 43(2):ehac544.2250

Sidney S, Go AS, Jaffe MG, Solomon MD, Ambrosy AP, Rana JS (2019) Association between aging of the US population and heart disease mortality from 2011 to 2017. JAMA Cardiol 4(12):1280–1286

Singh Dhindsa D, Khambhati J, Schultz WM, Tahhan AS, Quyyumi AA (2020) Marital status and outcomes in patients with cardiovascular disease. Trends Cardiovasc Med 30(4):215–220

Stanojević G, Miljanović D, Doljak D, Ćurčić NB, Radovanović MM, Malinović-Milićević S, Hauriak O (2019) Spatio-temporal variability of annual PM2.5 concentrations and population exposure assessment in Serbia for the period 2001–2016. J Geogr Inst "Jovan Cvijić" SASA 69(3):197–211

Stanojević G, Malinović-Milićević S, Ćurčić NB, Radovanović M, Radivojević A, Popović T, Ćurčić S (2023) An assessment of the multidimensional drivers and determinants of public risk perception of and behaviors related to exposure to air pollution in Serbia. Sustainability 15(24):16901

Stanojević G, Spalević A, Kokotović V, Stojilković J (2014a) Does Belgrade (Serbia) need heat health warning system? Disaster Prev Manag 23(5):494–507

Stanojević G, Stojilković J, Spalević A, Kokotović V (2014b) The impact of heat waves on daily mortality in Belgrade (Serbia) during summer. Environ Hazards 13(4):37–41

Stanojević G, Malinović-Milićević S, Brđanin E, Milanović M, Radovanović MM, Popović T (2024) Impact of domestic heating on air pollution—extreme pollution events in Serbia. Sustainability 16(18):7920

Statistical Office of the Republic of Serbia (2024) Age and sex. Data by settlements

Statistical Office of the Republic of Serbia (n.d.-a) Municipalities in the Republic of Serbia (2005–2008), Municipalities in Serbia (2009–2010), Municipalities and regions in the Republic of Serbia (2011–2022)

Stojilković Gnjatović J (2023) Starosna struktura stanovništva Srbije–koje su promene zabeležili popisi 2002, 2011. i 2022? Sociološki pregled 57(3):792–822

Tanjug (2023) Nakon 100 dana rada bez prekida Kardiohirurgija UKC Niš nema više liste čekanja

Todorović MN, Radenković MB, Rajšić SF, Ignjatović LJM (2019) Evaluation of mortality attributed to air pollution in the three most populated cities in Serbia. Int J Environ Sci Technol 16:7059–7070

United Nations (1999) Resolution 1244

Wang T, Li Y, Zheng X (2024) Association of socioeconomic status with cardiovascular disease and cardiovascular risk factors: a systematic review and meta-analysis. J Public Health (Berl) 32:385–399

World Health Organization (2021) Cardiovascular diseases (CVDs)

World Heart Federation (2023) World heart report 2023: confronting the world's number one killer. Geneva, Switzerland

Landscape as Heritage: Teaching and Learning Essays Using Geospatial Technologies

Laura Soares, Elsa Pacheco, and Paulo Lemos

Abstract Landscape is one of the concepts whose definition—non-consensual and variable in the epistemological context of Geography—is assumed to be the visible and perceived space, constituting the tangible expression of the dynamic combination of physical, biological and anthropic elements over time. In this sense, it brings together the past and present, revealing the history of socio-economic, political and cultural transformations that occurred in conjunction with environmental characteristics. Thus, it integrates a legacy bequeathed by the societies that built it and supports current changes. The main objective of this paper is to present and discuss work projects developed by Portuguese Geography students from two levels of education—basic/secondary and Geography undergraduate studies—based on the perspective of landscape as heritage. The unifying idea of all the works lies in a compromised/integrated perspective of natural and cultural heritage, which we must promote and preserve to enhance the appreciation and development of territories. Project goals and methods have been adapted to meet specific characteristics and needs of each educational level, particularly concerning the selection of appropriate geospatial technologies.

Keywords Heritage education · Landscape · Geotechnologies · Active learning methodologies

---

L. Soares (✉)
Faculty of Arts and Humanities, Geography and Spatial Planning Research Centre (CEGOT), University of Porto, Porto, Portugal
e-mail: lauras@letras.up.pt

E. Pacheco
Faculty of Arts and Humanities, Transdisciplinary Research Center for Culture, Space and Memory (CITCEM), University of Porto, Porto, Portugal
e-mail: elsap@letras.up.pt

P. Lemos
Geography and Spatial Planning Research Centre (CEGOT), Porto, Portugal
e-mail: paulomclemos@outlook.com

L. Soares et al. (eds.), *Compromised Geography*, Key Challenges in Geography,
https://doi.org/10.1007/978-3-032-05201-8_11

Introduction

Education for the twenty-first century is one of the most discussed topics in today's Education System. Its foundations date back to the UNESCO report co-ordinated by Delors (1996), to which we can associate, more recently, the 2030 Agenda for Sustainable Development, namely its fourth goal, which aims to "[to] ensure inclusive and equitable quality education and promote lifelong learning opportunities for all" (United Nations 2020, p. 32). If, in the words of Nelson Mandela (2003), 'education is the most powerful weapon which you can use to change the world',[1] we can assume that only a quality education can guarantee the fulfilment of all the other SDGs (Soares et al. 2023).

So, the legislation and regulations of the Portuguese Education System (ES), at the level of compulsory education (12 years), have been experiencing changes with the need to modify/adapt curricula and define contents that respond to the problems faced by society. Teaching–learning theories must put students at the centre of the educational process—defending their autonomy in knowledge, skills and attitudes 'construction' (Constructivism)—and should apply (geo)technologies since they are part of our daily lives and must be regarded as essential teaching resources. (PASEO 2017; Decretos-Lei n.º 54 e 55 de 2018a, b; Despacho n.º 5908/2017; Portaria n.º 181/2019). But always reflecting on 'why, how and when' to use them: although they are essential to prevent schools from entering into a legitimacy crisis in the face of the technological environment that prevails in society (Lynch et al. 2013; Whalley et al. 2011; Lejla et al. 2014), (geo)technologies should be seen as support tools and not as the goal in themselves (Puertas-Aguilar et al. 2021).

In addition, some authors consider that if education is increasingly dependent on the digital world, the use of technology in schools is something that is applied but without much reflection. To this are added socio-cultural, political and economic difficulties in the ES (Selwyn 2017; Walshe and Healy 2021), as well as the difficulty of controlling data and information sources that students use, given their constant reliance on social networks (Hepburn 2018). However, if correctly selected and controlled, they can be positively incorporated (Williams 2022). Portugal's regulatory framework also includes the National Strategy for Citizenship Education (ENEC 2017), which defines the areas in which students should develop skills "(…) that enable them to intervene in the life and history of individuals and societies, make free and informed decisions on natural, social and ethical issues, and have the capacity for developing active, conscious and responsible civic participation". (PASEO 2017, p. 10, translated).

In this context, projects at a local scale must be favoured. These can be made through *Place Based Learning* (Stevenson 2008; Gruenewald 2008; Van Eijck and Roth 2010; Zandvliet 2012; Johnson et al. 2020)—in which the problems and realities of the student's daily life are defined and discussed—*Problem Based Learning* (De

[1] Mandela speech made in July 2003 at the inauguration of the *Mindset Network*, a non-profit organisation founded to improve education and health in Africa.

Graaf and Kolmos 2003; Loyens e 2008; Cachinho 2012; Yew and Goh 2016)—fostering meaningful learning that is difficult to achieve when abstract thematic content is introduced, since it is decontextualised from real life, proposing solutions through project work; and *Project-Based Learning* (Ravitz 2008; Tamim and Grant 2013; Kokotsaki et al. 2016)—which then is presented to the school community and the surrounding society. In other words, integrated teaching–learning methodologies/strategies are advocated, which involve the "(…) plural and responsible participation of everyone in the construction of themselves as citizens and of fairer and more inclusive societies, within the framework of democracy, respect for diversity and the defence of human rights" (ENEC, ob, cit., p.3, translated).

These norms and legislation of the Portuguese ES are combined with the Essentials Learning (EL), guiding documents for the knowledge/skills and attitudes that all students should learn in each discipline by year of schooling, thus establishing the contents/concepts to be apprehended in the various disciplinary areas. Within the scope of the themes and sub-themes included in the Geography EL, concepts like *Landscape* and *Heritage* are directly or indirectly involved in the knowledge, skills and attitudes considered, combined with the strategic teaching actions suggested to comply with PASEO.

In the context of Higher Education, within the framework of bachelor's and master's degrees, despite the differences between the Geography Study Plans of each University, we can also consider that the two concepts targeted by our study are always present in the two major branches into which this science is artificially subdivided: Physical Geography and Human Geography. Both cover subjects that allow the specialisation into sub-areas to simplify the complexity of our study object: the Earth in its inextricable interactions between natural, cultural and social environments.

While we aim to avoid an overly theoretical stance, it is important to justify the choice of our study theme—landscape as heritage—and to define the relevant concepts before we present and discuss the compromised Geography research conducted by our students.

Landscape as Heritage: Integrated Perspective

If we think about the meaning of 'heritage' from a juridical perspective, we will enter a discussion that takes on different contours and meanings, questioning "(…) whether heritage should be faced as a set of legal positions or rather as a set of belongings (…) in which case [heritage] is regarded in a material or natural sense" (Sequeira 2013: 298). However, the most common meaning in Family Law is associated with the idea of the possession of an object or a set of goods seen as an inheritance from the 'father' transmitted to his descendants. It can also be seen from a more global point of view as "(…) the inheritance of a community in its multidimensionality (…) a set of cultural assets, material or immaterial, which have passed down from generation to generation (…) together with a proportional awareness to conserve and value them" (Martins 2006, n.p.). Regardless of the type of heritage, natural or cultural, material or immaterial, it is in the landscape that they are best expressed, creating what we can call "visible or perceived space" which Wagner and Mikesell (1962, p. 36) consider being "(...) a concrete and characteristic product of the complicated interaction between a particular human community, encompassing specific cultural preferences and potentials, and a specific set of natural circumstances [constituting] a legacy of a long period of natural evolution and many generations of human endeavour".

It is in this sense that we can speak of "Landscape as Heritage". In other words, a landscape that results from the articulation between nature and culture, illustrating their evolution over time, mutually dependent on the influence of biogeophysical constraints or opportunities and successive socio-economic, cultural and technological interventions (Domingues 2001). This is in line with the concept of "Integrated Cultural Landscape", as envisaged by Panizza and Piacente (2003), which we have chosen to call "Landscape as Geocultural Heritage" (Soares et al. 2018).

Several other definitions can be mentioned emphasising the importance of promoting and protecting heritage, such as the *World Heritage Convention of 1972*. This convention, which in 1994 presented the *Global Strategy for the definition of the World Heritage List*, incorporates the world's cultural and natural diversity of outstanding universal value. They are a source of identity and socio-cultural cohesion, requiring a holistic governance system that can promote the development of the communities that inhabit different areas. This development is associated with a wide range of activities linked to (geo)tourism, which, by exploiting the values of (geo)heritage, means that it can be seen as a natural, cultural and economic resource and used as such (Reynard 2005, 2007, 2008; Brilha and Carvalho 2010). Similarly, the International Charter on Cultural Tourism, adopted by the International Council on Monuments and Sites (ICOMOS 2002:4), considers that

> (…) Heritage is a broad concept and includes the natural as well as the cultural environment. It encompasses landscapes, historic places, sites and built environments, as well as biodiversity, collections, past and continuing cultural practices, knowledge and living experiences. It records and expresses the long processes of historic development, forming the essence of diverse national, regional, indigenous and local identities and is an integral part of modern life. It is a dynamic reference point and positive instrument for growth and change. The particular heritage and collective memory of each locality or community is irreplaceable and an important foundation for development, both now and into the future.

The very definition of Geopark, expressed by the *UNESCO Global Geoparks Network* (Network, G.G., 2010), is in line with an integrated perspective of (geo)heritage in its connection with (geo)tourism, considering that the concept of Geopark applies to a "(…) geographical area where geological heritage sites are part of a holistic concept of protection, education and sustainable development [but that should consider] the whole geographical setting of the region, and shall not solely include sites of geological significance".

From these brief reflections, emerged a conceptual scheme that would be applied in the projects developed by our students at both levels of education (Fig. 1).

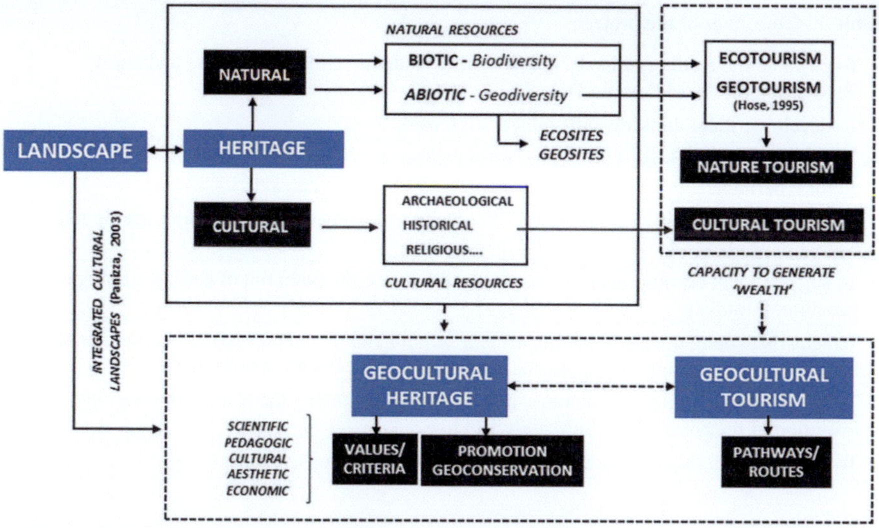

Fig. 1 The landscape as an integrated natural and cultural heritage

The Projects: Objectives, Methodologies and Results

Basic and Secondary School Examples: Knowing Your Place—The Heritage that Marks the Evolution of Landscape

Considering the various sub-themes that integrate the Geography EL of basic and secondary education (BSE), several projects were developed by students of these disciplinary levels, guided by master students in Geography Teaching (MGT) and supervised by BSE and MGT teachers. The first project aimed to "discover" the natural and cultural heritage that characterises the parish where the schools are located and where most students live, focusing on promoting this area through geotourism. Different objectives were considered when defining the work project (Table 1)—applied in the parish of Bonfim, in Porto city—which involved the application of different methodologies/strategies subdivided into several work steps (Fig. 2).

Besides involving the students in exploring their home territory in a guided and more exhaustive way—many of them recognising that they didn't know the area they walk through every day—the analysis of the produced materials led them on a space–time journey, which allowed them to understand the historical–geographical evolution of this eastern parish of Porto. In this context, we highlight four aspects, illustrated by some of the cartography created:

1. Analysing the urban fabric and the distribution of heritage, it is clear that the study area expanded from south to north, with the oldest heritage concentrated in the former, while the latter (north) clearly shows a subsequent evolutionary period

Table 1 Objectives of the project

• To promote the reading of landscape, fostering students' ability to observe and enjoy discovering the 'world' around them
• To develop critical thinking and collaborative work
• To contribute towards the knowledge, preservation and promotion of the local natural and cultural heritage
• To recognise the positive contribution of heritage to society and the economy, seeing it as and endogenous resource
• To suggest local development strategies in order to use the potential of heritage through geocultural tourism
• To sensitise young people to the importance of publicising and preserving heritage, promoting links with local communities (citizens in general and local authority leaders)
• To help students' build their memory and identity, recognising the historical–geographical evolution of the areas where they study and live
• To value local heritage and identity as support for global consciousness

(mainly from the mid-twentieth century), with buildings constructed in continuous strips along wide aligned roads, for housing and commercial purposes, public and private services, education and green spaces for recreation and leisure. The separation between these two sections of the parish, was recognised by the students and marked on the cartography (Fig. 3) based on architectural features, urban morphology (with emphasis on the design of the road network and the typology of residential buildings) and distinct functional contexts. Assertively, students identified the south of the parish as the city of the nineteenth century that grew as a result of the (…) mass arrival of the northern population fleeing the enormous problems of Portuguese rurality (…) to such an extent that "(…) in the second half of the nineteenth century' Bonfim became the most populous parish in the city" of Porto (Pinto 2011–2012: 1, 4).

2. Today, there are few traces of the industries (mainly textiles and, later, tobacco production units) that proliferated in south Bonfim during the nineteenth century, being responsible for the attraction the parish exerted over migrants, motivating the growth of its population. This was the aspect that most surprised the students, especially after consulting selected bibliography, since, according to Magalhães (1988), in 1852 there were 150 establishments in this area, employing 2207 workers. On the other hand, as Pinto (ob.cit. p. 12) points out referring to the industrial survey of 1881 "(…) in the parish of Bomfim there are streets and streets in which every house is a factory". This Bonfim has nothing to do with today's parish.
3. The third aspect relates to the traditional housing of the workers, the so-called "islands", which some of the students knew mainly from the requalification projects that some of them underwent, becoming Local Lodging. These "islands" are perhaps the clearest remnant of Bonfim's industrial character in the nineteenth century, as this is where the migrant workers lived. As described very clearly by Pinto (2015),

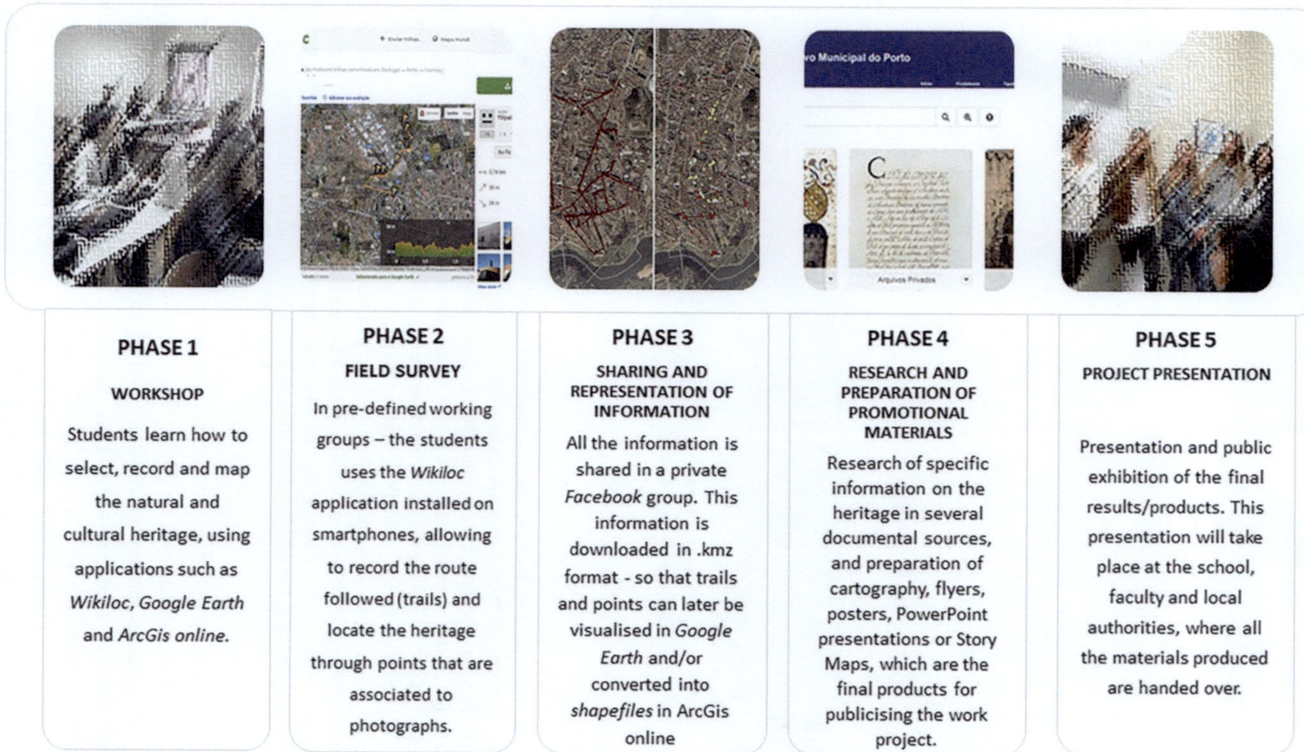

Fig. 2 Methodological framework: phases of the work project

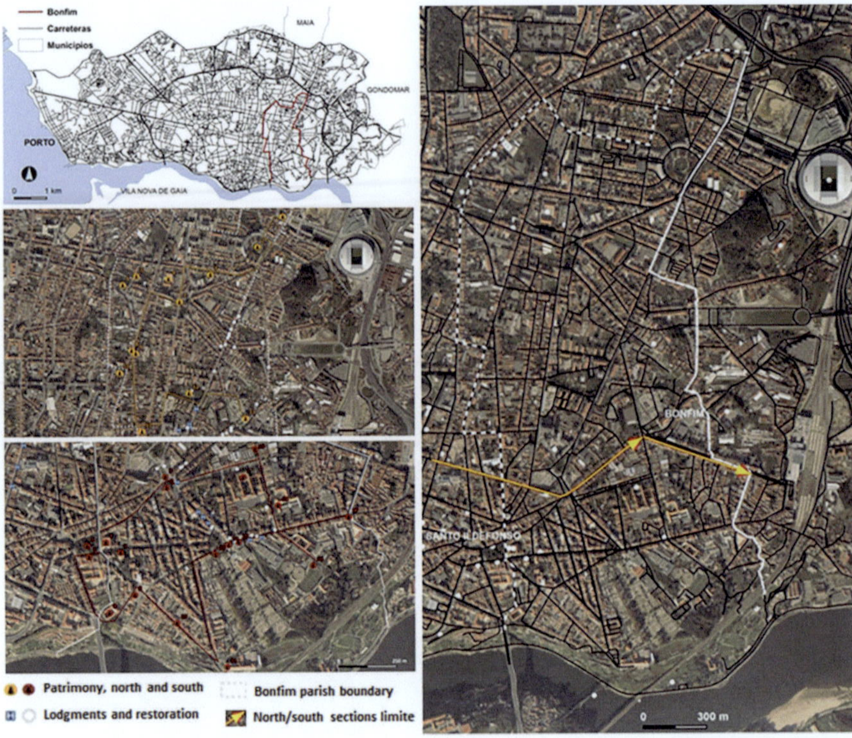

Fig. 3 Localisation of the patrimony in the north and south of the parish of Bonfim

> The 'islands' were lines of houses, usually small and single storey, built at the back of the deep plots of bourgeois houses (...) and sometimes on plots free of any occupation (ob.cit., p. 6). The houses had very small areas, their façades had only one door and one window, with a 4-metre front (...) that allowed the creation of a significant number of houses, plot by plot, which helped to accommodate the population surplus (…) they were made of low-quality materials and did not require ventilation, running water, electricity or sewage. The houses only had one independent wall and, in their interior, just three micro-rooms (living room, kitchen and bedroom), sometimes also with a small attic, a kind of drawer above the kitchen and bedroom, a cramped space where the younger ones slept. When the toilets existed, were at the end of the yard or in the middle of the corridor and shared by all the inhabitants (ob.cit., p. 7).

4. Finally, another of the students' highlights was the discrepancies and evident socio-economic and cultural inequality between rich and poor, typified, in particular, by the palaces built after 1860 in front of the most important streets, "(…) of remarkable size that boasted quality materials, including imposing balconies and elaborate Baroque artistic work, and remarkable gardens where palm trees, japonica trees, araucarias and other exotic species dotted the landscape" (Pinto 2011–2012: 14). Today, some of these buildings are occupied by public institutions, such as the Palacete dos Cirne, which is the headquarters of the Bonfim

Parish Council, or the Palacete de S. Lázaro, which houses the Faculty of Fine Arts of the University of Porto.

Together with the production of cartography (in ArcGis), posters, flyers to publicise the parish's heritage and its presentation in public spaces (namely at the Parish Council itself, with the presence of the mayors), the work project carried out by the basic and secondary school students, in collaboration with the master students and their teachers, allows us to highlight several aspects:

1. The adopted methodology (Project Based Learning), involved fieldwork, consultation of bibliography and the use of geotechnologies. All this allowed the acquisition of scientific knowledge, namely those related to the contents of the Essential Learnings of the school years involved, using a motivating and interactive approach, in which the students, although supported, worked autonomously. So, they feel that their ideas area valued and that, after all, the theoretical concepts of Geography involve them in their day-to-day lives.
2. The project subject, centred on the natural and cultural heritage of their living area, allowed students to obtain a more consistent view of their territory, enhancing local identity, culture and daily life elements, applying different research tools and establishing a bridge between the students, the school and society.
3. Understanding the space in which they live allows them to gradually approach regional, national and global analysis.
4. Working directly in the territory, applying an integrated perspective of heritage as a resource for endogenous development, helps students become active citizens in its community promotion and valorisation, as well as learning is more meaningful (Ausubel 1968).
5. Their astonishments and conclusions allowed them to realise that the parish of Bonfim, like the entire geographical space, has different spatial and temporal dynamics, so that the area they thought they knew so well acquired a new meaning… after all, the contents of the school curriculum have real expression and learning is, after all, an activity that empowers them.

Based on Michael Young (2009), Windsor and Kriewaldt (2023: 3) emphasise that powerful knowledge "(…) is as subject-specific, coherent, conceptual knowledge derived from disciplines that will enable students to make decisions and become action competent in a way that will influence their and others lives positively".

Geography Undergraduation. Geocultural Landscape: Integrated Perspective Environment-Culture

In the final year of the Geography course at the Faculty of Arts and Humanities of the University of Porto, students must develop a research project whose themes, proposed by several professors according to their speciality, they are free to choose. Two of the themes are related to geoheritage, but from two perspectives:

1. The first one, based on the integrated nature-culture vision of landscape, aims to develop geotourism itineraries, primarily focused on identifying and selecting the geocultural heritage in previously defined spatial contexts (districts or municipalities), to boost their territorial development and preserve the community's memory and identity traits.

The second, which we call "The story told by landscape", aims to analyse the natural constraints on the distribution of settlements over time, based on the archaeological heritage of different chronologies, from the Palaeolithic to the Roman period. Cross-referencing the sites where the archaeological remains are located with a set of natural variables and applying the logistic regression methodology, we define a Predictive Model in which the areas where human occupations of different chronologies are most likely to occur are identified and compared with the most populated contemporary places.

(Geo)tourism Itineraries

The work steps and methodology applied in this project involve the following:

(i) Inventorying and locating natural and cultural heritage, drawing up a georeferenced database with several attributes (Table 2), namely its classification according to its geocultural type and size (scale), its scientific, educational, cultural, aesthetic and economic values, location (coordinates) and naming.
(ii) This database, when transposed to ArcMap and associated with the basic information, made it possible to cross-reference the different variables that characterise the territory, using 3D Analyst and Spatial Analyst, producing maps of hypsometry, slopes, slope exposure, geology, land use and road network, the last one supporting the definition of itineraries (trails). Once the geodatabase is loaded into ArcMap, the spatial distribution of the heritage is analysed, and all the sites of interest are then characterised in a descriptive/qualitative way, using specific bibliography, cartography and the features identified during fieldwork.
(iii) The third step involves identifying tourism support infrastructures and defining thematic itineraries—designed for target people with specific interests—or integrated itineraries.

Drawing up the geodatabase is the most time-consuming stage of the project, especially for sites of natural interest (Fig. 4), because it involves fieldwork, analysing the maps drawn up and researching geological and geomorphological literature on the area under study. In addition, the classification of geosites must comply with the thematic inventory typology defined in Portugal for sites of geological interest (Brilha et al. 2005) and geomorphological interest (Pereira et al. 2006). Sites of cultural interest are easier to identify since there is already classified and georeferenced information on specific websites, especially considering patrimony classified as national monuments and buildings of municipal and public interest, archaeological sites and civil, military and religious architecture.

Table 2 Extract of simplified natural inventory with some attributes of geoforms

Name	Municipality	Geoform	Classification	Dimension	Latitude	Longitude
Bacia de Celorico	Celorico da Beira	Graben	Tectonic geoforms	Area	40.658346	– 7.350586
Cântaro Magro	Manteigas	Nunatak	Glacial and periglacial geoforms	Isolated place	40.326301	– 7.599282
Covão da Ametade	Manteigas	Cirque	Glacial and periglacial geoforms	Isolated place	40.328269	– 7.587899
Lagoa Comprida	Seia	Lagoon, erratic blocks, striations	Glacial and periglacial geoforms	Area	40.360538	– 7.641890
Nave de Santo António	Manteigas	Moraine accumulations	Glacial and periglacial geoforms	Area	40.283193	– 7.607345
Ribeira de Candeeira	Manteigas	Suspended valley	Glacial and periglacial geoforms	Area	40.343083	– 7.574236
Serra da Marofa	Fig. Castelo Rodrigo	Quartzite synclinal	Residual geoforms	Area	40.863591	– 6.992926
Depressão de Seia	Seia	Graben	Tectonic geoforms	Area	40.437673	– 7.696895
Poço do Inferno	Manteigas	Knick point	Relief and river drainage Iberian Massif	Isolated place	40.372473	– 7.516431
Tintinolho	Guarda	Tor	Granite geoforms	Area	40.573112	– 7.288348

Itineraries can be drawn up for one or several days, and they can also be designed for tourists with particular interests, connected, for example, to the values (e.g., scientific, cultural, aesthetic) attributed to the heritage identified in each area (Raynard 2007; Guimarães and Mariano 2015). They can be planned to use different types of transport, with stop-offs and short walking routes, or include only footpaths. Regardless of these characteristics, the itineraries should encourage contact with the local population, valuing intangible patrimony and activities linked to commerce and services. As a result, tourism support infrastructures should be incorporated on routes to develop the local economy, favouring accommodation in Guest Houses and Rural Tourism establishments (Country houses and Agrotourism), Tourist Villages or the Pousadas de Portugal network that occupy notable properties or monuments. It is also possible to include in the itinerary's places selling handicrafts and typical products associated, for example, with gastronomy and oenology (e.g., meals and wine tastings in cellars). The cartography produced by the students using ArcMap is often

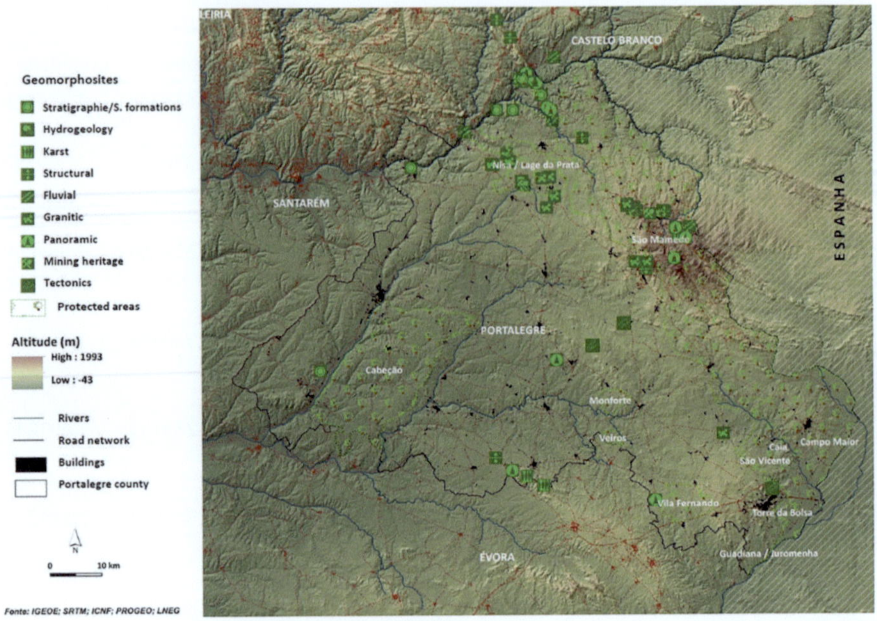

Fig. 4 Geo(morpho)sites of Portalegre district

integrated into flyers using applications such as Canva (a free-to-use online graphic design tool), which frames the information mentioned above and photographs taken during fieldwork.

The example shown in Fig. 5 represents an itinerary located on the eastern edge of the Portalegre district, on the border with Spain, between the municipalities of Castelo de Vide and Arronches. It covers around 110 km and is a 2-day journey by car or van—although it includes some short pedestrian trails—with an overnight stay in Marvão, a walled town that forms part of the municipality with the same name. Situated at the top of the Serra do Sapoio, a residual relief of Ordovician quartzites with an NW–SE orientation (included in the Serra de S. Mamede Natural Park) with a maximum elevation of 868 m, the town has an outstanding landscape over the extensive flat area around it, which is composed of schists and granites. The route includes several geocultural sites, from geosites of geological, geomorphological, fluvial and mining interest to archaeological and architectural cultural heritage (religious, civil and military), three of which classified as National Monuments—namely the Castle of Marvão, the Roman Ruins of the City of Ammaia and the Cruzeiro da Estrela—and four as Properties of Public Interest, of which stands out the entire urban settlement located within the perimeter of the Walls of Marvão.

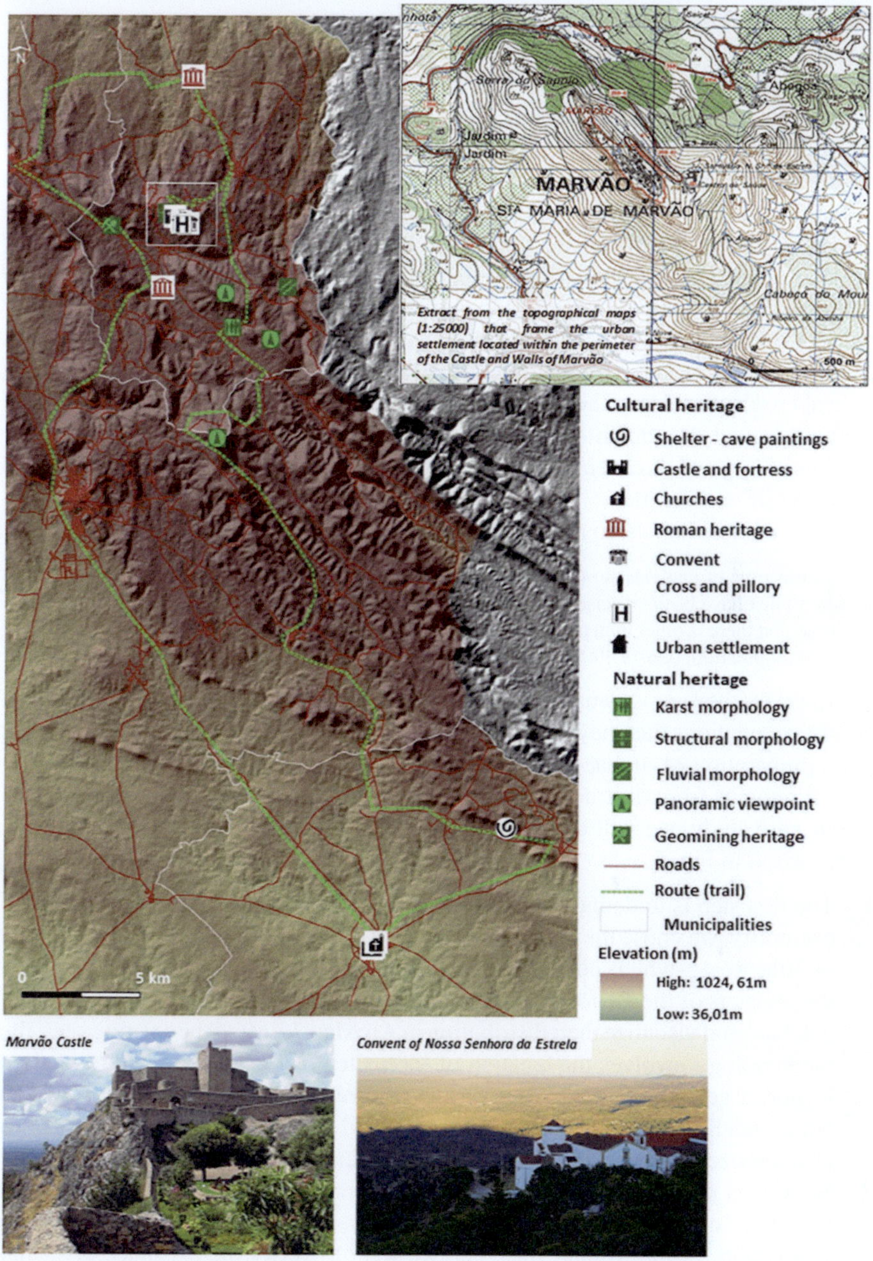

Fig. 5 Route around the Marvão area

The Story Told by Landscape

The last example of our projects was also developed during the final year of the undergraduate Geography program. In this case, the "story told by landscape" has the following objectives: (1) to evaluate the 'natural' constraints associated with the distribution of settlements, through the analysis of archaeological remains from different periods; (2) to define a predictive model (PM) "(…) *a technique that seeks to predict the probability of encountering a phenomenon in unsampled areas based on knowledge gained from sampled areas, may provide insight into a suspected pattern* (Conolly and Lake 2008: 145–146). In other words, it allows us to assess/model the probability of occurrence of a given phenomenon depending on various factors, in our case corresponding to the heritage of a given chronology, considering several terrain variables. These studies are more complex in terms of conceptual and methodological design. In our project, we apply *Logistic Regression Model*—the most usual type of Archaeological PM—a multivariate statistical model developed within a GIS environment.

Many archaeologists do not take these PMs very seriously, because

> (...) this type of analysis is often based on trends in archaeological thinking (New Archaeology) that have been criticised by more recent approaches (post-Processualism) (...) which argue that there is other than physical-environmental factors which constrain the settlements of the population. (Costa 2009, p. 40)

However, they are very useful in understanding the spatial distribution patterns of human occupations that may have existed, making it possible to guide research work (e.g., preparation of archaeological potential maps) and protect possible archaeological remains from destruction (Noronha Vaz 2008). To simplify the presentation of our case study, we will detail the work steps involved in the MP development, summarised in Fig. 6.

1. The first step is to define the chronological and spatial context of the model. To exemplify, we chose the archaeological sites of Roman occupation in the district of Vila Real—which correspond to the dependent variable of the MP—using the registers of the *Portal do Arqueólogo*,[2] allowing us to create a georeferenced database and visualize their distribution in ArcMap. Located the existing sites, the methodology involves the creation of 'non-sites', which correspond to random locations generated by ArcMap—using the *create random points tool*—, "(…) not coinciding with any known archaeological site, which makes it possible to characterize all the territory under study" (Soares et al. 2010: 184).
2. Next, it is necessary to define the physical conditions (independent variables) which, by characterizing the area under study, may have influenced the choice of location for the Roman sites, i.e., we want to verify the existence of spatial patterns common to archaeosites from this period. In this context, it's important to note the diversity and number of variables used by various authors (Verhagen 2007; Noronha Vaz 2008; Costa 2009; Vaughn and Crawford 2009; Graves

[2] https://arqueologia.patrimoniocultural.pt/

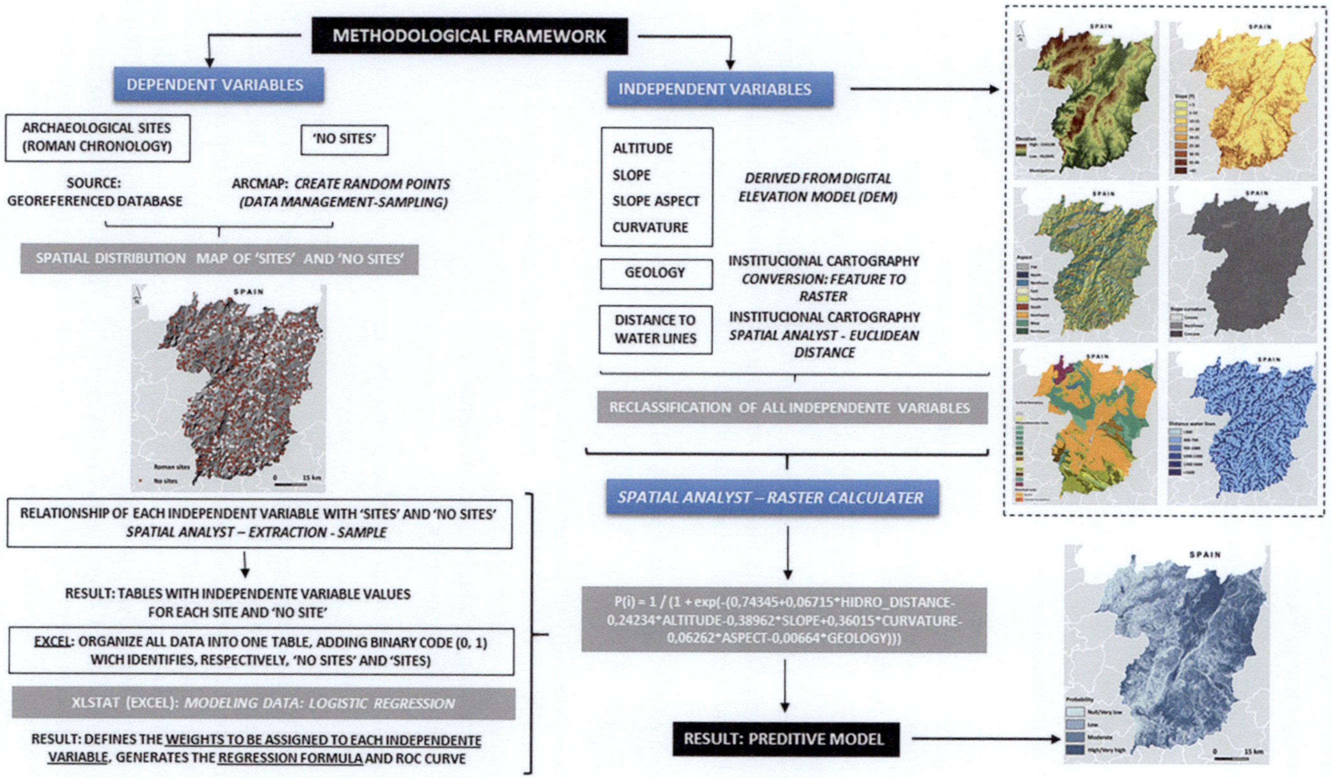

Fig. 6 Work phases involved in the creation of PM, using the Logistic Regression model

2011; Costa 2019), although determining the coefficients assigned to each one by applying logistic regression—using XLSTAT, statistical software that works as a complement to Excel—makes it possible to define the most significant ones. In our case study, we considered six variables: altitude, slopes, slope exposure, curvature, geology and distance to water lines. These variables are all represented in raster format: the first four were derived from a Digital Elevation Model (DEM); geology corresponds to the conversion from vector to raster format of institutional geological cartography (National Energy and Geology Laboratory); distance to water lines was determined using *Spatial Analyst's Euclidean Distance* tool, taking into account the hydrographic network represented in institutional topographic cartography (Army Geospatial Information Center).

3. The next stage involves crossing the sites and 'non-sites' with the independent variables, extracting the respective value for each point of the dependent variables (*extraction sample—Spatial Analyst*). This generates two tables merged into a single one (using Excel), identifying the values of the independent variables corresponding to the dependent ones, identified by a binary code (0— 'non-sites'; 1—sites).
4. Using Excel XLSTAT extension, the *Logistic Regression tool* is then applied to the previous table, obtaining the weighting coefficients for each independent variable, and then incorporated into the PM equation:

$$P(i) = 1/1 + exp[-(a + ß1X1i + ß2X2i + \cdots + ßnXni)] \quad (1)$$

where:
$P(i)$—probability of occurrence of an event
exp—raises e (Euler's number, approximately 2.72) to the value in brackets
a—intercept value of y
$ß$—regression or weighting coefficients
X—independent variables.

This equation is applied in ArcMap using the *raster calculator tool*, resulting in the Predictive Model.

The result allows us to observe the distribution of the different probabilities of occurrence of Roman heritage in the district under study (Fig. 7a), verifying that the PM obtained has a prediction rate of 70.6% (area under the curve). By associating this distribution with the variables of the physical environment, we can see, in global terms, that the areas with the highest probability of occurrence are located along the NNE-SSW alignment that practically divides the district in two—corresponding to the Verín-Régua-Penacova late-Variscan left fault (De Vicente and Vegas 2009)—where the Tâmega and Corgo rivers are incised (Fig. 7b). The archaeological sites are linked to the watercourses, as well as to the alluvial plains located near them, but also to the tops of the reliefs that border, to the east, the vast tectonic depression that extends from Chaves to Vila Real.

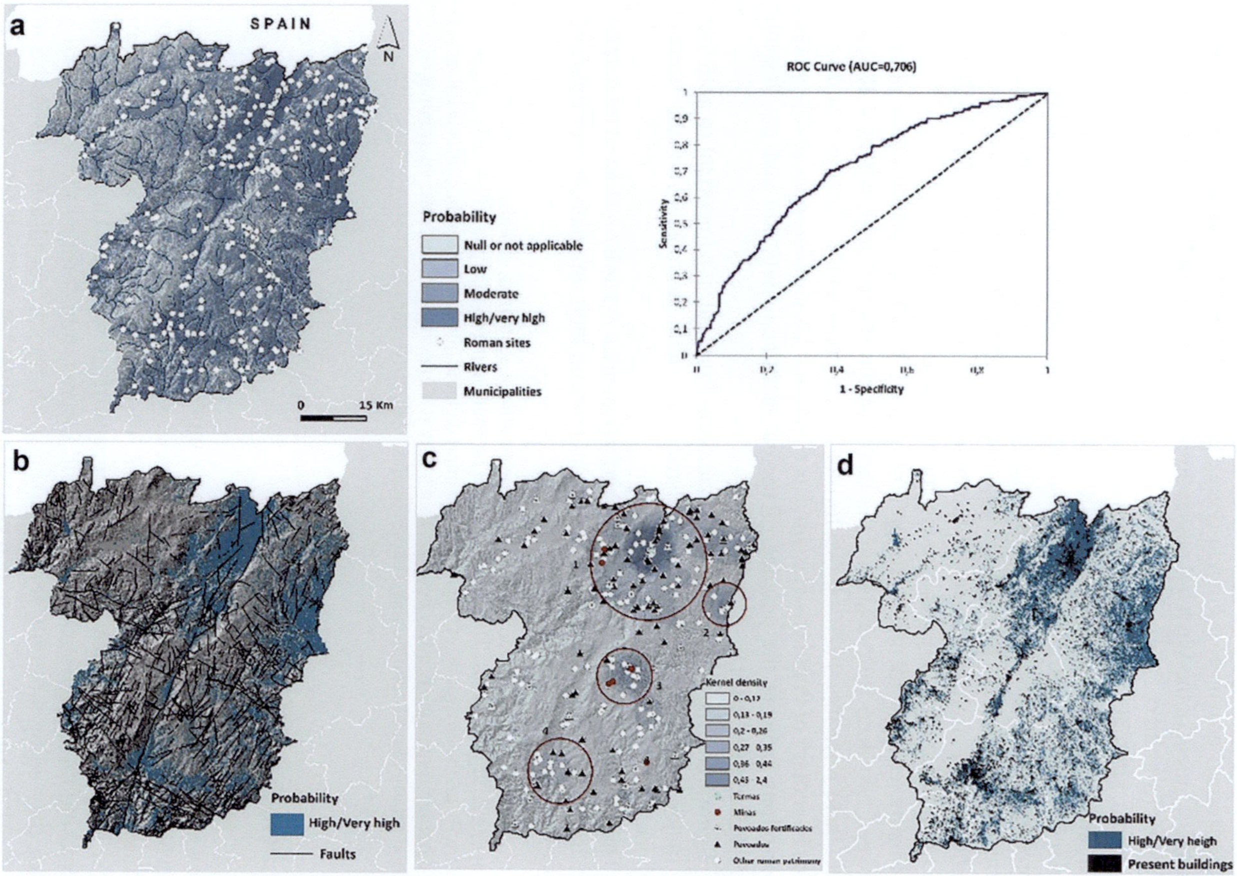

Fig. 7 Predictive model. **a** Probabilities of occurrence of Roman patrimony and Roc Curve; **b** Areas of high/very high probability of occurrence; **c** Kernel density of Roman sites; **d** Present-day territory occupation compared with high/very high probability of occurrence of Roman sites

The Kernel density analysis[3] (Fig. 7c) reveals that the highest concentration of heritage sites is in the northeastern part of the district, as would be expected. Indeed, circle 1 corresponds to *Aquae Flaviae*, one of the most significant urban centres in the Roman province of Galicia (Pérez Losada 1998). This area is also where a notable segment of the Via XVII passes through, a Roman road listed in the Antonine Itinerary (Soutinho 2024), recognised as the oldest in the northwest of the Iberian Peninsula, stretching from Braga to Astorga (Spain).

In this area, it is also important to note circle 3, where the Roman Mine Complex of Tresminas is found, in the surroundings of the mining village. This complex represents one of the most significant explorations of gold and other minerals and is currently classified as a Property of Public Interest. The gold-rich deposits, which also contain silver, are found in metasedimentary rocks and/or granites. This highlights the importance of geology as a key factor influencing our MP.

Finally, circle 4 identifies the present area of Vila Real, situated on a plateau between two deep valleys and on the banks of the River Corgo, where at Roman time, there were nine habitats, two fortified settlements and the famous and mysterious Sanctuary of Panóias—classified as a National Monument since 1910—whose inscriptions indicate the practice of sacred sacrificial, initiation and purification rituals (Redentor 2021).

Highlighting the impact of Romanization in Portugal, it's important to note that the major settlements in the present district align with the most significant Roman archaeological sites.

Thus, although the location of archaeological sites is hardly restricted by natural constraints, these must be considered to understand their articulation over time and space. However,

> (…) the landscape is also a space of appropriations, meanings, cultural references and symbolic significations. These elements can be challenging to interpret when studies focus strictly on the environment, neglecting the unpredictable human choices that influence it, as these are rooted in specific contexts. That's why it's important to articulate environmental and socio-cultural conditioning factors (Soares et al. 2010: 186).

Conclusion

As we mentioned in the introduction of this study, the skills needed to succeed in what we refer to as the knowledge society—both today and in the future—are different from those required in the past. Therefore, nowadays education system must be prepared to meet the learning needs of its students. This is essential for increasing their motivation and academic success, transforming the school into a place they enjoy, not only for socializing with friends but also for the desire to learn. Many students "(…) attend classes by obligation, without actively engaging

[3] The Kernel density, was calculated trough ArcGis pro, considering the shapefile of roman sites in a neighbourhood around those features.

in the basic activities, which is expressed through attitudes and behaviours ranging from mere apathy to behaviours/actions that disrupt the normal functioning of the educational environment, ultimately contributing to their academic failure" (Pezzini and Szymanski 2008, s/p).

In this sense, educational guidelines must adapt to the present socio-cultural context—in which knowledge and techniques are evolving at a vertiginous rhythm—by applying active teaching–learning methodologies that place students at the centre of the process, aiming for their autonomy, critical thinking and the development of attitudes and values that appeal to emotional intelligence—promoting the formation of citizens committed to the sustainable and sustained growth of the communities where they live and work. Thus, in Portugal, the most recent legislation/norms and educational referential have undergone several changes, providing more effective responses to the demands of society and young people. But never ignore the fact that each learning situation and each educational project is unique and special and must be contextualised, relativised and defined according to the specific conditions of the students, the resources involved, the scientific contents, the time available and other specificities of each pedagogical situation.

We have seen that the use of geotechnologies as teaching–learning tools has become essential in education, supported by legislation that guides the system. These tools align with project-based methodologies, which benefit from their application at the scale of the students' experience, enhancing meaningful learning by transforming complex and abstract geographic concepts into knowledge that is more easily apprehended.

The examples we present, covering basic, secondary and university education, focus on the study of landscapes study as heritage/patrimony. They illustrate the value of geotechnologies as a motivating factor for students, the new millennium learners. However, it's important to remember that the application of geotechnologies should not be seen as an end use in itself. Geotechnologies are tools that facilitates the input, representation, access, analysis and output of data. Additionally, the validity of the results relies entirely on the accuracy of the information that the user inputs and their ability to interpret it within the context of scientific knowledge.

In fact, in the case studies presented, geotechnologies—progressively more complex—are an essential tool for collecting and representing the necessary information, specifically the application of GIS. By allowing the combination of databases and their visualization in space, GIS supports the interpretation of territorial dynamics, as they enable the integration of different variables organised across multiple layers of information.

Its ability to articulate diverse data in different formats and store it interactively enhances analytical procedures, leading to an interpretation that, when combined with scientific and empirical knowledge, helps us, in the project works presented, to understand the heritage that shapes the evolution of landscape. This understanding is achieved through an integrated environment-culture perspective and, in this sense, reflects 'the story told by landscape'.

References

Ausubel DP (1968) Educational psychology: a cognitive view. Holt, Rinehart and Winston, New York

Brilha J, Andrade C, Azerêdo A, Barriga FJ AS, Cachão M, Couto H, Terrinha P (2005) Definition of the Portuguese frameworks with international relevance as an input for the European geological heritage characterisation. Episodes J International Geoscience 28(3): 177-186

Brilha J, Carvalho A (2010) Geoconservação em Portugal: Uma Introdução. In: Neiva JM et al (eds) Ciências geológicas: ensino, investigação e sua história, vol II, pp 435–441. https://hdl.handle.net/1822/10572

Cachinho H (2012) Criando experiências de aprendizagem significativas: do potencial da Aprendizagem Baseada em Problemas. El Hombre Máquina 40:58–67. https://www.redalyc.org/pdf/478/47826850007.pdf

Conolly J, Lake M (2008) Geographical Information Systems in Archaeology, 3rd edn. Cambridge University Press, Cambridge

Costa A (2009) A Aplicabilidade dos SIG e das Imagens de Satélite da Identificação de Áreas com Potencial Arqueológico: Estações Arqueológicas da Idade do Ferro. Dissertação de Mestrado. Faculdade de Letras da Universidade do Porto

Costa L (2019) Aplicação de Modelos Preditivos Na Investigação da Evolução dos Padrões de Povoamento No Neolítico do Baixo Tejo. Master's thesis, Universidade do Porto. https://repositorio-aberto.up.pt/bitstream/10216/124843/2/371106.pdf

De Graaf E, Kolmos A (2003) Characteristics of problem-based learning. Int J Eng Educ 19(5):657–662. https://research.tudelft.nl/en/publications/characteristics-of-problem-based-learning

Decreto-Lei n.º 54/2018, de 6 de julho. Diário da República n.º 129/2018, Série I de 2018-07-06, 2918–2928. https://diariodarepublica.pt/dr/detalhe/decreto-lei/54-2018-115652961

Decreto-Lei n.º 55/2018, de 6 de julho. Diário da República n.º 129/2018, Série I de 2018-07-06, 2928–2943. https://diariodarepublica.pt/dr/detalhe/decreto-lei/55-2018-115652962

Delors J (1996) Learning: the treasure within. In: Report to UNESCO of the international commission on education for the twenty-first century. https://precog.com.br/bc-texto/obras/ue000009.pdf

Despacho n.º 5908/2017, de 5 de julho. Diário da República n.º 128/2017, Série II de 2017-07-05, p 13881–13890. https://diariodarepublica.pt/dr/detalhe/despacho/5908-2017-107636120

De Vicente G, Vegas R (2009) Large-scale distributed deformation controlled topography along the western Africa–Eurasia limit: Tectonic constraints. Tectonophysics 474(1-2):124-143

Domingues A (2001) A paisagem revisitada. Lisboa. Finisterra XXXVI 72:55–66. https://doi.org/10.18055/Finis1621

ENEC (2017) Estratégia Nacional de Educação para a Cidadania. Direção-Geral da Educação. https://www.dge.mec.pt/estrategia-nacional-de-educacao-para-cidadania

Graves A (2011) Practical variational inference for neural networks. Proc Adv Neural Inf Process. Syst 24

Gruenewald D (2008) The best of both worlds: a critical pedagogy of place. Environ Educ Res 14(3):308–324. https://doi.org/10.3102/0013189X032004003

Guimarães T, Mariano G (2015) Uso de Trilhas como Estratégia de Geoconservação. Anuário Inst Geocienc 38(1):40–73. https://doi.org/10.11137/2015_1_40_53

Hepburn H (2018) Social media 'catastrophic for teachers. Times Educational Supplement. https://www.tes.com/magazine/archive/social-media-catastrophic-teachers

ICOMOS (2002) International cultural tourism charter. Principles and guidelines for managing tourism at places of cultural and heritage significance. International Council on Monuments and Sites. International Cultural Tourism Committee. https://drive.google.com/file/d/1_raDTw-4cPJTKx5Gv78QKTSR_ysghjO5/view

Johnson M, Sprowles A, Goldenberg K, Margell S, Castellino L (2020) Effect of a place-based learning community on belonging, persistence, and equity gaps for first-year STEM students. Innov High Educ 45:509–531. https://doi.org/10.1007/s10755-020-09519-5

Kokotsaki D, Menzies V, Wiggins A (2016) Project-based learning: a review of the literature. Improv Sch 19(3):267–277. https://doi.org/10.1177/1365480216659733

Lejla A, Bexheti B, Betim H (2014) An analysis of social media usage in teaching and learning: the case of SEEU. http://www.europment.org/library/2014/venice/bypaper/OLA/OLA-14.pdf

Loyens S, Magda J, Rikers R (2008) Self-directed learning in problem-based learning and its relationships with self-regulated learning. Educ Psychol Rev 20:411–427. https://doi.org/10.1007/s10648-008-9082-7

Lynch K, Bednarz B, Boxall J, Chalmers L, France D, Kesby J (2013) E-learning for geography's teaching and learning spaces. In: Active learning and student engagement, pp 116–130. https://doi.org/10.4324/9781315879376

Michael, Young (2009) (2009) Education globalisation and the 'voice of knowledge' Journal of Education and Work 22(3) 193-204. https://doi.org/10.1080/13639080902957848

Magalhães M (1988) A indústria do Porto na primeira metade do século XIX. In: Revista da Faculdade de Letras da Universidade do Porto. Geografia. Ser. I, vol 4, pp 111–155. https://ojs.letras.up.pt/index.php/geografia/article/view/7841/7193

Network GG (2010) Guidelines and criteria for national geoparks seeking UNESCO's assistance to join the Global Geoparks Network (GGN). http://www.globalgeopark.org/uploadfiles/2012_9_6/ggn2010.pdf

Noronha Vaz E (2008) GIS from a cultural heritage perspective: when past and future colide. Master's thesis, Universidade Nova de Lisboa (Portugal)

PASEO (2017) Perfil do Aluno à Saída da Escolaridade Obrigatória. Direção Geral da Educação. https://www.dge.mec.pt/perfil-dos-alunos

Panizza M, Piacente S (2003) Geomorfologia culturale. Pitagora Ed., Bologna

Pérez Losada F (1998) Cidades e aldeias na Galiza romana: uma proposta de classificação hierárquica do habitat galaico-romano. O Arqueólogo Port 16:157–174

Pereira D, Pereira P, Alves MIC, Brilha J (2006) Inventariação temática do património geomorfológico português. Publicações da Associação Portuguesa de Geomorfólogos 3:155-159

Pezzini CC, Szymanski MLS (2008) Falta de desejo de aprender: causas e consequências. Portal Dia-a-dia Educação, Paraná

Pinto J (2011–2012) A expansão (sub)urbana no Porto Romântico. O caso da freguesia do Bonfim. Percursos Ideias 3&4—2ª série. https://percursoseideias.iscet.pt/wp-content/uploads/2022/03/PI_2011-12_N03-04_TR01.pdf

Pinto JR (2015) As ilhas do Porto. Ilhas do Porto: levantamento e caracterização. Porto: Domus Social 5-18

Portaria n.º 181/2019, de 11 de junho. Diário da República n.º 111/2019, Série I de 2019–06–11, 2954–2957. https://diariodarepublica.pt/dr/detalhe/portaria/181-2019-122541299

Puertas-Aguilar MÁ, Álvarez-Otero J, de Lázaro-Torres ML (2021) The challenge of teacher training in the 2030 agenda framework using geotechnologies. Educ Sci 11(8):381. https://doi.org/10.3390/educsci11080381

Ravitz J (2008) Project based learning as a catalyst in reforming high schools. Buck Institute for Education. https://files.eric.ed.gov/fulltext/ED540113.pdf

Redentor A (2021) A paisagem sagrada de Panóias: (com)textos de um santuário rupestre. In: Martínez Caballero S, Santos Yanguas J (Coord.) Paisajes sagradas de la Antigüedad en el valle del Duero. Atas del Coloquio, pp 195–215. ISBN 978846979419

Reynard E (2005) Géomorphosites et paysages. In: Géomorphologie: relief, processus, environnement, vol 3, pp 181–188. https://doi.org/10.4000/geomorphologie.338

Reynard E (2007) A method for assessing the scientific and additional values of geomorphosites. Geogr Helv 62(3):148–158. https://doi.org/10.5194/gh-62-148-2007

Reynard E (2008) Scientific research and tourist promotion of geomorphological heritage. Geogr Fis Dinamica Quaternaria 31:225–230. https://serval.unil.ch/resource/serval:BIB_9448D137D568.P001/REF.pdf

Selwyn N (2017) Education and technology: key issues and debates. Bloomsbury, London. ISBN 978-1-3501-4554-2 (hbk)

Sequeira E (2013) Da objectividade jurídica do património. Direito Justiça 1(Especial):297–326. https://doi.org/10.34632/direitoejustica.2013.9883

Soares L, Costa A, Gomes A (2010) Geografia, Arqueologia e Sistemas de Informação Geográfica: exemplos prospectivos de articulação. Estudos do Quaternário. Rev APEQ, III J Quaternário 171–188. https://hdl.handle.net/10216/73995

Soares L, Pacheco E, Lucas J (2018) (Geo)diversidade, cultura e património: uma leitura integrada da paisagem. CEM Cultura, Espaço & Memória 4:157–175. https://ojs.letras.up.pt/index.php/CITCEM/article/view/4815/4497.

Soares L, Pacheco E, Lemos P (2023) Escola imaginária: ensaios (s) didático (s) para uma educação livre, equitativa e de qualidade. In: Didática da Geografia para uma Cidadania Territorial: uma bússola para um mundo em profunda transformação: Atas do X Congresso Ibérico de Didática da Geografia, pp 90–113. https://hdl.handle.net/10216/147545

Soutinho P (2024) Vias Romanas em Portugal. www.viasromanas.pt

Stevenson R (2008) A critical pedagogy of place and the critical place (s) of pedagogy. Environ Educ Res 14(3):353–360. https://doi.org/10.1080/13504620802190727

Tamim S, Grant M (2013) Definitions and uses: case study of teachers implementing project-based learning. Interdisc J Prob-Based Learn 7(2):3. https://doi.org/10.7771/1541-5015.1323

Yew E, Goh K (2016) Problem-based learning: an overview of its process and impact on learning. Health Prof Educ 2(2):75–79. https://doi.org/10.1016/j.hpe.2016.01.004

Van Eijck M, Roth WM (2010) Towards a chronotopic theory of "place" in place-based education. Cult Sci Edu 5:869–898. https://doi.org/10.1007/s11422-010-9278-2

Vaughn S, Crawford T (2009) A predictive model of archaeological potential: an example from northwestern Belize. Appl Geogr 29(4):542–555. https://doi.org/10.1016/j.apgeog.2009.01.00

Verhagen P (2007) Predictive models put to the test. Case studies in ar-chaeological predictive modelling. ASLU 14:115–168. ISBN (print) 9789087280079

Wagner PL, Mikeseil MW (Eds.) (1962) Readings in Cultural Geography. Chicago: University of Chicago Press

Walshe N, Healy G (2021) Geography education in the digital world. Routledge, London. ISBN 9780367224479

Whalley W, Saunders A, Robin A, Buenemann M, Sutton P (2011) Curriculum development: producing geographers for the 21st century. J Geogr High Educ 35(3):379–393. https://doi.org/10.1080/03098265.2011.589827

Williams R (2022) An overview of social media in education.https://doi.org/10.9734/bpi/costr/v7/4525A

Windsor S, Kriewaldt J (2023) School geography's critical role for a more sustainable future: powerful knowledge and praxis. Soc Sci 12(11):585. https://doi.org/10.3390/socsci12110585

United Nations (2020) Sustainable development goals—decade of action. Retrieved from https://www.un.org/sustainabledevelopment/decade-of-action/

Zandvliet D (2012) Development and validation of the place-based learning and constructivist environment survey (PLACES). Learn Environ Res 15:125–140. https://doi.org/10.1007/s10984-012-9110-x

Human Geography

The Conceptualization of Justice and Equity in Transport and Mobility Planning

Andreia Monteiro, Miguel Saraiva, and Teresa Sá Marques

Abstract Urban changes resulting from mobility solutions based on the use of private transport have had a particularly strong impact on the most vulnerable groups and highlight issues of injustice and inequality in transport and mobility. Today, society is moving toward finding sustainable solutions for transport and mobility/accessibility patterns, yet the existence of such issues compromises the development of adequate policies. In recent years, socio-economic, environmental, and political changes have fomented academic research and have seen the issue of justice in transport planning starting to appear in political agendas. Yet knowledge of what exactly this entails, and how to materialize it, is still scarce. Thus, this chapter aims to offer a theoretical reflection, revisiting some of the classic theories of justice to identify how they can, or have been translated into the (geographical) challenges of current transport and mobility planning. Building from utilitarianism, Rawls' egalitarianism, Sen's capability approach, and "the right to the city," a discussion is made of concepts as: Lucas' transport and social justice; Martens' transport justice; Sheller's mobility justice; van Wee's accessibility equity; Pereira's transportation equity; or Karner's transportation equity and justice. Ultimately, it is expected that this debate helps to better integrate the different discourses (equity and justice) into transport planning processes.

Keywords Theories of justice · Transportation equity · Inequalities · Approaches

A. Monteiro (✉) · M. Saraiva · T. S. Marques
Faculty of Arts and Humanities of the University of Porto, Centre of Studies in Geography and Spatial Planning (CEGOT), University of Porto, Porto, Portugal
e-mail: andrmmonteiro@gmail.com

M. Saraiva
e-mail: mmsaraiva@letras.up.pt

T. S. Marques
e-mail: tmarques@letras.up.pt

L. Soares et al. (eds.), *Compromised Geography*, Key Challenges in Geography,
https://doi.org/10.1007/978-3-032-05201-8_12

Introduction

Urban changes, such as city growth, congestion, and long-distance travel, which are the historical result of planning and designing mobility solutions mostly based on the use of private transport, have had a particularly strong impact on the most vulnerable groups and highlight issues of injustice and inequality in transport and mobility (Enright 2019; Grengs 2018).

In recent years, at the same time that societies around the world faced socio-economic, geo-political, environmental, migration/refugees, and indeed identity crisis (e.g. Paris Agreement, United Nations 2015a), a paradigm of a more sustainable mobility, eco-friendly, and supporting better quality of life permeated urban agendas (European Commission 2011, 2013; OECD 2004). In this confluence, evidences of injustice and inequality became more manifest—as cities compromise to reach short- and then long-term solutions, there is a risk that some may "left behind."

Closely aligning with the goals of the 2030 Agenda for Sustainable Development, to reduce inequalities (SDG 10) and make cities inclusive, safe, resilient, and sustainable (SDG 11) (United Nations 2015b), increased scientific research has started to address inequalities in transport and mobility (Farber et al. 2011; Pereira et al. 2017; Tiznado Aitken et al. 2023; Verlinghieri and Schwanen 2020), as have policy agendas, for example, that established in flagship 9 of the Sustainable and Smart Mobility Strategy: "making mobility fair and just for all" (European Commission 2020).

However, despite the importance of these issues within the contemporary society, there is still little conceptual clarity about what "justice" means in transport and mobility (Pereira et al. 2017) but also about how to define and how to integrate the concept of equity into planning processes (Linovski et al. 2018). Consequently, this chapter addresses a very specific challenge of compromised geography and intends to dig into classic theories of justice to offer a debate on how, and in what way, current transport and mobility planning can build upon that legacy to assist in the transformation of more sustainable and equitable societies.

Section 2 presents the major question: how the concepts of justice and equity are or should be regarded within transport and mobility planning. Section 3 revisits the classic theories and principles of justice. Section 4 makes the leap, pondering current planning approaches for transport, their inheritance of the principles of justice and how they can be used, or shaped, into responding to the current societal challenges. Finally, Sect. 5 closes the paper.

Justice and Equity in Transport and mobility—An Overview

According to Verlinghieri and Schwanen (2020), a set of socio-economic developments, like neoliberalization, global urbanization, climate change, and health pandemics, as well as the traditional conceptualization of transport systems oriented toward efficiency and economic development, created an ideal situation—a "perfect storm," which placed equity and justice at the center of debates in the field of transport and mobility.

In the traditional conceptualization of transportation systems, the focus was mainly on managing traffic. There was little consideration of how non-drivers were served or how external costs (congestion, accidents, pollution) deriving from planning decisions could result in unfair outcomes (Litman 2020). Indeed, nondrivers continue to face the greatest disadvantages in their mobility and accessibility, as many cities were built and developed for the use of private transport in mind (Grengs 2018).

The urban mobility model is, however, clearly changing (Javier-Gutiérrez and Arias-Molinares 2022), and justice in transport emerges as a fundamental political and ethical issue in this transition process (Enright 2019). There is no doubt that including issues of justice in the planning and design of transportation systems is essential to building sustainable futures (Harada 2023); because, quite simple, "an unjust transition is not sustainable" (Williams and Doyon 2019, p. 144).

Despite the growing importance that these issues have acquired in recent years for decision makers, scholars but also the public in general, there is still "no single overarching definition of justice" in the context of transport (Pereira et al. 2017, p. 171). It is a concept "difficult" to define (Bret et al. 2016, p. 80), that "is the object of a wide variety of interpretations" (Brennetot 2011, p. 115)[1], so no single meaning can be attributed to it (Beyazit 2011). Martens argues that "there has been little systematic effort to define what justice may require in the domain of transport" in contrast to what happens in other key domains of government intervention, such as health, housing, and education (2020, p. 383).

Concept of Equity

Although they are not exact synonyms and refer to different concepts in certain contexts, the terms equity, fairness, ethics, and justice are commonly used interchangeably in academic literature (Martens 2012; Pereira et al. 2017; van Wee 2022). The term equity is also often, albeit incorrectly, used interchangeably with the term equality (Carleton and Porter 2018). Equity implies an ethical and moral dimension with an underlying value judgment, whereas equality refers to the distribution of a given good (e.g. accessibility) independently of a moral judgment (Marques et al. 2016; van Wee and Geurs 2011). Thus, like the justice, the concept of equity has also been understood in different ways (Pereira et al. 2017), leading to inevitable conceptual confusions (Lewis et al. 2021).

Equity is intrinsically associated with the "allocation and distribution of resources." It is related to the "idea of distributive justice and the work developed by Rawls" in which the "existence of differentiated individual needs is admitted and

[1] Translation by the author.

the importance of the entire population accessing more effective equality of opportunities is highlighted" (Marques et al. 2016, p. 173) (see Footnote 1). Furthermore, because it implies a moral judgment, it becomes a much broader concept than the simple distribution of resources (van Wee and Geurs 2011); it becomes, in the words of Litman (2020, abs.) "the fairness with which impacts (benefits and costs) are distributed."

Even so, as it is rooted in ethical and moral issues, the debate on possible definitions, as well as its practical implications in the context of transport and mobility, remains open (Bruzzone et al. 2023). Several scholars have proposed ways to characterized it (Levinson 2010; Litman 2020; Thomopoulos et al. 2009), but the division between vertical and horizontal equity seems to be the most frequently used (Bruzzone et al. 2023). Horizontal equity (fairness or egalitarianism) assumes that individuals or groups considered equal in needs and capabilities should be treated similarly (Delbosc and Currie 2011; Litman 2020). Vertical equity (social justice, environmental justice, or social inclusion) assumes that individuals or groups with specific needs should receive favorable treatment to offset general social inequalities (Delbosc and Currie 2011; Litman 2020). When applied to the transport context, horizontal equity should refer to the availability of a balanced transport offer that is appropriate to the travel needs of the entire population, while vertical equity should refer to the level of adequacy of transport offers according to the specific needs of certain population groups (Silver et al. 2023).

However, these two perspectives can be contrasting (Delbosc and Currie 2011), so their effective application may lead to inconsistencies that accentuate disparities, as a decision that seems equitable using one criterion may be inequitable using another (Camporeale et al. 2019; Litman 2020). In the case of horizontal equity, treating everyone equally implies that existing social inequalities are not adequately observed (Camporeale et al. 2019). In the case of vertical equity, as disadvantaged individuals or groups are prioritized over others, this inherently implies that not everyone is being treated equally, and therefore, a situation that should be equitable becomes unequal (Camporeale et al. 2019; van Wee and Geurs 2011).

Litman has argued that "there is no single correct way to evaluate transportation equity. It is generally best to consider various perspectives, impacts and analysis methods" (2020, p. 2). Nonetheless, most of the literature that addresses equity in the context of transportation seems to place greater emphasis on the concept of vertical equity (Carleton and Porter 2018). These are mostly empirical studies, dedicated to describing and explaining the existing disparities between population groups based on a variety of transportation dimensions (such as kilometers traveled or level of accessibility) (Martens 2020). Furthermore, there seems to be no consensus on the methodological approaches to be applied (Schwanen 2020), as well as a lack of instruments to evaluate the results of planning policies in relation to justice, which naturally implies difficulties in verifying whether they provide "a just transit system" (Nazari Adli et al. 2019, p. 88).

Distributive Justice

Justice is generally understood and organized into three main domains (Davoudi and Brooks 2014; Pereira et al. 2017; Williams and Doyon 2019). Legal justice concerns how people are treated and what rights and entitlements should be recognized and implemented. Distributive justice concerns how benefits (resources and opportunities) and costs (harms and risks) are distributed within society. Finally, procedural justice relates to the processes and procedures of decision-making and governance. In the context of transport and mobility, the most widely considered and used domain is that of distributive justice (Pereira and Karner 2021; Schwanen 2020).

In highly mobile contemporary societies, the distribution of transport services is a matter of utmost importance (Martens 2012) and a major concern for transport planners and policy makers (Camporeale et al. 2019). In this context, the biggest challenge becomes how to operationalize it (Di Ciommo and Shiftan 2017). Three intrinsically related questions have been put forward by scholars (Bruzzone et al. 2023; Di Ciommo and Shiftan 2017; Pereira and Karner 2021): (1) What benefits and burdens are distributed?; (2) To whom (population and social groups) these benefits and burdens are being distributed?; and (3) What principle or criterion determines the fairest distribution? Indeed, if there is no doubt that the planning and design of transport services should be based on principles of justice, the main question is which principles they should be based on (Martens 2020). Various scholars have suggested that the answers lie in the theories of justice found in the disciplines of political philosophy, political science, and political science (Hananel and Berechman 2016; Martens 2012, 2020; Nazari Adli et al. 2019; Pereira et al. 2017), which are reviewed in the next section.

Theories and Principles of Justice

Utilitarianism

Initially worked on by Jeremy Bentham and further developed by John Stuart Mill, utilitarianism has been arguably the most influential theory of justice (Castro 2009). Its fundamental principle is the maximization of human well-being or the total "utility" of society as a whole, with utility being defined as anything that produces happiness or satisfaction (Adli 2020; Sandel 2009; Thomopoulos et al. 2009). Therefore, anything that maximizes this "utility" or well-being is considered fair (Pimentel 2014).

Another fundamental assumption of this theory is to assign equal weights to well-being and common interest, since all individual preferences are considered to have the same weight and are taken into account together with those of everyone else (Adli 2020; Sandel 2009) In a logic of consequentialism, the utilitarian perspective considers that the judgment of actions or policies must be morally right or wrong depending on their effects (Adli 2020; Pimentel 2014).

Because utilitarianism is fundamentally concerned with maximizing outcomes regardless of the form of the resulting distribution (Humberto 2023), many of its critics argue that the vulnerability of this theory lies in its failure to respect individual rights (Sandel 2009). It is considered legitimately justifiable if an action or policy results in maximizing the well-being of a greater number of people even if that means that some are excluded (Pereira et al. 2017).

This means that utilitarianism is the theory underlying cost–benefit analysis (CBA) and has been predominant in traditional transport performance evaluations (Lucas et al. 2016; Tabascio et al. 2024; Thomopoulos et al. 2009). By considering all benefits and costs, as far as possible, in monetary terms, it allows for an easy comparison of alternatives (van Wee and Geurs 2011). However, limitations of this method include equity issues, as not all impacts can be monetized, especially social ones, and therefore, it does not observe the loss of well-being of certain groups or regions, focusing only on aggregate well-being (Di Ciommo and Shiftan 2017; Lucas et al. 2016; Thomopoulos et al. 2009).

Egalitarianism and Ralswian Principles

Egalitarian theories have as their basic principle the equal treatment of all people (Adli 2020), and therefore, the best situation is one where there is the smallest possible difference in the level of well-being, or any other benefit between groups or individuals (Humberto 2023).

Of the various egalitarian theories, the "justice as fairness" developed by John Rawls and presented in his work *A Theory of Justice* (2013) has become an inevitable reference for egalitarian liberalism (Castro 2009) and is considered the most influential theory of justice in modern moral philosophy (Sen 2010).

Rawls proposes a basic structure of society oriented toward redistributive justice of "primary goods" (rights, freedoms and opportunities, income and wealth, and self-respect) that aims to reduce inequalities and result in benefits for all (2013). It is based on two central principles of justice that he considers would be chosen in the "original position" under the "veil of ignorance." The first is the principle of maximum equal freedom and the second is subdivided into the difference principle (inequalities are only fair if they result in the greatest possible benefits for those least benefited) and the principle of fair equality of opportunity.

These two principles, according to Rawls, reflect a more general conception of justice that can be summarized as: "All social values (…) should be equally distributed, unless an unequal distribution of some or all of these values would result in benefits to all."(2013, p. 69) (see Footnote 1).

According to van Wee and Geurs (2011), although Rawls does not identify "accessibility" as a primary social good, they consider that a basic level of access to some destinations can be understood as a "primary social good" and, therefore, from a

Rawlsian perspective, transport policies should aim to distribute and improve accessibility levels for all and, under the difference principle, focus on improving accessibility levels for the least advantaged groups in order to be fair (Adli 2020; Pereira et al. 2017).

Sufficientarianism and Capability Approaches

Sufficientarianism also assumes a perspective of distributive justice, but its priority is to guarantee "that each of us has enough to cover their basic needs" (Adli 2020, p. 24). Sufficientarians seek to improve the situation of the less privileged, however this concern is only considered if these individuals are in a position below the established sufficiency threshold. And once this threshold is reached, inequalities cease to be a concern "everyone has enough to live above a sufficiency threshold" (Humberto 2023, p.3).

One of the most relevant sufficientarian theories is the Capability approach developed by Amartya Sen, in his work *The Idea of Justice* (2010). Building from Rawls, Sen proposes a change from "primary goods" to "capabilities," with capabilities being understood as "what people are actually able to do and to be" (Nussbaum 2003, p. 33).

Sen's motivation is to present a theory of justice that can serve as a basis for practical reasoning and should include ways of assessing how to reduce injustice, rather than aiming only to characterize perfectly just societies. Therefore, he argues that principles of justice should not be defined in normative terms, but rather in terms related to the lives and freedoms of the people involved (Sen 2010) Thus, justice is based on "people's ability and freedom to make their own decisions" (Marques and Maia 2022, p. 343) (see Footnote 1), recognizing the diversity of individuals' characteristics (e.g., preferences, values, and needs), but also the social structures and constraints that affect individuals' abilities to convert resources into real functioning (Pereira et al. 2017).

Consequently, an equitable distribution of resources is insufficient as it does not consider the diverse needs and preferences of individuals. Different levels of resources would thus be needed to achieve the same level of functioning capacity (Nussbaum 2003; Sen 2010). The capability approach may clearly be more appropriate because, in addition to taking into account how resources are distributed, it also considers how they affect people's opportunities differently, according to each person's personal characteristics, aspirations, and choices (Vecchio and Martens 2021).

The Right to the City and Spatial Justice

The "right to the city" presented by Henri Lefebvre in his book of the same name is announced as a demand for "*the right to urban life*, transformed, renewed" (Lefebvre 2012, p. 119, emphasis in the original) (see Footnote 1). It is effective through the strengthening of citizenship, which implies the city's integration and participation capacities and the right to appropriation of space and time, with the aim of creating a fair, beautiful, and democratic city. According to Harvey (2008, p. 23), the right to the city "is a right to change ourselves by changing the city," but this transformation must be carried out through the exercise of a collective power to lead and remodel urbanization processes, constituting a common and not an individual right (Lefebvre 2012).

Like Lefebvre's right to the city, Edward Soja's notion of spatial justice also addresses pertinent questions about who controls urbanization processes and the interconnection between social justice and spatial justice (Enright 2019). Thus, the right to the city constitutes a representation of the construction of space collectively and of the interests of those who are currently excluded from the "center" of value, creativity, decision, and vitality. It implies a human capacity to participate in planning institutions to transform urban life against all forms of segregation, exploitation, and hierarchy (Enright 2019). Spatial justice or injustice, according to Soja, "places intentional emphasis on the spatial or geographic aspects of justice and injustice," (see Footnote 1) because the political organization of space and the normal development of an urban system are sources of inequality and injustice (Soja 2016, p. 101). This implies considering as a starting point "the fair and equitable distribution in space,"(Soja, 2016, pp.101-102) (see Footnote 1) not only of the resources considered important in society but also of the opportunities that allow for their use.

David Harvey, in his work *Social Justice and the City*, also discussed the geographies of inequalities and social justice and highlighted the relevance that the concepts of space and place have in the organization of social life (Sarmento 2012). His theorizing and analysis on social justice are carried out from a critical spatial perspective (Adli 2020). Although he does not specifically use the expression "spatial justice," Harvey presents his version of the spatiality of justice using Bleddyn Davies' concept, "territorial justice" (Soja 2016); which he defined "as a socially just distribution that is justly arrived at" (Adli 2020, p. 28).

In turn, Peter Marcuse (2016, p. 107) (see Footnote 1) states that "spatial justice is part of the objective of every planning policy" and considers the perspective of distributive justice necessary in a normative stance in planning, despite not being sufficient. Because, according to him, it does not address the causes of injustice, which are structural. Thus, he considers that "a good city should not be limited to distributive equity; must support the full development of each and every individual" (Marcuse 2016, p. 108) (see Footnote 1); and focuses its approach on spatial injustice that is fundamentally caused in two ways: "the lack of freedom" due to "involuntary confinement" (segregation, ghettoization) and the "unequal distribution of resources in space" (p. 111) (see Footnote 1).

According to Marques and Maia (2022, p. 343) (see Footnote 1), "the notion of spatial justice presupposes an integrated perspective of social justice and territorial-based policy that assesses redistribution systems and degrees of access, but also levels of participation in decision-making."

Current Transport Planning Approaches Based on the Principles of Justice

According to Martens et al. (2022), in recent decades two important changes have been occurring in transport planning research, practice, and political priorities. The first consists of focusing on accessibility as a planning objective (Handy 2020) and the second concerns issues of equity and justice (Martens 2021; Pereira and Karner 2021), combining the various conceptions and the different principles of justice presented in the past (reviewed in the previous section). The goal seems to establish a new planning vision for cities aligned with current societal challenges, even though the plurality of proposals indicates, once more, that what exactly "justice" in transport means, and how that can be achieved, remains somewhat elusive.

Transport and Social Justice by Karen Lucas

Lucas (2006, 2012; 2016) advocates a social justice perspective on transportation influenced by "environmental justice" and "social exclusion," and which will be implemented through accessibility planning. Accessibility planning should be done more collaboratively, between local transportation planners, land use planners, and local communities, so that decision makers have informed knowledge not only about the areas where accessibility is poorest, but also about the perspectives of the people who live there regarding existing accessibility barriers. And thus consider the needs and concerns of disadvantaged groups whose demand for transport is conditioned by various factors, such as financial inability or fears for personal safety.

Social exclusion can be defined as the deprivation of goods and services and the inability to participate in activities that are fundamental for a good quality of life (Levitas et al. 2007). So, in this case, social exclusion highlights the causal interactions between intrinsic factors of individuals (notably age, gender, race and disability) and between external factors related to socio-spatial contexts, such as lack or inadequate supply of transport services, absence of basic local services, lack of supply in the labor market, cultural influences, migrations, and legislative structures (Lucas 2006).

Lucas states that transport and social exclusion cannot depend solely on an agenda focused on transport. Indeed, accessibility planning, being necessary to meet the travel needs of socially excluded people, should be broadly integrated into policies, programs, and objectives in other areas such as land use, housing, health, education and welfare, local regeneration, and regional development (Lucas 2006, 2012). Thus, to overcome inadequate levels of services and avoid social exclusion resulting from lack of access to employment, education, health, and other basic amenities, it is necessary to comply with basic standards of inclusion in transport (Lucas 2012). The use of the concept of sufficiency can contribute to the development of policies that guarantee a minimum level of accessibility to the main destinations (Lucas et al. 2016).

Thus, Lucas's perspective on transportation justice can be understood as a combination of principles from the theories of the right to the city, spatial justice, and sufficientarianism.

Transport Justice by Karel Martens

Martens (2006, 2016, 2020) notes that traditionally the design and conception of transport systems are governed by guidelines of economic growth and efficiency. Therefore, in his book *Transport Justice* (2016), he presents an alternative approach to transport planning based on the principles of justice, with the conviction that this will lead to the construction of fairer systems.

Martens' approach to transport justice is people-centered and shifts its focus to the notion of accessibility. He considers that the dynamic process of production of the transport good has generated injustices in transport and therefore, using Michael Walzer's Spheres of Justice,[2] argues that transport justice deserves the creation of a separate transport sphere to prevent the financial aspect from becoming (completely) the dominant force in the production of accessibility.

Martens explores various theories and approaches to social justice to subsequently define a set of principles of justice for the domain of transportation and concludes that "a fair transportation system provides all persons with a sufficient level of accessibility under most, but not all, circumstances." (2016, p.14).

While transport justice theory cannot determine what level of accessibility would be sufficient, it seeks to develop guidelines from which stakeholders can determine what they consider sufficient accessibility and what is insufficient accessibility.

In short, transport justice is not just about providing minimal solutions for the disadvantaged, but about providing a sufficient sense of accessibility for everyone. Which, according to Martens (2020, p. 384, emphasis in the original), "is basically

---

[2] In Sect. 2 of this chapter, the Theories of Justice most referred to in the literature were considered.

Michael Walzer's Spheres of Justice is a distributive theory of social goods (and not abstract basic goods) with socially constructed meaning and therefore can be different in different societies. Walzer, based on the "theory of goods," develops the concept of "distributive spheres" that are characterized by the distribution of distinct (and non-normal) goods through a distributive principle other than free exchange and autonomously from the distribution of other goods. *In* Martens, K. (2012). Justice in transport as justice in accessibility: applying Walzer's 'Spheres of Justice' to the transport sector. *Transportation, 39*(6), 1035–1053. https://doi.org/10.1007/s11116-012-9388-7

a notion of freedom, as a sufficient level of accessibility enables people to *select* the activities they want to engage in and thus enables them to give direction to their lives."

Mobility Justice by Mimi Sheller

Sheller (2018a, 2018b, 2020) sees mobility justice as a fundamental political and ethical issue, crucial for the "transition to more environmentally sustainable and socially just mobility" (2018b, p. 17).

In her book *Mobility Justice: The Politics of Movement in an Age of Extremes* (2018a, p. 34), the author argues that we need to move beyond current discussions of transportation equity, which are substantially focused on distributional outcomes and do not understand "the full multi-scalar and entangled dimensions of mobility justice". Alternatively, it develops a more comprehensive theory of mobility justice, albeit based on distributive concerns, including accessibility and elements of spatial justice, but which presents broader alternatives. And that is why it considers that mobility justice has five types of justice as fundamental components: distributive, deliberative, procedural, restorative, and epistemic. Distributive justice focuses on the best distribution of benefits (accessibility) and burdens (externalities), prioritizing the most disadvantaged. Deliberative and procedural justice analyzes which rights and prerogatives should be considered and implemented, as well as the processes and procedures for participation in policy formulation and decision-making. Restorative justice involves recognizing the causes that gave rise to current and ongoing historical inequalities, as well as redressing them. Finally, epistemic justice relates to the conception of new knowledge, continuous adaptation, and understanding of different scales.

Sheller indicates that a complete theory of mobility justice necessarily involves addressing certain elements of injustice, namely injustices related to individual factors, other inequalities structurally distributed in historical contexts; the struggles for "the right to the city" and the public sphere, for freedom of movement and the exercise of the right to assembly; mobilizations related to borders, migrations and other types of transnational mobility; and the circulation of goods, resources, and information.

Sheller considers that the various theories of justice do not deal with the spatial and geographical dimensions of mobility, and that "spatial justice" is therefore important and useful, even if it is necessary to go beyond debates just about transport.

Accessibility Equity by Bert Van Wee

Van Wee (2022; 2011) considers that accessibility levels present differences that are impossible to avoid, since it is not possible "to provide equal levels of access for all people to all destinations" (2022, p. 1), develops a conceptual model in which he integrates the factors that influence and dominate ethical concerns regarding accessibility. It is based on the ethical principles of sufficientarianism (all people should have a certain minimum level of accessibility that is considered "sufficient" to satisfy basic needs) and the egalitarian view (differences in levels of accessibility between people or groups are unfair in themselves).

The conceptual model proposed by van Wee comprises the four components that had already been presented by Geurs and van Wee (2004), namely, the land use system, the transportation system, the temporal component, and the individual component. According to him, the importance of the first three components is inherent to the definition of accessibility considered, while the individual component is differentiated, since accessibility levels are influenced by contextual and individual factors (socioeconomic and demographic variables, attitudes, preferences, and needs, as well as capabilities and social networks).

van Wee finds this model useful not only from a theoretical perspective, but also for assessing which factors to include when applying ethical principles and making moral judgments about equity in accessibility. This is exemplified by the issue of low accessibility, which can be the result of preferences and voluntary choices, such as choosing a place to live in more distant areas.

Transportation Equity by Rafael H. M. Pereira

According to Pereira (2021; 2017), equity in transportation incorporates issues of distributive justice arising from the distribution of the benefits and burdens of transportation in society among social, economic, and governmental institutions. In this sense, he develops a perspective of distributive justice that combines the principles of Rawls' egalitarianism and capability approaches.

This perspective considers that accessibility should be the focus of transport policies and planning, since some minimum level of accessibility to the main destinations necessarily constitutes a fundamental capacity of people to ensure the satisfaction of essential needs. And for the distribution of accessibility to be fair, it must respect the fundamental rights and freedoms of individuals, which can never be violated or sacrificed to the detriment of improving the accessibility levels of others; seek to reduce inequalities of opportunity, in particular by giving priority to vulnerable groups; define minimum accessibility standards to key destinations that must be safeguarded through social and transport policies; and only limit higher levels of accessibility to social groups and transport modes when marginal improvements imply losses for base groups.

For Pereira, equity in transportation is a fundamental part of incorporating justice, because although "transport justice goes beyond distributive concerns" it "cannot be achieved without equity" (2021, p. 272).

Pereira believes that the added value of this perspective consists in the fact that it includes in a balanced and flexible way not only the universalist concerns of justice, especially the guarantee of basic rights, the satisfaction of essential needs and the promotion of equal opportunities, but also the understanding of issues related to basic needs and minimum levels resulting from the specific cultural and historical characteristics of each context.

From Transportation Equity to Transportation Justice by Alex Karner

Karner et al. (2023; 2020; 2024), like Sheller and Pereira, also argues that equitable distributions are necessary but not sufficient to achieve justice in mobility, since justice implies a broader set of considerations than just equitable outcomes. Therefore, it proposes a transformative and reformist approach to transport planning structures and processes that lead to the unequal distribution of multiple transport externalities, drawing on critical perspectives in environmental justice studies and the movements of activist groups and nongovernmental organizations.

His perspective on transportation Justice is close to Sheller's mobility justice; as it also considers that the perspective of justice must also include, in addition to distributive components, deliberative, procedural, restorative, and epistemic elements.

In short, it advocates a shift toward justice in transport that consists of a society-centered approach, based on the active participation of stakeholders to be able to influence decisions, identifying and redressing the injustices of a structural redistribution of resources across society, linked to wider political movements and realizing transformative changes to achieve true justice.

Final Considerations

According to Karner et al. (2020), the two perspectives that regularly emerge in the literature and practice of transport planning are "transportation equity" and "transportation justice."

The equity approach is widely used by actors with concerns regarding the distribution of benefits and burdens associated with transport services and the mitigation of inequalities. This means, investing in transport reform to improve social opportunities (equity).

The justice approach aims to contribute to transforming social structures (justice) through collaborative participatory processes and greater integration of different policies. This means a transformative approach to transport planning processes, focusing on perspectives that are more focused on society's aspirations. The aim is to correct injustices by promoting the redistribution of resources and services. Here, transport policies are just one component of a broader and more complex transformative process.

In fact, the proposals from the different authors we visited in this chapter bring this confirmation. From the perspective of equity, according to Pereira and van Wee's approaches, emphasizing concerns about mitigating inequalities in the distribution of benefits and burdens based on egalitarian and sufficient principles of justice. From the perspective of justice, whether in transport or mobility, as discussed by Lucas, Martens, Karner, and Sheller, oriented toward broader and more comprehensive concerns of justice, in addition to distributive concerns, relying on environmental justice, social justice, spatial justice, and the right to the city (with the exception of Martens).

Approaches from the perspective of equity are necessary and present an empirically feasible perspective. Approaches from the perspective of jus-tice, considering their scope, become more complex and difficult to implement that is why compromised geography continues to seek solutions to build more sustainable and fair cities. In this sense, this research will seek to operationalize these two concepts in the future with a view to contributing to the development of transportation policies that promote greater spatial justice.

Acknowledgements This work is funded by the Foundation for Science and Technology (FCT) in a Doctoral Scholarship Program under reference 2023.04647.BD

References

Adli SN (2020) Inclusion of social justice in public transit planning: Development of a systematic framework for accessibility evaluation. Doctoral thesis, The University of Auckland. https://researchspace.auckland.ac.nz/docs/uoa-docs/rights.htm

Beyazit E (2011) Evaluating social justice in transport: lessons to be learned from the capability approach. Transp Rev 31(1):117–134. https://doi.org/10.1080/01441647.2010.504900

Brennetot A (2011) Les géographes et la justice spatiale: généalogie d'une relation compliquée. Ann Géog 678(2):115–134. https://doi.org/10.3917/ag.678.0115

Bret B, Gervais-Lambony P, Hancock C, Landy F (2016) Justicia e injusticias espaciales

Bruzzone F, Cavallaro F, Nocera S (2023) The definition of equity in transport. Transp Res Procedia 69:440–447. https://doi.org/10.1016/j.trpro.2023.02.193

Camporeale R, Caggiani L, Ottomanelli M (2019) Modeling horizontal and vertical equity in the public transport design problem: a case study. Transp Res Part A: Policy Pract 125:184–206. https://doi.org/10.1016/j.tra.2018.04.006

Carleton PR, Porter JD (2018) A comparative analysis of the challenges in measuring transit equity: definitions, interpretations, and limitations. J Transp Geogr 72:64–75. https://doi.org/10.1016/j.jtrangeo.2018.08.012

Castro RA (2009) As teorias da justiça e o igualitarismo liberal contemporâneo como disfarce da dominação capitalista. Códigos da violência nas democracias globalizadas. In: Revista do curso de mestrado em ciência jurídica da Fundinopi/Faculdade Estadual de Direito do Norte Pioneiro

Davoudi S, Brooks E (2014) When does unequal become unfair? Judging claims of environmental injustice. Environ Planning A: Econ Space 46(11):2686–2702. https://doi.org/10.1068/a130346p

Delbosc A, Currie G (2011) Using Lorenz curves to assess public transport equity. J Transp Geogr 19(6):1252–1259. https://doi.org/10.1016/j.jtrangeo.2011.02.008

Di Ciommo F, Shiftan Y (2017) Transport equity analysis. Transp Rev 37(2):139–151. https://doi.org/10.1080/01441647.2017.1278647

Enright T (2019) Transit justice as spatial justice: learning from activists. Mobilities 14(5):665–680. https://doi.org/10.1080/17450101.2019.1607156

European Commission (2011) WHITE PAPER roadmap to a single European transport area—towards a competitive and resource efficient transport system. Brussels. Retrieved from https://eur-lex.europa.eu/legal-content/EN/TXT/?uri=celex:52011DC0144

European Commission (2013) Communication from the commission to the European Parliament, The Council, The European Economic and Social Committee and The Committee of the regions together towards competitive and resource-efficient urban mobility. European Commission, Brussels. Retrieved from https://eur-lex.europa.eu/legal-content/EN/TXT/?uri=celex:52013DC0913

European Commission (2020) Communication from The Commission to The European Parliament, The Council, The European Economic and Social Committee and The Committee of The Regions sustainable and smart mobility strategy—putting European transport on track for the future. Brussels. Retrieved from https://eur-lex.europa.eu/legal-content/EN/TXT/HTML/?uri=CELEX:52020DC0789

Farber S, Páez A, Mercado RG, Roorda M, Morency C (2011) A time-use investigation of shopping participation in three Canadian cities: is there evidence of social exclusion? Transportation 38(1):17–44. https://doi.org/10.1007/s11116-010-9282-0

Geurs KT, van Wee B (2004) Accessibility evaluation of land-use and transport strategies: review and research directions. J Transp Geogr 12(2):127–140. https://doi.org/10.1016/j.jtrangeo.2003.10.005

Grengs J (2018) On the way but not there yet: making accessibility the core of equity planning in transportation. In: Norman K, Kathryn Wertheim H (eds) Advancing equity planning now. Cornell University Press, pp 127–148. https://doi.org/10.7591/9781501730399-009

Hananel R, Berechman J (2016) Justice and transportation decision-making: the capabilities approach. Transp Policy 49:78–85. https://doi.org/10.1016/j.tranpol.2016.04.005

Handy S (2020) Is accessibility an idea whose time has finally come? Transp Res Part D: Transp Environ 83:102319. https://doi.org/10.1016/j.trd.2020.102319

Harada T (2023) Mobility justice and sustainable futures. Aust Geogr 54(4):425–431. https://doi.org/10.1080/00049182.2023.2271615

Harvey D (2008) The right to the city. New Left Rev 53:23–40

Humberto M (2023) How to translate justice theory into urban transport metrics? Synchronic assessment of Latin American cities based on equality, priority and sufficiency. J Transp Geogr 110:103630. https://doi.org/10.1016/j.jtrangeo.2023.103630

Javier-Gutiérrez JC, Arias-Molinares D (2022) Nuevas formas de Mobilidad en la ciudad. In: Rio Fernandes JA, González RCL, Beltrào Sposito ME (eds) El mundo visto de las ciudades. Tirant Humanidades, pp 583–591

Karner A, London J, Rowangould D, Manaugh K (2020) From transportation equity to transportation justice: within, through, and beyond the state. J Plan Lit 35(4):440–459. https://doi.org/10.1177/0885412220927691

Karner A, Bills T, Golub A (2023) Emerging perspectives on transportation justice. Transp Res Part D: Transp Environ 116:103618. https://doi.org/10.1016/j.trd.2023.103618

Karner A, Pereira RHM, Farber S (2024) Advances and pitfalls in measuring transportation equity. Transportation. https://doi.org/10.1007/s11116-023-10460-7

Lefebvre H (2012) O direito à cidade/Henri Lefebvre; trad. Rui Lopo; apres Carlos Fortuna. Estúdio e Livraria Letra Livre

Levinson D (2010) Equity effects of road pricing: a review. Transp Rev 30(1):33–57. https://doi.org/10.1080/01441640903189304
Levitas R, Pantazis C, Fahmy E, Gordon D, Lloyd E, Patsios D (2007) The multi-dimensional analysis of social exclusion
Lewis EOC, MacKenzie D, Kaminsky J (2021) Exploring equity: how equity norms have been applied implicitly and explicitly in transportation research and practice. Transp Res Interdisc Perspect 9:100332. https://doi.org/10.1016/j.trip.2021.100332
Linovski O, Baker DM, Manaugh K (2018) Equity in practice? Evaluations of equity in planning for bus rapid transit. Transp Res Part A: Policy Pract 113:75–87. https://doi.org/10.1016/j.tra.2018.03.030
Litman T (2020) Evaluating transportation equity: guidance for incorporating distributional impacts in transportation planning. Victoria Transport Policy Institute. www.vtpi.org/equity.pdf. 31 Jan 2025
Lucas K (2006) Providing transport for social inclusion within a framework for environmental justice in the UK. Transp Res Part A: Policy Pract 40(10):801–809. https://doi.org/10.1016/j.tra.2005.12.005
Lucas K (2012) Transport and social exclusion: where are we now? Transp Policy 20:105–113. https://doi.org/10.1016/j.tranpol.2012.01.013
Lucas K, van Wee B, Maat K (2016) A method to evaluate equitable accessibility: combining ethical theories and accessibility-based approaches. Transportation 43(3):473–490. https://doi.org/10.1007/s11116-015-9585-2
Marcuse P (2016) La justicia espacial: consecuencia pero también causa de la justicia social (H. Pons, Trans.). In: Bret B, Gervais-Lambony P, Hancock C, Landy F (eds) Justicia e Injusticias Espaciales (1 edición en español ed.). UNR Editora
Marques TS, Maia C (2022) Retratos de la injusticia espacial en Porto: de la persistencia temporal a los impactos de las crisis. In: Fernandes J, González R, Sposito ME (eds) El mundo visto de las ciudades. Tirant Humanidades, pp 341–353
Marques TS, Santinha G, Mejías LMF (2016) Equidade. In: Dicionário de Geografia Aplicada—Terminologia da análise, do planeamento e da gestão do território
Martens K (2006) Basing transport planning on principles of social justice. Berkeley Planning J 19:1–17. https://doi.org/10.5070/BP319111486
Martens K (2012) Justice in transport as justice in accessibility: applying Walzer's 'Spheres of Justice' to the transport sector. Transportation 39(6):1035–1053. https://doi.org/10.1007/s11116-012-9388-7
Martens K (2016) Transport justice: designing fair transportation systems. Routledge
Martens K (2020) How just is transportation justice theory? The issues of paternalism and production: a comment. Transp Res Part A: Policy Pract 133:383–386. https://doi.org/10.1016/j.tra.2020.01.012
Martens K (2021) Equity considerations in transport planning. In: International encyclopedia of transportation. pp 154–160. https://doi.org/10.1016/b978-0-08-102671-7.10634-7
Martens K, Singer ME, Cohen-Zada AL (2022) Equity in accessibility. J Am Plann Assoc 88(4):479–494. https://doi.org/10.1080/01944363.2021.2016476
Nazari Adli S, Chowdhury S, Shiftan Y (2019) Justice in public transport systems: a comparative study of Auckland, Brisbane, Perth and Vancouver. Cities 90:88–99. https://doi.org/10.1016/j.cities.2019.01.031
Nussbaum M (2003) Capabilities as fundamental entitlements: Sen and social justice. Fem Econ 9(2–3):33–59. https://doi.org/10.1080/1354570022000077926
OECD (2004) Communicating environmentally sustainable transport: the role of soft measures. OECD Publishing, Paris. https://doi.org/10.1787/9789264106642-en
Pereira RHM, Karner A (2021) Transportation equity. In: In: Vickerman R (ed) International encyclopedia of transportation, vol 1. Elsevier, pp 271–277
Pereira RHM, Schwanen T, Banister D (2017) Distributive justice and equity in transportation. Transp Rev 37(2):170–191. https://doi.org/10.1080/01441647.2016.1257660

Pimentel JT (2014) Uma estrutura metodológica para avaliação de intervenções sociais baseada na ideia de justiça de Amartya Sen. Tese de Doutoramento, Universidade Federal do Rio Grande do Sul. http://hdl.handle.net/10183/109258
Rawls J (2013) Uma teoria da justiça/John Rawls; trad Carlos Pinto Correia, 3rd edn. Editorial Presença
Sandel MJ (2009) Justiça: o que é fazer a coisa certa? 6rd edn. [Tradução Heloísa Matias e Maria Alice Máximo, 2012]. Civilização Brasileira
Sarmento J (2012) David Harvey e a geografia cultural. In: Geografia Cultural: uma antologia. Editora da Universidade do Estado do Rio de Janeiro, pp 344
Schwanen T (2020) Low-carbon mobility in London: a just transition? One Earth 2(2):132–134. https://doi.org/10.1016/j.oneear.2020.01.013
Sen A (2010) A ideia de justiça/Amartya Sen; trad Nuno Castello-Branco Bastos. Almedina
Sheller M (2018a) Mobility justice: the politics of movement in an age of extremes. Verso Books
Sheller M (2018b) Theorising mobility justice. Tempo Soc 30:17–34. https://doi.org/10.11606/0103-2070.ts.2018.142763
Sheller M (2020) Mobility justice. In: Büscher M, Freudendal-Pedersen M, Kesselring S, Grauslund Kristensen N (eds) Handbook of research methods and applications for mobilities. Edward Elgar Publishing, pp 11–20. https://doi.org/10.4337/9781788115469.00007
Silver K, Lopes A, Vale D, da Costa NM (2023) The inequality effects of public transport fare: the case of Lisbon's fare reform. J Transp Geogr 112:103685. https://doi.org/10.1016/j.jtrangeo.2023.103685
Soja EW (2016) La ciudad y la justicia espacial (H Pons, trans). In: Bret B, Gervais-Lambony P, Hancock C, Landy F (eds) Justicia e injusticias espaciales (1 edición en español ed.). UNR Editora
Tabascio A, Tiznado-Aitken I, Higgins C, Farber S (2024) Incorporating equity into transit performance measures: a disaggregated bus route level approach. Case Stud Transp Policy 17:101256. https://doi.org/10.1016/j.cstp.2024.101256
Thomopoulos N, Grant-Muller S, Tight MR (2009) Incorporating equity considerations in transport infrastructure evaluation: current practice and a proposed methodology. Eval Program Planning 32(4):351–359. https://doi.org/10.1016/j.evalprogplan.2009.06.013
Tiznado Aitken I, Vecchio G, Mora R, Gonzalez L, Marshall C (2023) Planning for accessibility: the divide between research and policy in the promotion of equitable mobility. Area Develop Policy 1–23.https://doi.org/10.1080/23792949.2023.2261530
United Nations (2015a) Paris agreement. United Nations
United Nations (2015b) Transforming our world: the 2030 Agenda for sustainable development. (A/RES/70/1). United Nations Retrieved from https://docs.un.org/en/A/RES/70/1
van Wee B (2022) Accessibility and equity: a conceptual framework and research agenda. J Transp Geogr 104:103421. https://doi.org/10.1016/j.jtrangeo.2022.103421
van Wee B, Geurs K (2011) Discussing equity and social exclusion in accessibility evaluations. Eur J Transp Infrastr Res 11(4). https://doi.org/10.18757/ejtir.2011.11.4.2940
Vecchio G, Martens K (2021) Accessibility and the capabilities approach: a review of the literature and proposal for conceptual advancements. Transp Rev 41(6):833–854. https://doi.org/10.1080/01441647.2021.1931551
Verlinghieri E, Schwanen T (2020) Transport and mobility justice: evolving discussions. J Transp Geogr 87:102798–102798
Williams S, Doyon A (2019) Justice in energy transitions. Environ Innovation Societal Transitions 31:144–153. https://doi.org/10.1016/j.eist.2018.12.001

Looking at Maps: Reflecting on the Role of Cognitive Cartography in the Cartographic Design Process

María Zúñiga Antón

Abstract This chapter explores the theoretical foundations and current developments in cognitive cartography, focussing on its implications for map design and user interaction. The first section analyses cognitive processes involved in map reading—particularly attention, perception, and cognitive load—highlighting how these are shaped by user-related factors (such as age, experience, or environmental conditions) and map-related factors (such as visual complexity and task demands). The second part addresses methods for evaluating optimal map use, combining subjective tools like user questionnaires with objective techniques such as eye-tracking. This dual approach allows for a deeper understanding of how users cognitively engage with maps, informing more effective an user-centred cartographic design. By bridging cognitive science and cartographic practice, this chapter contributes to the development of evidence-based strategies that enhance spatial understanding and communication.

Keywords Cognitive cartography · Cognitive load · Map design · Perception · Eye tracking

Maps: More Than just Pretty Pictures

Cartographic representation is geography's natural language for explaining occurrences in the world, with maps serving as its preferred tool (Montello 2002). Since the mid-twentieth century, it has been widely accepted that the primary function of maps is linked to communication, which directly depends on the explicit or implicit decisions made by the cartographer in designing and producing the map (Robinson 1952).

There are **multiple perspectives** (Fig. 1) **through which the process of understanding cartographic reading** can be framed (Mocnik 2022), depending on the

M. Z. Antón (✉)
Department of Geography. Environmental Sciences Institute (IUCA), University of Zaragoza, Zaragoza, Spain
e-mail: mz@unizar.es

L. Soares et al. (eds.), *Compromised Geography*, Key Challenges in Geography,
https://doi.org/10.1007/978-3-032-05201-8_13

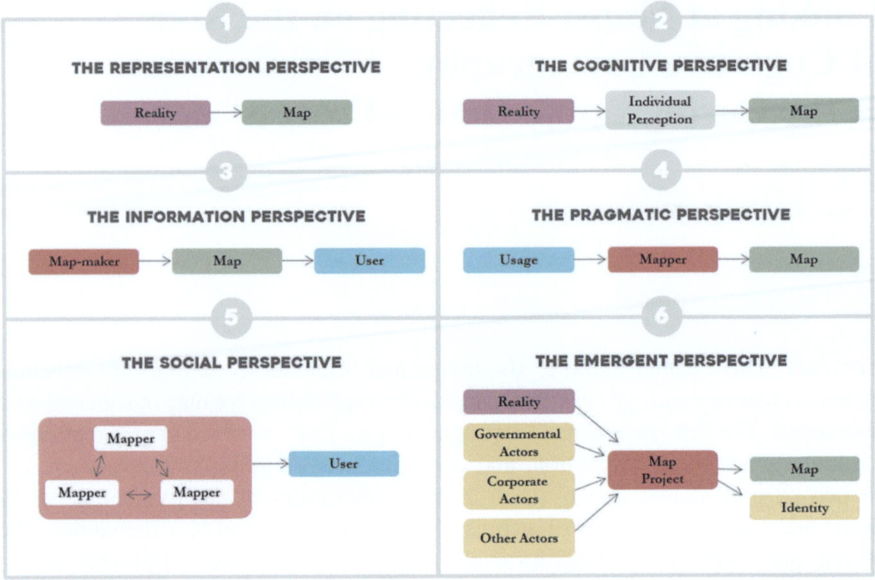

Fig. 1 Global framework perspectives on understanding cartographic reading (Mocnik 2022)

standpoint adopted: (1) the representation perspective, in which the map adheres to reality; (2) the cognitive perspective, which posits that individual perception of maps follows systematic patterns; (3) the information perspective, where maps are considered a means of communication; (4) the pragmatic perspective, wherein the map's appearance is determined by its intended use; (5) the social perspective, which acknowledges collaboration among cartographers to align their efforts; and (6) the emergent perspective, which accounts for the influence of external agents in rationalising design decisions in cartography. These perspectives, in any case, are not mutually exclusive but rather compatible with one another.

Within the context of this chapter, the informational perspective is examined in depth to establish a connection between cognitive science and cartographic design.

From Sender to Symbol: The Cartographic Communication Chain

The information perspective aligns the phases of the cartographic process with the terminology of Communication Theory or SMCR model (Source–Message–Channel–Receiver) (Berlo 1999) based on the premise that it effectively explains the encoding, transmission, and interpretation of geographic information in a simple, intuitive, and sufficiently comprehensive manner.

Building on SMCR, a workflow sequence is proposed (Fig. 2), enabling a logical and coherent integration of the two traditional spheres of cartography—the production and use of maps—as established in classical models (Koláčný 1969). Additionally, a third sphere related to map design is incorporated (Zúñiga Antón 2009). Organising all elements from various conceptual frameworks into a single scheme is not straightforward; however, it has been possible to delineate the position of each phase within the global framework. This working model specifically refers to the cartographic process associated with thematic maps, understood as those designed to analyse the spatial variations of one or more themes that can be geographically located.

This perspective, which conceptualises **cartography as a communication process**, assumes that two active agents participate in the existence of a map: the mapmaker, responsible for designing and producing the map, and the map-user, who interprets it. Both agents hold equal importance and give meaning to the map itself (Caquard 2015). The design of a map should be guided primarily by the correspondence between the characteristics of the represented data and the chosen symbology (Lambert and Zanin 2023). At the same time, it is also essential to consider user experience to enhance the overall design (Montello 2002).

The first assumption derives from *Sémiologie Graphique*, defined by Bertin in the late 1960s (Bertin 1967). In Europe—particularly in France and Spain—it became a sort of cartographic dogma that has shaped the practice of generations of cartographers since the 1970s (Palsky 2017). It cannot be considered a conventional cartography manual, as it encompasses the broader field of graphic expression, analysing a

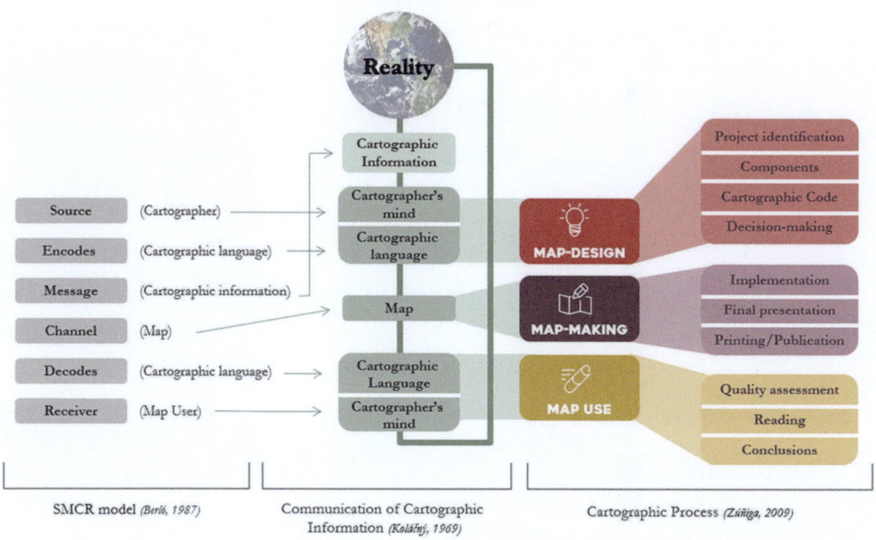

Fig. 2 Correspondence between communication theory (SMCR), communication of cartographic information and the cartographic process. *Source* Own elaboration based in Berló (1987), Koláčný (1969) and Zúñiga (2009)

set of graphical methods that Bertin regarded as a *monosemic system*, where each sign carries a single, predefined meaning. Until then, 'good representation' was primarily expected to be complete and accurate. However, with Bertin, the priority shifted towards the effectiveness of communication, taking into account the importance of the mental cost of perception, as defined by Zipf.

This connects to the second assumption, which assigns an active role to the map user, particularly regarding how the human mind processes maps—activating specific cognitive processes. These processes include both low-level perception, which deals with basic sensory input, and high-level cognition, which involves attention, perception, learning, and memory (Montello 2002). Perception is particularly relevant in this context, as it refers to the cognitive process of recognition, interpretation, and meaning-making, enabling individuals to form judgments based on sensory input from their physical and social environment. This process also involves learning, memory, and attention (Vargas 1994).

Perception is considered subjective—it varies from person to person—selective, as individuals, sometimes unconsciously, choose to perceive certain elements while ignoring others, and temporary, as it processes information over a short period (Etece 2023).

The **Perception Process** consists of three stages (Fig. 3): sensory stimulation and selection, organisation, and interpretation. Although individuals are rarely aware of passing through these stages separately, they fundamentally shape how we construct mental images of the world around us.

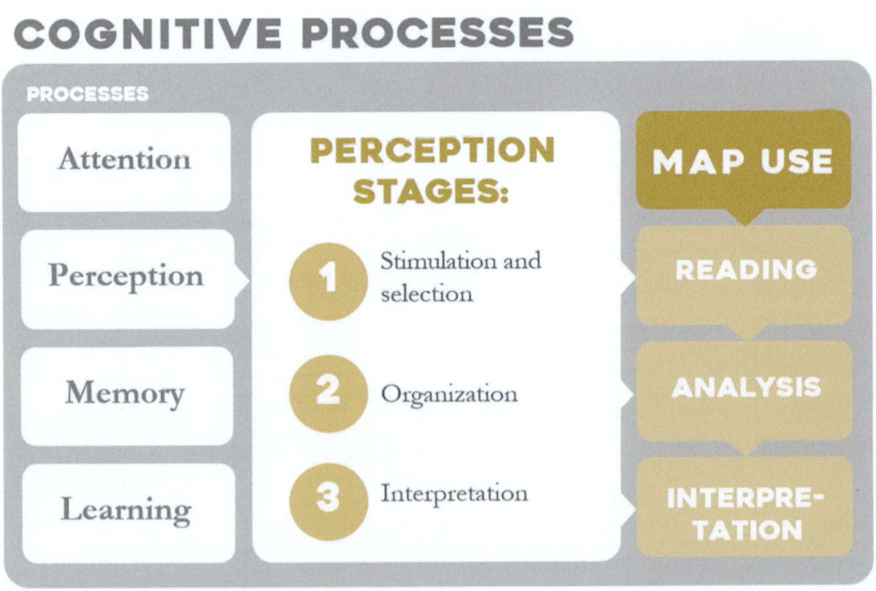

Fig. 3 Correspondence between global cognitive processes and their application in cartographic use. Author's own elaboration based on Vargas (1994) and Rock and Palmer (1990)

These processes, when applied to map use, function as follows: in **Sensory Stimulation and Selection stage** individuals perceive only a limited portion of the stimuli in front of them. The selection of which part of the stimulus to focus on is neither random nor entirely conscious. Three key factors influencing this selection process have been identified: needs, interests, and physiological limitations (Green et al. 2023). Individual characteristics also play a role in this selection. In cartographic design, this phase highlights the importance of visual hierarchy, which relies on size, colour, and the positioning of map elements. Ultimately, this process involves structuring and prioritising map content to enhance communication (Llasera 2023). It corresponds to the most fundamental stage of map reading and decoding.

The **Organisation stage** involves processing stimuli and fitting them into a recognisable pattern. The way we understand this process is rooted in Gestalt Theory, which explains how external stimuli are translated into mental images, emphasising that perception is a complex interaction between external input and internal cognitive processes (Rock and Palmer 1990). When applied to map reading, this phase consists of connecting visual elements with stored knowledge—engaging working memory to process what is seen while linking it to pre-existing schemas in long-term memory (such as spatial patterns, cartographic design strategies, and knowledge about the subject and location represented). In essence, this is the stage where the human mind attempts to understand how the map functions. It corresponds to the analytical phase, focussing on identifying spatial patterns and trends within the map.

Finally, the **Interpretation stage** is where meaning is assigned to the perceived content (Green et al. 2023). It represents the final step in which the previously selected and organised stimuli are given context. Individual experiences and personal values influence this process (Etece 2023). In relation to cartography, this is the phase where the user fully comprehends the map's function and representation. At this point, the map can be read, evaluated, and used to describe spatial patterns and trends. Depending on the user's prior knowledge, conclusions can be drawn not only about the causes behind a spatial distribution but also about its potential consequences. This phase corresponds to interpretation, knowledge integration, and the deduction of conclusions regarding spatial distributions.

Linking these stages of perception to cartographic reading, it is crucial to consider both initial visual attention, which plays a significant role in the sensory stimulation and selection phase, and cognitive load, which is particularly relevant in the organisation and interpretation phases.

Inside the Map-Reader's Mind: Exploring Cognitive Cartography

The use of the concepts outlined in the previous section has led to the development of the conceptual framework known as **Cognitive Cartography**. This is understood as a subdiscipline that encompasses both the application of cognitive theories and methods to the understanding of maps, and the use of maps to understand cognitive processes (Kiik et al. 2017).

Cognitive cartography is structured around three **main lines of research** (Montello 2002): (1) map-design research, which focuses on understanding how maps are read in order to improve their design, making them more efficient, effective, and visually appealing; (2) map-psychology research, which investigates human cognition using maps as stimuli—though not necessarily with the goal of improving map design; and (3) map-education research, which aims to enhance teaching and learning processes through the use of maps.

Ultimately, the objective is to identify the **cognitive load** required by a map—that is, the mental effort necessary to extract and utilise the information encoded within it (Kiik et al. 2017). This concept is framed within the **Cognitive Load Theory** (CLT), which posits that when users read a map, they engage their working memory (WM)—which is limited in capacity—while attempting to connect it with their long-term memory (LTM), which is theoretically unlimited (Baddeley 1986). Cognitive load, in turn, is influenced by both factors related to the map user and those inherent to the cartographic document itself (Fig. 4). Building upon earlier classifications (Bunch and Lloyd 2006), the following categorisation has been developed.

Among the **user-related factors** are the **user profile**, which includes individual characteristics such as sex, age, and prior experience in map reading or familiarity

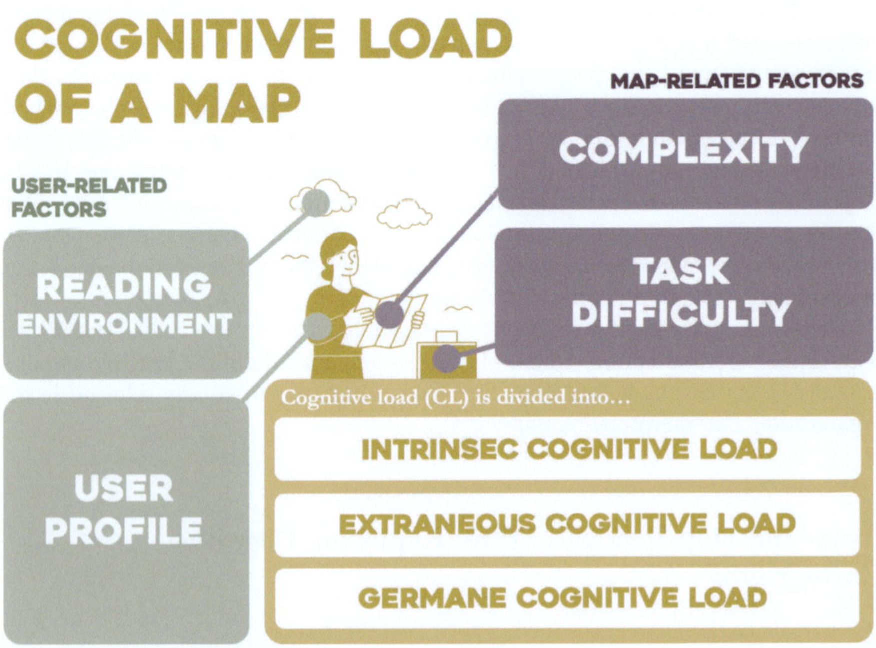

Fig. 4 Cognitive load: summarise of factors and components. Author's own elaboration based on Paas et al. (2003) and Bunch and Lloyd (2006)

with the specific region or theme depicted, as well as the **reading environment**, referring to the conditions under which the map is used and how these affect each individual.

Map-related factors include the complexity of the map, since each map has a specific level of complexity depending on the design decisions made during its creation, and the difficulty of the task, understood as the level of cognitive demand required by the nature of the specific task the user is expected to perform.

In addition, cognitive load (Fig. 4) consists of three **key components:** intrinsic, extraneous, and germane cognitive load (Paas et al. 2003). **Intrinsic cognitive load** refers to the demand placed on working memory (WM) due to the interaction between the stimuli generated by the specific map being read and the long-term memory (LTM), particularly in terms of whether existing mental schemas can be utilised. **Extraneous cognitive load** relates to the cognitive resources required to understand the instructions associated with a specific map-reading task. Finally, **Germane cognitive load** concerns the WM resources needed to acquire, practise, and automate reading strategies and schemas, which are later consolidated in LTM.

Making Maps Work: Gauging Cognitive Load and Optimal Use

One of the key tasks of the cognitive cartography approach to map use is to identify what constitutes optimal use and an appropriate level of cognitive load. In this regard, it is essential to deepen our understanding of the cognitive process in order to improve map design. Achieving this requires exposing users to different cartographic models to assess both their efficiency (the speed at which maps can be read) and their effectiveness (the accuracy with which tasks are completed) (Kiik et al. 2017). It is also important to consider the user's reaction, understood as their level of satisfaction with the map, and retention, or their ability to recall the information they have read (Song et al. 2022).

There are three main methodologies used to collect information on how users complete assigned tasks (Kiik et al. 2017; Ooms et al. 2015). The first involves directly surveying or interviewing participants. The second is the *'Thinking Aloud Method'*, where users are asked to verbally express the steps they are taking as they interpret the map. The third involves eye tracking, which monitors the movement of participants' eyes as they read the map. It will be explored in greater depth as this technique has regained considerable attention over the last decade.

Following the Gaze: Eye Tracking as a Cartographic Ally

The concept of eye tracking refers to a set of technologies that allow for the monitoring and recording of how a person visually explores a given scene or image—in particular, which areas draw their attention, for how long, and in what sequence (Hassan Montero and Herrero Solana 2007).

The use of this technology dates back to the late nineteenth century (Płużyczka 2018). Since then, there have been significant advances in sophistication, precision, and user comfort. Early devices were invasive for participants, requiring contact with the eye via specialised lenses. Later, true *eye trackers* emerged, operating through electrooculography, photooculography, or videooculography—the latter, particularly infrared videooculography, being the most widely used today (Kiefer et al. 2017; Krassanakis and Cybulski 2019).

The first applications of eye tracking in cartography drew on research from other fields such as art, photography, and advertising, which had developed recording techniques and task types applicable to map design and interpretation (Steinke 1987). A key milestone was the 1970 'Map's User' conference held at Queen's University in Kingston, Ontario, which marked the beginning of a research period in cartography using eye tracking techniques. The work of Jenks (1973, 1975) during this period proved influential. As an example (Fig. 5), one can observe the earliest instances of eye tracking on maps (White 2018). Research gradually became more complex, both in terms of experimental design (task construction) and data processing (eye tracking metrics).

However, from the 1980s onwards, research on map design lost prominence. This shift was partly due to the growing focus on geographic information systems (GIS) and also because eye tracking experiments were expensive and time-consuming, often producing limited results in return (Montello 2002).

In recent years, though, the concepts of cognitive cartography have regained relevance, echoing the interest they received during the 1970s (Caquard 2015). This renewed attention stems from both empirical and critical branches of geography, which have recognised the importance of placing cognitive processes at the centre of the cartographic experience. The rapid evolution of geo-technologies and their integration into digital environments has produced a paradigm shift. The widespread adoption of online cartography (Webmapping) and the democratisation of location technologies through tools such as Google Maps and OpenStreetMap (OSM) has led to a high dependence on navigation systems for everyday tasks.

This shift raises concerns about the potential weakening of spatial mental schemas—those lasting spatial memories stored in long-term memory—since location information is now often processed through digital applications rather than through direct interaction with space (Caquard 2015). This change challenges the traditional, *indexical* nature of spatial understanding—based on personal, grounded experience—by replacing it with a *non-indexical* approach that offers a generic worldview to all users. This concern reinforces the idea of cartography as a process

Fig. 5 Example of eye tracking in map reading conducted by Jenks (1973). *Source* White (2018)

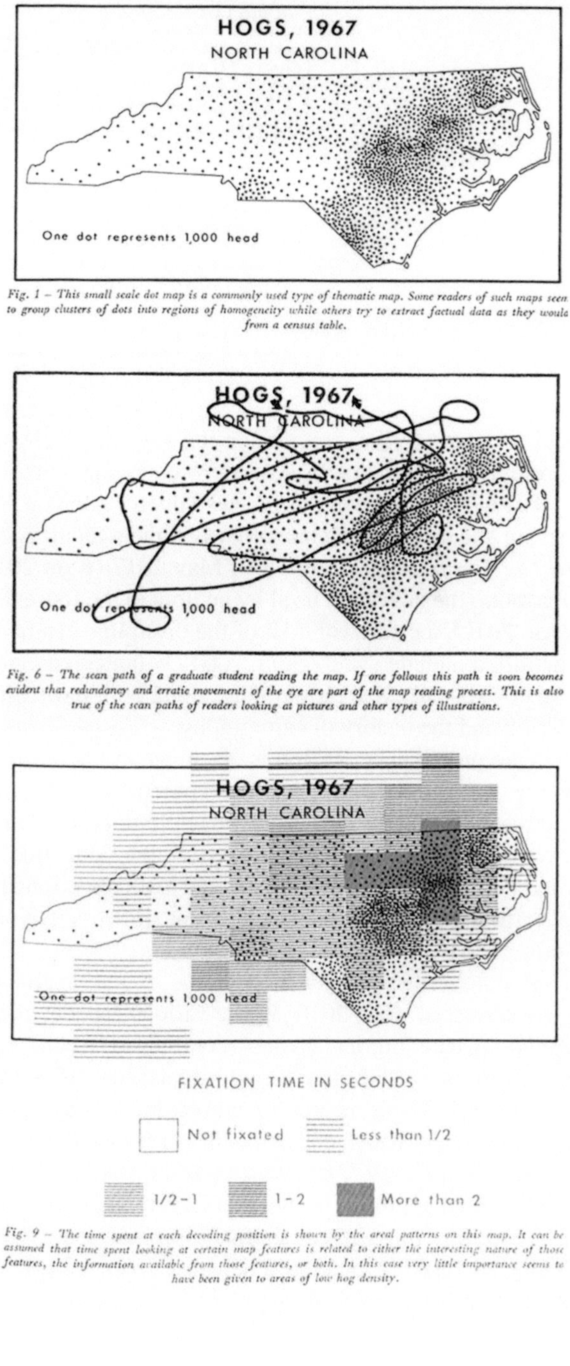

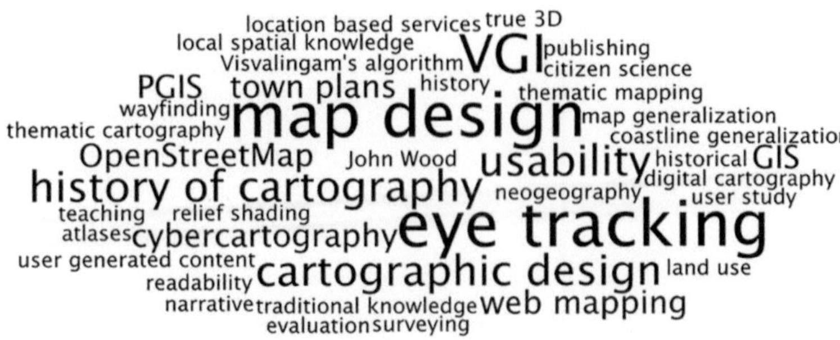

Fig. 6 Keyword cloud (excluding generic terms), The Cartographic Journal 2005–2016. *Source* Kent (2017)

and provides fertile ground for the revival of cognitive cartography, especially through the integration of post-representational perspectives (Roberts 2012).

Further indicators of this renewed interest include the growing importance of map design and eye tracking techniques. A meta-analysis by Alexander Kent, published in *The Cartographic Journal* in May 2017 (Kent 2017), highlights this trend. By analysing the keywords used by authors in the journal between 2005 and 2017 (Wu et al. 2012), a clear evolution of the discipline can be observed (Fig. 6). Key topics identified include eye tracking, VGI (Volunteered Geographic Information), OpenStreetMap, and cybercartography, along with more established themes such as map design and the history of cartography.

The prominence of terms such as *eye tracking*, *usability*, *evaluation*, and *surveying*—with the first rivalling the frequency of *map design*—indicates a growing interest in evaluating the usability of cartographic documents and ensuring their effectiveness from the user's perspective. At the same time, concepts such as VGI, OpenStreetMap, citizen science, PGIS, and town plans reflect the ongoing democratisation of cartography and its incorporation into processes of participation and governmental transparency at the local level.

A notable increase in the volume of publications related to *eye tracking* has been observed since 2010. While various aspects of cartographic design—such as encoding, labelling, and layout—continue to be constant subjects of study, the current trend leans towards renewed cognitive aspects linked to the processes of map production and use. These include the employment of dynamic and interactive cartography across different geovisualisation platforms and the use of navigation tools on a range of (Kiefer et al. 2017; Krassanakis and Cybulski 2019; Punde et al. 2017).

As a final argument in support of the rising importance of Cognitive Cartography, it is worth noting the collaborative process led by the International Cartographic Association (ICA) to define a new research agenda (Griffin et al. 2017). Ten themes were identified through ten keywords, one of which refers to Usability, encompassing concepts such as user-centred map design, usability test, visual perception of map and spatial thinking, understanding and cognition (ICA 2025).

Additionally, geospatial databases are now more accessible than ever, owing to technological advances in data storage, processing, and distribution, resulting in an exponential increase in cartographic output (Ooms et al. 2014). Moreover, digital cartography, particularly online mapping, has become the primary means through which maps are consumed today.

This situation necessitates the use of emerging technologies to incorporate greater levels of user interaction and to embrace new conceptualisations that support more effective communication—potentially reframing cartographic design as a form of visual narrative, adopting the Anglo-American term Visual Storytelling (Song et al. 2022). In this context, visual storytelling provides an entry point for hybridisation in cartography, bridging technology and practice, product and process, and design and critique, while also opening up new and valuable avenues for transdisciplinary research and design (Roth 2021).

Design Meets Data: A Dual Approach to Measuring Map Use

As previously mentioned, it is essential to be able to identify what constitutes optimal use and an appropriate level of cognitive load in the process of map reading. This requires exposing users to different cartographic models to evaluate both aspects (Kiik et al. 2017; Song et al. 2022). In order to establish whether a map facilitates such optimal use and maintains a manageable cognitive load, it becomes necessary to assess how users perform when interacting with the map. To assess the outcomes of map use, the following elements are proposed for evaluation: (1) effectiveness: the accuracy with which a task associated with the map is completed; (2) efficiency: the speed at which a specific map can be read; (3) user reaction: the level of satisfaction expressed by the user regarding the map; and (4) retention: the user's ability to recall the information previously read.

To gather information on map reading and cognitive load, two complementary techniques are proposed: (1) a **subjective approach**, based on participant questionnaires, and (2) an **objective approach**, based on eye tracking data that records participants' eye movements while reading the map. Ultimately, both approaches must necessarily converge in order to obtain a comprehensive understanding of the user's cognitive experience during map interaction.

Methodological Proposal for a Subjective Approach Based on Questionnaires

Subjective data can be collected using questionnaires that include open-ended questions of varying length, which require participants to write responses freely, without predefined answer choices and closed-ended questions, which solicit brief, specific, and limited responses. In some cases, Likert-scale items will be used—psychometric tools where respondents indicate their level of agreement or disagreement with a statement, using an ordered, unidimensional scale (Matas 2018).

To assess cognitive processes and user interaction during and after map reading, a structured questionnaire is proposed. The instrument is designed to capture both performance and subjective experience, and is composed of several sections, to be administered at two key moments: (1) while the participant is reading the map and immediately afterwards; and (2) in a follow-up session to evaluate long-term retention.

Initial Questionnaire

The initial questionnaire comprises the following four sections:

- **Section A: Participant Profile**

  This section collects basic demographic and experiential data, including sex, age, prior experience in map reading, and familiarity with the specific region or theme represented in the map.
- **Section B: Task-based Multiple-Choice Questions**

  This section contains approximately ten closed-ended questions designed to be answered while observing the map. The questions focus on map-reading tasks such as identifying, comparing, ordering, or extracting information.

  Example: For a thematic map of the city of Zaragoza showing the distribution of ageing rates by city block, a typical question might be:

  '*The elderly population in the Las Delicias neighbourhood is: (a) Between 0 and 12%; (b) From 12 to 14%; (c) From 14 to 18%; (d) Over 18%*'.
- **Section C: Questions on Map Content (Post-reading, without map)**

  Administered immediately after the map is no longer visible, this section includes both open-ended and closed-ended questions aimed at evaluating the participant's understanding and recall.

  Example:

  Open-ended: '*Explain the map in your own words*'.

  Closed-ended: '*The Historic Centre and the University area have the same percentage of elderly population: (a) Agree; (b) Neither agree nor disagree; (c) Disagree*'.
- **Section D: User Evaluation of the Map**

  This section assesses the participant's subjective evaluation of the map through closed-ended items rated on a five-point Likert scale.

  Example:

1. 'Would you say you understood the map'?
2. 'Would you say you liked the map'?
3. 'Is the design of this map complex'?
4. 'Is it difficult to read this map'?
5. 'Is it difficult to answer the specific questions'?

Follow-up Questionnaire (Retention Assessment)

To evaluate long-term retention, participants may be invited to complete a simplified follow-up questionnaire approximately two months after the initial map reading. This version retains the core structure of the initial questionnaire but focuses specifically on memory and perceived understanding.

- **Section A: Recall of Map Content**
  This section includes both open-ended and closed-ended questions aimed at evaluating retained information.
  Example:
  Open-ended: '*Explain the map in your own words*'.
  Closed-ended: '*The Historic Centre and the University area have the same percentage of elderly population: (a) Agree; (b) Neither agree nor disagree; (c) Disagree*'.
- **Section B: Self-Assessment of Retention**
  This section gathers participants' subjective perceptions of their memory and understanding using Likert-scale items.
  Example:

1. 'Would you say you understood the map'?
2. 'Would you say you liked the map'?
3. 'Would you say you remember the information shown on the map'?

This combined structure allows for the integration of immediate cognitive and affective responses with delayed assessments of retention, providing a more holistic understanding of the map-reading experience and its cognitive demands.

Methodological Proposal for an Objective Approach Based on Eye Tracking Techniques

Objective information will be gathered through the registration of eye-tracking metrics. The function of eye trackers is to determine, based on ocular rotation, where the person directs their central vision—therefore, identifying the visual area being perceived most clearly at any given moment (Jacob 1995). The rest of the visual field is processed through peripheral vision, which does not allow for precise identification of objects or elements, but does enable the detection of changes—such as motion—that guide or capture attention. In fact, the movement of the mouse pointer itself is often a source of visual distraction during eye tracking tests (Goldberg and Wichansky 2002).

The fundamental assumption underlying eye tracking is that it provides a manifest measure of attention (Glenn Cummins 2017). The continuous monitoring of the gaze point reveals attentional eye movements that cannot be directly measured by the researcher (Jacob 1995). Eye movements occur in any visual processing activity as automatic responses that compensate for head or environmental movements, allowing the stabilisation of the retinal image on a fixed point.

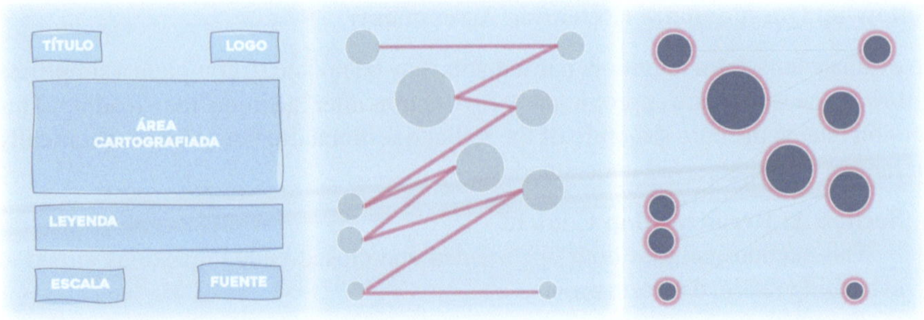

Fig. 7 Eye tracking structure: areas of interest (left) and basic eye movements: saccades (centre) and fixations (right). *Source* Carmen Bentué Martínez

Depending on their function, eye movements can be classified as follows:

(1) automatic compensatory movements that stabilise the retinal image by adjusting to head and environmental movements, allowing voluntary fixation on a specific point; (2) voluntary movements that shift the point of fixation across the visual field (*saccades, fixations*), or that follow moving objects (*pursuit* and *vergence* movements); and (3) micromovements associated with ocular fixation, including *tremors*, *microsaccades*, and *drifts*.

The two primary **types of eye movements** are saccades and fixations (Fig. 7). When we visually scan a scene, the eyes do not move smoothly but instead perform rapid jumps to redirect the visual axis to a new position (lasting 30–120 ms). These movements are known as **saccades** (Jacob 1995). During a saccade, visual perception is effectively suppressed—even though we are not consciously aware of this lapse. Between saccades, the eyes pause briefly in a relatively stable position (lasting 100–300 ms), allowing the focussed area to be seen clearly, this movements are **fixations**.

Eye tracking devices and software translate eye movements into what are known as eye tracking metrics. Most current software includes functionality to segment the visual stimulus into Areas of Interest (AOIs), enabling the extraction of specific metrics for each. These AOIs often correspond to the space occupied by individual map elements, although in some cases a grid-based segmentation is used, which tends to have less cartographic significance.

Eye trackers are typically categorised according to the type of hardware support into two main groups: mobile and remote devices (Punde et al. 2017). **Mobile eye trackers**, or head-mounted devices (usually in the form of glasses), record eye movements at short range (Fig. 8). These allow participants to move freely and are therefore commonly used in studies requiring tasks to be performed in natural environments. **Remote or screen-based eye trackers** (Fig. 9) record eye movements from a distance. These are the most widely used devices today. They are placed on or integrated into a screen that displays the visual stimuli with which participants interact. The system tracks the eyes within a designated area (*headbox*), while allowing sufficient freedom of movement for effective tracking.

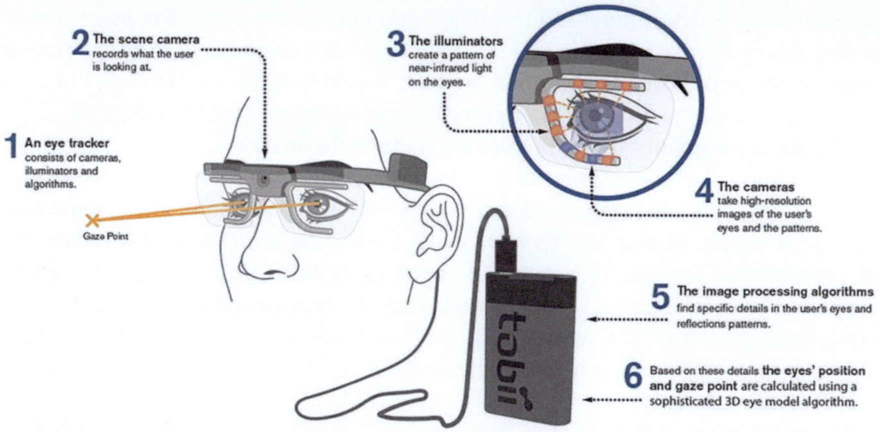

Fig. 8 Components of a mobile eye tracker (Tobii 2023)

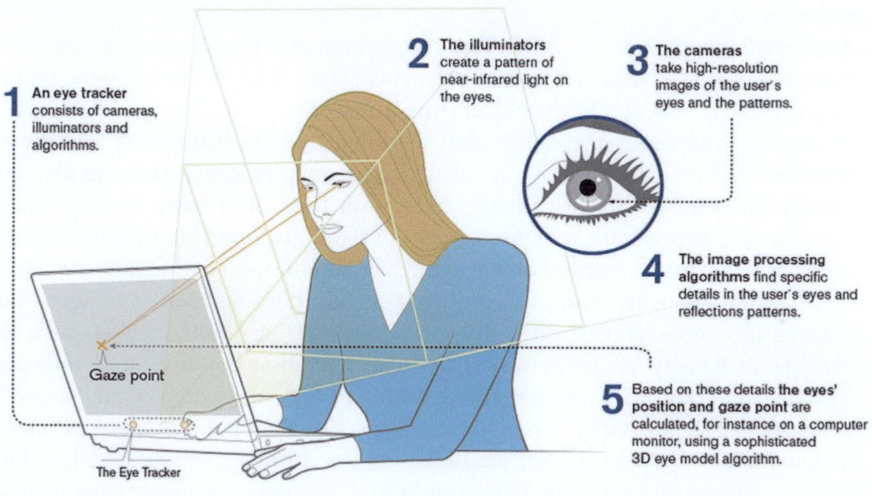

Fig. 9 Components of a screen-based eye trackers (Tobii 2023)

With few exceptions, the operational principles of these systems are broadly similar. They generally consist of one or more cameras, infrared light sources, and computing capabilities. The device emits infrared light (via *pupil illuminators*) to illuminate the eye and produce visible reflections. High-resolution cameras embedded in the device capture images of the eye, identifying the reflections on the cornea and pupil. Using advanced image processing and machine learning tools, the angle between the corneal and pupillary reflections is calculated. Based on this vector, the gaze direction is determined and translated into gaze points—the data points representing where the participant is looking. Contemporary devices and software include mechanisms to perform a calibration process, which generates a 3D model of each participant's gaze, ensuring greater precision and accuracy in data collection.

The most commonly used eye tracking metrics for analysing the graphical properties of a visual scene are outlined below, alongside examples of how these may be interpreted in cartographic evaluation experiments. It is important to emphasise that eye tracking, in essence, merely reveals where an individual looks; thus, the interpretation of the resulting data must be made within context (Glenn Cummins 2017). In the case of cartography, this includes the experimental design and objectives, the map's properties, which result from the sequence of decisions made during the cartographic process, and the profile and characteristics of the user undertaking the observation.

Eye tracking metrics (Glenn Cummins 2017; Kiik et al. 2017; Rayner 1998)**:**

- **Dwell Time**: This refers to the total duration of time spent observing a map. In an experiment involving either free viewing or task resolution, it may be useful to measure the time taken to complete the task or the proportion of time spent on different AOIs.
- **Fixation Duration on an AOI**: This metric indicates how long a specific area was observed. The duration of a fixation is considered to be related to the depth of cognitive processing of the object (Rayner 1998).
- In relation to saccades, the most widely used metric is **the Number of Fixations**, recorded as the count of fixations made during map reading. This can also be expressed as a percentage of total observation time. In a basic thematic map free-viewing experiment, a higher number of fixations might be expected on key elements such as the mapped area and the legend, followed by the title and other auxiliary components. However, an unexpectedly high number of fixations on a given area may suggest difficulty in comprehension. Other metrics, such as the spatial density of fixations or the average fixation duration, may indicate visual search efficiency. A more concentrated pattern of fixations could suggest efficient search behaviour, while greater dispersion may point to a less efficient visual strategy. Another useful metric is the time to first fixation, based on the premise that the quicker a user fixates on an AOI, the more effectively its graphical properties capture visual attention.
- Saccade metrics can also be expressed as counts or percentages relative to total observation time, in addition to their characteristic measurement—**saccade length**, generally expressed in degrees. One may infer that a certain layout configuration is more efficient than another if it reduces the **total number or amplitude of saccades**. For example, in a cartographic composition featuring multiple maps, if each map's legend is located close to the corresponding map, the user might read the content more quickly than if all legends were grouped elsewhere.
- **Scanpath**: This refers to the sequence of eye movements (saccades and fixations) across a stimulus over a given time. It has both a beginning and an end, and therefore a measurable length. It may be of interest to analyse how this sequence varies

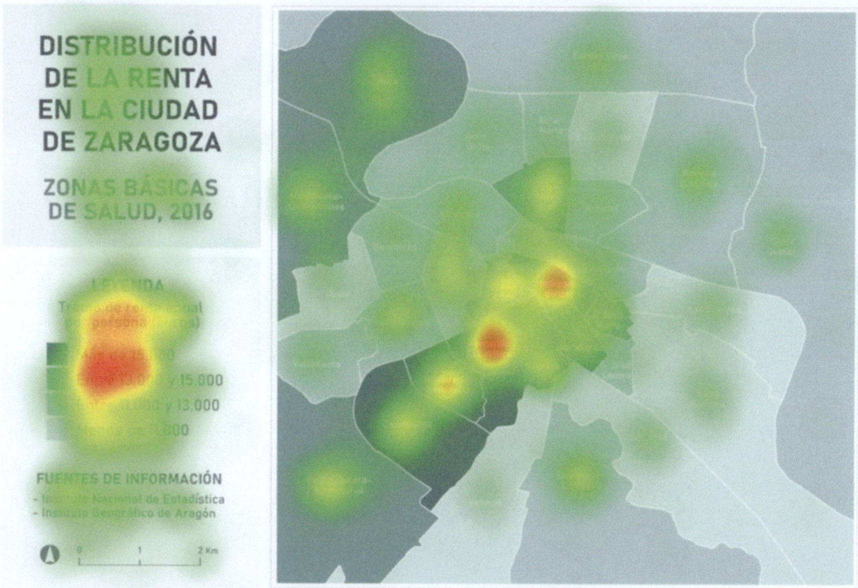

Fig. 10 Example of a heat map generated from eye tracking data. *Source* Carmen Bentué Martínez

according to different layout configurations, as well as based on user profiles. On one hand, one might expect a scanpath consistent with the visual hierarchy principles upon which the layout was based. On the other hand, users with a specific profile (e.g., more experience in map reading) might exhibit more regular and ordered sequences.

Finally, it is worth mentioning the **graphical representations** that some software tools can generate in order to summarise user behaviour visually (Hassan Montero and Herrero-Solana 2007). The most widely recognised of these are heat maps (Fig. 10), which aggregate gaze points and fixations to show a gradient between areas of higher and lower attention. It is also possible to obtain a graphical representation of fixations and saccades, where the larger the circle, the more time the eye spends on that point (Fig. 11). In these visualisations, the number inside each circle indicates the temporal sequence in which that point was visited, allowing researchers to reconstruct the order of visual exploration.

Overall Assessment of Map Reading

To fully understand how users engage with a map, it is necessary to move beyond isolated metrics and adopt an integrated evaluation framework. This section proposes a comprehensive approach that synthesises both subjective and objective data—combining questionnaire responses, eye-tracking metrics, and contextual variables

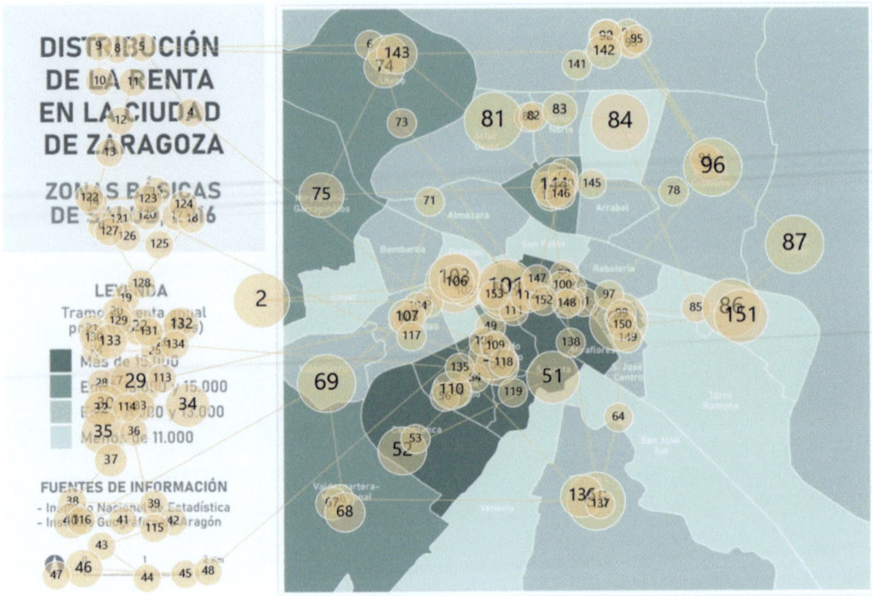

Fig. 11 Example of a graphical representation of fixations and saccades on the map generated from eye tracking data. *Source* Carmen Bentué Martínez

related to map design and user profile. By bringing together these dimensions, the goal is to assess not only the cognitive effort required to read a map but also the outcomes of that effort in terms of effectiveness, efficiency, satisfaction, and user reaction. This holistic perspective enables a more robust evaluation of cartographic products, guiding future improvements in design and usability.

To **calculate overall cognitive load**, three specific measures are selected: total reading time, number of fixations, and average fixation duration. In addition, it is possible to calculate variables associated with the factors that influence cognitive load. These can be categorised as map-related and user-related factors. Map-related factors include complexity, which is calculated based on the response to the question 3 in Section D—'Is the design of this map complex'?—as answered by both the user and the person who designed the map. Task difficulty is calculated as the average of the responses to questions 4 and 5 in the same section. User-related factors are also considered. The reading environment is evaluated using a 1–5 scale to reflect the conditions of tranquillity or distraction during the map-reading process. The user's profile is derived from the information collected in Section A, including age, gender, level of education, experience in reading maps, and familiarity with the region or topic represented.

The **outcomes of map use are measured** through responses to the questionnaire and through selected eye tracking metrics. Effectiveness is assessed by the scores obtained from the questions in Sections B and C, and is validated by the response to question 1 in Section D, which asks whether the participant felt they had understood the map. Efficiency is calculated using eye tracking data, taking into account the total time spent viewing the map freely, the average time taken to complete the tasks, and the sum of both measures. User reaction is gauged through question 2 of Section D, which asks whether the participant liked the map. Retention is assessed in two phases: short-term retention is based on the responses to Sections B and C of the questionnaire completed immediately after viewing the map, while long-term retention is measured by a simplified questionnaire administered two months later.

Once the cognitive load of a map has been calculated, along with the identification of its influencing factors and an analysis of map-reading outcomes, it becomes possible to assess whether the design of the map was appropriate. From this assessment, potential improvements can be identified, and directions for refinement can be proposed.

Unfinished Maps: Reflections, Challenges, and Future Routes

This chapter began with the idea that maps are geography's natural language since they are powerful tools for communicating spatial phenomena. Rooted in the information perspective, we examined how cognitive cartography builds on this foundation by integrating knowledge from cognitive science, psychology, and perception theory into the cartographic design process.

By understanding maps as communication artefacts involving both a producer and a user, cognitive cartography offers a richer framework for map evaluation and improvement. The exploration of perception, attention, memory, and especially cognitive load, reveals how users interact with maps not merely as static visuals but as dynamic cognitive experiences.

We have proposed a dual methodological approach by combining subjective surveys and objective eye tracking techniques to assess map effectiveness, efficiency, user satisfaction, and user retention. These tools not only allow us to quantify cognitive effort but also open pathways for adaptive and inclusive design, tailored to diverse user profiles and contexts.

As cartography evolves in a digital and participatory age, it becomes ever more necessary to reconsider the role of the map user, not as a passive reader, but as a co-constructor of meaning. Future research in cognitive cartography should continue to bridge design and cognition, promoting evidence-based guidelines that respond to human mental processes while embracing new forms of visual storytelling.

Ultimately, by placing the user at the centre of the cartographic process, we not only refine our maps, we redefine what it means to truly read them. This work is supported by the project "Digital maps in the classroom: innovation in the learning of geographical competencies through eye-tracking techniques" (code PID2024-155735OB-I00), funded by the State Research Agency (Spain).

References

Baddeley A (1986) Working memory. Claredon Press

Berlo DK (1987) El proceso de la comunicación. Introducción a la teoría y a la práctica, Buenos Aires, Librería El Ateneo, 173 pp

Bertin J (1967) Sémiologie graphique. Les diagrammes—Les reseaux—Les cartes. Gauthiers-Villars

Bunch RL, Lloyd RE (2006) The cognitive load of geographic information. Prof Geogr 58(2):209–220

Caquard S (2015) Cartography III: a post-representational perspective on cognitive cartography. Prog Hum Geogr 39(2):225–235. https://doi.org/10.1177/0309132514527039

Etece EE (2023) Percepción—Concepto, etapas y componentes. https://concepto.de/percepcion/

Glenn Cummins R (2017) Eye tracking. In: The international encyclopedia of communication research methods. Texas Tech University, USA, pp 1–9

Goldberg JH, Wichansky AM (2002) Eye tracking in usability evaluation: a practitioner's guide. In: To appear in: Hyönä

Green K, Fairchild R, Knudsen B, Lease-Gubrud D (2023) Introduction to communication. Minnesota State Colleges and Universities

Griffin AL, Robinson AC, Roth RE (2017) Envisioning the future of cartographic research. Int J Cartography 00(2):1–8. https://doi.org/10.1080/23729333.2017.1316466

Hassan Montero Y, Herrero Solana V (2007) Eye-tracking en interacción persona-ordenador. No Sólo Usabilidad: Rev Multidisciplinar Sobre Diseño Personas Tecnología. http://www.nosolousabilidad.com/articulos/eye-tracking.htm

ICA (2025) Structure of the research Agenda. https://icaci.org/research-agenda/structure/

Jacob RJK (1995) Eye tracking in advanced interface design

Jenks GF (1973) Visual integration in thematic mapping: fact or fiction? Int Yearbook Cartography 13:27–35

Jenks GF (1975) The evaluation and prediction of visual clustering in maps symbolized with proportional circles. In: Davis J, Maccullaugh M (eds) Display and analysis of spatial data. Willey, New York, pp 311–327

Kent A (2017) Mapping cartography: towards some insights. Cartogr J 54(2):103–105. https://doi.org/10.1080/00087041.2017.1318471

Kiefer P, Giannopoulos I, Raubal M, Duchowski A (2017) Eye tracking for spatial research: cognition, computation, challenges. Spat Cogn Comput 17(1–2):1–19. https://doi.org/10.1080/13875868.2016.1254634

Kiik A, Nyström M, Harrie L (2017) Cartographic design matters—a comparison of thematic polygon design. Cartogr J 54(1):24–35. https://doi.org/10.1080/00087041.2016.1147191

Koláčný A (1969) Cartographic information—a fundamental concept and term in modern cartography. Cartogr J 6:47–49

Krassanakis V, Cybulski P (2019) A review on eye movement analysis in map reading process: the status of the last decade. Geodesy Cartography 68(1):191–209. http://journals.pan.pl/dlibra/publication/edition/110001

Lambert N, Zanin C (2023) Practical handbook of thematic cartography: principles, methods, and applications. CRC Press

Llasera J (2023) Jerarquía visual. https://imborrable.com/blog/jerarquia-visual/

Matas A (2018) Diseño del formato de escalas tipo Likert: un estado de la cuestión. Rev Electrónica Inv Educativa 20(1):38–47

Mocnik FB (2022) Why we can read maps. Cartogr Geogr Inf Sci. https://doi.org/10.1080/15230406.2022.2127911

Montello DR (2002) Cognitive map-design research in the twentieth century: theoretical and empirical approaches. Cartogr Geogr Inf Sci 29(3):283–304

Ooms K, De Maeyer P, Fack V (2014) Study of the attentive behavior of novice and expert map users using eye tracking. Cartogr Geogr Inf Sci 41(1):37–54

Ooms K, De Maeyer P, Fack V (2015) Listen to the map user: cognition, memory, and expertise. Cartogr J 52(1):3–19. https://doi.org/10.1179/1743277413Y.0000000068
Paas F, Renkl A, Sweller J (2003) Cognitive load theory and instructional design: recent developments. Educ Psychol 38(1):1–4
Palsky G (2017) ¡La Semiología gráfica de Jacques Bertin cumple cincuenta años! VisionsCarto. https://visionscarto.net/semiologia-grafica-bertin
Płużyczka M (2018) The first hundred years: a history of eye tracking as a research method. Appl Linguist Papers
Punde PA, Jadhav ME, Manza RR (2017) A study of eye tracking technology and its applications. In: 1st international conference on intelligent systems and information management (ICISIM), pp 86–90. https://doi.org/10.1109/ICISIM.2017.8122153
Rayner K (1998) Eye movements in reading and information processing: 20 years of research. Psychol Bull 124(3):372
Roberts L (2012) Mapping cultures: place, practice, performance. Springer
Robinson AH (1952) The look of maps. University of WIsconsin Press
Rock I, Palmer S (1990) The legacy of Gestalt psychology. Sci Am 263(6):84–91
Roth RE (2021) Cartographic design as visual storytelling: synthesis and review of map-based narratives, genres, and tropes. Cartogr J 58(1):83–114. https://doi.org/10.1080/00087041.2019.1633103
Song Z, Prestby T, Roth RE, Iverson A, Houtman L, Gao S (2022) Visual storytelling with maps: an empirical study on story map themes and narrative elements, visual storytelling genres and tropes, and individual audience differences. Cartographic Perspect 2022(100):10–44. https://doi.org/10.14714/CP100.1759
Steinke TR (1987) Eye movement studies in cartography and related fields. Cartographica: Int J Geogr Inform Geovisualization 24(2):40–73. https://doi.org/10.3138/J166-635U-7R56-X2L1
Tobii (2023) How do Tobii eye trackers work? https://connect.tobii.com/s/article/How-do-Tobii-eye-trackers-work?language=en_US
Vargas LM (1994) Sobre el concepto de percepción. Alteridades 4(8):47–53
White TM (2018) The George F. Jenks map collection. Cartographic Perspect 91:114–121
Wu B, Xiao H, Dong X, Wang M, Xue L (2012) Tourism knowledge domains: a keyword analysis. Asia Pac J Tourism Res 17(4):355–380
Zúñiga Antón M (2009) Propuesta cartográfica para la representación y análisis de la variable población mediante Sistemas de Información Geográfica: el caso español. https://zaguan.unizar.es/record/4149

Urban–Rural Migration—An Opportunity or a Threat for Rural Areas in Poland

Monika Wesołowska and Anita Kulawiak

Abstract The chapter presents the conclusions of part of the investigation conducted within the research project: 'Rural–urban knowledge transfer—models of interdependence,' developed by a group of researchers from Poland. In its fundamental-theoretical assumptions, the project refers to the contemporary discourse on the socio-cultural turn in geography. The essence of the project is the identification and description of the broadly understood interactions between old and new inhabitant. Ultimately, the research aims to develop a model of interdependence in the urban–rural transfer of knowledge and information. The findings presented are the result of literature studies and interviews conducted with key respondents, civil society groups migrants, and field observation. The study's starting point was to identify and define the types of people who decide to move to the countryside in Poland and, in this context, to try to present the profile of the Polish counterurbanite. In the light of the literature, the difference between the suburbanite and the counterurbanite is based on cultural distinctiveness and a different lifestyle, which they lead and prefer in a rural environment. The research undertaken by the authors confirmed this regularity.

Keywords Poland · Urban–rural migration · Suburbanite · Counterurbanite · Knowledge transfer

Introduction

Since the 1960s, there has been an increased interest in urban-to-rural population migration in the United States and Western Europe. The observed trend has resulted in changes in rural and urban populations and has significantly increased researcher interest in this issue. An analysis of the number of papers published in English through

M. Wesołowska (✉)
Maria Curie Skłodowska University, Lublin, Poland
e-mail: monika.wesolowska@mail.umcs.pl

A. Kulawiak
University of Łódź, Łódź, Poland
e-mail: anita.kulawiak@geo.uni.lodz.pl

L. Soares et al. (eds.), *Compromised Geography*, Key Challenges in Geography,
https://doi.org/10.1007/978-3-032-05201-8_14

the *Google Scholar* research engine, using the keyword *counterurbanization* in the title of the text, showed that while 64 papers were published between 1960 and 1990, between 1991 and 2024, the number rose to 123, an increase of almost 200%. Among the first to address the problem of urban–rural migration in terms of the process of counterurbanization were Berry (1976) and Beale (1974, 1976), who analysed the issue concerning the United States. Their contribution was to observe and describe the phenomenon of the so-called population *turnaround* (*population turnaround*, *population reversal*), consisting of a decrease in population in central (urban) areas and an increase in peripheral ones (Grzeszczak 2000: 373). Subsequently, the topic has been pursued in Western European countries, especially in the UK, where not only many empirical studies on the subject have been produced, but also discussions have been undertaken to develop a theoretical and methodological basis for the issue of counterurbanization (e.g. Fielding 1982, 1983; Cloke and Goodwin 1992; Cloke 1985; Champion and Atkins 1996; Halfacree 1994). The importance of the UK in developing this topic is also evidenced by the fact that, as pointed out by Dej and Zajda (2017), these studies often represent multi-year research programmes conducted by the same authors (e.g. Champion 1987; Boyle et al. 2013). In Poland, as in other countries of Central and Eastern Europe, despite the great interest in migration from the city to the peripheral countryside, research in this area was practically absent, so the occurrence of this process, its scale, and characteristics are not well recognized to this day. Much more extensively, the topic of urban–rural migration has been developed only from the perspective of the suburbanization process. Recently, an attempt has been made by Wesołowska and Kulawiak (2025) to organize the research topics addressed in this area in geographical terms. It is also worth adding that moving from the city to the countryside is spreading so widely that it has attracted the interest of academics and the media world. Proof *of this is* the appearance on *the* Polish market of a specialist magazine entitled '*Poza miastem' (Out of Town*), aimed mainly at people considering moving to the countryside, as well as the massive popularity of *a series of* reportages on Polish television entitled '*Daleko od* miasta' (*Far from the city*), which present 'city dwellers' who have decided to move to the countryside.

Considering the literature research, there are many reasons why peripheral villages are gaining in popularity. It is worth noting, however, that slightly different ones are mentioned in the English-language literature (USA, UK), and slightly different ones are articulated in Polish or more widely in Central and Eastern Europe. Researchers only agree that counterurbanization cannot be explained by just one isolated factor and that the same factors can, depending on the situation, either foster or inhibit counterurbanization tendencies (Llleris 1990; Grzeszczak 2000).

In English-speaking countries, researchers most often identify the causes of counterurbanization as economic determinants—distinguishing between economic cyclical factors and economic structural factors—in addition to technological innovations and direct or indirect governmental policies, followed by socio-cultural and environmental factors (Champion 1987; Frey 1987, 1988; Kontuly 1998; Kontuly and Dcarden 1998). In the Polish literature, however, greater importance than *strictly* economic factors are attributed to factors which, as Sant and Simons (1993) put it, have a demand (consumption) form, under which individual decisions of people looking for places to live with specific environmental values are hidden. In this regard, safety, the friendliness of the rural population, recreational opportunities and the beauty of the landscape, as well as remoteness from civilization and the practice of healthy lifestyles, are most frequently indicated in the literature (Ołów-Wachowicz 2020; Dolińska et al. 2023). Thus, what mainly attracts new settlers to the Polish countryside located peripherally to the city is related to the broadly understood 'concept of rurality', under which there is a desire to realize the dream of living in a natural and uncontaminated by the 'dirt' of civilization environment, defined by traditions, authenticity, attachment to the land and custom, and patriotic devotion and sacrifice. Indeed, considering research to date, counterurbanites are a social group strongly influenced by idealized images, where the countryside is treated as a source of pure and natural truths. This fact was pointed out, among others, by Laskowska-Otwinowska writing: '…in groups of Novosadniks it was the imagination that first became the driving force behind behaviour of a contesting nature, including escape to the countryside, and over time also the material for the realization of social activities of an ideological, educational and artistic nature…' (Laskowska-Otwinowska 2008: 213). In the case of this group of migrants, it is, therefore, counterurbanization, which the literature refers to as *lifestyle-driven counterurbanization.*

As mentioned in the summary, this article presents conclusions from part of the research conducted by a group of researchers from Poland as part of the research project of the National Science Center 'Transfer of knowledge between the countryside and the city—models of interdependence'. The essence of the project is to identify the mutual influence of old and new residents of the countryside, including users of second homes. The final effect of the project will be an attempt to construct the title model.

The project set four specific objectives: i.e. identify and diagnose groups of newcomers and forms of their migration to the countryside; identify categories and types of knowledge subject to transfer; and determine the dominant directions and intensity of knowledge transfer and their consequences. This chapter presents the results of the research related to the realization of the first objective, including an attempt to answer four questions:

1. *What types of groups of people are migrating to the countryside?*
2. *What forms does this migration take?*
3. *Where are migrants concentrated in Poland?*
4. *What are the characteristics of this rural space?*

Therefore, the chapter focuses firstly on discussing the socio-cultural characteristics of migrants who have decided to move from the city to a non-metropolitan countryside and, secondly, on identifying the areas and characteristics that counterurbanites choose as their new residence. The literature studies and their research led the authors to reflect on the positive and negative aspects (opportunities and threats) related to the process of counterurbanization, from the perspective of rural areas. This issue seemed interesting to the authors because, despite researchers' ever-growing interest in counterurbanization, it is mainly considered from an urban perspective. The 'rural' perspective, including in the Polish literature, has not yet been extensively studied.

Research Methods and Source Materials

Various research materials were obtained through two research methods to achieve the stated objectives. The presented findings are the result of both an extensive literature search, including official statistical data, social research, a field inventory, questionnaire, and in-depth interviews, with key respondents (public sector—local government, important rural institutions), civil society groups (e.g. NGOs), migrants (leaders, new farmers). The survey was addressed to a purposive sample of respondents adequate to the research subject (indigenous people living in the village and newcomers). The material was collected during research as part of the already referred project funded by the National Science Centre (UMO-2021/41/B/HS4/02055) and was conducted between 2023 and 2024 on a group of adult residents of the villages selected for the study who had moved from town to village within at least the last 5 years. A total of 40 casual interviews and 379 questionnaire interviews were conducted. Then, based on bibliography and our research, the socio-cultural profile of counterurbanites and the motives driving the population to non-agglomeration villages were characterized.

According to the survey, the group of counterurbanites is made up of middle-aged people, i.e. between 40 and 59 years of age (41%), and, in second place, relatively young people, i.e. between 26 and 49 years of age (24%). Occasionally, the newcomers also include pensioners, i.e. people over 65, but it is worth mentioning that they have already reached retirement age while living in the countryside. More than 60% of respondents moved out of the city to the countryside more than 5 years ago, and 64% of them from a large city (with more than 100,000 inhabitants). A reasonably high occupational position also characterizes the surveyed group. Most are well-educated people (specialists in various fields, e.g. engineers, doctors, lawyers, scientists). More than half of the respondents (52%) declared a university degree and another (nearly 40%) a secondary education. One in three respondents is self-employed.

The villages selected for the study are shown in Fig. 1. They are located in six voivodeships: Mazowieckie, Lubelskie, Łódzkie, Kujawsko-Pomorskie, Dolnośląskie, and Pomorskie. Their selection was purposeful and dictated by two conditions. Firstly, the examined villages are located within the territory of the voivodeships from which the researchers carrying out the above-mentioned research project originate. Secondly, only those units located outside the zone of direct influence of large cities were considered, in this case, the capitals of the voivodeships in which they were located.

In turn, statistical data from the Central Statistical Office in Poland for 2002–2021 was used to identify areas of high interest to Polish migrants. Spatial analyses covered 1769 rural municipalities and rural areas of urban–rural municipalities located outside

Fig. 1 Villages surveyed as part of the 'Urban–rural knowledge transfer—interdependence models' project. *Source* Own elaboration

the agglomeration areas designated by the Central Statistical Office.[1] Then, based on

[1] The CSO's rural delimitation was developed using the typology of Functional Urban Areas (FUAs). FUAs include cities and their commuting zones. The delimitation of rural municipalities (rural municipalities and rural areas in urban-rural municipalities—TERYT unit type symbols 2 and 5, respectively) considered the impact of large cities on them. The delimitation includes level 1, which refers to the location of the rural area to FUAs of cities with at least 150,000 inhabitants or FUAs of provincial cities; level 2, which refers to the value of population density of rural areas. The delimitation distinguishes the following groups of areas:

1. agglomeration—rural areas located within the FUAs of provincial cities or the FUAs of other cities with at least 150,000 inhabitants:

- high-density agglomeration—meets both the conditions for agglomeration areas and the condition of population density higher than the average population density for Poland.

- low-density agglomeration—meet both the conditions for agglomeration areas and the condition of population density equal to or lower than the average population density for Poland.

2. Non-agglomeration—rural areas outside the FUAs of provincial cities or the FUAs of other cities with at least 150,000 inhabitants:

literature and research, an attempt was made to assess the opportunities and threats brought by migration from cities to non-agglomeration rural areas.

The authors are aware that the present research, based only on limited empirical material, does not constitute a complete treatment of the assumed problem or an exhaustive description of it. However, the collected material may provide a starting point for further in-depth research.

Background of the Study in Polish Literature

The topic of urban–rural peripheral (non-metropolitan) migration in Polish geographical literature is a relatively new research issue, as it only appeared on a broader scale in the second half of the 1990s with the change of social and economic foundations in Poland and the period of transformation (Więcław-Michniewska 2006). This does not mean, however, that the issue of urban–rural migration had not appeared in the literature before. On the contrary, it has quite a long tradition, originating from settlement geography studies. It should be noted that research in this area was conducted almost in parallel with the geography of rural and urban settlements. The functionalist current in the geography of rural settlements made a particular contribution to this research. This is because, within its framework, the concept of the 'suburban village' was developed (Straszewicz 1954, 1985; Liszewski 1987a, b), including those of a recreational and touristic nature which became a contribution to the initiation of research into tourism space (Matczak 1985 after: Wójcik 2013), and with it and the impact of tourism migration on permanent settlement in these areas (Matczak 1986, 1991; Dziegieć 1987; Szkup 2003). In this early period, geographical studies of peri-urban villages focussed primarily on quantitative approaches, including the analysis of their spatial and functional structure, with the possible addition of life quality aspects to the research (Liszewski 1987a, b, 2005; Jakóbczyk-Gryszkiewicz 1991, 1998; Wójcik 2008, 2013; Włodarczyk 2009). Within urban studies, on the other hand, the process of urban–rural migration has been analysed as an element of the suburbanization process, focussing attention mainly on its quantitative spatial aspects, including consideration of its intensity, size, and location (Ilnicki 2002a, b). A popular area of research in this field has been, as with rural areas, also morphological, settlement, and functional transformations of these areas in the context of the migration process (Kowalewski et al. 2018; Kajdanek 2011, 2012; Miszewska

- High-density non-agglomeration areas—meet both the conditions for non-agglomeration areas and the condition of population density higher than 1/3 of the population density for Poland.

Low-density non-agglomeration areas meet both the conditions for non-agglomeration areas and the condition of population density equal to or less than 1/3 of the population density for Poland.

The provincial and other cities group included twenty-five centers with more than 150,000 inhabitants. The population density in 2023 was 119.9 persons/km2, and the threshold of 1/3 of the average population density was 40.0 persons/km^2.

https://stat.gov.pl/statystyka-regionalna/jednostki-terytorialne/delimitacja-obszarow-wiejsk ich-dow-/

1985), as well as issues related to the change in the population of suburban areas (Prochownik 1975).

The 1990s brought a fundamental turnover in research on urban–rural migration, which resulted not only from the increase in the number of people deciding to move to the countryside, but also from a particular cultural transformation that took place in Polish geography at that time, consisting of a shift from positivist approaches to post-structural approaches and human geography. Paying more attention to society as a source of change in contemporary rural and urban areas significantly expanded and enriched the previous scientific output concerning the issue of migration. The aftermath of this change has been works describing the process of urban–rural migration from the perspective of the local communities involved in the process, including those exceptionally devoted to the study of the relationship between new and old villagers, their lifestyles, cultural difference, their living conditions and quality of life, as well as the types of settlers and their motives for moving to the countryside. These works have been written mainly in the fields of sociology and cultural anthropology (Kajdanek 2011, 2012), as well as geography (Szkup 2003; Wojciechowska 2009; Szmytkowska and Masik 2005) and formed the basis for drawing a picture of Polish suburbanization.

On the wave of achievements of foreign authors, the first studies on counterurbanization in Poland appeared in the 1970s and 1980s (Kukliński 1978; Węcławowicz 1979; Korcelli 1984). However, these isolated works did not arouse Polish researchers' interest, probably due to the low interest of urban residents in moving to the peripheral countryside. This is because, during the communist period in Poland, the ideology of the primacy of the city, industry, and the working class over the countryside was in force, and therefore, for years, it was the escape to the city that was a kind of social advancement (Zdrowicka-Wawrzyniak 2022). This statement is confirmed by the results of a study by, among others, Kupiszewski et al. (1998) on the redistribution of the population between 1984 and 1994, where it is stated that: '…in Poland so far, no signs of counterurbanization can be detected…' (Kupiszewski et al. after Grzeszczak 2000: 387). A unique contribution to the initiation of research on the process of counterurbanization in Poland, including in geography on a slightly broader scale, was made by Grzeszczak's studies (1996, 1998, 2000). This author was the first to present to the Polish reader the course of counterurbanization in various countries in the 1970s and 1980s and to discuss the development of views on the spread and persistence of this process, its essence, and conditions. Following him, research in this area was undertaken by, among others, sociologists and anthropologists which consider the process of counterurbanization mainly through the prism of social relations between old and immigrant groups of inhabitants and their consequences for rural areas (Kwiatkowska 2005; Laskowska-Otwinowska 2008; Wrona 2020; Dej and Zajda 2016, 2017; Wieruszewska 1992; Sztandara 2003), as well as individual geographers (Bednarek-Szczepańska and Janc 2025; Wesołowska and Kulawiak 2025).

However, these topics, especially in the field of geography, are still overshadowed by the issue of the suburbanization process. As a result, the scale and characteristics of the process of counterurbanization remain little recognized. As pointed out by

Dej and Zajda (2016), mentions of this topic appear only on broader descriptions of migration processes and are not supported by in-depth research (Szymańska and Biegańska 2011; Heffner 2016; Degórska 2017; Czarnecki 2019; Ruczkowski 2022; Michalska-Żyła 2017).

Who is a Counterurbanite? A Profile Analysis in the Light of Literature and Own Research

Paying attention to the socio-cultural profile of migrants is important because, as an analysis of the available literature on the subject indicates, migrants who move from the city to a village located far from it (beyond the commuting hinterland) manifest slightly different characteristics than those who chose to live in its periphery (Laskowska-Otwinowska 2008; Wrona 2020; Mitchell 2004, 2019; Tammaru et al. 2023; Wesołowska and Kulawiak 2025). Therefore, the literature distinguishes two categories of such people—suburbanites—described as representatives of suburban culture and include residents of villages on the urban periphery (O'Connor 1985; Nicolaides 1993; Williamson 2008) and *counterurbanites (novosadniks, vioszczanians)* who have settled far beyond the city's sphere of influence (peripheral village), but because of their philosophy and specific lifestyle are called representatives of alternative culture or counterculture, which means that they represent 'a set of cultural patterns autonomous from the dominant culture' (Kurczewska 2004; Burszta and Kuligowski 2005; Laskowska-Otwinowska 2008; Wrona 2020).

Considering the literature and our research, these groups represent different lifestyles and a completely different socio-cultural profile. Consequently, they also differ in their motives for living in a village and their attitudes towards the village and the local community (Table 1). What the newcomers from the city have in common, regardless of the place where they have settled, is that for them the countryside is an acquired, transactional commodity, a bought idea for life, a consumption space (Wesołowska and Kulawiak 2025). Compared to the Indigenous rural population, they are also people who are much better educated, younger, and have already achieved specific professional and material stability. It is worth mentioning that in terms of age, this is the opposite trend to that in Western Europe, where already in the 1980s, mainly pensioners prevailed among urban migrants, reinforcing, as noted by Sofranko et al. (1982) among others, the unfavourable demographic situation in rural areas.

Who, then, are the counterurbanites (neo-settlers) in the light of our research and the literature? Although contemporary literature provides more and more descriptions of this social group, it is challenging to outline their socio-cultural profile unambiguously. They are often characterized through the prism of the need to cultivate traditional values derived from rural culture. The common denominator linking these studies, especially on the Polish ground, is the recognition that this group manifests many of the characteristics inherent in the 'ethos of the peasant intelligentsia', which,

Table 1 Types of urban–rural migrants in the light of the literature on the subject

Suburbanita	Counterurbanite (Novosadnik)
They represent the culture of the suburbs, the embodiment of dominant consumerism	Representatives of alternative culture (counterculture) reject consumerist lifestyles
A resident of a former communist bloc pushed out of the city by uncomfortable living conditions	Moving to the countryside was an escape from the dominant culture and social patterns of the communist period, not politically driven or determined by poor living conditions
Personal role model identified with corporate employee (seeks satisfaction by fitting in, does not engage in competition)	They are pursuing their dreams of running a rural farm or escaping corporate jobs
In the villages, they are looking for a large, secluded home and proximity to wildlife	They are looking for simplicity, naturalness, self-sufficiency, and archetypes in the countryside
The group least culturally integrated compared to other villagers does not meet the criteria for a social group	The group strives to integrate into the local community and seeks to preserve the traditional cultural heritage of the village
They are in constant contact with the city due to the need to meet their many needs in terms of education, work, leisure, culture	This migration mainly concerns artistic circles, which locate festivals, workshops, and art exhibitions in the rural space. They also realize themselves in artistic activities based on the countryside's natural resources, such as pot making and sculpting
A man who seeks to put down roots in a place but is afraid of the effort and relationships involved	They speak with nostalgia about the disappearance of traditional rural lifestyles associated with agriculture and community
A whole series of meanings are attributed to them, i.e. consumerism, corporatism, and individualism	Fascination with ecology, agro-tourism, and the preservation of traditional ways of life and local cultural identities

Source Wesołowska and Kulawiak (2025)

in a nutshell, was characterized by anti-industrialism, anti-urbanism and an orientation towards the cultivation of regionalisms, including the study of folk culture, a conviction of the need to recreate and preserve local traditions, as well as the need to awaken local patriotism (Laskowska-Otwinowska 2008: 205). Counterurbanites build their system of values based on a worldview borrowed from Polish national culture, inspired by its components primarily derived from peasant tradition. Consequently, they are close to values traditionally identified with the countryside, such as nature, simplicity, harmony, self-sufficiency, and living in harmony with the natural order of the world, i.e. everything that stands in opposition to the cosmopolitan city.

Moving to the countryside is, therefore, primarily for this group, a search for a 'more humane way of life', as well as a release from the obligations of the big city. Researchers also often emphasize that this is a group that remains strongly influenced by the 'myth of the Arcadian countryside' (Halfacree 1994, 2007; Kwiatkowska 2005), and therefore, by moving to the countryside, they try not only to rescue its traditions and identity, which is expressed, among other things, by undertaking activities in the rural environment based on the natural and cultural resources of the countryside, but also to create their life in the countryside based on its idealized image. In the literature describing this issue, it is often emphasized that this group is particularly susceptible to assimilating images of the countryside, which are suggested to them by the media, advertising, or mass media, and based on these images, they create their own 'rural idyll' (Halfacree 1994). This makes them form a social group and a community of a particular character. Analysing New Orchard groups in Poland, Sztandara (2003) noted that although their actions are an exhibition of individual feeling, of individual expression, they have a community character. As she concludes, these communities, built on similarity and a sense of closeness, base their existence on various combinations of three slogans: ecology, community, and spirituality (Sztandara 2003: 48). These general observations were also confirmed in the present research. In their light, counterurbanites are people looking for, above all, in the countryside: contact with nature and wildlife, a respite from the ideology and rhythm of life imposed by the city, and inspiration to realize their own needs and desires, most often in the artistic dimension. The exemplary responses of respondents indicate this:

- *"...The countryside is freedom. Contrary to appearances, because you live in a small, closed group, the countryside gives you more freedom than the city. Living here, you feel like you are not restricted by much but can create your life freely."*
- *"The city creates culture and new ideas but poisons itself with these ideas, destroying people's lives. The city is noisy and edgy; everything must be done on time now. It is some sick organization. In the countryside, life is different; it flows with a different rhythm. It is just more concrete than everything. It is more for the sake of life."*

The counterurbanites selected for the study are also people who need to make the environment in which they live more valuable and thus a need to actively work for its benefit, as demonstrated, for example, by the founding of the Museum of History and Tradition (village of Kamieńczyk) or the establishment of the 'Handicrafts' Association operating at the Lavender Living Museum (Nowe Kawkowo). In their case, migration to the countryside was not infrequently dictated by a desire to fulfil the dream of running one's own business in a rural environment. It is worth adding that this does not mean any business, but one firmly rooted in the local environment and draws on its resources. An example of combining these 'rural, unique aspects' and a passion that was realized when moving to the countryside is an activity related to the production of lavender (Nowe Kawkowo), breeding dogs (Kamieńczyk), horses or alpacas (Siedlątków). The need to realize the values adhered to by counterurbanites, which, as Wrona (2020) puts it, can be divided into personal (contact with nature, harmony, authenticity, desire for self-realization) and social (desire to repair the world, save the Earth or build community relations) is also indicated by the fact that this social group, not autochthony, are also the majority of initiators of organic farms, including running agrotourism or professionally involved in organic farming.

An example is the statement of one respondent: "... *I have been living here for about 7–8 years. It is a plot of land that belonged to my current wife's parents. I met my wife at university, and when we graduated, we decided to move and live there. The plot was unoccupied with an old house. We renovated it. I come from France, where healthy, organic food is vital. I wanted to build an organic farm in Poland without fertilizers or pesticides. At first, it was for me, but over time, the farm grew, and now I also sell some of my products on the market in Piotrków Trybunalski...*"

The characteristics of counterurbanites cited both in the light of the literature and our research indicate that, in some respects, they form a relatively homogeneous and unique social group. Interestingly, contemporary counterurbanites combine what distinguished previous waves of migration to the countryside in Poland. Thus, this indicates a particular consolidation of the counterurbanite image and its relative permanence in Polish society. As Kuligowski (2007) pointed out, people with unique characteristics have always migrated to the Polish peripherally located countryside. Indeed, the first wave of migration, i.e. in the 1970s, was made up of people, mainly artists, who were troubled by the political, economic, and social system of the time, set by the rules of the People's Republic of Poland. For these people, escaping to the countryside was synonymous with a 'relative return to normality' and, at the same time, an opportunity to express their views, most often in artistic form in the form of festivals, workshops, or exhibitions. On the other hand, the second wave, which fell at the turn of the 1990s, was dominated by people fascinated by organic farming and agro-tourism and was driven by a desire to defend traditional ways of life and local cultural identities. The third wave, on the other hand—which continues nowadays—is made up of people with strongly defined views who consciously reject the consumerist *way* of *life (way of life)* (Kuligowski 2007, after: Laskowska-Otwinowska 2008).

Geography of Counter-Urbanization in Poland

The results of statistical analyses indicate an increase in the population of rural areas in Poland observed in recent years due to migration processes (Szymańska and Biegańska 2011). Increased movement of the population from urban to rural areas has been observed for at least two decades; from 2002 to 2021, it concerned almost 2.3 million people (Fig. 2) and included approximately 36% of all movements of people within the territory of Poland. In addition, as many as 69.5% of Poles surveyed by the Institute of Rural and Agricultural Research of the Polish Academy of Sciences (IRWiR PAN) declare a desire to live in the countryside or in a small town, which indicates that this process may still gain momentum over time. According to further estimates by the Central Statistical Office in Poland, 2 million people will move from cities to rural areas by 2035.

The enormous scale of migration from cities to non-agricultural rural areas was visible in the last 20 years in western Poland, particularly in the Dolnośląskie and Wielkopolskie Voivodeships, in southern Poland, in the Małopolskie Voivodeship, and in the northern Poland Voivodeships (Zachodniopomorskie and Pomorskie) (Fig. 3). In many municipalities in the voivodeships mentioned above, more than

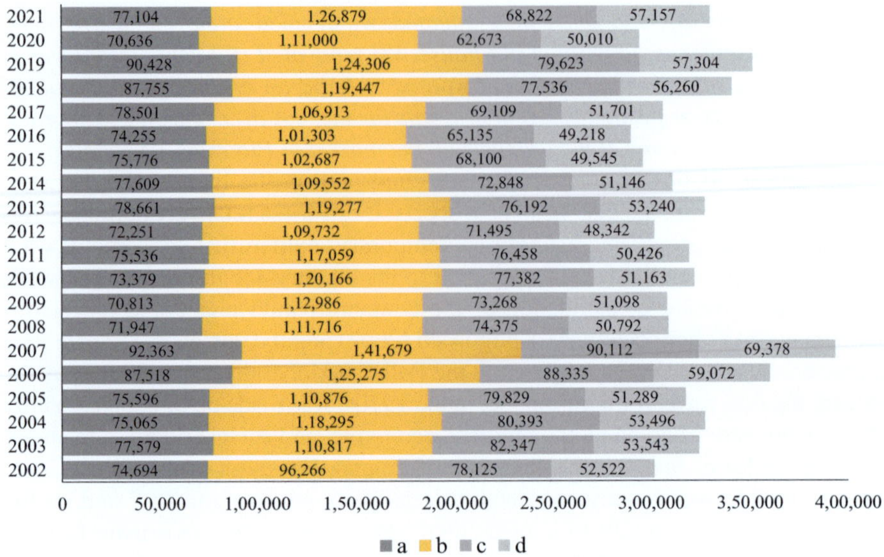

Fig. 2 Internal migration of Poland's population for permanent residence by direction 2002–2021: **a**—rural-to-urban migration; **b**—migration from urban to rural areas; **c**—migration from urban to urban, **d**—migration from the rural-to-rural areas. *Source* Own elaboration based on: https://stat.gov.pl/obszary-tematyczne/ludnosc/migracje-wewnetrzne-ludnosci/migracje-wewnetrzne-ludnosci-na-pobyt-staly-wedlug-wojewodztw-w-latach-1974-2023,3,1.html

1200 people appeared as a result of internal migration from the city over the two decades. The process of counterurbanization becomes even more evident when we analyse population movements among the people living in individual municipalities (Fig. 4). In many municipalities in western and southern Poland, particularly in the West Pomeranian, Lower Silesian, and Silesian Voivodeships, for every 1000 people, 100 come from the city.

When considering the nature of these inflow areas, they are primarily communes with high landscape values—located near the sea or in mountainous areas—confirming the observations on the directions of population movements to non-agglomeration villages in Poland. Counterurbanites choose areas with attractive, high landscape values, often located near mountains or the sea, perceiving them as 'better' and 'safer' areas that enable a change in their lifestyle. This confirms research by other authors that the process of moving people from the city to the countryside is mainly determined by non-economic factors (Geyer and Kontuly 1993; Gkartzios and Scott 2009; Halfacree 2008; McManus 2022; Mitchell 2004, 2019). The most common authors mention safety, the friendliness of the villagers, recreational opportunities and the beauty of the landscape (Sofranko et al. 1982), infatuation with the 'idyllic' image of the countryside (Palak 2016), remoteness from civilization (Bartoš et al. 2008; Republic 2008), and the practice of a healthy lifestyle (Ołów-Wachowicz 2020; Dolińska et al. 2023). Motives related to increasing the quality of life, related

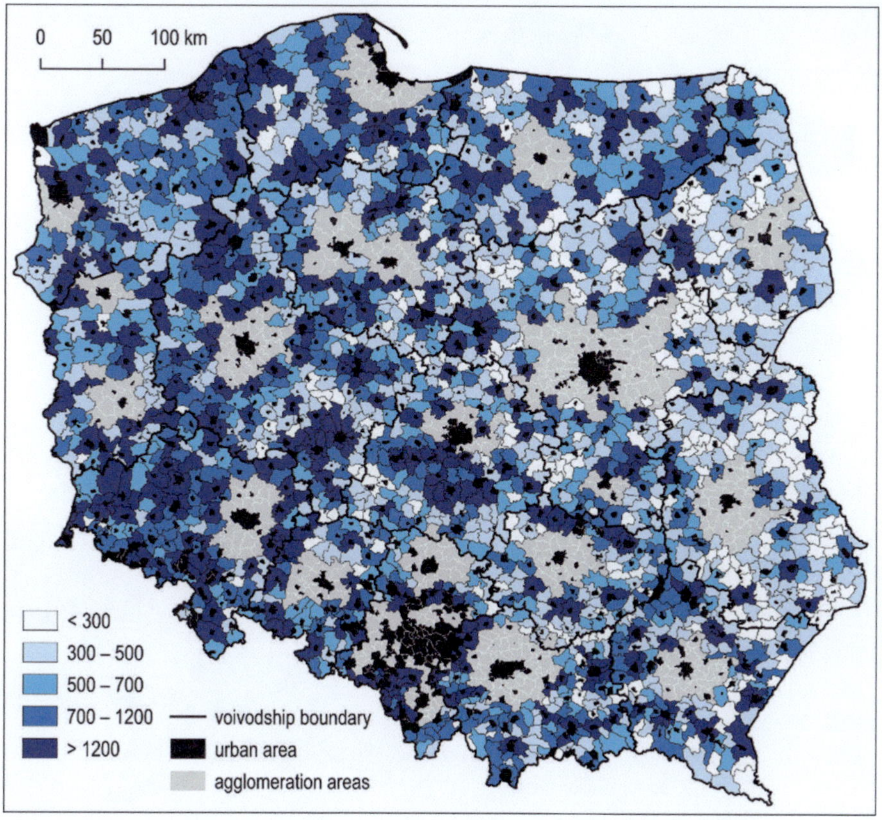

Fig. 3 Urban registrations in rural non-agglomeration municipalities in Poland 2002–2021 (persons). *Source* Own elaboration based on CSO data

to the quality of the natural and socio-cultural environment, are the most important (Beyers and Nelson 2000; Bijker et al. 2012; Brown and Schafft 2002; Marcouiller et al. 2002; Foulkes and Newbold 2008; Laitos and Ruckriegle 2013; Rainer 2019).

Discussion and Conclusion

In Poland, living 'away from the city' has recently been seen in the literature as the fulfilment of dreams of an idyllic life. This is because the countryside is commonly associated with nature, the land, the farm, tradition, seclusion, and social consolidation. Who does not dream of a cottage in the woods, a piece of their lawn, and the opportunity to walk in solitude? Today, when the synonym for luxury is peace, freedom from corporate jobs, the cost of working 'to the max' without interruption, urban traffic jams, and the big-city rush, living in the countryside offers the hope of

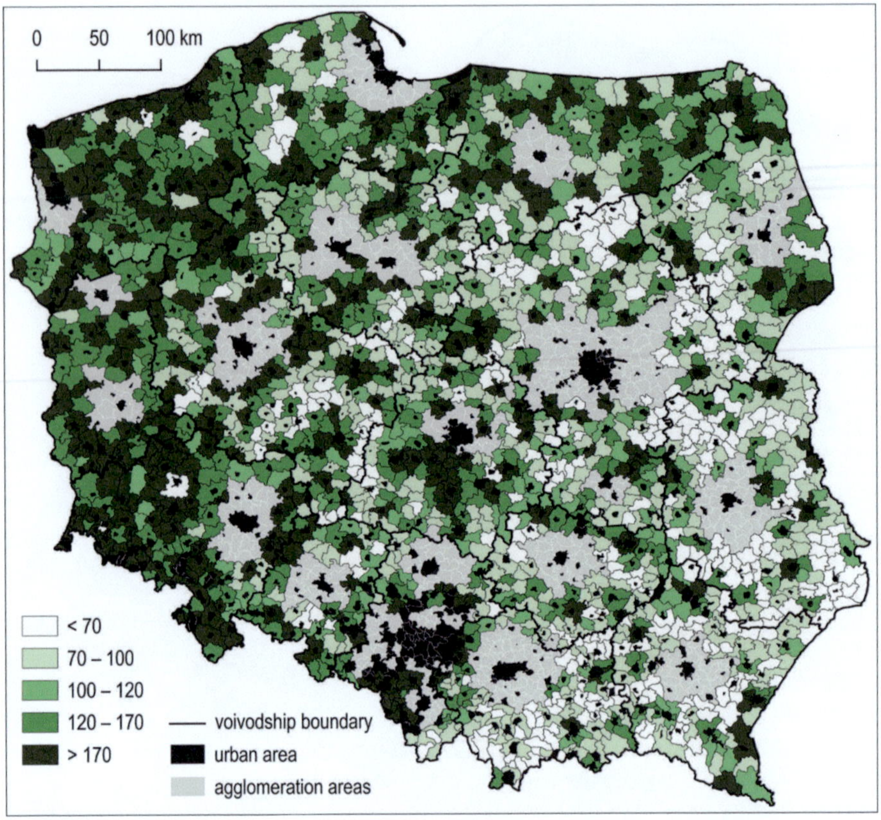

Fig. 4 Registrations from urban to rural non-agglomeration municipalities in Poland 1000 inhabitants (2002–2021). *Source* Own elaboration based on CSO data

stopping and meeting loved ones. However, those tired of the city and the associated pace of life fall into two groups. On the one hand, there are the dreamers, and on the other, those who are turning their dreams into reality, meaning they have decided to change their current residence. In the light of official statistics, the latter are increasingly numerous. It is worth mentioning that in Poland, since 2000—for the first time since the Second World War—the influx of people from the cities to the countryside has been more significant than the outflow of people from the countryside to the city. This also raises questions in the literature and answers who they are.

This article also fits into this rhetoric and reveals that the Polish counterurbanites form not only a cohesive group but also a community that is united not only by a similar socio-cultural profile identified with the 'middle class', characterized by a clear residential preference for rural areas—as they share a strong cultural affinity with the 'rural idyll' and a set of values, said to be from the past (Thrift 1989), but also guided by their perceptions of the countryside, they seek to preserve, enhance, and cultivate this 'idyll', which is reflected, among other things, in the form of economic activities they set up—mainly based on the natural and local resources of the countryside, as well as engaging in and initiating activities related to the propagation of the cultural and historical identity of these places. This observation is also confirmed by the research of Laskowska-Otwinowska, who stated: '…the centres of the Novosad communities in Poland appear in a way that makes one see in them a layer of the intelligentsia, a mixture of national and international, conservative and oppositional characteristics…' (Laskowska-Otwinowska 2008: 238). In the literature, it is also emphasized that what the newcomers from the city have in common is courage because they have dared like no one else, as Zdrowicka-Wawrzyniak writes, "…dared to live in their way and fulfil their dreams. To cut themselves off from what is known and stable, to put everything on the line, to confront their weaknesses, to endure the grumblings of their families and the disbelief of their friends, to rub shoulders with the great loneliness, to live in silence, to constantly learn and to push their limits. That is to say, to step out of one's comfort zone (Zdrowicka-Wawrzyniak 2022: 160).

The research undertaken also led the authors to reflect on the effects, both positive and negative, of the process of counterurbanization as seen from the perspective of the Polish countryside. This process undoubtedly significantly impacts the countryside and contributes to its positive changes. Considering literature studies, one of the key aspects of counterurbanization and counterurbanites that researchers most frequently articulate is the generation of broadly understood rural development. This aspect is particularly relevant in a situation where questions about the condition of the Polish countryside in the context of changes in its socio-economic foundations forced by the shift away from traditional agriculture towards multifunctional development are topical (Bański 2014; Stanny 2009). As Borkowski (2001) pointed out, outlining a vision for the Polish countryside—while the search for development opportunities within agriculture is minimal (developed agricultural policy and food surpluses), the countryside still appears to have a profound but unrealized development potential. This potential is influenced, among other things, by the admission from the cities of a population with its sources of income and seeking healthy lifestyle conditions, a place for new ventures due to raw materials from agriculture, attractive surroundings, lower taxes, bonding, sentiment, or the shaping of new tourist values (Borkowski, after: Zdrowicka-Wawrzyniak 2022: 156).

With their ideology and lifestyle, the counterurbanites seem to realize this plan and thus also ideally activate dormant potential. Our research also confirmed this thesis, meaning the Polish countryside is gaining from this migration. The respondents, who represented various professions (agro-tourism hosts, guesthouses, atypical crops, breeding or activities), argued that there is great potential in the countryside, and therefore moving to the countryside—apart from making dreams of one's own home surrounded by greenery come true—is also a good idea for professional fulfilment, and thus for the revival of its mostly dormant potential. Thus, the emergence of counterurbanites can be regarded as a qualitative change in the countryside, as it creates a new economic, social, and professional structure based on development factors other than the traditional agricultural ones, i.e. a place for leisure activities, a place for modern services, tourist and recreational functions, cultural functions, etc. The creation of new economic activities is also often a source of local products, contributing not only to the promotion of the village but also to the promotion of its assets by referring to and using local resources. By becoming active in the rural environment, counterurbanites thus show an alternative development path for the local community or activate it through a multiplier effect. In their activities, counterurbanites often make use of local resources, both natural and social, thanks to which they interact with the local market by creating added value to the rural environment in which they operate in the form of, among other things, new jobs, the creation of new, informal social networks or social exchange (relations, ties), which in the future may result in new economic activities. In addition, the promotion of a healthy lifestyle and the education of rural society through the creation of various cultural and educational initiatives (e.g. Museum of History and Tradition in Kamieńczyk, 'Handicrafts' Association at the Lavender Living Museum in Nowy Kawkowo) cannot be overestimated. From this point of view, counterurbanites represent an opportunity to preserve the local identity of the village and its traditions.

The emergence of a new social stratum often also gives rise to the need to expand a village's existing technical and social infrastructure and changes its physiognomic character, which increases not only its functional but also its aesthetic assets. In the light of research, Polish non-metropolitan villages are often described through the prism of fewer employment opportunities, less accessibility to health care and educational or training institutions, and worse equipment in the road, water supply, or sewage infrastructure (Michalska-Żyła 2017; Stanny 2013; Murawska 2012). Consequently, village development, public space, new investments, or broadly understood needs in this area are often the declared topics of conversation between villagers and newcomers, often leading to a change in this situation and integration between old and new villagers. Indeed, such an initiative often requires collective action, which forms associations and activates itself and its neighbours.

The image of counterurbanization presented in most Polish literature emphasizes its positive impact on the countryside and its inhabitants. The authors mostly agree that counterurbanization is undoubtedly one factor contributing to 'village renewal' in the broad sense of this concept. Moreover, this does not only apply to its productive sphere but, above all, to its cultural and social sphere, and its expression is the cultivation of rural traditions, the protection of cultural assets, and the integration of the rural community. Indeed, the 'village renewal' programme promotes organic and biodynamic farming and draws attention to protecting the rural landscape. As a result of 'village renewal' programmes, it is also noticeable that people are returning to the rural areas they inhabited years ago (before migration to the city), often returning 'to their roots' or rebuilding the demographic structure (Halliday and Coombes 1995; Rodríguez et al. 2002; Gonda-Soroczyńska 2006; Ni Laoire 2007; Damyanovic and Reinwald 2014). These processes are most pronounced in the United States and Western Europe and are referred to in the literature as the counterurbanization cascade, new turnaround, rural renaissance, or rural rebound (Dahms 1995; Hudson 1989). It may also lead, contrary to the views of some researchers (e.g. Weekley 1988), to the inhibition of the 'rural depopulation' process, which, until recently, was so characteristic of non-metropolitan villages and constituted a significant barrier to its development.

In the mainly foreign literature, where counterurbanization has a much longer research tradition, authors warn against merely positively assessing it and articulating its drawbacks. One of them is the pressure of tourist colonization, the consequence of which is both the loss of the original character of the village and the devastation of its natural environment. As Wesołowska (2006) pointed out, many environmental threats from recreational settlement are noted in areas subject to tourist colonization pressure. The main threat is the pollution of soil and groundwater associated with the lack of connection to the water supply and sewage system. In addition to the one discussed, there are numerous examples of dense settlements of 'second homes' where, near each other, there are lavatories with absorption pits and dug wells using groundwater, resulting in the co-occurrence of sources of water abstraction and foci of water pollution (Wesołowska 2006). In addition, Sant and Simons (1993) have also highlighted the commercialization of this space, resulting in the treatment of the environmental features of the countryside as a commodity. As a result, they argue that amenities that were once freely available now must be bought often at inflated prices (Sant and Simons 1993: 114).

The claim that counterurbanization is leading to a rural revival is also questioned. As the aforementioned Sant and Simon write, there is no evidence of widespread growth in employment within the village itself. As a rule, the opposite is accurate, and growth is concentrated in small towns and other centres rather than in rural areas. In addition, the research experience also reveals that counterurbanites, through their activities, do not revive traditional lifestyles but create new ones with strong urban connotations. Their activities, which most often take the form of hobby farms or communal alternatives, as the above-mentioned authors conclude, are merely a form of escapism, and their emergence is the result not of a rural tradition but of innovative urbanism (Sant and Simons 1993: 119). Thus, there is no shortage of opinions in the foreign literature that counterurbanization merely extends the urbanization process, including creating its new forms. However, it may lead to fundamental changes in the contemporary settlement pattern, where the lowest-order centres will cease to fulfil their role as service facilities for the higher-order centres. Thus, they will retain their rural character, but only in physiognomy, but their social and economic bases are likely to be less spatially connected than in earlier systems (Sant and Simons 1993).

Bearing in mind the influence of the media on public opinion, as well as the 'culture of the Polish peasant and the Polish countryside', which is firmly ingrained in the Polish national consciousness, it can be assumed that for a long time to come to the myth of the positive links between urbanization and rural idyll will persist in Polish society. In turn, this will influence Polish researchers, who will subconsciously search for and draw a vision of the main positive aspects of this relationship. The question remains: To what extent does this image coincide with the real face? Undoubtedly, the answer requires further in-depth research. This chapter concludes with an attempt, based on the literature on the subject and my observations, to verbalize the strengths, weaknesses, opportunities, and threats of non-agglomeration rural areas

Table 2 SWOT analysis

Strengths	**Weaknesses**
– Polyfunctional economic structure – Creation of new jobs – Creation of new economic entities – Integration of residents – Strengthening of the village brand (recognition) – Community activation – The emergence of regional products – Village renewal in its aesthetic dimension – Development of technical and social infrastructure – Increased wealth of the local community – Strengthening the human and social capital of villages	– Local conflicts (between old and new villagers) – The difference in world view between new and old villagers – The difference in socio-demographic characteristics between migrant and old residents – Insufficient exploitation of the potential of the opportunities offered by the countryside (for the development of agro-tourism, recreation, etc.) – Unfavourable service structure – Community ageing – Lack of adequate knowledge of the financial resources and conditions created by the state to support investments
Opportunities	**Threat**
– Rural development – Population growth – Emergence of a new social group—educated, economically prosperous, creative (gentrification of the countryside) – Development of tourism infrastructure, organic farms – Conservation of the rural heritage – Increased wealth of the local community – More excellent opportunities to benefit from rural development capital measures, i.e. EU grants, state budget funds, etc. – Progressive modernization and development of social infrastructure facilities (educational, cultural, sports, etc.)	– Environmental devastation – Commercialization of space (limited opportunities for small businesses, increase in land prices) – Loss of village identity (globalization of villages) – Loss of rank (importance) in the regional and national settlement network – Increasing passive and demanding social attitudes – Mismatch between the professional qualifications of the local community and the requirements of the new labour market – Deterioration of safety due to an increase in the number of people, greater anonymity, and an increase in traffic on the roads

Source Own elaboration based on literature and own observations

from the point of view of the impact of counterurbanization on their space and society (Table 2).

References

Bański J (2014) Perspektywy rozwoju polskiej wsi—wybrane zagadnienia. Wieś i Rolnictwo 4(165):13–25. https://doi.org/10.53098/wir.2014.4.165/01

Bartoš M, Kušová D, Těšitel J, Kopp J, Novotná M (2008) Amenity migration in the context landscape. J Landscape Ecol 1(2):5–21. https://doi.org/10.2478/v10285-012-0006-3

Beale CL (1974) Rural development, population and settlement prospects. J Soil Water Conserv 29(1):23–27

Beale CL (1976) A further look at non-metropolitan population growth since 1970. Am J Agr Econ 58:953–958. https://doi.org/10.2307/1239998

Bednarek-Szczepańska M, Janc K (2025) Konsekwencje miejsko-wiejskiego transferu wiedzy z perspektywy interakcji pomiędzy starymi i nowymi mieszkańcami wsi. Czasopismo Geograficzne 96(1):27–50. https://doi.org/10.12657/czageo-96-02

Berry BJL (1976) Urbanization and countrurbanization. Sage Publications, Beverly Hills

Beyers W, Nelson PB (2000) Contemporary development forces in the nonmetropolitan West: new insights from rapidly growing communities. J Rural Stud 16:459–474

Bijker R, Haartsen T, Strijker D (2012) Migration to less-popular rural areas in the Netherlands: exploring the motivations. J Rural Stud 28:490–498

Borkowski J (2001) Obszary wiejskie—niewykorzystany potencjał rozwojowy. w: Kolarska-Bobińska L, Rosner A, Wilkin J (eds) Przyszłość wsi polskiej. Wizje, strategie, koncepcje. Warszawa

Boyle P, Halfacree K, Robinson V (2013) Exploring contemporary migration. Rout-ledge, New York

Brown D, Schafft K (2002) Leaving Athens: narratives of counterurbanisation in time of crisis. J Rural Stud 32:158–167

Burszta W, Kuligowski W (2005) Sequel. Dalsze przygody kultury w globalnym świecie. Wyd. MUZA, Warszawa

Champion AG (1987) Recent changes in the pace of population deconcentration in Britain. Geoforum 18:379–401

Champion T, Atkins D (1996) The counterurbanisation cascade: an analysis of the 1991 Census Special Migration Statistics for Great Britain. Department of Geography Seminar Paper 66, University of Newcastle upon Tyne

Cloke P (1985) Counterurbanisation: a rural perspective. Geography 70(1):13–23. http://www.jstor.org/stable/40571490

Cloke P, Goodwin M (1992) Conceptualizing countryside change: from post-fordism to rural structured coherence. Trans Inst Br Geogr 17(3):321–336. https://doi.org/10.2307/622883

Czarnecki A (2019) Urbanizacja kraju i jej etapy. In: Halamska M, Stanny M, Wilkin J (red.) Ciągłość i zmiana. Sto lat rozwoju polskiej wsi, t 1. Instytut Rozwoju Wsi i Rolnictwa, Warszawa, pp 51–76

Dahms FA (1995) "Dying villages", 'counterurbanization' and the urban field—a Canadian perspective. J Rural Stud 11(1):21–33

Damyanovic D, Reinwald F (2014) The "comprehensive village renewal programme in burgerland" as a means of strengthening the social capital in rural areas. European Countryside 1:18–35

Degórska B (2017) Urbanizacja przestrzenna terenów wiejskich na obszarze metropolitarnym Warszawy. Prace Geograficzne 262:179

Dej M, Zajda K (2016) Kontrurbanizacja i jakość życia na wsi. Acta Universitatis Lodziensis, Folia Sociologica 57:51–66. https://doi.org/10.18778/0208-600X.57.04

Dej M, Zajda K (2017) Wizerunek kontrurbanizacji w Polsce na przykładzie reportaży "Daleko od miasta". Cechy migrantów, motywy migracji i jej konsekwencje dla wiejskiej przedsiębiorczości. Acta Universitatis Lodziensis, Folia Sociologica 63:29–43. https://doi.org/10.18778/0208-600X.63.03

Dolińska A, Jończy R, Śleszyński P, Rokitowska-Malcher J, Ptak M (2023) Migracje z miast na wieś. Determinanty oraz wybrane konsekwencje w wymiarze ekonomicznym i przestrzenno-środowiskowym (na przykładzie strefy podmiejskiej Wrocławia). Wydawnictwo Uniwersytetu Ekonomicznego we Wrocławiu, Wrocław

Dziegieć E (1987) Przemiany osadnictwa wiejskiego pod wpływem turystyki i wypoczynku w świetle literatury. Turyzm 3:7–32

Fielding AJ (1983) Counterurbanization in Western Europe: recent empirical and theoretical contributions to the debate. Paper presented at the Anglo-Dutch migration symposium, Socstcrberg, The Netherlands, September 14–16
Fielding AJ (1982) Counterurbanization in Western Europe. Prog Plan 17(1):1–52. https://doi.org/10.1016/0305-9006(82)90006-X
Foulkes M, Newbold KB (2008) Poverty catchments: migration, residential mobility, and population turnover in impoverished rural Illinois communities. Rural Sociol 73:440–462
Frey WH (1987) Migration and depopulation of the metropolis: regional restructuring or rural renaissance? Am Sociol Assoc 52(April):240–257
Frey WH (1988) The re-emergence of core region growth: a return to the metropolis? Int Reg Sci Rev 11(3):261–267
Geyer H, Kontuly T (1993) A theoretical foundation for the concept of differential urbanisation. Int Reg Sci Rev 17:157–177
Gkartzios M, Scott M (2009) Residential mobilities and house building in rural Ireland: evidence from three case studies. Sociol Rural 50:64–84
Gonda-Soroczyńska E (2006) Sieć osadnicza Europy na przykładzie Niemiec, Francji i Holandii. In: Sieć osadnicza jako przedmiot badań, E.Bagiński (red.), Oficyna Wydawnicza Politechniki Wrocławskiej, Wrocław, pp 163–184
Grzeszczak J (1996) Tendencje kontrurbanizacyjne w krajach Europy Zachodniej. Prace Geograficzne IGiPZ PAN 167, Warszawa
Grzeszczak J (1998) Przesunięcie miasto-wieś w przemyśle krajów Unii Europejskiej. Zeszyty IGiPZ PAN 55, Warszawa
Grzeszczak J (2000) Kontrurbanizacja—idea i rzeczywistość. Przegląd Geograficzny 72(4):373–393
Halfacree K (1994) The importance of 'the rural' in the constitution of counterurbanization: evidence from England in the 1980s. Sociologis Ruralis 34:164–189. https://doi.org/10.1111/j.1467-9523.1994.tb00807.x
Halfacree K (2007) Trial by space for a "radical rural": introducing alternative localities, representations and lives. J Rural Stud 23:125–141
Halfacree K (2008) To revitalise counter-urbanisation research? Recognising an international and fuller picture. Popul Space Place 14(6):479–495
Halliday J, Coombes M (1995) In search of counterurbanisation: some evidence from devon on the relationship between patterns of migration and motivation. J Rural Stud 1(4):433–446
Heffner K (2016) Proces suburbanizacji a polityka miejska w Polsce. In: Marszał T (ed) Miasto-region-gospodarka w badaniach geograficznych. W stulecie urodzin Profesora Ludwika Straszewicza. Wyd. Uniwersytetu Łódzkiego, Łódź, pp 75–110
Hudson P (1989) Change and adaptation in four rural communities in New England, NSW. Aust Geogr 20:54–64
Ilnicki D (2002a) Przekształcenia wybranych elementów struktur demograficznych miejscowości charakterystycznych dla strefy podmiejskiej Wrocławia. Bull Geogr Socio-Econ Series 1:202–216
Ilnicki D (2002b) Przemiany w strukturach demograficznych w miejscowościach strefy podmiejskiej Wrocławia. Przykład Mirków i Bielany Wrocławskie. Biuletyn Geograficzny IG PAN 1:201–214
Jakóbczyk-Gryszkiewicz J (1991) Zróżnicowanie urbanizacji wsi w strefie podmiejskiej Łodzi. Acta Uniwersitatis Lodziensis. Folia Geographica 13:75–99
Jakóbczyk-Gryszkiewicz J (1998) Przeobrażenia stref podmiejskich dużych miast. Studium porównawcze Warszawy, Łodzi i Krakowa. Wyd. Uniwersytetu Łódzkiego, Łódź
Kajdanek K (2011) Pomiędzy miastem, a wsią. Suburbanizacja na przykładzie osiedli podmiejskich Wrocławia. Zakład Wydawniczy NOMOS, Kraków
Kajdanek K (2012) Suburbanizacja po polsku. Zakład Wydawniczy NOMOS, Kraków
Kontuly T (1998) Contrasting the counterurbanisation experience in European nations. In: Migration into rural areas: theories and issues, pp 61–78

Kontuly T, Dcarden B (1998) Regionale Umverteilungsprozesse der Bevölkerung in Europa seit 1970. Informationen Zur Raumentwicklung 11(12):713–722
Korcelli P (1984) The turnaround of urbanization in developed countries. In: Population distribution, migration and development. In: International conference on population 1984. United Nations, New York, pp 349–372
Kowalewski A, Markowski T, Śleszyński P (ed) (2018) Studia nad chaosem przestrzennym. Studia KPZK 182, 1–3
Kukliński A (1978) Polityka regionalna w perspektywie porównawczej. Przegląd Geograficzny 50(I):3–11
Kuligowski W (2007) Antropologia współczesności. Wiele światów jedno miejsce. Universitas. Kraków
Kupiszewski M, Durham H, Rees P (1998) Internal migration and urban change in Poland. Eur J Popul 14(3):265–290
Kurczewska J (2004) (ed) Oblicza lokalności. Tradycja i współczesność. Wydawnictwo Instytutu Filozofii i Socjologii PAN, Warszawa
Kwiatkowska O (2005) Nowoosadnictwo w Lucimiu, czyli historia pewnego eksperymentuartystyczno-społecznego. Etnografia Polska 49(1–2):147–167
Laitos JG, Ruckriegle H (2013) The problem of amenity migrants in North America and Europe. Urban Lawyer 45(4):849–914
Laskowska-Otwinowska I (2008) Globalne przepływy kulturowe a obecność nowoosadników na wsi polskiej. Muzeum Archeologiczne i Etnograficzne, Łódź
Liszewski S (1987a) Geneza i rozwój osadnictwa wypoczynkowego w otoczeniu Łodzi. Turyzm 3:33–54
Liszewski S (1987b) Strefa podmiejska jako przedmiot badań geograficznych. Próba Syntezy. Przegląd Geograficzny 59(1–2):65–79
Liszewski S (2005) Przestrzeń turystyczna w ujęciu podmiotowym. Przyczynek do dyskusji o przestrzeni w geografii. In: Maik W, Rembowska K, Suliborski A (eds) Geografia jako nauka o przestrzeni, środowisku i krajobrazie. Podstawowe idee i koncepcje w geografii. T.1. ŁTN, Łódź, pp 50–60
Llleris S (1990) Counterurbanization north-western revisited: the new map of population distribution in central and Europe. Norsk Geografisk Tidsskrift 44(I):39–52
Marcouiller DW, Clendenning JG, Kedzior R (2002) Natural amenity-led development and rural planning. J Plan Lit 16:515–542
Matczak A (1985) Funkcja wypoczynkowa strefy podmiejskiej Łodzi. In: Straszewicz L (ed) Pojęcia i metody badań strefy podmiejskiej. Acta Uniwersitatis Lodziensis. Folia Geographica 5:299–312
Matczak A (1986) Budownictwo letniskowe w strefie podmiejskiej Łodzi. Acta Universitatis Lodziensis, Folia Geographica 7:137–165
Matczak A (1991) Przemiany przestrzenne wsi położonych w strefie podmiejskiej Łodzi na przykładzie rejonu Grotnik. Acta Universitatis Lodziensis. Folia Geographica 14:33–46
McManus P (2022) Counterurbanisation, demographic change and discourses of rural revival in Australia during COVID-19. Australian Geographer, Published online. Accessed 28 Feb 2022
Michalska-Żyła A (2017) Zróżnicowanie jakości życia w gminach wiejskich województwa łódzkiego. Prace Naukowe Uniwersytetu Ekonomicznego We Wrocławiu 477:172–181. https://doi.org/10.15611/pn.2017.477.17
Miszewska B (1985) Przemiany morfologiczne osadnictwa wiejskiego w strefie podmiejskiej Wrocławia. Metoda analizy. Acta Universitatis Lodziensis. Folia Geographica 5:103–122
Mitchell CJA (2004) Making sense of counterurbanisation. J Rural Stud 20(1):15–34
Mitchell CJA (2019) The patterns and places of counterurbanisation: a 'macro' perspective from Newfoundland and Labrador, Canada. J Rural Stud 70:104–116
Murawska A (2012) Zmiany w poziomie i jakości życia ludności na obszarach wiejskich w Polsce. J Agribus Rural Dev 3(25):169–180

Ni Laoire C (2007) The 'green, green grass of home'? Return migration to rural Ireland. J Rural Stud 23:332–344
Nicolaides B (1993) In search of the good life: community and politics in working-class Los Angeles. 1920–1955. Columbia University, Washington
O'Connor CA (1985) Sorting out the suburbs: patterns of land use, class and culture. Am Q:382–394
Ołów-Wachowicz A (2020) Zmęczeni miastem. Nowi mieszkańcy. Nowy styl życia. Kierunek zmian 3:17–20
Palak M (2016) Nowe oblicza przedmieść. Socjologiczne studia suburbanizacji w Polsce na przykładzie Rzeszowa. Wydawnictwo Uniwersytetu Rzeszowskiego, Rzeszów
Prochownik A (1975) Urbanizacja wsi, czy deruralizacja? Czasopismo Geograficzne 46(4):399–405
Rainer G (2019) Amenity/lifestyle migration to the Global South: driving forces and socio-spatial implications in Latin America. Third World Q 40(7):1359–1377. https://doi.org/10.1080/01436597.2019.158729
Republic C (2008) Amenity migration in the context landscape. Landscape Ecol:5–21
Rodríguez V, Egea C, Nieto JA (2002) Return migration in Andalusia, Spain. Int J Popul Geogr 8(3):233–254
Ruczkowski P (2022) Suburbanizacja oraz kontrurbanizacja jako wyzwania dla gmin. Studia Prawnoustrojowe 57:443–461. https://doi.org/10.31648/sp.7999
Sant M, Simons PL (1993) The conceptual basis of counterurbanisationi: critique and development. Aust Geogr Stud 31:113–126
Sofranko AJ, Fliegel FC, Glasgow N (1982) Starsi migranci z miast na obszarach wiejskich: problemy i perspektywy. Int J Aging Human Dev 1982–1983 16(4):297–309. https://doi.org/10.2190/vlgm-hb2u-q5w1-98qt
Stanny M (2009) Zróżnicowanie poziomu rozwoju obszarów wiejskich w Polsce a problem realizacji polityki spójności. Wieś i Rolnictwo 4(145):246–257. https://doi.org/10.53098/wir.2009.4.145/15
Stanny M (2013) Przestrzenne zróżnicowanie rozwoju obszarów wiejskich w Polsce. Instytut Rozwoju Wsi i Rolnictwa PAN, Warszawa
Straszewicz L (1954) Strefa Podmiejska Łodzi. Przegląd Geograficzny 26(4):182–197
Straszewicz L (1985) Strefa podmiejska. Pojęcia i definicje. In: Straszewicz L (ed) Pojęcia i metody badań strefy podmiejskiej. Acta Uniwersitatis Lodziensis. Folia Geographica 5:7–16
Szkup R (2003) Kształtowanie podmiejskiej przestrzeni wypoczynkowej. Przykład zachodniego sektora strefy podmiejskiej Łodzi. Wydawnictwo Uniwersytetu Łódzkiego, Łódź
Szmytkowska M, Masik G (2005) Społeczne aspekty suburbanizacji w obszarze metropolitarnym Trójmiasta. In: Lorens P (ed) Problem suburbanizacji. Wydawnictwo Urbanista, Warszawa, pp 71–80
Sztandara M (2003) Sposoby na ludowość, Powrót do źródeł i idea działań alternatywnych. In: Materiały pokonferencyjne. Tradycja z przyszłością, Wyd. Towarzystwa Naukowego, Sandomierz
Szymańska D, Biegańska J (2011) Fenomen urbanizacji i procesy z nim związane. Studia Miejskie 4:13–38
Tammaru T, Kliimask J, Kalm K, Zalıte J (2023) Did the pandemic bring new features to counterurbanisation? Evidence from Estonia. J Rural Stud 97:345–355
Thrift N (1989) Editorial: new times and new spaces? The perils of transition models. Environ Plann d: Soc Space 7:127
Weekley I (1988) Rural depopulation and counter urbanisation: a paradox. Area:127–134
Wesołowska M (2006) Rozwój budownictwa mieszkaniowego a przemiany przestrzenne wsi województwa lubelskiego. Studia Obszarów Wiejskich 10, Warszawa
Wesołowska M, Kulawiak A (2025) Profil społeczno-kulturowy kontrurbanity w soczewce studiów literaturowych i badań własnych. Czasopismo Geograficzne 96(1):47–68. https://doi.org/10.12657/czageo-96-03
Węclawowicz G (1979) [recenzja:] B.J.L. Berry (ed.) Urbanization and counterurbanisalion. Przegląd Geograficzny 51(2):364–366

Wieruszewska M (ed) (1992) Odnowa wsi między mitem, a nadzieją. IRWiR PAN, Warszawa
Więcław-Michniewska J (2006) Krakowskie suburbia i ich społeczność. Instytut Geografii i Gospodarki Przestrzennej Uniwersytetu Jagiellońskiego, Kraków
Williamson T (2008) Sprawl, spatial, spatial location and politics: how ideological identification. Am Polit Res 36:903–933
Włodarczyk B (2009) Przestrzeń turystyczna. Istota, koncepcje, determinanty rozwoju. Wydawnictwo Uniwersytetu Łódzkiego, Łódź
Wojciechowska J (2009) Procesy i uwarunkowania rozwoju agroturystyki w Polsce, Wydawnictwo Uniwersytetu Łódzkiego, Łódź
Wójcik M (2008) Przemiany społeczno-gospodarcze wsi w okresie transformacji ustrojowej. Wydawnictwo Uniwersytetu Łódzkiego, Łódź
Wójcik M (2013) Przemiany społeczno-przestrzenne osiedli wiejskich. Studium przypadku łódzkiego Obszaru Metropolitarnego, Wydawnictwo Uniwersytetu Łódzkiego, Łódź
Wrona A (2020) Przeprowadzki z miast do wsi: spotkanie dwóch kultur i co z niego wynika dla lokalnych społeczności. Górnośląskie Studia Socjologiczne. Seria Nowa 6:123–144
Zdrowicka-Wawrzyniak M (2022) Życie z dala od miasta. O popularności serii dokumentalnej Daleko od miasta. Zeszyty Wiejskie 28:151–165. https://doi.org/10.18778/1506-6541.28.06

UNESCO Creative Cities in the Inland Centre of Portugal

Pedro Vaz Serra and Cláudia Seabra

Abstract With a view to promoting and cooperating with and between cities that identify and embrace creativity among the factors likely to lead to urban development with sustainability criteria, the UNESCO Creative Cities Network (UCCN) was founded in 2004. Three UNESCO Creative Cities are in the central interior of Portugal. Due to their location, these cities have specific characteristics and have become areas that have seen a systematic decrease in the resident community, progressive ageing of the population, and more difficulty in maintaining and retaining younger people. We propose a comprehensive approach, drawing up a SWOT analysis for each case, a summary of the most relevant economic activity, and how they have incorporated the status of Creative Cities into everyday life and cultural programming. Suggestions for development and/or improvement will be made, bearing in mind how each city's specificities can achieve competitive and sustainable advantages in and through it. The data will be collected over five years, between 2019 and 2023, and obtained from official and secondary sources. We aim to contribute to a more detailed knowledge of the reality associated with Creative Cities, particularly those inland with challenges and limitations that they want to face and overcome. Contributions to science, management, and operators, public and private are envisaged.

Keywords UNESCO creative cities network · Creative cities in the Inland centre of Portugal · Tourism and creative cities · SWOT analysis of creative cities

P. Vaz Serra (✉) · C. Seabra
Faculty of Arts and Humanities of University of Coimbra, Geography and Spatial Planning Research Centre (CEGOT), University of Coimbra, Coimbra, Portugal
e-mail: pedrovazserra@gmail.com

Faculty of Arts and Humanities of University of Coimbra, Geography and Spatial Planning Research Centre (CEGOT), University of Coimbra, Coimbra, Portugal

C. Seabra
e-mail: cseabra@uc.pt

L. Soares et al. (eds.), *Compromised Geography*, Key Challenges in Geography,
https://doi.org/10.1007/978-3-032-05201-8_15

Introduction

The UNESCO Creative Cities Network (UCCN) was created with the strategic goal of adopting cultural and creative activities as the basis for urban development criteria and enhancing cooperation at national and international levels (UCCN 2024). After its first 20 years, it will have 350 cities representing more than 100 countries by 2024 (UCCN 2024).

Through integration into the UCCN, its members are linked to the sharing of good practices and the development of multilateral actions that include public entities, private operators, and civil society (UCCN 2024).

Among the objectives of this convergence between cities are (i) the affirmation of an attitude tending to create, produce, distribute, and publicise cultural activities, goods, and services; (ii) the development of creative and innovative hubs, with more opportunities for the various actors involved; (iii) improving accessibility to cultural activities, through the integration of more vulnerable citizens; (iv) the adoption of cultural value and creative action in sustainable development plans (UCCN 2024).

There are seven creative areas covered by the network: (i) literature; (ii) design; (iii) film; (iv) crafts and folk arts; (v) music; (vi) media arts; and (vii) gastronomy (UCCN 2024).

For UNESCO, the UCCN represents a strategic partnership (UCCN 2024). In addition to being a space for reflection that uses creativity to support and accelerate the sustainability associated with city development, it also makes an important contribution to implementing the 2030 Agenda for Sustainable Development, namely its Goal 11—Sustainable Cities and Communities (United Nations 2015; UCCN 2024). This goal aims to ensure the presence of values and dimensions associated with inclusion, security, resilience, and sustainability in these places (United Nations 2015).

The Creative Cities in the Inland Centre of Portugal

In the Inland Centre of Portugal, three cities make up the UCCN, Idanha-a-Nova, *City of Music*, since 2015; Covilhã, *City of Design*, since 2021; and Castelo Branco, *City of Crafts and Folk Arts*, since 2023 (Municipio de Idanha-a-Nova 2024b; Municipio da Covilhã 2024b; Municipio de Castelo Branco 2024b).

These three Creative Cities of the Inland Centre are all located in the same district, which has Castelo Branco as its capital.

Any of these cities is less than 200 kms from Coimbra and less than 300 kms from Lisbon or Porto, and all these cities are remarkably close to Spain, which proximity makes it possible to attract new audiences.

Let us look at the SWOT analysis of each city in question.

Idanha-a-Nova SWOT Analysis

In Idanha-a-Nova and as a strength, ancient music has had a structural element in its annual programme lines since 2001, with various music groups and festivities throughout the municipality (see Table 1).

This city has a resident population of 8,355 (2021), 14% less than in 2011, and a purchasing power per capita of 68.5 (2021), compared to Portugal (100), and unemployment of 8.6% (2022), compared to Portugal (4.6%), what represent weaknesses (see Table 1).

As a UCCN member and concerning opportunities, Idanha-a-Nova can highlight cultural assets on a global platform and innovate through the exchange of knowledge and experiences. Under the tourist activity (see Table 1), the average stay in tourist accommodation has good indicators: 2.1 (2022), compared to 1.8 (Centre of Portugal).

As threats, the imbalance in regional development (interiority), the difficulty in retaining and attracting qualified young people, and the ageing population are the most significant challenges (see Table 1).

In short, with the vulnerabilities typical of an inland city in contemporary Portugal, Idanha-a-Nova has in the UCCN a platform that, associated with a more comprehensive growth and development strategy, can constitute an opportunity for local affirmation.

Table 1 Idanha-a-Nova SWOT analysis

Strengths	*Weaknesses*
• **Ancient music** as a **structural element** in its annual programme lines since 2001 (14 years before the integration into the UCCN) • Various **music groups** and **festivities** throughout the municipality	• Resident **population**: 8355 (2021), 14% less than in 2011 • **Purchasing power** *per capita*: 68.5 (2021), Portugal (100) • **Unemployment**: 8.6% (2022), Portugal (4.6%)
Opportunities	*Threats*
• Highlight **cultural assets** on a **global platform** • Cultivating **innovation** through the exchange of **knowledge** • **Average stay** in tourist accommodation: 2.1 (2022), compared to 1.8 (Centre of Portugal)	• **Imbalance** in **regional development** (interiority, in a country very centralised in Lisbon) • Difficulty in retaining and attracting qualified **young people** • **Ageing** population

Note Município de Idanha-a-Nova (2024a, b), Pordata (2024)

Table 2 Covilhã SWOT analysis

Strengths	*Weaknesses*
• **Three museums** (Arte Sacra, Covilhã and Lanifícios, the latter a multi-nuclear museum) and three **art galleries** • **Routes** of Urban Art, *Art Nouveau*, Discover Covilhã, Jewish Quarter, Industrial Heritage, and the Wool Route	• Resident **population**: 46,455 (2021), 10,3% less than in 2011 • **Purchasing power** *per capita*: 86.9 (2021), Portugal (100)
Opportunities	*Threats*
• The *New Hand Lab* is dedicated to **industrial** and **creative tourism** and is a **benchmark** in the field of art and design • **Culture** and **creativity** as ecosystem values for the **development** of the city and the region	• **Imbalance** in **regional development** (interiority, in a country very centralised in Lisbon) • The **average stay** in tourist accommodation without significant evolution: 1.8 (2022), compared to 1.6 (2001)

Note Município da Covilhã (2024a, b), Pordata (2024)

Covilhã SWOT Analysis

Covilhã has a tradition linked to textiles, with three museums—Arte Sacra, Covilhã, and Lanifícios, the latter a multi-nuclear museum—and three art galleries, and routes of Urban Art, Art Nouveau, Discover Covilhã, Jewish Quarter, Industrial Heritage, and the Wool Route, which correspond to its strengths (see Table 2).

A resident population of 46,455 (2021), 10.3% less than in 2011, and a purchasing power per capita of 86.9 (2021) compared to Portugal (100), constitute its main weaknesses (see Table 2).

The Covilhã´s opportunities are related to the New Hand Lab dedicated to industrial and creative tourism and a benchmark in the field of art and design, and culture and creativity as ecosystem values for the development of the city and the region (see Table 2).

The imbalance in regional development (interiority) and the average stay in tourist accommodation without significant evolution—1.8 (2022), compared to 1.6 (2001)—constitute threats to be considered (see Table 2).

In short, Covilhã can also have, as a UCCN member, a vehicle for growth and development, if this participation is included in relevant and complementary actions that enhance investment and employment.

Castelo Branco SWOT Analysis

Castelo Branco, the capital of the district and the largest city among those considered, strategically, sees creativity as an element of the region's development, as a means of creating value and retaining human resources, and has 10 tourist routes related

Table 3 Castelo Branco SWOT analysis

Strengths	*Weaknesses*
• Strategically, it sees **creativity** as an element of the region's **development,** as a means of **creating value** and retaining **human resources** • It has 10 **tourist routes** related to tangible and intangible **heritage**, leisure, nature, and natural resources	• Resident **population**: 52,272 (2021), 6,8% less than in 2011 • **Unemployment**: 5.1% (2022), Portugal (4.6%)
Opportunities	*Threats*
• Intensify the definition and implementation of **public policies** in **creativity** and **innovation** in the **municipal** context • The **average stay** in tourist accommodation showing **satisfactory** progress: 1.9 (2022), compared to 1.6 (2001)	• **Imbalance** in **regional development** (interiority, in a country very centralised in Lisbon) • **Difficulty** in retaining and attracting qualified **young people**

Note Município de Castelo Branco (2024a, b), Pordata (2024)

to tangible and intangible heritage, leisure, nature, and natural resources, which constitute its strengths (see Table 3).

However, as a weakness, it has been losing its resident population—52,272 (2021), 6.8% less than in 2011—with an unemployment of 5.1% (2022), compared to Portugal (4.6%).

The city, as a UCCN member, can intensify the definition and implementation of public policies in creativity and innovation in the municipal context. The average stay in tourist accommodation showing satisfactory progress: 1.9 (2022), compared to 1.6 (2001).

Like the other two cities analysed, Castelo Branco also has an imbalance in regional development (interiority, in a country very centralised in Lisbon), and difficulty in retaining and attracting qualified young people, which constitute threats in the present and the future (see Table 3).

In short, the recent integration into the UCCN may represent a commitment that, in a virtuous way, accelerates local and regional growth and development, with an impact on investment and employment and, therefore, on attracting and retaining young people and increasing tourist flows.

Economic Activity

As we can see at Table 4, the service companies dominate in all three cities, followed by industrial and agricultural companies.

We can conclude, about economic activity, that, although services prevail, agriculture and industry are significant in Idanha-a-Nova. Also, in Covilhã, industry

Table 4 Companies based in the municipality (2022)

Companies based in the municipality (2022)			
Sectors	Cities		
	Idanha-a-Nova	Covilhã	Castelo Branco
Agriculture	23,2%	3,2%	10,3%
Industry	21,4%	30,4%	13,3%
Services	55,4%	66,4%	76,4%

vines, olive groves, cereals, and fresh fruit

cereals, fruit, vegetables, and olive oil; cattle, sheep and goats are raised

textiles, construction, wood sawmills and metalworking

pulp and wood

commercial activity and tourism services: extreme sports and sports linked to snow, hunting and fishing are on the increase

Note Pordata (2024)

continues to have some expression, although much smaller than services, around half. In turn, Castelo Branco is, fundamentally, a city of services (see Table 4).

How Has the Brand (Creative City) Been Defined by the City?

All the cities analysed, through dedicated teams inserted in their respective municipalities, outlined an action programme that includes multiple initiatives, where they seek to lay the foundations for the future, but extolling the characteristics and potential that, in some cases, come from the past (see Table 5).

Concerning the main relating activities, in Idanha-a-Nova, more than 20 traditional music groups, most of which play the *adufe* instrument; international ancient music courses (CIMA) offer classes, masterclasses, and lectures at historic sites; and *Boom Festival*, one of the most important electronic music festivals in the world, with more than 40,000 people from 150 countries, in an area of 140 hectares, for ten days, 24 h a day with the performance of 800 artists (Municipio de Idanha-a-Nova 2024a, b).

In Covilhã, since 2021 and annually, the Jardim das Artes welcomed thousands of visitors to FIADA, the International Trade Fair for Crafts, Design and Other Arts; the Covilhã Creative Week, a diverse programme that includes a range of free and simultaneous cultural activities open to the community; *WOOL—Covilhã's Urban Art Festival*, is a remarkable series of interventions spread around the city's historic area, to revitalise run-down areas and promote forgotten spaces; and the *New Hand*

Table 5 How has the brand (creative city) been defined by the city?

Idanha-a-Nova	Covilhã	Castelo Branco
Making **creativity** an essential element of local economic and social development	Strengthening **public policies** in the field of urban creativity and contemporary culture	Creating reference spaces for **innovation** and **value creation**
Strengthening **local capacity** and training cultural agents in business	Establish a new urban cartography for the **creative** community and **cultural industries**	Creation of the *Creativity Factory*, to housing projects in the **creative industries** area
Highlighting **cultural assets** on a global platform	Catalysing creative potential at all levels of **urban development**	Creativity to **generate value** and retain **human resources**
Sharing **knowledge** between cultural groups around the world	Empowering **citizens** to contribute to sustainability and citizenship	Preservation and valorisation of **heritage**
Cultivating **innovation** through the exchange of knowledge and experiences	Seminars and creative residencies to promote **synergies** between stakeholders	Strong commitment to the **cultural offer**
Promoting cultural products on the **national** and **international** markets	Design as an enabler of collective and participatory **research**	
Create new **opportunities** for cooperation with other cities	Expanding **design** as a creative tool in the **urban** management process	

Note Município de Idanha-a-Nova (2024a, b), Município da Covilhã (2024a, b), Município de Castelo Branco (2024a, b)

Lab is a space that promotes creativity, innovation, and entrepreneurship through the realisation of ideas, products, and initiatives (Municipio da Covilhã 2024a, b).

In Castelo Branco, through the Embroidery Route, it is possible to view all the stages of production of the materials and the execution of the embroidery itself; a children's book with an educational game on the theme of Castelo Branco's embroidery was published; and the Creativity Factory was built to encourage the emergence of new projects in the creative and innovative industries, so that they become generators of development in the cultural, artistic, economic, and technological spheres (Municipio de Castelo Branco 2024a, b).

Promotion and Communication in the (Digital) Media as a Creative City

Idanha-a-Nova and Covilhã are well disseminated on digital channels (see Table 6). Castelo Branco's delay may be because it has only been part of the network since October 2023.

The number of followers on the various social networks is exceptionally low. On the contrary, Google references are significant, especially about Castelo Branco (see Table 7).

However, Facebook is the most expressive social network to Idanha-a-Nova and Covilhã (see Table 7).

As for the main content of the messages disseminated on social networks, it is the cultural dimension that deserves special emphasis (see Table 8).

There is little interaction between social media users and the publications made (see Table 9).

Table 6 Creative cities' digital presence

Digital presence		Cities		
		Idanha-a-Nova	Covilhã	Castelo Branco
Web site	Institutional (own)	✓	✓	✗
	Municipality	✓	✓	✓
	Tourism Centre of Portugal	✓	✓	✗
	UNESCO Creative Cities Network	✓	✓	✗
Social networks	Instagram	✓	✓	✗
	Facebook	✓	✓	✗
	X	✗	✗	✗
Google	References	✓	✓	✓

Note Municipio de Idanha-a-Nova (2024a, b), Municipio da Covilhã (2024a, b), Municipio de Castelo Branco (2024a, b), Turismo Centro de Portugal (2024), UCCN (2024), Instagram (2024), Facebook (2024), X (2024), Google (2024); *Legend* (*) Not only the reference of the nomination but also several paragraphs about the city, its characteristics and goals and contributions as a Creative City

Table 7 Creative cities' digital presence: followers and references (January 2024)

Digital presence		Cities		
		Idanha-a-Nova	Covilhã	Castelo Branco
Social networks	Instagram	739	0	–
	Facebook	4400	1800	–
	X	–	–	–
Google	References	67,800	233,000	1,180,000

Note Instagram (2024), Facebook (2024), X (2024), Google (2024)

Table 8 Creative cities' digital presence: appeal of key messages on social media (January 2024)

Digital presence		Cities		
		Idanha-a-Nova	Covilhã	Castelo Branco
Social networks	Instagram	Cultural activities, such as concerts, shows, and exhibitions, directly related to music	Private account, with only 16 posts	–
	Facebook	Cultural activities, such as concerts, shows, and exhibitions, directly related to music	Activities related to design, as well as the cultural programme of the Municipality	–
	X	–	–	–

Note Instagram (2024), Facebook (2024), X (2024)

Table 9 Creative cities 'digital presence: interaction with social media users (likes, comments, and replies—January 2024)

Digital presence		Cities		
		Idanha-a-Nova	Covilhã	Castelo Branco
Social networks	Instagram	8–10 likes per post, without any comments	Private account	–
	Facebook	5–7 likes per post. without any comments	4–6 likes per post, without any comments	–
	X	–	–	–

Note Instagram (2024), Facebook (2024), X (2024)

From the 674, 15,459, and 7125 opinions for Idanha-a-Nova, Covilhã, and Castelo Branco, respectively, we see that the cities have the same appreciation from visitors, with 4 out of 5 ratings in all categories, except for Idanha-a-Nova concerning gastronomy, which has 3 out of 5 (see Table 10).

Table 10 Reputation of the cities among visitors (January 2024)

Platform		Cities		
		Idanha-a-Nova	Covilhã	Castelo Branco
Trip Advisor	Do			
	Stay			
	Eat			

Note TripAdvisor (2024)

Dimensions/Aspects that Might Be Improved or Leveraged, Conclusions, and Contributions

In this approach, we aim to share good practices and comparative analyses to provide a more realistic view of how the UCCN can serve as a factor in the territories´ economic and social growth and development.

Stepping up policies to attract investment and create jobs not only to enhance a better quality of life and well-being for the population but also to intensify actions to retain and attract young people are inevitable measures, based on the indicators analysed, for the cities of Idanha-a-Nova, Covilhã, and Castelo Branco.

Complementarily, the intensification of partnerships with other cities belonging to the UCCN and reinforcing the presence on digital channels, in general, and social networks in particular, are significant in increasing visibility and tourist flows.

From the analysis, we can conclude that integrating a municipality into the UCCN may not be sufficient for its growth and development. Complementary measures will be necessary, not only on the part of local, regional, and central authorities but also on the part of private initiatives in a joint effort, and by sharing common goals, actions can be outlined that promote the destination, with repercussions on its resident community and those who visit it.

However, the synergies that can arise from participation in the UCCN, in addition to its visibility and notoriety—national and international—can be a first step towards revitalising a territory's socio-economic activity and, in this way, an inducer to consider for the dynamics of the various sectors and the rejuvenation of its population.

Since integration into the UCCN is still a topic rarely addressed in the literature, this work contributes to greater clarification of the topic and points out paths for public decision-makers and private operators.

For science, as we do not know of any identical study, to date. The UCCN management will have access to new data on the current reality and future perspectives, as well as the Creative City team on-site. For operators, the observation of a reality susceptible to evolution implies assertive actions.

Acknowledgements This research was supported by the Geography and Spatial Planning Research Centre (CEGOT), financed by national funds through the Portuguese Foundation for Science and Technology (FCT) under reference UI/DB/04084/2025, and, also, by the FCT, financed by national and EU funds through the ESF, under reference UI/BD/154288/2022.

References

Facebook (2024) Facebook. https://www.facebook.com/?locale=pt_PT
Google (2024) Cidades Criativas da UNESCO
Instagram (2024) Instagram. https://www.instagram.com/
Município da Covilhã (2024a) Covilhã. https://www.cm-covilha.pt/
Município da Covilhã (2024b) Covilhã Cidade Criativa. https://covilhacriativa.com/

Município de Castelo Branco (2024a) Castelo Branco. https://www.cm-castelobranco.pt/
Município de Castelo Branco (2024b) Castelo Branco Cidade Criativa. https://www.cm-castelobranco.pt/investidor/territorio-dinamico-e-inovador/cidade-criativa/
Município de Idanha-a-Nova (2024a) Idanha-a-Nova. https://www.cm-idanhanova.pt/
Município de Idanha-a-Nova (2024b) Idanha-a-Nova city of music. http://cityofmusic.cm-idanhanova.pt/
Pordata (2024) Municípos—estatísticas, indicadores e dados. https://www.pordata.pt/Municipios
Tripadvisor (2024) Tripadvisor: over a billion reviews & contributions for hotels, attractions, restaurants, and more. https://www.tripadvisor.com/
Turismo Centro de Portugal (2024) Cidades Criativas da UNESCO. Turismo Centro Portugal. https://turismodocentro.pt/artigo/cidades-criativas-da-unesco-no-centro-de-portugal/
UCCN (2024) UNESCO creative cities network. https://en.unesco.org/creative-cities/home
United Nations (2015) General assembly resolution A/RES/70/1. Transforming our world, the 2030 agenda for sustainable development. https://sustainabledevelopment.un.org/index.php?page=view&type=111&nr=8496&menu=35
X (2024) X. O que está acontecendo/X. https://twitter.com/?lang=pt

Territorial Profiles of Unworthy Housing in Portugal

Teresa Sá Marques, Fátima Matos, Fernando Honório, Catarina Maia, Gabriela Ribeiro, Diogo Ribeiro, Paulo Conceição, and João Queirós

Abstract Housing is a fundamental human right and is crucial to quality of life and social well-being and is on international and national agendas (United Nations; European Commission). European agendas have highlighted the importance of improving access to housing and reducing housing deprivation. The literature on unworthy housing focuses mainly on identifying different types of unworthiness and developing comparative analyses between countries, highlighting also the need to develop integrated territorial approaches that can support the design of public policies. This chapter compensates for the lack of territorial reflection by conducting a spatial analysis of Portugal. We developed a methodological approach based on multivariate

T. S. Marques (✉) · F. Matos
Faculty of Arts and Humanities, Department of Geography, Geography and Spatial Planning Research Centre (CEGOT), University of Porto, Porto, Portugal
e-mail: tmarques@letras.up.pt

F. Matos
e-mail: fmatos@letras.up.pt

F. Honório · G. Ribeiro
Faculty of Arts and Humanities, University of Porto, Porto, Portugal
e-mail: fern.honorio@gmail.com

G. Ribeiro
e-mail: ribeirogabriela551@gmail.com

C. Maia · D. Ribeiro
Faculty of Arts and Humanities, Geography and Spatial Planning Research Centre (CEGOT), University of Porto, Porto, Portugal
e-mail: catarina.maia2@gmail.com

D. Ribeiro
e-mail: mrcotgeo@gmail.com

P. Conceição
Faculty of Engineering, Department of Civil and Georesources Engineering, CITTA, University of Porto, Porto, Portugal
e-mail: psc@fe.up.pt

J. Queirós
IS-UP, Faculty of Arts and Humanities, University of Porto, Porto, Portugal
e-mail: jplqueiros@gmail.com

L. Soares et al. (eds.), *Compromised Geography*, Key Challenges in Geography,
https://doi.org/10.1007/978-3-032-05201-8_16

analysis to identify territorial profiles based on different types of unworthy housing. The findings provide very useful information for the design of territorial housing policies.

Keywords Unworthy housing · Territorial profiles · Portugal

Introduction

Access to decent housing is a fundamental human right and an indicator of the advancement and quality of life of citizens. The right to decent housing is linked to the right to the city and to urban functions, in line with the 2030 Agenda (United Nations 2015), in particular Goal 11—Sustainable Cities and Communities. Accordingly, by 2030, countries must guarantee access to housing with adequate living conditions for all citizens, integrated into the urban fabric, with good access to infrastructure and basic facilities. This presupposes the need to promote more inclusive and sustainable urbanisation.

In 2016, the "New Urban Agenda", presented at the UN-Habitat III Conference, renewed the political commitment to decent housing and sustainable urban development. It reaffirmed the centrality of inner city housing as advocated by the United Nations, highlighting the importance of access to housing and service availability, which implies addressing discrimination in access to housing and increasing the protection of the most socially vulnerable people (United Nations—Habitat III 2017).

In recent decades, housing has been absent from political agendas in most European countries, including Portugal. In recent years, however, it has once again been brought to the public and academic debate, and to the political agenda. This growing concern is particularly evident at European level. In 2016, the Urban Agenda for the EU was presented, highlighting housing as one of the priority areas for intervention.

In 2017, the European Commission issued Recommendation (EU) 2017/761 on the European Pillar of Social Rights. Principle 19 addresses "housing and assistance for the homeless" and aims to promote access to social housing or housing assistance of good quality; the right to appropriate assistance and protection against forced eviction; and shelter and services to the homeless to promote their social inclusion (European Commission 2017, Chap. III, no. 19).

In January 2021, the European Parliament adopted the motion for a resolution on access to decent and affordable housing for all, in which it "calls on the Commission and the Member States to ensure equal access for all to decent housing, including clean and high-quality drinking water, adequate and equitable sanitation and hygiene, connection to sewage and water networks, a high quality indoor environment and to affordable, reliable, sustainable energy for all, thereby contributing to eradicating poverty in all its forms, protecting the human rights of disadvantaged households and supporting the most vulnerable groups, so as to protect their health and well-being" (European Parliament 2021, pp. 11–12).

In Portugal, according to Article 65 of the Constitution of the Portuguese Republic, "everyone has the right, for themselves and their family, to housing of adequate size, in conditions of hygiene and comfort, and which respects personal and family privacy" (República Portuguesa 2005, Article 65(1)). The State and local authorities are also responsible for guaranteeing this right and are obliged to "plan and implement a housing policy that is integrated into the general land-use planning instruments and supported by urbanisation plans that guarantee the existence of an adequate transport network and social facilities" (República Portuguesa 2005, Article 65(2a)).

The lockdown during the COVID-19 pandemic once again highlighted the importance of housing conditions. While some households were able to live comfortably during the lockdown periods due to good housing and neighbourhood conditions, others faced a much more difficult situation, such as those living in unhealthy housing (Matos et al. 2024b; Mendes 2020).

Specifically, this chapter addresses a key challenge for a Committed Geography Agenda, as it deals with housing as a fundamental right of individuals and a crucial issue in building more equitable cities. This study is part of an ongoing research project (Housing Project: Territorial identification of social vulnerability and unworthy housing[1]), aimed at identifying the different characteristics of unworthy housing in Portugal and helping inform public policies anchored in the territory that can respond more effectively to the challenges faced by different geographical contexts. The specific objectives of this chapter are to identify the territorial patterns of unworthy housing at municipal level and their determining factors. With these objectives in mind, Sect. 2 discusses the concept of unworthy housing, problematising different perspectives. Section 3 establishes the methodological approach used to identify different territorial problems with a view to characterising unworthy housing conditions. The approach we have developed considers a dwelling unworthy where any of the following habitability requirements are not adequately met: exclusion and insecurity; inaccessibility; inadequacy or unsuitability; overcrowding; unsanitary and uncomfortable. Section 4 presents the findings for Portugal at municipal level, identifying a territorial typology that illustrates different problems related to unworthy housing. Conclusions are drawn in Sect. 5, including indications for the future development of this research.

---

[1] This study was developed under the Science4Policy 2023 (S4P-23): annual science for policy project call, an initiative by the Centre for Planning and Evaluation of Public Policies in partnership with the Foundation for Science and Technology, funded by the Portuguese Recovery and Resilience Plan.

Besides the chapter authors, other members of the research team also contributed to the discussion, in particular Pedro Leão Neto, Maria Neto, Ana Miriam Rebelo (CEAU, Faculty of Architecture, University of Porto); Isabel Breda Vázquez, Jorge Afonso Rios (CITTA, Faculty of Engineering, University of Porto); Paula Guerra, Vanessa Rodrigues (IS-UP, Faculty of Arts and Humanities, University of Porto); Isabel Martins, Miguel Saraiva (CEGOT, Faculty of Arts and Humanities, University of Porto).

The Concept of Unworthy/worthy Housing

Adequate/worthy/unworthy housing has been a central topic in housing policy debates. The term "adequate housing" first appeared in 1966, in the International Covenant on Economic, Social and Cultural Rights, under which the State Parties recognised "the right of everyone to an adequate standard of living for himself and his family, including adequate food, clothing and housing, and to the continuous improvement of living conditions" (United Nations 1966, Article 11(1)). The United Nations (1991) defines worthy housing as providing adequate privacy, legal security, affordability, sufficient lighting and ventilation, adequate space, basic infrastructure and a location that is accessible to the workplace and essential services, all at a reasonable cost to the beneficiaries. Thus, the definition of 'unworthy housing' arguably applies to dwellings that do not meet these standards (United Nations 1991).

The UN has systematised this right to adequate housing based on seven dimensions (United Nations 2014): security of tenure, availability of services, materials and infrastructures, affordability, habitability, accessibility, location, and cultural adequacy. Recently, an eighth dimension—sustainability—has been proposed (Mazucatto and Farha 2023).

Recognising the diversity of contexts and factors that can contribute to housing exclusion, the European Federation of National Organisations Working with the Homeless (FEANTSA 2005, 2007) has defined homelessness under four categories: rooflessness, houselessness, insecure housing, and inadequate housing (Table 1). The ETHOS Light typology (2007) develops a harmonised definition of homelessness for statistical purposes, considering the following categories: people living rough (living in the streets or public spaces without a shelter that can be defined as living quarters); people in emergency accommodation (with a place to sleep, but temporary in institutions or shelters, e.g. homeless hostels, night shelters, women's shelters or refuge accommodation); people living in institutions (health care institutions, penal institutions); people living in non-conventional dwellings due to lack of housing (mobile homes, non-conventional buildings, temporary structures) and homeless people living temporarily with family and friends (due to a lack of housing).

In 2010, a European Union study (Stephens et al. 2010) revealed that housing policies can help to reduce, but not remove, the link between poverty and housing deprivation. Not all poor people experience housing deprivation, but it is more likely to occur among the poor and the most vulnerable groups (the homeless, the elderly, the sick, immigrants, etc.).

In 2016, the United Nations, in the Sustainable Development Goals of the 2030 Agenda, namely Goal 11 "Sustainable Cities and Communities", defined as one of its targets to ensure "by 2030, access for all to adequate, safe and affordable housing and basic services, and upgrade slums" (United Nations 2016). In general, there is great concern about the need to promote decent and adequate housing and to address housing deprivation, as evidenced by precarious housing, inaccessibility to housing, lack of basic services, and poor location of housing (United Nations 2016).

Table 1 Categories of living situations that constitute homelessness or housing exclusion (FEANTSA 2005)

Roofless	Houseless	Insecure	Inadequate
People living rough People in emergency accommodation	People living in accommodation for the homeless People living in women's shelter People living in accommodation for immigrants People are due to be released from institutions	People receiving long-term support (due to homelessness) People living in insecure accommodation People living under threat of eviction People living under threat of violence	People living in temporary/ non-conventional structures People living in unfit housing People living in extreme overcrowding

In 2018, the World Health Organization (WHO) considered that people living in substandard housing have at least four of the following problems: (i) leaking roofs; (ii) damp walls, floors and foundations; (iii) rotting window frames or floors; (iv) no bathtub or shower in the dwelling; (v) no flushing toilet inside the dwelling for the sole use of the family; (vi) the house is very dark (WHO 2018).

In "The Future of Cities", published by the Joint Research Centre (Vandecasteele et al. 2019), the European Commission points out that one of the main challenges for cities is to promote an adequate and affordable housing supply. The increase in housing prices in recent years, due to the growing role of financial actors in the urban housing market, the so-called financialisation of housing, has raised significant concerns in several European cities about the impact on adequate and affordable housing, as prices rise faster than household incomes and the availability of affordable housing is reduced (Marques et al. 2022; Vandecasteele et al. 2019).

In its literature review on housing deprivation, focusing on household surveys, Eurofound concluded that most studies examine three areas—basic facilities, structural problems and overcrowding—, although some include the first two, but not overcrowding, and others occasionally include accessibility or environmental problems (Eurofound 2016, pp. 11–27).

According to Clair et al. (2019), housing precariousness comprises four components: (1) accessibility; (2) tenure security (risk of eviction or foreclosure); (3) quality (dampness, poor ventilation, toxicity, house adequately cool in summer and warm in winter, housing with a toilet or shower, with a toilet for the exclusive use of the household and inside the house, leakage/humidity and overcrowding); (4) access to basic services (banking, postal services, transport, food and health services).

The European Parliament (2021) considers unworthy housing to be housing with safety problems, that is, unhealthy, of poor quality, unaffordable, energy inefficient or overcrowded, also including homeless people or those at risk of eviction.

The official EU measure of severe housing deprivation (Eurostat, EU Statistics on Income and Living Conditions survey—EU-SILC) includes five indicators: overcrowding; darkness (the dwelling is too dark, meaning that there is not enough daylight coming through the windows during the day); leakage (the dwelling has any of the following issues: leaking roof and/or damp ceilings, damp walls, floors or foundations and/or rotting window frames and doors); and lack of an indoor bath/shower and indoor flushing toilet for the sole use of the household. A household is only classified as deprived if it lacks these items, while the concept of "severe housing deprivation" presupposes overcrowding plus at least one of the other indicators of deprivation (Eurostat 2022a).

The concept of inaccessible and inadequate housing proposed by Eurofound (2023) encompasses four problem situations and four dimensions of analysis, involving different groups and social situations: (1) *housing exclusion,* which includes the homeless population, young adults living with their parents and without access to housing; (2) *housing insecurity*, families or individuals at risk of becoming homeless, experiencing delays or difficulties with payments and services, in a situation of eviction or at risk of eviction; (3) *housing cost problems*, families overburdened by housing costs or the cost of housing services; (4) *housing inadequacy*, which includes satisfaction with housing and place of residence, housing conditions, overcrowding, and problems/positive aspects associated with the place of residence.

Governments are also increasingly committed to this issue. The French government (*Loi no° 90-449 du 31 mai*1990*; Loi no° 2014-366 du 24 mars 2014—art. 34*) defines unworthy housing as accommodation used for residential purposes that is by its nature unfit for that purpose, as well as dwellings or buildings that expose their occupants to risks that could endanger their physical safety or health. The law also adds the concept of informal housing, which is defined as places or facilities for residential use, built without the right or title to the land on which they are located, without drinking water supply or sewer and rainwater collection systems, or without roads or public facilities that guarantee satisfactory access to health and safety.

According to French law (…) *mobilisation pour le logement et la lutte contre l'exclusion (Loi n° 2009-323 du 25 mars* 2009) precarious housing covers all housing situations that constitute a denial of the right to housing and compromise human dignity, including, in particular, unhygienic and unsuitable housing, buildings and facilities (health risk); housing and buildings containing lead (risk of lead poisoning); buildings in ruins, at risk (risk of insecurity).

The Spanish government, in the Law on the Right to Housing (*Ley 12/*2023*, de 24 de mayo, por el derecho a la vivenda, art. 3º*), defines unsuitable housing as a building or part of a building intended for habitation, that does not meet the minimum conditions required by the legislation in force, that is, dwellings that do not meet the requirements for surface area, number, size and characteristics of the habitable parts, those that have serious defects in their basic installations and equipment, and those that do not meet the minimum safety and universal accessibility requirements. The law also defines decent and adequate housing as that which, by virtue of its size, location, living conditions, universal accessibility, energy efficiency and use of renewable energy, with access to basic supply networks, meets the housing needs of families, in conditions that are affordable in relation to their financial resources, and which constitute their home, where they can live in dignity, respecting their privacy, and enjoy family or social relations, favouring the full development and social inclusion of people and the building habitability required.

In many countries, particularly in the English-speaking world, the issue of housing insecurity (loss of housing due to default on mortgage payments and, above all, eviction) has received increasing attention, as in the case of the work of the Eviction Lab in the US (https://evictionlab.org/), which monitors and maps eviction processes in the country and presents other analyses related to housing insecurity, loss of housing and homelessness (see also Desmond 2016). The issue of housing accessibility and the burden of housing costs is currently at the centre of most technical and scientific studies (Beer et al. 2016; Gabriel and Painter 2020; Galster and Lee 2021; Shamsuddin and Campbell 2021; Wetzstein 2017; Wood et al. 2017). Decent or adequate housing is that which does not represent an excessive cost for families, given their income—hence the proliferation of terms such as "affordable housing", or mobilising indicators such as "housing cost burden" (Eurostat 2022b, 2023, 2024).

In Portugal, 2017 marked a political turning point with the publication of the report of the UN Special Rapporteur on the Right to Adequate Housing (United Nations 2017). This report highlighted the serious housing shortage in Portugal, especially for young people aged between 20 and 29, and the sharp increase in the risk of homelessness. It also pointed out the particularly poor housing conditions of some communities living in Portugal, such as the Roma and people of African descent. The report's recommendations were fundamental to the policy strategies developed from 2017 onwards. The Portuguese government gave special prominence to the 1st Right Housing Programme in 2017, advocating the design of financial instruments to promote access to housing for people living in unworthy housing conditions (Alegra and Tulumello 2019; Matos et al. 2024a; Portuguese Government 2017).

In 2018, the Institute of Housing and Urban Rehabilitation (IHRU—*Instituto da Habitação e da Reabilitação Urbana*) conducted a survey on housing shortages on the basis of precarious housing situations on urban centres, namely shacks or precarious constructions; illegal encampments; campsites; rundown social housing estates; urban areas of illegal origin (AUGI); rundown urban areas; living in an area of risk, not including other situations of housing deprivation such as overcrowding; housing inadequacy related to the characteristics of the occupants, in particular people with reduced mobility; domestic violence situations; homelessness; among others included in some of the documents referred above.

As regards housing conditions, the Basic Housing Law (*Lei nº 83/2019, República Portuguesa* 2019) states in Article 9 that "(1) a dwelling is considered to be of an adequate size for its occupants where the area, the number of rooms and the available solutions for water, sanitation and energy supply are sufficient and do not give rise to situations of insalubrity, overcrowding or risk of promiscuity. (2) The law defines the minimum requirements for the qualification of dwellings, taking into account the number of rooms and surface area, as well as guaranteeing hygiene, health, comfort, safety and accessibility conditions. (3) There is a risk of housing promiscuity and inadequacy for its residents if it is not possible to guarantee differentiated sleeping and sanitary facilities to preserve people's intimacy and family privacy." The same law (Article 8) also considers persons or groups in situations of particular vulnerability, such as the homeless, minors who are victims of abandonment or mistreatment, victims of violence, or persons in a situation of vulnerability.

Decree-Law No. 37/2018 (*República Portuguesa* 2018), which created the 1st Right Housing Programme, aimed at putting an end to unworthy housing situations in Portugal, considers that people who do not have adequate housing live in the following situations:

(a) "precariousness, which includes people who are homeless, (…), as well as people who have no alternative housing solution to the place they use as their permanent residence, in particular when they have to vacate it due to the declaration of bankruptcy of household members or the owner of the property where the household lives, domestic violence situations, municipal urban development measures or the non-renewal of a lease agreement;
(b) unhealthy and insecure housing, cases in which the person or household lives in a place, whether built or not, that does not meet basic health, structural safety, watertightness and hygiene conditions or because they do not meet the minimum conditions of habitability;
(c) overcrowding, where the composition of the household and the number of rooms in the dwelling indicate an inadequate living space due to a lack of two or more rooms, as defined by the National Statistics Institute[2];
(d) inadequate housing, due to the conditions of the dwelling which are incompatible with the specific characteristics of the people who live in it, such as in the case of disabled people, in particular where the dwelling i) has barriers to access to the floor on which it is situated; and/or ii) the dimensions of the openings and interior areas prevent circulation and use adapted to the specific characteristics of the people who live in it" (Article 5). The latest amendment to this decree-law (*República Portuguesa* 2025) added another situation, people "overburdened by costs with rent or monthly mortgage loan payment, where this implies an effort rate of more than 40% of their average monthly income" (Article 5(e)).

Decree-Law No. 89/2021 (*República Portuguesa* 2021), regulating the provisions of the Basic Housing Law, considers people in a situation of actual housing need to be "those who do not have or are in actual risk of losing a home and who have no alternative housing" (Article 3).

Proposed Methodological Approach

The various documents cited above demonstrate the persistence and centrality of the issue of unworthy housing, but also the complex, multidimensional, contextual and evolving nature of the concept. A definition related to the unworthy housing conditions of families prevails, linked to the characteristics of the dwelling (safety,

[2] This provision was later amended to consider any situation of overcrowding, as defined by the National Statistics Institute (República Portuguesa, Decree-Law No. 81/2020, Article 5(c)).

health, equipment, size) and the legal relationship to the dwelling (security of tenure). Some definitions of housing precariousness/exclusion/insecurity also include issues of housing affordability (Clair et al. 2019; Eurofound 2023; Eurostat 2022b, 2023, 2024) and other aspects of housing security. The related debate, therefore involves several challenges:

- the diversity of situations and dimensions covered (individual, community), which poses the challenge of developing robust analytical frameworks capable of reflecting the relationship between the manifestations of problems and the evolution of structural vulnerability factors that restrict access to housing (Clair et al. 2019; Dewilde and Decker 2016; Debrunner et al. 2024; Kimhur 2023). Hence, the importance of taking into account the diversity of situations and recognising the central role of local contexts (Conceição 2023; Howell et al. 2019; Matos et al. 2020, 2024a; Tosi 2010);
- the diversity of situations and the need for greater local knowledge imply changes in objectives, in the design of public policies and their instruments, and in organisational models (Elsinga et al. 2020; Fijalkow 2022; Levy-Vroelant 2010; Matos et al. 2019), making it essential to combine knowledge of unworthy housing situations with institutional capacity building and reflection on the organisation of housing policies.

Amidst a structural crisis, with new contours emerging from the aftermath of the 2008 crisis and the COVID-19 pandemic, unworthy housing should be analysed from a relational perspective encompassing the various changing dimensions (Marques et al. 2021; Matos et al. 2024b; Moos 2019).

Given the diversity of the challenges, it is therefore essential to develop a robust analytical framework capable of reflecting the different manifestations of the problems and monitoring the evolution of the structural vulnerability factors that restrict access to housing. The approach we have developed considers a dwelling unworthy (precarious for some authors) where any of the following habitability requirements are not adequately met: (i) exclusion and insecurity; (ii) inaccessibility; (iii) inadequacy or unsuitability; (iv) overcrowding; (v) unsanitary and uncomfortable.

(i) *Housing exclusion and housing insecurity*

Housing insecurity and exclusion (Bolt and Czirfusz 2022; Eurofound 2023; Kadi and Musterd 2015) are manifest where people do not have access to housing (homelessness), or are unable to access housing due to personal insecurity (e.g. domestic violence) (FEANTSA 2005, 2007), or where housing can be a financial overburden (e.g. young adults living at home with their parents) (Bolt and Czirfusz 2022; Eurofound 2023). The increase in evictions and repossessions, where income is insufficient to meet contractual payments, has contributed to increased housing insecurity (Eurofound 2023), as has the proliferation of short-term or non-renewable tenancies (Silva 2019). This situation has seen many low-income families living in very precarious housing conditions (Carmo et al. 2015; Silva 2019).

(ii) *Unaffordable housing*

As rent and housing prices rise (European Parliament 2025; Eurostat 2022b, 2023, 2024; Lourenço and Rodrigues 2017; Marques et al. 2022; OECD 2025), families will have to spend more of their income to cover housing costs. Labour market flexibility, an increase in precarious working conditions and low-paying jobs will exacerbate the problem of finding affordable housing. These factors could lead to greater affordability problems among low-income households. Rising costs make housing increasingly precarious for a range of groups, primarily affecting those receiving lower incomes (Bolt and Czirfusz 2022).

Within the private rental sector, the highest risk of housing cost overburden is found in the informal segments. These exploitative forms of rental housing have emerged during recent and historical periods of housing crises across Europe, including Portugal (Bolt and Czirfusz 2022; Mendes 2022a).

The heavy burden of housing costs on household income can reduce financial capacity to meet maintenance costs, which can reduce comfort of the dwelling and threaten the structural sustainability of the building.

(iii) *Housing inadequacy or unsuitability*

Poor quality and inadequate housing conditions compromise minimum comfort, well-being and sometimes the health of individuals (WHO 2018). Buildings must ensure mechanical strength and stability, fire safety, environmental protection, noise protection, thermal insulation, ventilation, and functionality. It is therefore vital to consider the building's state of repair, including the roof, structure, external walls and window frames, as their poor condition can lead to the spread of damp, poor ventilation and poor lighting, factors that affect the health and quality of life of occupants (Borg 2015; Gibson et al. 2011; Shaw 2004; WHO 2011). The precariousness of construction, in particularly mobile homes, non-conventional buildings and temporary structures, worsens living conditions. The age of the building also affects housing conditions, as old, very small and unmaintained houses give rise to problems of functionality, watertightness and safety. In Portugal, the old rent freeze has contributed to the rundown state of a large number of residential buildings (Matos et al. 2019; Mendes 2022b).

(iv) *Overcrowded housing*

The level of occupancy of the dwelling and indoor comfort conditions can determine the comfort of the dwelling. In the case of poor occupancy or overcrowding, it is essential to analyse the severity of the situation and the deficit of habitable space (Cutts et al. 2011; Eurofound 2016, 2023), ensuring an appropriate balance between the household composition and the number of rooms, taking into account the size of the living space. Moreover, shared accommodation (occupied by more than one family unit) raises issues of personal intimacy and family privacy. These situations have a negative impact on physical and mental health (Pendall et al. 2012; Ruiz-Tagle and Urria 2022). Overcrowding is even more serious for the most vulnerable, such as families with young children and families of single elderly people, who are often disabled and have no access to services or family support.

(v) *Unsanitary and uncomfortable housing*

As regards minimum comfort and sanitary conditions, access to basic services must be guaranteed, namely water supply and sanitation, and minimum health conditions (sanitary facilities essential for hygiene) (WHO 2018). It is also necessary to ensure that homes provide thermal comfort to guarantee the health and well-being of individuals and families (WHO 2018). Physical accessibility to housing is also crucial, especially for an increasingly elderly and disabled population, including lifts in multi-storey buildings (WHO 2018).

This theoretical approach will structure the subsequent analysis. Our goal is to identify and spatially characterise unworthy housing in Portugal. To this end, a typology of problems is identified (Table 2), as well as the corresponding statistical indicators to be included, taking into account the statistical information available at municipal level (Table 3). Most of the indicators were obtained from the 2021 Population and Housing Census, published by the National Statistics Institute (INE). As for the homeless, the latest survey of this population, from 2023, from the National Strategy for the Integration of People Experiencing Homelessness (ENIPSSA) was taken into account. In terms of thermal comfort, beneficiaries of the social energy tariff provided by the Directorate-General for Energy and Geology (DGEG) for 2020 were included. Information on public water services and sanitation infrastructures was used from ERSAR (Water and Waste Services Regulatory Authority), ERSARA (Azores Water and Waste Services Regulatory Authority) and DREM (Madeira Regional Statistics Directorate) for 2020.

To discuss and select the indicators, a focus group was held involving the entire project team (16 researchers from geography, architecture, civil engineering, planning and sociology). We discussed the indicators best suited to measure the intensity of each type of vulnerability and assessed their spatial distribution and representativeness to select two (preferably) complementary indicators, suitable for assessing two levels of vulnerability:

- one that reflects the presence and/or intensity of each type of vulnerability;

Table 2 Typology of housing vulnerabilities

Housing exclusion and housing insecurity	Unaffordable housing	Inadequate or unsuitable housing	Overcrowded housing	Housing with poor sanitation and comfort
Homeless people Adults living at home with their parents	Families struggling to meet housing costs Families struggling to meet rental costs	Dwellings in rundown buildings Old buildings Precarious and temporary housing	Overcrowded housing Housing shared by two or more families	Unsanitary housing Housing with poor physical accessibility conditions Housing with thermal comfort problems

Table 3 Typology of housing vulnerabilities and related statistical indicators

	Vulnerabilities	Indicators (%)
Housing exclusion and housing insecurity	1. Individuals without access to housing	1. Homeless people
	2. Adults living with their parents	2. Households with adults (over 25 years old) living with their parents
Unaffordable housing	3. Difficulties in meeting housing costs	3. Number of years to buy a 72 m^2 house, considering an effort rate of 40% of the declared gross income of households in Quintile 1 4. Number of years to buy a 72 m^2 house, considering an effort rate of 40% of the declared gross income of households in Quintile 2
	4. Difficulties in paying the housing rent	5. Weight of rent in the gross monthly incomes of households in Quintile 1 (lower incomes) 6. Weight of rent in gross monthly income of households in Quintile 2
Inadequate or unsuitable housing	5. Rundown housing	7. Dwellings in rundown buildings
	6. Old houses	8. Dwellings in buildings constructed before 1981 9. Dwellings in buildings constructed before 1948
	7. Precarious and temporary housing	10. Households living in non-typical dwellings (precariously built, mobile, improvised and not built for living)
Overcrowded housing	8. Overcrowded homes	11. Overcrowded housing 12. Overcrowding with vulnerable residents
	9. Shared housing	13. Dwellings with two or more family units
Housing with poor sanitation and comfort	10. Unsanitary housing	14. Dwellings without a bath or shower 15. Dwellings without a public water supply 16. Dwellings not connected to a sewer system

(continued)

Table 3 (continued)

	Vulnerabilities	Indicators (%)
	11. Housing with poor accessibility conditions	17. Dwellings in buildings with four or more floors without a lift 18. Dwellings without a wheelchair-accessible entrance
	12. Thermal comfort problems	19. Dwellings without a heating system 20. Energy-supported housing

- and a more accurate level that reflects only the most serious situations within each vulnerability type.

To characterise dwellings, and in the few cases where this criterion could not be met due to a lack of compatible indicators, two or three indicators were selected to cover different aspects of the phenomenon, or only the most appropriate indicator was chosen.

At this stage, the aim was to produce a composite index for each dimension that would reflect the housing vulnerability of each territory as accurately as possible. In other words, the index is a statistical weighting of the most serious situations, without losing sight of the overall picture of each vulnerability.

The territorial profiles of unworthy housing were identified as follows:

1. Calculation of the Composite Index (CI) for each vulnerability dimension: (i) normalising the initial variables 0–100; (ii) calculating the new variable resulting from the average of the normalised initial variables; (iii) normalising the new variable 0–100 = Composite Index. This index makes it possible to give equal weight to all the dimensions, while maintaining the original variances. The indicators were not weighted.
2. The application of a Principal Component Analysis (PCA), with all active CIs to allow a classification that eliminates random and insignificant variances. PCA is a multivariate statistical method suitable for processing databases made up of several numerical variables.
3. Mixed classification, ascending hierarchical and mobile centres, given the large number of territorial units (municipalities), using Euclidean Distance as a measure of similarity and Ward's Method as an aggregation criterion. This method is the most common and guarantees a more appropriate division of the groups.
4. Selection of the number of classes to retain for analysis, based on the most significant splits in the dendrogram.
5. Description of the retained classes based on a statistical significance test (V-test—which translates the value of the deviation between the values of the variables observed in the class and the values observed in the data set).

6. Creation of a standard profile for each class by selecting the most characteristic variables (based on the highest V-test values).

Territorial profiles of unworthy housing in Portugal

Portugal's Unworthy Housing

Unworthy housing in Portugal was characterised taking into account the five types of vulnerability described above.

Housing exclusion and housing insecurity in Portugal have deteriorated considerably in recent years. Social inequalities and poverty levels have increased, reflected in the number of people who are unable to afford housing. According to the December 2023 Characterisation of Homeless People Survey (ENIPSSA 2024), more than 13,000 people are homeless.

Domestic violence was not included in our analysis due to a lack of data at municipal level. However, the data available for Portugal show an increase in domestic violence, 26,511 cases in 2021 and 30,389 in 2022 (Portal da Violência Doméstica). There were 475,138 households with adults (over 25 years old) living with their parents in 2021, corresponding to 11.5% of all families. Evictions have also increased in Portugal. According to data from the National Rental Centre, there were 2672 eviction requests in 2023, 17% more compared to 2022 (Diário de Notícias 2024).

As regards **unaffordable housing**, the weight of monthly rent in the gross monthly income of low-income families (Quintiles 1 and 2) is 61% and 41%, respectively. These figures show that families are struggling to gain access to housing. It takes 44 years for families in Quintile 1, and 30 years for families in Quintile 2, to buy a 72-square-meter house, considering a 40% effort rate. These figures highlight the inaccessibility of housing for the most economically vulnerable groups.

In terms of **inadequate or unsuitable housing**, 33% of dwellings in Portugal in 2021 were in rundown buildings, 14.4% were built before 1961, and 4042 were non-typical dwellings (mobile homes, non-conventional buildings or temporary structures).

In 2021, occupancy levels in Portugal revealed 527,855 (12.7%) overcrowded dwellings and 122,572 (3.0%) dwellings shared by two or more families.

An assessment of **sanitation and comfort** found that there are still 79,269 (2.0%) households without a bath or shower, 3% do not have a public water supply, and 13% are not connected to a sewer system. Physical accessibility to housing requires immediate attention, as 558,188 families (18%) live in buildings with four or more floors without a lift and 2,732,787 (66%) of buildings lack a wheelchair-accessible entrance.

Territorial Typology of Unworthy Housing

The analysis is structured around five areas of housing unworthiness: two focusing on affordability problems (housing exclusion; unaffordable housing) and three on unworthy housing conditions (inadequate or unsuitable housing; overcrowding; poor sanitation and comfort). Each typology takes into account two or three dimensions of vulnerability.

We identified seven territorial profiles of unworthy housing in Portugal (Table 4). The analysis in Table 4 allows us to characterise each of the seven classes identified, taking into account the dimensions of vulnerability considered (selection of the most characteristic variables, based on the highest V-test values).

The profiles identified were mapped by municipality (Fig. 1) to provide a territorial portrait of the different types of vulnerability to unworthy housing in Portugal. They reflect the diversity of the problems that affect different areas of the country and demonstrate the need to develop territory-based policies that focus on the different problems.

Profile A has a well-defined territorialisation. It is present in the Lisbon region (Lisbon and Cascais), in most of the Algarve municipalities, in some areas of Madeira (Funchal, Santa Cruz and Ponta do Sol) and in the Azores (Ponta Delgada). Two major vulnerabilities were identified: housing affordability and overcrowding.

People with low and medium incomes struggle to afford housing (buying or renting), as costs are very high in relation to household income and effort rates are at their highest. Housing exclusion is reflected in a large presence of 'homeless' people (especially in Lisbon).

Overcrowding is also high, often very serious (dwellings lacking 3 or more rooms), more so when overcrowding affects, for instance, vulnerable age groups (people

Table 4 Profiles for the seven classes (from A to G), using the most characteristic variables (based on the highest V-test values)

	Dimensions	A	B	C	D	E	F	G
i) Housing exclusion	Homelessness	5,2						
	Adult children living with parents		-2,9	-3,3		5,6	-3,1	
ii) Unaffordable housing	Unaffordable purchase price	**9,7**	**5,1**	**4,7**	**5,4**		-5,9	-7,0
	Unaffordable rent	**6,1**	**3,9**	**2,0**	**3,7**	3,0	-8,7	-3,4
iii) Inadequate or unsuitable housing	Rundown housing				2,6	4,0	-6,2	
	Housing in old buildings			-2,8	-3,7	-2,6	12,0	-2,2
	Precarious and temporary housing		**4,9**				2,6	-3,1
Overcrowded housing	Overcrowding	**8,7**	**2,2**	**3,6**	**5,9**	-2,1	-2,7	-7,7
	Overcrowding with vulnerable residents	7,9	2,7	3,6	3,8	-2,6		-7,6
	Shared housing	3,5			5,4	5,1	-6,1	-5,3
v) Housing with poor sanitation and comfort	Sanitation, water and bathroom issues			-4,3	-5,6	9,3		
	No lift or accessibility	2,9		11,2		-7,2	-2,3	2,9
	Thermal comfort issues				4,1	3,1	-8,9	

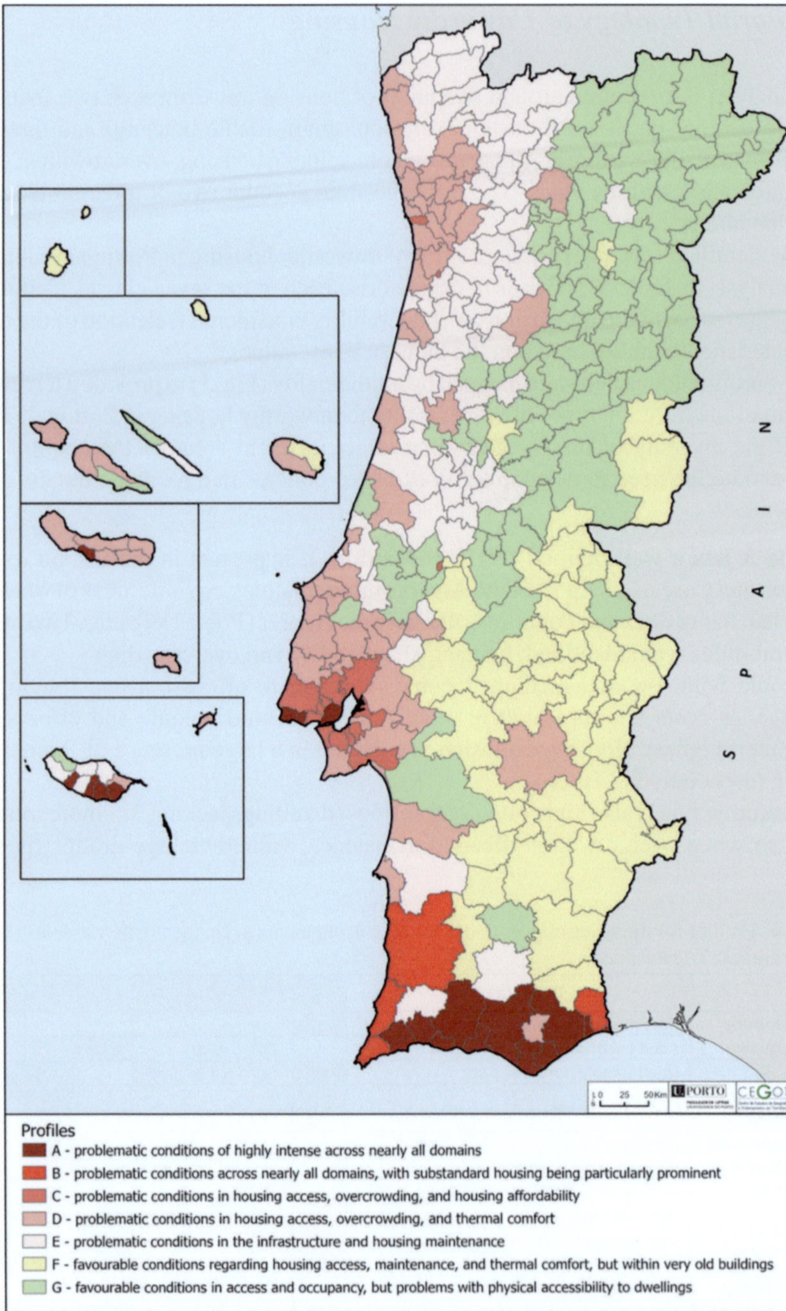

Fig. 1 Territorial profiles of unworthy housing

younger than 15 years of age or the elderly). The existence of small houses with few rooms and little space contributes to this situation.

Profile B is prevalent along the Alentejo coast and in the Algarve (Odemira, Aljezur, Vila do Bispo and also Vila Real de Santo António). The main vulnerability is the high cost of housing (on the sales and rental markets), making housing unaffordable for people with lower incomes.

On the other hand, there are problematic situations linked to a significant presence of precarious and temporary housing, with inadequate conditions and insecurity. Some of the most problematic situations at national level are concentrated in this territorial profile.

There are also worrying situations of overcrowding, affecting the most vulnerable age groups.

Profile C is found in most of the suburban municipalities of Lisbon and in the municipality of Porto. Here, the absence of a lift in buildings with four or more floors makes accessibility dramatically worse. These private and public buildings house an ageing population with more physical disabilities, who are often alone and unable to leave their homes.

Difficulty in affording housing is another significant vulnerability of this profile, albeit to a lesser extent than in profile A. In particular, the price of houses on the sales and rental markets are reaching levels beyond the reach of the lower-income social classes. As in the previous profiles, overcrowding is a very visible problem, but also to a lesser extent than in profile A. Overcrowding is sometimes severe (more than three rooms short, given the size of the household) and at vulnerable ages (situations involving elderly people).

On the positive side, the building stock is younger, with better living conditions and adequate sewer systems.

Profile D is more territorially extensive than the previous profiles, found in the second suburban area of Lisbon; in the neighbouring municipalities of Porto, extending to the north-west; in some medium-sized cities, namely Viana do Castelo, Vila Real, Aveiro, Viseu, Coimbra, Figueira da Foz, Leiria, Évora, and Sines; and in some municipalities in Madeira and the Azores.

Difficulties in accessing affordable housing remain high, especially for low and lower-middle-income groups. Overcrowding is also clearly problematic. In these contexts, there are dwellings with insufficient rooms for the size of the household, severe overcrowding, and overcrowding associated with vulnerable groups (especially the elderly). Housing shared by more than one family unit is also common, which may contribute to overcrowding.

The state of repair of buildings is also a problem, that is, they are in a serious state of disrepair and offer poor thermal comfort.

Profile E is prevalent in the extensive suburban sprawl along the north and central coast, where there is less pressure on the housing market. This is generally a more positive profile in terms of overall housing conditions compared to the other territorial profiles, but it reveals higher levels of rundown buildings and problems with water infrastructure and sewer systems.

Situations where adults continue to live with their parents were also identified, often because they are financially unable to emancipate themselves, and sometimes for cultural reasons (extended multi-generational families).

Newer buildings, the space available in the dwellings and better occupancy standards are of note. Nevertheless, it should be borne in mind that the profile is based on data from the 2021 census and that in recent years (2022, 2023 and 2024) the number of foreigners arriving in Portugal to take up residence has been almost as high as the number of residents in the area. As these regions are attractive because of the job opportunities they offer, the inflow of such a large number of people could have a negative impact on housing conditions.

Profile F is predominant in the Alentejo and occasionally extends to the interior of the Centre region. Each dimension has both positive and negative aspects.

As regards the physical condition of the dwellings, old dwellings (which strongly characterises this group) account for a large portion of the housing stock, but all are well maintained.

As regards occupancy, although the dwellings are small, occupancy figures and the number of rooms available are good. Often, one or two people, usually seniors, live in the dwellings.

No infrastructure problems were identified, namely related to thermal comfort and physical accessibility. However, the advanced age of owners and tenants, combined with small pensions, may make it difficult to maintain the good condition of the dwellings.

Profile G covers the northern and central inland municipalities. This profile reflects the highest level of housing worthiness, mainly due to better occupation conditions compared to the rest of the territory. Affordability is not an issue in this area. Effort rates are very low, as are rental costs relative to income.

No significant vulnerabilities were identified as regards the physical condition of dwellings.

Almost all the dimensions of vulnerability related to housing occupancy (capacity, size, number of rooms) are favourable. Generally speaking, these are municipalities in demographic decline, and there is a relatively low demand for housing, as opposed to along the coast. Overcrowding is therefore not an issue.

However, we believe that this situation may already have changed somewhat with the inflow the foreigners in the past three years. The occupancy profile will certainly tend to be more balanced.

In terms of comfort, access to the dwellings themselves is poor, particularly wheelchair access, which could be a major problem given the increasingly ageing population.

These findings provide vital information for the design of territorial policies aimed at addressing different spatial problems.

In summary, we consider that:

- policies focusing on citizens excluded from housing should be directed towards the areas with the highest concentration of homeless people, especially Lisbon and some municipalities in the Alentejo;
- policies to increase the supply of affordable housing for people with low incomes should focus on territorial profiles A, B, C and D, with a bias towards profile A;
- policies to minimise overcrowding problems should focus on profiles A, B, C and D, prioritising profile A;
- policies to improve the physical conditions of housing should be prioritised in profile E and D areas and should promote the conservation of buildings and improve the thermal conditions of dwellings. For profile E, greater attention should be paid to infrastructure issues that support housing, in particular sewer systems;
- policies to improve the accessibility of housing, namely for residents of buildings with four or more floors without a lift, should give priority to profile A and C areas;
- for profile F territories, the goal should be to reduce the number of precarious and temporary dwellings and to improve the functionality of dwellings in very old buildings;
- unworthy housing is virtually non-existent in profile G territories, but poor physical accessibility to housing, particularly for the elderly, needs to be addressed;
- greater coordination and the combination of policy instruments are needed in some territories/profiles, such as profile A.

Conclusion

The literature on unworthy housing mainly presents comparative analyses between countries, seeking to identify common problems and different specificities. This study took a more specific territorial approach, focusing on the municipalities of Portugal.

Based on the available literature, we offer a conceptual perspective structured around five aspects of unworthy housing: two focusing on affordability problems (housing exclusion; unaffordable housing) and three on unworthy living conditions (inadequate or unsuitable housing; occupancy levels; comfort conditions). To this end, a methodological approach was developed taking into account 12 dimensions of vulnerability, organised around a set of 20 indicators, and encompassing 308 municipalities, to characterise the territorial typology of unworthy housing. The data collected was used to identify and outline seven territorial profiles, each depicting how unworthy housing is expressed and identifying the different vulnerabilities and the intensity thereof. These resulting profiles provide guidelines for the design of policies to address the problems in each territory.

The results obtained at municipal level can be used to inform further territorial analyses. It would therefore be useful in the future to conduct a broader analysis (e.g. for statistical purposes) to gain a more specific territorial perspective and better characterise existing problems to support the development of an action programme focused on real intervention needs.

On the other hand, the literature suggests that territorial assessments of socio-economic vulnerability can help to better understand the spatial concentration of unsafe housing, as some social groups are more likely to be affected by precarious housing, such as low-income families, renters, the elderly, families with children, single-parent families, and minorities (Beer et al. 2016; Debrunner et al. 2024; Mckee et al. 2020). Socio-economic vulnerabilities appear as contextual conditions and are independent variables, while those related to the identification and characterisation of unworthy/substandard housing are dependent variables, as they are a reflection/consequence of multiple social dynamics (Debrunner et al. 2024).

Further study would therefore be useful to analyse the contexts and the extent to which they relate to the identified territorial profiles. This analytical advancement will have to be carried out at the municipal and for statistical purposes.

In parallel with the spatial typology work, the research also includes the development of visual methodologies—particularly artistic documentary photography—as a tool for documenting, interpreting, and critically engaging with the lived realities of unworthy housing. By making visible the everyday conditions experienced in the affected territories, this approach complements the statistical analysis and contributes to a deeper, situated understanding of each territorial profile, giving them a more human and tangible presence.

Acknowledgments This research received support from the Centre of Studies in Geography and Spatial Planning (CEGOT), funded by national funds through the Foundation for Science and Technology (FCT) under the reference INV210015—UID/04084—CEGOT—FB.

References

Alegra M, Tulumello S (2019) O estado da habitação: introdução ao dossier. Cidades, 38. http://journals.openedition.org/cidades/997

Beer A, Bentley R, Baker E, Mason K, Mallett S, Kavanagh A, LaMontagne T (2016) Neoliberalism, economic restructuring and policy change: precarious housing and precarious employment in Australia. Urban Studies 53(8):1542–1558. https://doi.org/10.1177/0042098015596922

Bolt G, Czirfusz M (2022) Precariousness and the financialization of housing. In: Münch S, Siede A (eds) PusH working paper series, working paper 2 (Version 09/2022). PusH Strategic Partnership. https://ec.europa.eu/programmes/erasmus-plus/project-result-content/89f55067-12f5-44e6-a74d-f940d84ded71/PusH_Working_Paper_2_Precariousness_and_Financialization_end.pdf

Borg I (2015) Housing deprivation in Europe: on the role of rental tenure types. Hous Theory Soc 32(1):73–93. https://doi.org/10.1080/14036096.2014.969443

Carmo R, Cachado R, Ferreira D (2015) Desigualdades em Tempos de Crise: vulnerabilidades Habitacionais e Socioeconómicas na Área Metropolitana de Lisboa. Revista Portuguesa de Estudos Regionais (40):5–22. https://doi.org/10.59072/rper.vi40.366

Clair A, Reeves A, McKee M, Stuckler D (2019) Constructing a housing precariousness measure for Europe. J Eur Soc Policy 29(1):13–28. https://doi.org/10.1177/0958928718768334

Conceição P (2023) Políticas locais de habitação: condições para uma agenda transformadora. In: Santos A, Branco Teixeira M, Valença P (eds) Políticas Locais de Habitação. Grupo Editorial Vida Económica, pp 289–301

Cutts DB, Meyers AF, Black MM, Casey PH, Chilton M, Cook JT, Geppert J, Ettinger de Cuba S, Heeren T, Coleman S, Rose-Jacobs R, Frank DA (2011) US Housing insecurity and the health of very young children. Am J Public Health 101(8):1508–1514. https://doi.org/10.2105/ajph.2011.300139

Debrunner G, Hofer K, Wicki M, Kauer F, Kaufmann D (2024) Housing precarity in six European and North American cities: threatened by the loss of a safe, stable, and affordable home. J Am Plann Assoc 90(4):610–626. https://doi.org/10.1080/01944363.2023.2291148

Desmond M (2016) Evicted: poverty and profit in the American city. Crown Publishers

Dewilde C, De Decker P (2016) Changing inequalities in housing outcomes across Western Europe. Hous Theory Soc 33(2):121–161. https://doi.org/10.1080/14036096.2015.1109545

Diário de Notícias (2024) Pedidos de despejo aumentam 17%, a maioria por incumprimento

Elsinga M, Hoekstra J, Sedighi M, Taebi B (2020) Toward sustainable and inclusive housing: underpinning housing policy as design for values. Sustainability 12(5):1920. https://doi.org/10.3390/su12051920

ENIPSSA (2024) Inquérito Caracterização das Pessoas em Situação de Sem-Abrigo. https://www.enipssa.pt/documents/10180/11876/Inqu%C3%A9rito+Caracteriza%C3%A7%C3%A3o+das+Pessoas+em+Situa%C3%A7%C3%A3o+de+Sem-Abrigo+-+31+de+dezembro+2023+-+Quadros/bc4e2eb8-31ba-4aa4-984b-dbeccaaddf22

Eurofound (2016) Inadequate housing in Europe: costs and consequences. Publications Office of the European Union. https://publications.europa.eu/resource/cellar/5187e545-78ab-11e6-b076-01aa75ed71a1.0001.02/DOC_1

Eurofound (2023) Unaffordable and inadequate housing in Europe, Publications Office of the European Union. https://www.eurofound.europa.eu/system/files/2023-05/ef22024en.pdf

European Commission (2017) Recommendation (EU) 2017/761 of 26 April 2017 on the European pillar of social rights. https://eur-lex.europa.eu/legal-content/EN/TXT/PDF/?uri=CELEX:32017H0761

European Parliament (2021) European Parliament resolution of 21 January 2021 on access to decent and affordable housing for all (2019/2187(INI)). https://www.europarl.europa.eu/doceo/document/TA-9-2021-0020_EN.pdf

European Parliament (2025) Rising housing costs in the EU: the facts. https://www.europarl.europa.eu/topics/en/article/20241014STO24542/rising-housing-costs-in-the-eu-the-facts-infographics

Eurostat (2022a) EU statistics on Income and Living Conditions (EU-SILC) methodology–definition of dimensions. https://ec.europa.eu/eurostat/statistics-explained/index.php?title=EU_statistics_on_income_and_living_conditions_(EU-SILC)_methodology_-_definition_of_dimensions#Household_type

Eurostat (2022b) Housing in Europe—2022 interactive edition. https://ec.europa.eu/eurostat/cache/digpub/housing/index.html?lang=en

Eurostat (2023) Housing in Europe—2023 interactive edition. https://ec.europa.eu/eurostat/web/interactive-publications/housing-2023

Eurostat (2024) Housing in Europe—2024 interactive edition. https://ec.europa.eu/eurostat/web/interactive-publications/housing-2024

FEANTSA (2005) ETHOS: European typology of homelessness and housing exclusion. https://www.feantsa.org/download/ethos2484215748748239888.pdf

FEANTSA (2007). ETHOS light: European typology of homelessness and housing exclusion. https://www.feantsa.org/download/fea-002-18-update-ethos-light-0032417441788687419154.pdf

Fijalkow Y (2022) The notion of housing need in France: from norms to negotiations (19th-21st centuries). Urban Plann 7(1):197–206. https://doi.org/10.17645/up.v7i1.4706

Gabriel S, Painter G (2020) Why affordability matters. Reg Sci Urban Econ 80:103378. https://doi.org/10.1016/j.regsciurbeco.2018.07.001

Galster G, Lee KO (2021) Housing affordability: a framing, synthesis of research and policy, and future directions. Int J Urban Sci 25(sup1):7–58. https://doi.org/10.1080/12265934.2020.1713864

Gibson M, Petticrew M, Bambra C, Sowden AJ, Wright KE, Whitehead M (2011) Housing and health inequalities: a synthesis of systematic reviews of interventions aimed at different pathways linking housing and health. Health Place 17(1):175–184. https://doi.org/10.1016/j.healthplace.2010.09.011

Governo Português (2017) Para Uma Nova Geração de Políticas de Habitação: sentido estratégico, objetivos e instrumentos de atuação [For a New Generation of Housing Policies: strategic direction, objectives and instruments of action]. https://www.portugal.gov.pt/download-ficheiros/ficheiro.aspx?v=%3D%3DBAAAAB%2BLCAAAAAAABAAzNjWzAADDhmRABAAAAA%3D%3D

Howell KL, Mueller EJ, Wilson BB (2019) One size fits none: local context and planning for the preservation of affordable housing. Hous Policy Debate 29(1):148–165. https://doi.org/10.1080/10511482.2018.1476896

IHRU (2018). Levantamento Nacional das Necessidades de Realojamento Habitacional [National Survey of Housing Resettlement Needs]. https://www.portaldahabitacao.pt/documents/20126/58203/Relatorio_Final_Necessidades_Realojamento.pdf/d67e9ed8-93aa-d65a-eb36-b586389b2295?t=1550355947650

Kadi J, Musterd S (2015) Housing for the poor in a neo-liberalising just city: still affordable, but increasingly inaccessible. Tijdschr Econ Soc Geogr 106(3):246–262. https://doi.org/10.1111/tesg.12101

Kimhur B (2023) A capability approach to evaluating well-being and equality in housing: clear conceptual difference but unclear practical difference? Hous Theory Soc 40(2):192–218. https://doi.org/10.1080/14036096.2022.2149617

Levy-Vroelant C (2010) Housing vulnerable groups: the development of a new public action sector. Int J Hous Policy 10(4):443–456. https://doi.org/10.1080/14616718.2010.525051

Ley 12/2023, de 24 de mayo, por el derecho a la vivenda, Boletin Oficial del Estado, nº 124, de 25 de mayo de 2023. https://www.boe.es/buscar/pdf/2023/BOE-A-2023-12203-consolidado.pdf

Lourenço R, Rodrigues P (2017) House prices in Portugal—what happened since the crisis? Banco de Portugal. https://www.bportugal.pt/en/paper/house-prices-portugal-what-happened-crisis

Marques TS, Matos FL, Maia C, Ferreira M, Ribeiro D (2021) Habitação, uma questão central no ordenamento do território e um desafio para o processo de descentralização. In: Oliveira FP (Coord) Descentralização e Ordenamento do Território - Atas do Encontro Anual da Ad Urbem. Almedina, Coimbra, pp 145–162. ISBN: 9789724097305

Marques TS, Saraiva M, Matos FL, Maia C, Ribeiro D, Ferreira M, van Heerden S (2022) Property investment and housing affordability in Lisbon and Porto. EUR 30993 EN. Publications Office of the European Union, Luxembourg. https://doi.org/10.2760/55085. https://hdl.handle.net/10216/140111

Matos FL, Marques TS, Saraiva M, Maia C, Ribeiro D (2019) Housing in Portugal after the economic crisis: challenges for a new policy. In: Bargelli E, Heitkamp T, Simón-Moreno H, Vásquez Varela C (eds) Housing policy and tenure types in the 21st century a southern European perspective. Pisa University Press, Pisa, pp 53–72. ISBN: 978-88-3339-250-9. https://hdl.handle.net/10216/122657

Matos FL, Marques TS, Maia C, Ribeiro D (2020) O acesso à habitação e as políticas públicas. In: Remoaldo PC, Caldeira MJ, Teles V, Scalabrini EB, Fernandes JAR (eds) XII Congresso da Geografia Portuguesa—Geografias de Transição para a Sustentabilidade, Livro de Atas, pp 70–75. https://hdl.handle.net/10216/127140

Matos FL, Marques TS, Maia C, Ribeiro D (2024a) Habitação e dinâmicas de segmentação socioespacial na Área Metropolitana do Porto. In: Robaina I, López G, Cáceres C (Coord.) Geografía y segregación socioespacial urbana. Una mirada desde Iberoamérica. Universidad de Burgos, Burgos, pp 213–238. ISBN: 978-84-18465-63-5. https://doi.org/10.36443/9788418465635

Matos FL, Marques TS, Maia C, Ribeiro D, Varady D (2024b) Stuck at home: Housing and neighborhood conditions during the COVID-19 pandemic, a case study of Lisbon and Porto metropolitan areas. J Urban Aff. https://doi.org/10.1080/07352166.2024.2323527
Mazucatto M, Farha L (2023) The right to housing, a mission-oriented and human rights-based approach. UN Habitat, Council on urban Initiatives, Working Paper 2023/01. https://make-the-shift.org/wp-content/uploads/2023/06/Right-to-Housing-Missions-Final-DIGITAL.pdf
McKee K, Soaita A, Hoolachan J (2020) "Generation rent" and the emotions of private renting: self-worth, status and insecurity amongst low-income renters. Hous Stud 35(8):1468–1487. https://doi.org/10.1080/02673037.2019.1676400
Mendes L (2020) How can we quarantine without a home? Responses of activism and urban social movements in times of COVID-19 pandemic crisis in Lisbon. Tijdschr Econ Soc Geogr 111(3):318–332. https://doi.org/10.1111/tesg.12450
Mendes L (2022a) Mercado de Arrendamento em Portugal: crónica de uma morte anunciada. Boletim Goiano de Geografia 42(01). https://doi.org/10.5216/bgg.v42.70100
Mendes L (2022b) The dysfunctional rental market in Portugal: a policy review. Land 11(4):566. https://doi.org/10.3390/land11040566
Moos M (2019) Housing today. In: Moos M (ed) A Research agenda for housing. Edward Elgar, pp 2–13
OECD (2025) Affordable housing database. https://www.oecd.org/content/oecd/en/data/datasets/oecd-affordable-housing-database.html
Pendall R, Theodos B, Franks K (2012) Vulnerable people, precarious housing, and regional resilience: an exploratory analysis. Hous Policy Debate 22(2):271–296. https://doi.org/10.1080/10511482.2011.648208
Portal da Violência Doméstica, Indicadores Estatísticos. https://www.cig.gov.pt/area-portal-da-violencia/portal-violencia-domestica/indicadores-estatisticos/
República Portuguesa (2005) Constituição da República Portuguesa [Portuguese Constitution]. https://www.parlamento.pt/Legislacao/Documents/constpt2005.pdf
República Portuguesa (2018) Decreto-Lei n.º 37/2018, de 4 de junho. Cria o 1.º Direito—Programa de Apoio ao Acesso à Habitação [Decree Law no. 37/2018, of June 4, Creates the 1st Right - Support Program for Access to Housing]. Diário da República n.º 106/2018, Série I de 2018–06–04, pp 2348–2364
República Portuguesa (2019) Lei n.º 83/2019, de 3 de setembro. Lei de Bases da Habitação [Law no. 83/2019, Basic Housing Act], Diário da República n.º 168/2019, Série I, pp 11–33
República Portuguesa (2021) Decreto-Lei n.º 89/2021, de 3 de novembro. Regulamenta normas da Lei de Bases da Habitação relativas à garantia de alternativa habitacional, ao direito legal de preferência e à fiscalização de condições de habitabilidade [Decree-Law no. 89/2021, Rules of the Basic Housing Act, establishing the obligations of public entities in order to guarantee adequate housing alternatives]. Diário da República n.º 213/2021, Série I de 2021–11–03, pp 15–22
República Portuguesa (2025) Decreto-Lei 44/2025, de 27 de março. Aprova um regime especial de comparticipação destinado a determinadas soluções habitacionais e altera o Decreto-Lei n.º 37/2018, de 4 de junho, que cria o 1.º Direito—Programa de Apoio ao Acesso à Habitação [Decree-Law no. Approves a special co-payment scheme for certain housing solutions and amends Decree-Law no. 37/2018, of June 4, which creates the 1st Right - Support Programme for Access to Housing]. Diário da República nº 61/2025, Série I de 2025–03–27
République Française. Loi nº 2009-323 du 25 mars 2009 de mobilisation pour le logement et la lutte contre l'exclusion, Journal Officiel de La République Française, nº 0073 du 27 mars 2009. https://www.legifrance.gouv.fr/download/pdf?id=AQcz-NkfCylW_OQ3FEMf6E1tUE4pff_NWtPY0T-2KIM=
République Française. Loi nº 90-449 du 31 mai 1990. Visant à la mise en œuvre du droit au logement. Version actuelle—Loi n.° 2014-366 du 24 mars 2014—art. 34 (V) Pour l'accès au logement et un urbanisme rénové) Journal Officiel de La République Française, nº 0072

du 26 mars 2014. https://www.legifrance.gouv.fr/download/pdf?id=7Wb5kNdWdwPFhrotL9FHWVsDFihSq-tW46KWa2ISZzs

Ruiz-Tagle J, Urria I (2022) Household overcrowding trajectories and mental well-being. Soc Sci Med 296:114051. https://doi.org/10.1016/j.socscimed.2021.114051

Shamsuddin S, Campbell C (2021) Housing cost burden, material hardship, and well-being. Hous Policy Debate 32(3):413–432. https://doi.org/10.1080/10511482.2021.1882532

Shaw M (2004) Housing and public health. Annu Rev Public Health 25:397–418. https://doi.org/10.1146/annurev.publhealth.25.101802.123036

Silva R (2019) Crise e desigualdade habitacional: como (não) se vive em Lisboa. In: Santos AC (Coord.) A Nova Questão da Habitação em Portugal. Actual, pp 233–257

Stephens M, Fitzpatrick S, Elsinga M, van Steen G, Chzhen Y (2010) Study on housing exclusion: welfare policies, labour market and housing provision. European Commission, Brussels. https://researchportal.hw.ac.uk/en/publications/939983ce-dd31-4da2-b842-4430be5021cc

Tosi A (2010) Coping with diversity. Reflections on homelessness in research in Europe. In: O'sullivan E, Busch-Geertsema V, Quilgars D, Pleace N (eds) Homelessness research in Europe. FEANTSA, Brussels, pp 221–236

United Nations—Habitat III (2017) New urban agenda. https://habitat3.org/the-new-urban-agenda/

United Nations (1966) International covenant on economic, social and cultural rights, resolution 2200A (XXI) of 16 December 1966. https://www.ohchr.org/sites/default/files/Documents/ProfessionalInterest/cescr.pdf

United Nations (1991) Committee on economic, social and cultural rights. CESCR General Comment No. 4: the right to adequate housing (Art. 11 (1) of the Covenant). https://www.refworld.org/pdfid/47a7079a1.pdf

United Nations (2014) The right to adequate housing, fact sheet No. 21/Rev.1. https://unhabitat.org/the-right-to-adequate-housing-fact-sheet-no-21rev1

United Nations (2015) Resolution adopted by the General Assembly on 25 September 2015—transforming our world: the 2030 Agenda for Sustainable Development. https://docs.un.org/en/A/RES/70/1

United Nations (2016) Sustainable development goals (SDG). https://www.un.org/sustainabledevelopment/

United Nations (2017) Report of the Special Rapporteur on adequate housing as a component of the right to an adequate standard of living, and on the right to non-discrimination in this context. Mission to Portugal. https://docs.un.org/en/A/HRC/34/51/Add.2

Vandecasteele I, Baranzelli C, Siragusa A, Aurambout JP (ed) (2019) The future of cities. Publications Office of the European Union, Luxembourg. ISBN 978-92-76-03847-4. https://doi.org/10.2760/375209

Wetzstein S (2017) The global urban housing affordability crisis. Urban Studies 54(14):3159–3177. https://doi.org/10.1177/0042098017711649

WHO (2011) Environmental burden of disease associated with inadequate housing. https://www.who.int/publications/i/item/9789289057899

WHO (2018) WHO Housing and health guidelines. World Health Organization, Geneva. https://www.who.int/publications/i/item/9789241550376

Wood GA, Smith SJ, Cigdem M, Ong R (2017) Life on the edge: a perspective on precarious home ownership in Australia and the UK. Int J Hous Policy 17(2):201–226. https://doi.org/10.1080/14616718.2015.1115225

Physical and Environmental Geography

Landscapes of Samtskhe-Javakheti: Intersection of Natural and Cultural Elements with Thematic Database Analysis

Roman Maisuradze, Tamar Khardziani, and Gia Chilingarashvili

Abstract This Paper examines the distribution of megalithic and rock-cut structures, analysing their spatial and temporal characteristics in Samtskhe-Javakheti, Georgia. A new landscape map, compiled by us in 2024, delineates 43 classification units. Information on rock-cut and megalithic structures was sourced from field studies, large-scale topographic maps, and the online portal *Memkvdreoba.ge*. Thematic classifiers were developed to group landscapes based on the following criteria: (a) Hypsometric levels; (b) Relief forms; (c) Relief types; (d) Ecosystems. The study identified and assessed 77 rock-cut monuments and 218 megalithic structures across administrative and landscape units. Key findings include (I) Rock-Cut Monuments. About 89% are associated with volcanic landscapes. Concentrations are within the middle mountainous belt, particularly in steppe, meadow-steppe, and semi-humid forest-shrub ecosystems. (II) Megalithic Structures. Concentrate on the Javakheti Plateau. Approximately 85% are associated with volcanic relief, with 55% situated in the middle mountains and 43% in the subalpine belt. Their distribution aligns predominantly with steppe ecosystems, especially middle mountain and subalpine steppe zones. This research highlights the interplay of natural and cultural landscapes in Samtskhe-Javakheti, offering a foundation for further exploration of this region's rich heritage.

Keywords Landscapes · Rock-Cut monuments · Megalithic structures · Samtskhe-Javakheti · Ecosystems

---

R. Maisuradze (✉) · T. Khardziani · G. Chilingarashvili
Ivane Javakhishvili Tbilisi State University, Tbilisi, Georgia
e-mail: romani.maisuradze@tsu.ge

T. Khardziani
e-mail: tamar.khardziani@tsu.ge

G. Chilingarashvili
e-mail: giochilingarashvili1@gmail.com

L. Soares et al. (eds.), *Compromised Geography*, Key Challenges in Geography,
https://doi.org/10.1007/978-3-032-05201-8_17

Introduction

When studying and assessing historical and cultural landscapes, one of its key components is the monuments of immovable cultural heritage. They exist in the form of historical, architectural and archaeological monuments and represent a characteristic element for a specific landscape unit, reflecting the appearance and features of the natural and cultural landscape of a particular period. This work aims to study the distribution of immovable cultural heritage monuments within natural landscapes. This component represents one of the cultural elements of the natural landscape, and precisely, their harmonisation creates the integrity of the natural-cultural landscape. Such a phenomenon has been studied in different landscape conditions and is often referred to as a cultural landscape (Sauer 1948). Such components are one of the most important parts of the landscape's history as a long-term trace of human activity, which indicates that such buildings and associated lifestyles represented a rather deep structure and became a kind of inseparable part of the territory (Braudel and Wallerstein 2009). Among the cultural elements in Samtskhe-Javakheti, our study region, we selected megalithic settlements and rock-cut dwellings, which often repeat the contours of local traditional folk architecture.

Such cultural heritage further emphasises the value of folk architectural heritage (Battaini-Dragoni 2008; Sangiorgi 2008). The consideration of the landscape as a space that brings together single environmental components, is important for the selection of the cultural landscape as a research area and subject, since our investigation objects are primarily related to the main foundation of the landscape—geological and relief structures, which gives the study area its special significance, and also determined the types of residential buildings and lifestyle of the population (Maisuradze et al. 2021). Which, on the one hand, speaks of an interesting and valuable network of buildings and settlements (Berdzenishvili 1988; Gaprindashvili 1966; Tatishvili 1970).

Megalithic buildings and their remains are scattered worldwide and have been one of the well-tried ways for humans to adapt to their environment, defend themselves and cope with harsh conditions for millennia. Such structures are often associated with landscapes related to Quaternary volcanism or post-glacial times, and in some cases, both factors are combined. In addition, the study region is mountainous. It is not unfamiliar with the climatic fluctuations that accompanied the climatic changes that occurred in the Northern Hemisphere during the Late Pleistocene, Holocene, or even the Middle Ages (Mann 2002; Matthews and Briffa 2005). Therefore, the distribution of megalithic structures can also play a kind of indicator role, although, in addition to it, the analysis of the distribution of rock-cut residential structures is no less important. This type of architecture and cultural heritage is one of the categories of attention and protection (Charter 1931, 1964; Khoshtaria et al. 2019), which are somewhat unique systems in the study area, but at the same time reveal quite a lot of similarity and identity with the system of similar structures in the Mediterranean-Anatolia-Western Asia to consider the experience of their protection or inclusion in modern economic activities should be a timely and necessary step.

This is a prerequisite for the reasonable development of the historical and cultural landscape and the proper socio-economic functioning.

In the article, we consider landscape typological units as the space of distribution and accumulation of the above-mentioned forms of cultural heritage and combine the issues of landscape mapping and classification with the scales of their distribution. Therefore, we rely on existing examples of landscape classifications and discuss their different approaches. In addition to the classifications considered in the process of landscape mapping of the region (Javakhishvili 1947;Saneblidze et al. 1964; Jakeli et al. 1970; Beruchashvili 1979), we tried to introduce new additional classification elements, after which we diversified the hierarchy of landscape typological units. The different approaches to landscape classification can include a geological-relief, climatic, oroclimatic or ecosystem component. Therefore, the study analyses the distribution and concentration of megalithic residential-defensive and rock-cut monuments. We conducted an analysis of the altitudinal stratification of the landscape, relief types and landforms, as well as ecosystem types, which allowed us to analyse the distribution within which typology the objects selected for the study are characterised and what is the cultural significance of each typological unit in the interpretation of the historical-cultural landscape.

Study Area

Our research area is the Samtskhe-Javakheti region, one of Georgia's regional administrative regions (Fig. 1). Samtskhe-Javakheti is in southern Georgia and unites six administrative municipalities (Akhaltsikhe, Adigeni, Aspindza, Akhalkalaki, Borjomi, and Ninotsminda) as well as three historical provinces: Samtskhe and Javakheti. The border between Samtskhe and Javakheti was conditional. It was often referred to as one historical region. At the same time, Tori mainly corresponds to the present-day Borjomi municipality, except for the Tabatskuri cave in Borjomi municipality, which is historically located in Javakheti. The total area of the study region is 6412.8 km^2 (National Statistics Office of Georgia 2014), and the population is 160.5 thousand inhabitants (National Statistics Office of Georgia 2014, Census of the Republic of Georgia).

The region covers the upper section of Mtkvari River valley, which starts from the Tashiskari Narrows and is located between the Meskheti, Arsiani, Erusheti, Gekdagi, Javakheti, Samsari and Trialeti ranges. The main tributaries of the Mtkvari are the Kvabliani and Tavparavnistskali, which, in turn, are the main hydrographic objects of the Akhalikhi Basin and the Javakheti Plateau. The region is mountainous: its lowest point is near the Tashiskari Narrows (720 m.) and the highest is Mt. Didi Abuli (3301 m), which means that a sizeable hypsometric range characterises it. The amount of atmospheric precipitation in the Khertvisi-Aspindza section drops below 400 mm. In comparison, on the Javakheti Plateau, it varies between 500 and 700 mm, and in the Akhaltsikhe Basin, from 400 to 800 mm, by the increase in hypsometric steps. On the Javakheti and Samsari ranges, it varies between 800 and 900 mm, and

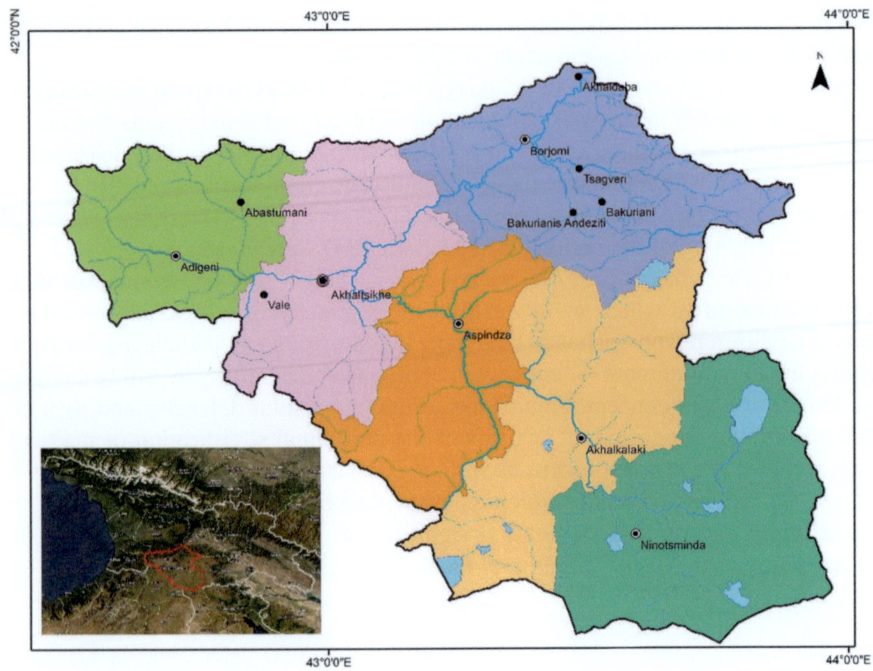

Fig. 1 Study area

in the Trialeti range it reaches 1000 mm, while in the Adjara-Imereti and Arsiani ranges, it varies between 1000 and 1200 mm, and in the Themuri part, it even exceeds this value (Gogishvili 1964).

The maximum air temperature is recorded in the Akhaltsikhe Basin, when it reaches 38°–41°C, in July–August, while the absolute minimum is recorded near Lake Kartsakhi, at −36°. In general, the Javakheti Plateau and the Samsari-Javakheti ridges are characterised by a very continental climate and cold winters, and on average are below −8°–12° (Gavasheli 1964). Forest brown, meadow brown, and mountain black soils are characteristic; in some places, they are eroded and primitive, and mountain-meadow soils are in the subalpine and alpine belts. The vegetation cover is quite compact, with semi-arid semi-shrublands, shrublands, steppe and mesophyllous broadleaf, mixed and coniferous forests, subalpine forest and meadow, mountain swamps and alpine herbaceous vegetation (Nakhutsrishvili 2012).

Part of the wetlands and mountain forests are included in the Javakheti National Park, which belongs to the protected areas of Javakheti. In addition to Javakheti, the Borjomi-Kharagauli National Park is under protection, part of which is also in the study area and includes the territories of both Borjomi and Akhaltsikhe and Adigeni municipalities. In addition, there is the Nedzvi reserve, which includes the Nedzvura River valley in Borjomi municipality.

Materials and Methods

Materials

To create a landscape map, we used large-scale topographic maps at 1: 25,000 and 1: 50,000 scales. We differentiated the relief component to identify valleys, plateaus, canyons, hydrographic units, lakes, swamps and other features. We also used topographic maps to determine the vertical distribution of landscapes, including the lower and middle mountain and subalpine belt.

For the geological maps, we used maps compiled in the last century for geological prospecting purposes. In addition to minerals, their main content is geological structures, lithology and age. From the existing geological maps, several sheets were available at a scale of 1: 200,000, covering the territory of Samtskhe-Javakheti (Gamkrelidze 1940; Gamkrelidze 1945; Gamkrelidze 1952; Kandelaki 1995). With their help, we clarified the characteristic lithological structure of landscape units.

Aerial and satellite images. To prepare the map, we used materials available in the framework of free access and satellite images available on various online portals (Google Earth, Napr.gov.ge). Satellite photos were taken from 2006 to 2007, and then in different periods. They were taken for forestry and cadastral arrangements, which were periodically carried out on the territory of Georgia.

Field description materials. We collected these materials during several expeditions, including field description forms, notes and photographs, and specific landscape views. The forms contain information about landscape-ecological features taken with high precision (GPS coordinates show the exact location of their description), and information is provided about ecosystem type, soil, vegetation and forest, and an assessment of floristic structure is described.

In addition to the information on geological structures, seismicity and seismic hazard data were also collected for megalithic and rock-cut defensive and residential buildings, since the study region is a reasonably active zone in this respect. One of the most significant building threats is seismic activity (Matcharashvili et al. 2000), which is also visible on seismic zoning maps (Javakhishvili 2012).

In addition to the coordinates taken in the field, we used information about the settlements on large-scale (1: 2500) topographic maps, which provided us with some information about their location as well as archaeological material, which grants information about the structure, distribution and architectural aspects of similar monuments (Narimanishvili et al. 2022; Narimanishvili 2019).

Methodology

Identification and mapping of geographical features forming the landscape. Preparing a map of geographical features forming the landscape involved several stages. Initially, we used orographic units on topographic maps (1:25,000, 1:50,000) as an essential element of climate formation. After that, we divided the orographic features according to the following characteristics: (1) The height of the ridge above sea level; (2) Assessment of the importance of the ridge as an orographic and climatic barrier, which in turn was divided into three categories: high, medium and low value.

After that, we used large-scale topographic maps and satellite images in QGIS to distinguish the primary relief forms. Since South Georgia is characterised by mountainous terrain, we selected relief forms, that differ from the forms of steep mountain slopes. We chose a working scale of 1: 10,000 to make the relief forms perceptible and clear. We distinguished the following forms: (1) Mountain plateaus; (2) intermountain valleys. We classified mountain plateaus according to a double classification:

(A) According to the height belt, the plateaus are stepped in nature and include different belts: (a) Middle mountain plateaus; (b) Upper mountain plateaus; (c) High mountain plateaus.
(B) According to the genesis of the relief form: (a) Structural; (b) Volcanic; (c) Volcanic-paleoglacial. Like mountain plateaus, a vertical steppe layout also characterises valleys and is climatically characterised by a sub-mountainous, intermediate, and upper mountain climate. Therefore, we have divided mountain valleys into three categories: (a) Lower Mountain, (b) Middle Mountain, and (c) Upper Mountain.

Ecosystem approach: Landscape units may contain an ecosystem that has been developing as a natural ecosystem for a long time. We have followed a single spatial approach, which implies forming one distinctive and most important type of ecosystem within the landscape, characteristic of a particular spatial unit. The work was carried out at a scale of 1: 10,000, although the working scale ranged from 1: 3000 to 1: 5000 for the differentiation of hydrographic units and wetlands. We have identified different types of ecosystems that extend across mountain valleys, plateaus and ridge slopes. If any ecosystem is fragmented, this is due to long-term human agricultural activities in each landscape. We have identified the ecosystem component of the landscape using satellite imagery. We have prepared its vector image and have identified agricultural and permanent residential areas in the database. However, this is a lower level in our typological classification of the landscape. At this stage, the map highlights natural landscapes that can be divided into sectors based on their socio-cultural activities.

We described and recorded the historical and cultural components of the landscape in the field, defining its geographical parameters, distribution features, importance and purpose, architectural type, and distinctiveness. In addition to field notes, we used the information posted on the online portal memkvidreoba.org.ge. Using the portal, we specified the cultural monuments' location, significance, and age. Based on the available information about the two types of cultural heritage selected by us—(1) megalithic dwellings and defensive structures and (2) rock-cut dwellings, defensive and cult structures-, we prepared a geo-information system and conducted a landscape analysis of their distribution, the results of which are presented in the form of thematic maps (Figs. 2, 3 and 4) and tables.

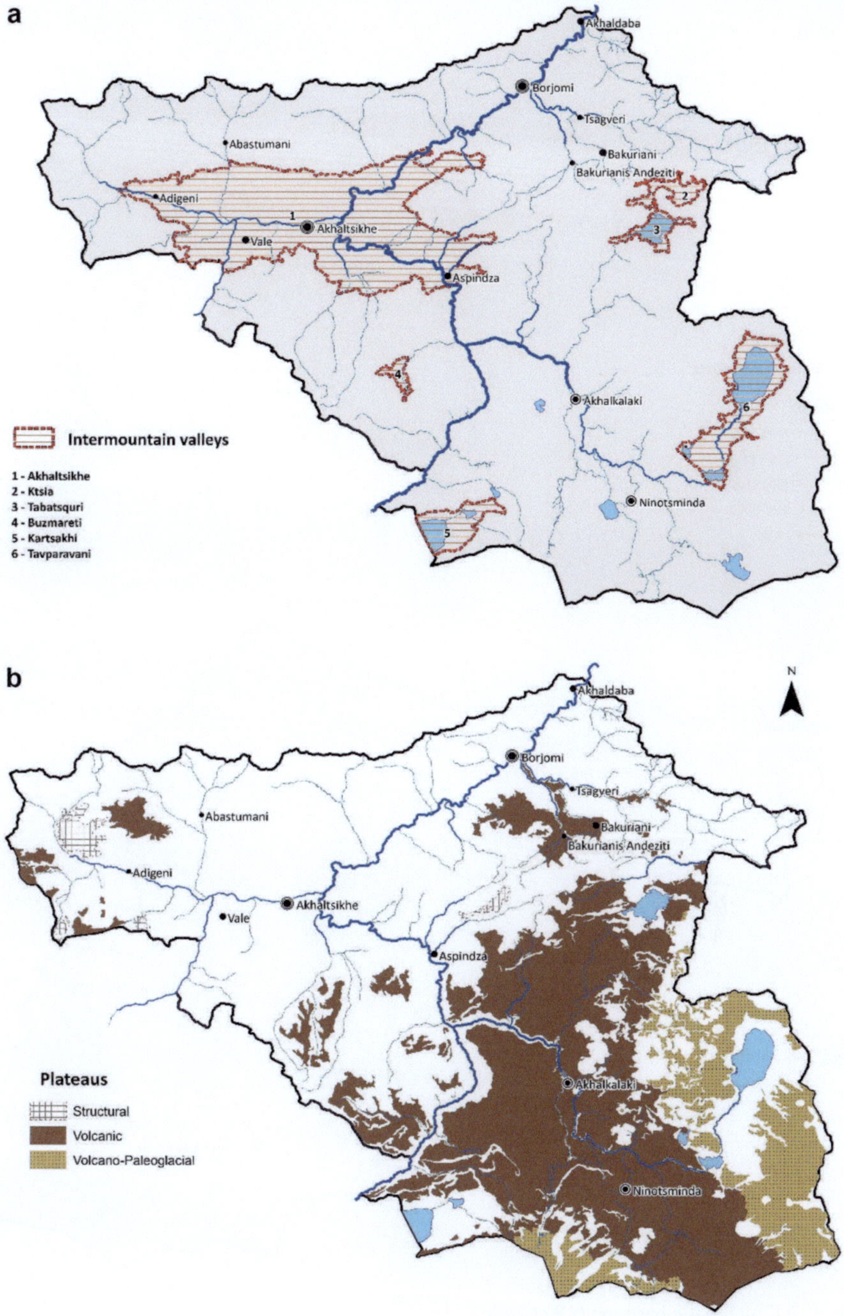

Fig. 2 The main landforms: **a** Intermountain valleys; **b** Plateaus

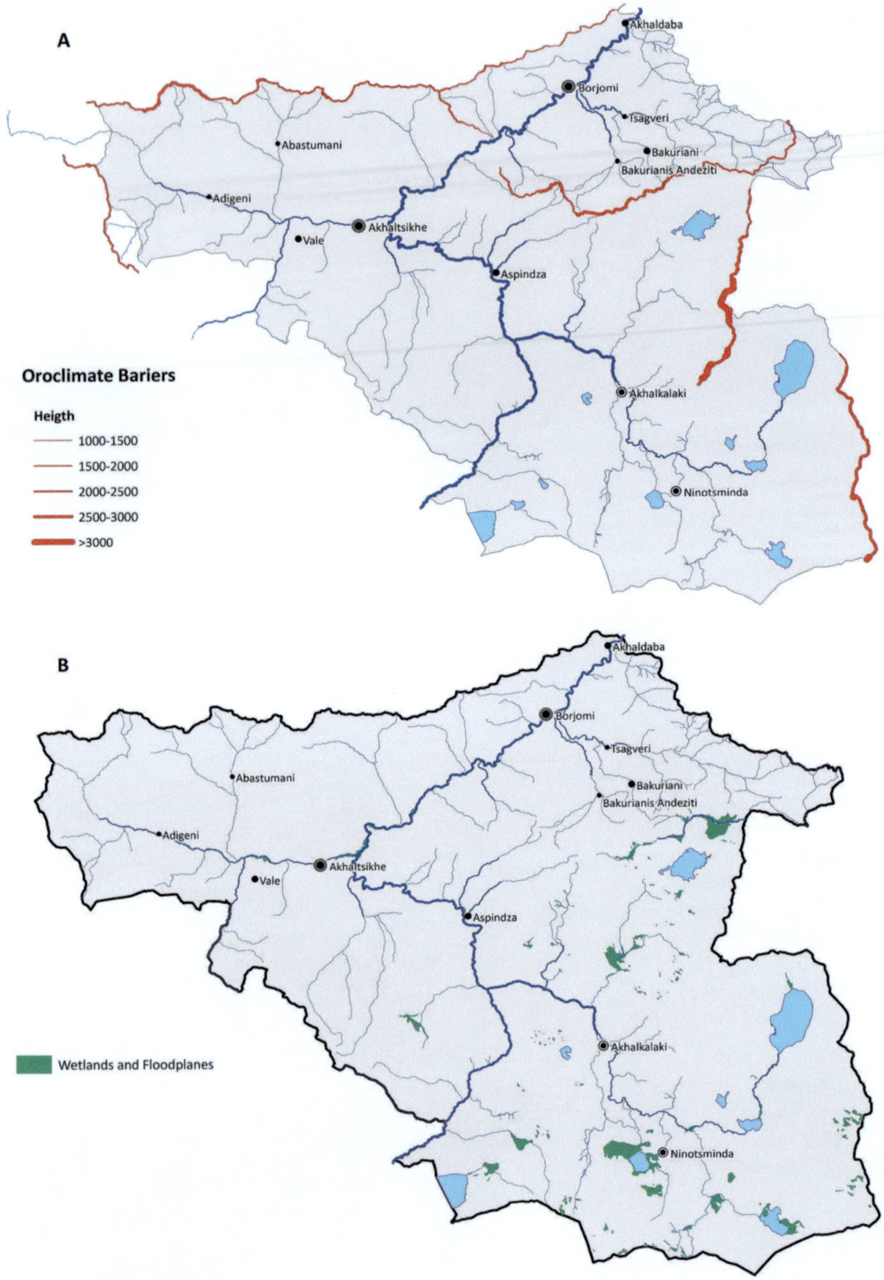

Fig. 3 **a** Oroclimatic barriers; **b** Wetlands

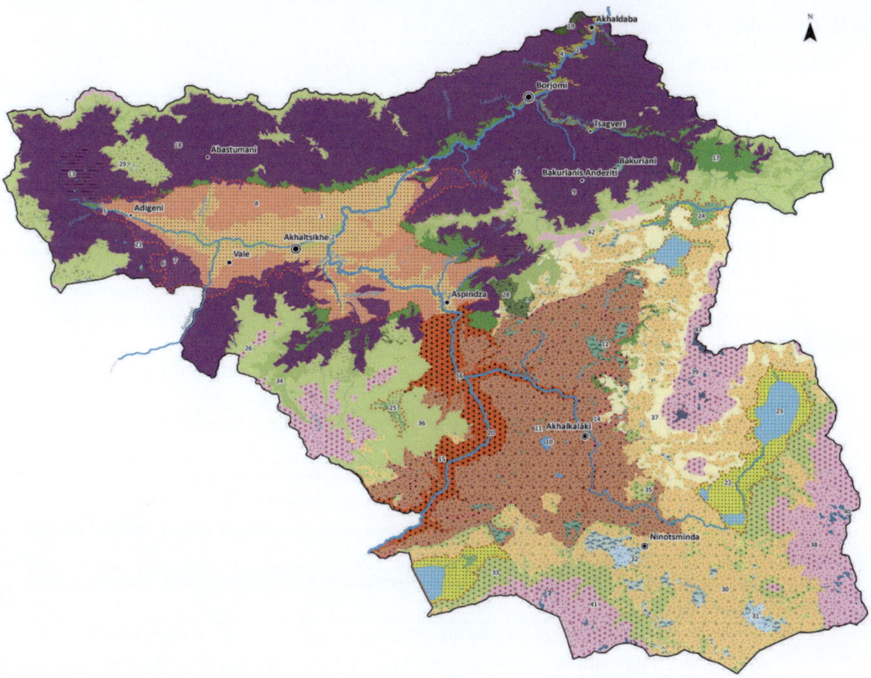

Fig. 4 Landscape map of Samtskhe-Javakheti (2024), typological units of the landscape

Results

We chose a landscape approach as the framework for the research, which was based on the distribution and concentration of selected categories of cultural heritage monuments. The landscape map was based on highlighting the spatial aspect of its cultural elements, such as in what space the above-mentioned objects were concentrated. The methodology for landscape mapping and classification, which is new and implies the consistency of the component elements in its classification, considered the identification of climatic, hypsometric steps, relief forms and genetic types, emphasising the ecosystem component.

Therefore, we identified the following criteria in the landscape classification: (A) hypsometric step; (B) Main relief forms; (C) Genetic relief types; (D) Ecosystems; € Moisture content and hydrothermal characteristics. These components are shown in Fig. 4, which we prepared in 2024, where 43 typological units are differentiated (Table 1).

Based on the GIS database we prepared and analysed, it was found that more than 200 megalithic monuments in the study region are under the status of cultural heritage or included in the potential heritage database. The distribution of administrative municipalities shows that more than half of them are in the Akhalkalaki municipality, and more than ¼ are found in the Ninotsminda municipality. Aspindza

Table 1 Landscapes of study area

Landscape Types	
	1. Lower mountain landscapes, erosive-accumulative, extra-humid, sub hydromorphic, with floodplain forest and meadow
	2. Landscapes of lower mountain valleys, erosive-accumulative, extra-humid, sub hydromorphic, with floodplain forest and meadow
	3. Lower mountain valleys, erosive-accumulative, transitional from semihumid to semiarid, meadow-steppe, shrubland and forest-shrubland
	4. Lower mountain erosive-denudative, semi-humid, with oak and hornbeam forests
	5. Middle mountain valleys, erosion-accumulative, transitional from humid to extra humid, subhydromorphic, with floodplain forest and meadow
	6. Middle mountain valleys, erosion-accumulative, humid, with coniferous, beech-coniferous and pine-coniferous forests
	7. Middle mountain valleys, accumulative, hydromorphic in transition from humid to extra humid, covered with a water surface
	8. Middle mountain valleys, erosive-accumulative, transitional from semi-humid to semi-arid, with steppe, meadow-steppe, shrubland and forest-shrubland
	9. Middle mountain plateaus, volcanic, humid, with coniferous, beech-coniferous and pine-coniferous forests
	10. Middle mountain plateaus, volcanic, extrahumidic, hydromorphic, covered with a water surface
	11. Volcanic landscapes of middle mountain plateaus, transitional from Semihumid to Semiarid, steppe and meadow-steppe, with planted pine forest in some places
	12. Volcanic landscapes of middle mountain plateaus, extrahumid, subhydromorphic, with mountain swamps and wetlands.
	13. Middle mountain plateau landscapes with erosive-denudation, humid, coniferous, beech-coniferous and pine-coniferous forests
	14. Middle mountain volcanic landscapes, transitional from semi-humid to semi-arid, steppe and meadow-steppe, with planted pine forest in some places
	15. Middle mountain volcanic landscapes, transitional from Semihumid to Semiarid, steppe, shrubland and forest-shrubland
	16. Middle mountain volcanic and erosion-accumulation landscapes with sub hydromorphic floodplain forests and meadows transitioning from humid to extra humid
	17. Middle mountain erosion-denudation landscapes, humid, beech and mixed deciduous forests
	18. Middle mountain erosion-denudation landscapes, humid, with coniferous, beech-coniferous and pine-coniferous forests
	19. Middle mountain, erosive-denudation landscapes, humid, Colchian forests with beech and chestnut
	20. Middle mountain accumulative landscapes, extra humid, hydromorphic, covered with the water surface

(continued)

Table 1 (continued)

	21. Middle mountain landscapes, accumulative, transitional from humid to extra humid, subhydromorphic, lakes and wetlands
	22. Upper mountain valleys, volcanic and erosive-accumulative, semi-humid, steppe and meadow-steppe
	23. Upper mountain valleys, volcanic and erosive-accumulative, extra humid, hydromorphic, covered with the water surface
	24. Upper mountain valleys, volcanic and erosive-accumulative, extra humid, sub hydromorphic, with mountain swamps and wetlands
	25. Upper mountain valleys, erosive-accumulative, semi-humid, steppe and meadow-steppe
	26. Upper mountain valleys, accumulative, extra humid, sub hydromorphic, with mountain swamps and wetlands
	27. Upper mountain, accumulative, extra humid, sub hydromorphic, with mountain swamps and wetlands
	28. Upper mountain plateaus, volcanic, humid, with subalpine forests
	29. Upper mountain plateaus, volcanic, semi-humid, with meadows and meadow-steppe
	30. Upper mountain plateaus, volcanic, semi-humid, with steppe and meadow-steppe
	31. Upper mountain plateaus, volcanic, extra humid, hydromorphic, covered with the water surface
	32. Upper mountain plateaus, volcanic, extra humid, subhydromorphic, with mountain swamps and wetlands
	33. Upper mountain volcanic, semi-humid, with subalpine meadows and steppe meadows
	34. Upper mountain erosive-denudative, humid, with subalpine forest
	35. Upper mountain volcanic and denudation-paleoglacial, humid, with subalpine forest
	36. Upper mountain erosive-denudative, humid, with subalpine meadows
	37. Upper mountain, erosive-denudative, semi-humid, with subalpine meadows and meadow-steppe
	38. High mountain plateaus with volcanic and denudation-palaeoglacial, humid, with alpine meadows
	39. High mountain volcanic and denudation-paleoglacial, extra humid, hydromorphic, covered with the water surface
	40. High mountain plateaus, volcanic and denudation-palaeoglacial, extra humid, subhydromorphic, with mountain swamps and wetlands
	41. High mountain volcanic and denudation-palaeoglacial, humid, with alpine meadows
	42. High mountain denudation-palaeoglacial, humid, with alpine meadows
	43. High mountain volcanic, humid, subnival

Fig. 5 **a** Megalithic settlement of Gogichaantghele, near Chobiskhevi (Borjomi municipality); **b** Remains of a megalithic structure in the Buzmareti valley (Aspindza municipality)

municipality is characterised by a relatively high rate—17.51%, while in the Adigeni and Akhaltsikhe municipalities, similar types of structures and their remains are not recorded. Perhaps their traces in the Arsiani Range and Erusheti Mountains have not yet been sufficiently identified, as on the Fersati volcanic plateau, which is also characterised by geological structures rich in building material for megaliths. In addition, some of them may be covered with vegetation to such an extent that it is difficult to identify them under the forest cover. This version is evidenced by the appearance of megaliths in Borjomi municipality, and not in the area of Lake Tabatskuri, which is located within the volcanic ridges and intermountain depressions, but near the village of Chobiskhevi (Fig. 5a), which is in the antecedent valley of the Mtkvari River and is less related to the system of volcanic plateaus and ridges. As for the rock-cut residential and monastic buildings (Fig. 5b), 77 of such complexes have been described in the study region. These objects are characterised by their sharp spatial concentration—more than 85% are concentrated in the Aspindza municipality. In contrast, in the Akhaltsikhe-Akhalkalaki municipalities, their distribution is small and slightly exceeds 5%; their number is even smaller in the Borjomi municipality, and they are not described at all in the Adigeni and Ninotsminda municipalities.

The distribution of cultural heritage monuments by relief types reveals the following patterns: megalithic defensive and residential structures are predominantly found in volcanic landscapes. Some of these monuments also extend into areas characterised by volcanic, tectonic, and erosion-accumulative relief. More than 85% of megalithic structures are associated with landscapes formed from volcanic activity.

These landscapes primarily consist of volcanic rocks such as massive andesites, dacites, basalts, and other high-strength materials. They are often dotted with boulders and cobblestones that accumulate at the base of volcanic ridges, forming cobblestone deposits. The construction material for the megalithic fortresses and dwellings is sourced locally, which is essential for adapting to harsh climatic conditions and for military defense.

Furthermore, megalithic systems are mainly linked to mountain plateaus, with nearly 60% of these structures located in such landforms. On the other hand, megalithic structures are less associated with mountain caves, since the natural depressions in these areas have low defensive and surveillance capabilities. Table 2 illustrates the relationship between megalithic structures and the types of relief.

The situation differs regarding residential, agricultural and monastic complexes carved into rock. About 92% of these complexes are found in volcanic landscapes. This trend is particularly evident in the Mtkvari Canyon, specifically in the Aspindza-Oloda section, where well-known cave systems like Vardzia, Vani Kvab, Gelsund, Fia, Oloda, and others are located. These caves are directly related to the area's distribution of Neogene, Miocene-Pliocene and Quaternary volcanic tuffs and tuff breccias. As a result, nearly 95% of these cave systems are associated with gorges, most of which are canyon-like in appearance (Table 3).

The distribution of megalithic and rock-cut residential and defensive structures varies across different hypsometric levels. Megalithic structures, including residential and defensive types, are not found in the lower mountain belt. Over half of these structures are in the middle mountain forest belt and a small part is located in the alpine belt. A significant concentration—more than 45%—is located in the upper mountain subalpine belt, which suggests that, at the time of their construction, the climate in the subalpine-alpine forest belt was probably no cooler than it is today.

With a slight difference, rock-cut residential, monastic, and defensive systems are mainly situated in the middle mountain belt, indicating that their distribution is

Table 2 Megalithic dwellings and defensive structures by relief types

Megalithic dwellings and defensive structures by relief types of landscape	Number of objects described	%
Erosive-Accumulative	4	1.83
Erosive-Denudative	26	11.93
Volcanic	177	81.19
Volcanic, tectonic and erosive-accumulative	9	4.13
Denudative and paleoglacial	2	0.92
Total	218	100

Table 3 Rock-cut dwellings, defensive and cult structures by relief types

Rock-cut dwellings, defensive and cult structures by relief types of landscape	Number of objects described	%
Erosive-Accumulative	4	5.19
Erosive-Denudative	2	2.60
Volcanic	69	89.61
Volcanic, tectonic and erosive-accumulative	2	2.6
Denudative and paleoglacial	0	0
Total	77	100

closely linked to specific landscape features. Table 4 provides more detailed information about their hypsometric distribution. Regarding their age, most megalithic structures lack precise dating, so further research is needed. However, some dating suggests that many of these structures date back to before 4000 BC, and some may be even older. Notably, some megalithic settlements exhibit multi-layered construction, indicating that they were occupied for centuries or even millennia (Fig. 6). As for the rock-cut systems, more than half remain undated, while nearly half of those dated originate from the eleventh to twelfth centuries. This period corresponds to the transformation of the Vardzia complex into a royal residence. The lack of preserved remains from earlier settlements probably contributes to the dating challenges, since important restorations, reconstructions, and urban transformations took place during this period.

Along with examining hypsometric steps and types of relief, we also tried to assess the distribution of our cultural heritage according to ecosystem types. In the landscape map, we identified 15 significant categories of ecosystems within natural landscapes, serving as these landscapes' most comprehensive external indicators. It should be noted that megalithic and rock-cut monuments are not evenly distributed across all areas. Tables 4 and 5 provide more detailed information about their distribution and concentration (Table 6).

These tables illustrate that the distribution of megalithic defensive and residential monuments corresponds to nine different types of ecosystems (Fig. 7.). Notably, a significant concentration of these monuments is found in two primary ecosystems: (1) Steppe and meadow-steppe, occasionally featuring planted pine forests. (2) Subalpine steppe and meadow-steppe. These ecosystems are adjacent to one another and exhibit a hypsometric difference, resulting in their similar characteristics.

At altitudes between 1900 and 2000 m, the environment transitions into subalpine steppe and meadow-steppe ecosystems. Approximately 76% of megalithic residential and defensive monuments are predominantly situated within the steppe and meadow-steppe ecosystems. When considering additional formations within the steppe ecosystem, 34 monuments (over 15%) are categorised within the semi-humid and semi-arid forest-shrubland. More than 4% of the monuments are found in the subalpine meadow and meadow-steppe ecosystems.

Table 4 Megalithic dwellings and defensive structures and rock-cut dwellings, defensive and religious structures. According to hypsometric steps of landscape

Megalithic dwellings and defensive structures	Number of objects described	%	Rock-cut dwellings, defensive and cult structures	Number of objects described	%
Lower mountain	0	0	Lower mountain	4	5.19
Middle mountain	119	54.59	Middle mountain	72	93.51
Upper mountain	97	44.50	Upper mountain	1	1.30
High mountain	2	0.92	High mountain	0	0
Total	218	100	Total	77	100

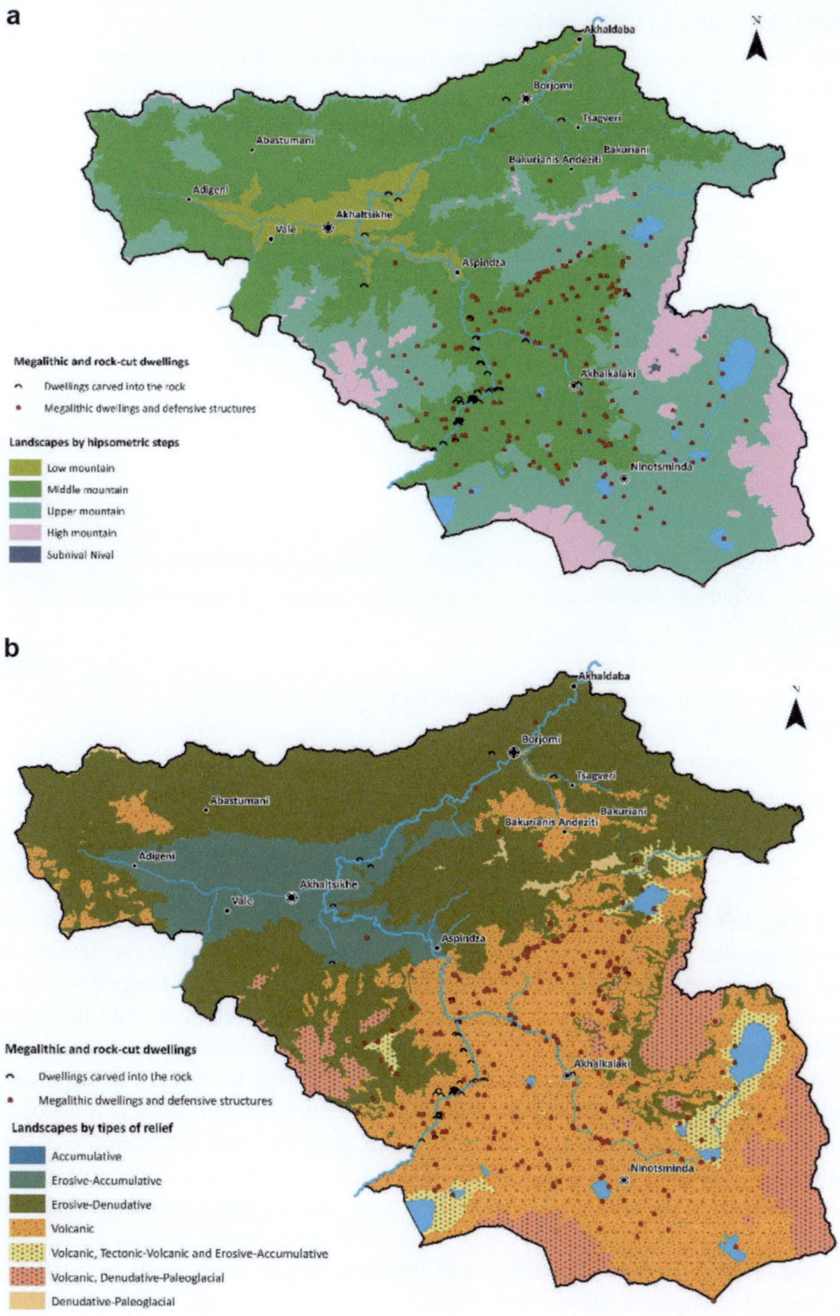

Fig. 6 Megaliths and rock-cut structures, **a** according to hypsometric belts. **b** According to landforms

Table 5 Megalithic dwellings and defensive structures by ecosystem types

Megalithic dwellings and defensive structures by ecosystem types of landscape	Number of objects described	%
Oak and oak-hornbeam forest	1	0.46
Steppe and meadow-steppe, with planted pine forest in places	83	38.07
Steppe, meadow-steppe, semi-humid and semi-arid forest-shrubland	34	15.60
Coniferous, beech-coniferous and pine-coniferous forests	3	1.38
Subalpine meadow	1	0.46
Subalpine meadow and meadow-steppe	9	4.13
Subalpine steppe and meadow-steppe	83	38.07
Mountain swamps and wetlands	2	0.92
Alpine meadows	2	0.92
Total	218	100

Table 6 Rock-cut dwellings, defensive and cult structures by ecosystem types

Rock-cut dwellings, defensive and cult structures by ecosystem types of landscape	Number of objects described	%
Oak and oak-hornbeam forest	1	1.30
Steppe, meadow-steppe, semi-humid and semi-arid forest-shrubland	73	94.81
Coniferous, beech-coniferous and pine-coniferous forests	2	2.60
Subalpine meadow and meadow-steppe	1	1.30
Total	77	100

Altogether, this indicates that over 95% of megalithic residential and defensive complexes are linked to the steppe and meadow-steppe ecosystems. In contrast, rock-cut residential and defensive structures are characteristic of fewer ecosystems, primarily concentrated in four types. However, it is important to note that nearly 95% of these structures are found in the steppe, meadow-steppe, semi-humid, and semi-arid forest-shrubland ecosystems. When examining the overall distribution of both megalithic and rock-cut defensive and residential monuments, a total of 107 complexes are identified within this ecosystem type. In total, 295 complexes across both categories have been documented, which indicates that more than 36% of them are concentrated in this specific ecosystem type, making these landscapes a kind of "hot spot."

Furthermore, in both the steppe and meadow-steppe ecosystems, as well as in the separate subalpine steppe and meadow-steppe ecosystems, this concentration exceeds 28%, thus designating them as exceptional in this context.

Today, accurately classifying the purpose of megalithic structures is challenging. Many of these structures have been destroyed and currently exist in a damaged state,

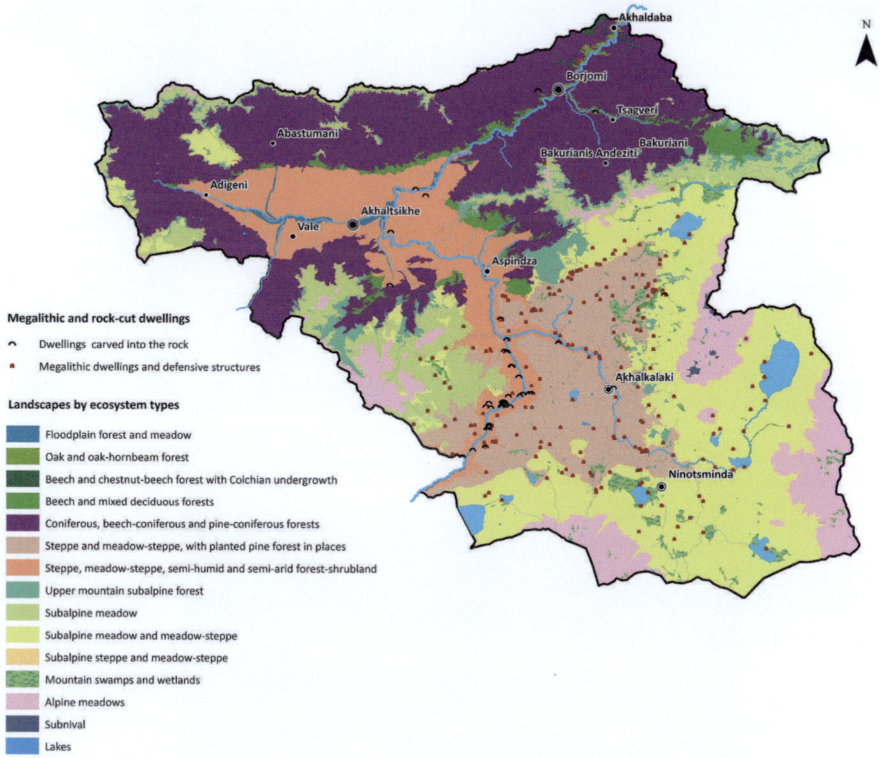

Fig. 7 Megalithic and rock-cut structures in the background of ecosystems

making it difficult to determine whether they served as a city, a wall, or a fortification. Some megaliths are spread over a large area, but the limited scope of archaeological research has not yet clarified their distribution.

In the village of Dzveli (Fig. 8a), the megaliths cover a reasonably expansive area. However, their distribution is complicated by the fact that they are located near an active settlement. In addition to villages, there are agricultural settlements and partially volcanic cairns nearby, which complicate the task of distinguishing the structures from natural volcanic formations (Fig. 8b).

In contrast, rock-cut dwellings present fewer challenges, as their boundaries are more clearly defined, making it easier to understand their purpose and function (Fig. 8c).

The landscape assessment focuses on the concentration of cultural monuments and aims to identify distinctive landscape units that can be considered "hot spots" in a positive sense. Our compiled map distinguishes 43 typological units of the landscape, evaluated according to this criterion. This practical approach highlights the cultural components within these units, which serve as focal points indicating the richness of cultural heritage monuments in the area. Using this method, we can

Fig. 8 Megalithic and rock-cut structures: **a** megalithic structure in the village of Dzveli, **b** megalithic structure in the Sulda area, **c** rock-cut settlement in the Chachkari area

define a typological unit of landscape as a space that corresponds to a cultural and historical landscape.

Although this may differ in terms of landscape classification, it reveals a relationship among these units. In this instance, the natural landscape aligns harmoniously with the historical-cultural landscape. The map we have prepared illustrates this spatial classification and zoning effectively (Fig. 9).

Several landscapes with significant historical and cultural importance have been identified on the map, designating a critical zone. This area is a landscape museum, essential for archaeological work and research, and is rich in artefacts. It provides a valuable opportunity for gathering information and reconstructing life from centuries and even millennia ago, which is vital for museum initiatives.

In addition to traditional museum activities, these landscapes have a high potential for developing open-air museums and living city and village museums. The richness of these sites enhances the landscape's potential by harmonising traditional and

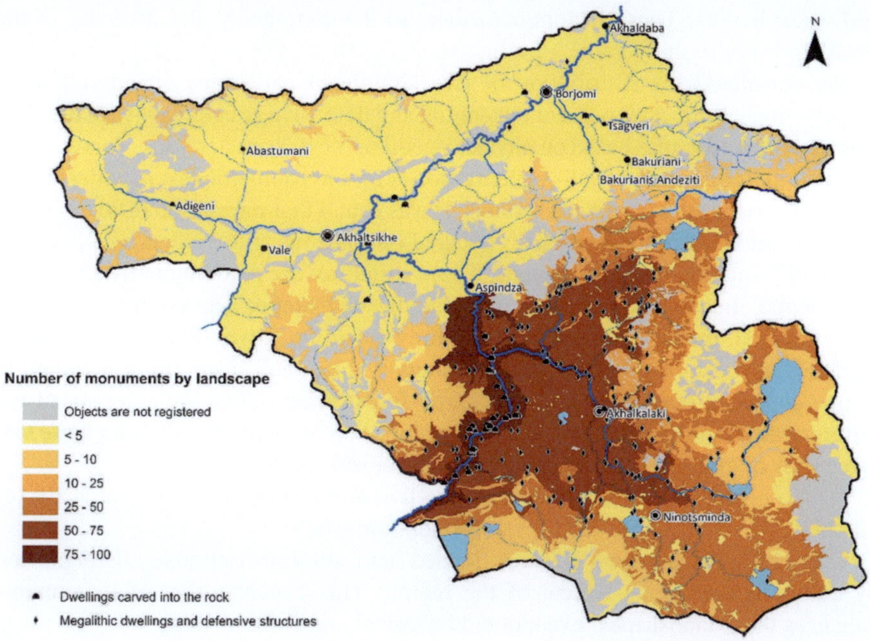

Fig. 9 Landscapes by the number of megalithic and rock-cut residential and defensive monuments

innovative developments in the area. As a result, their value increases due to the wealth of information they contain.

These landscapes become living chronicles, changing our perception of them. They are more than mere land; they are narrators of ancient history and culture, preserving stories and artefacts that reflect the experiences of past inhabitants. People living in such a space actively participate in its culture and history.

This transformation makes these areas even more appealing, laying the foundation for further development. Ultimately, this approach contributes to preserving the most valuable aspects of the landscape, presenting excellent prospects for our research area.

Conclusion

The differentiation of landscape typological units is a complex process that plays a significant role in scientific research, spatial development, and assessment. The study region is characterised by intricate and contrasting features, making it essential to consider the landscape classification hierarchy carefully. When differentiating and assessing natural and historical-cultural landscapes, we focused on key components such as hypsometric gradient, geological structure, relief, climate, vegetation cover, and water bodies. These factors contribute to the complexity and diversity of the landscapes in the study area.

We correlated the distribution of two significant categories of cultural landscapes—megalithic and rock-cut residential and defensive structures—with the classification of landscapes based on physical-geographical characteristics. This analysis revealed their distribution and concentration across hypsometric levels, relief forms, genetic types of terrain and ecosystems. It was found that megalithic residential and defensive monuments are primarily located on plateaus and are distributed within the middle and upper mountain belts. These structures are predominantly associated with steppe, meadow-steppe, and subalpine meadow-steppe ecosystems.

In contrast, rock-cut dwellings are mainly found in canyon-like valleys and are primarily associated with the middle mountain belt. Their distribution is linked to steppe and semi-humid to semi-arid shrub-forest ecosystems. We categorised the differentiated landscapes based on the concentration of monuments within these two categories of cultural heritage, distinguishing seven classes:

(1) No monuments described; (2) Up to five monuments; (3) Between 5–10; (4) 10–25; (5) 25–50; (6) 50–75; (7) 70–100 monuments.

Classes 6 and 7 represent the most "loaded" and culturally rich areas, distinguished by cultural monuments typical of the region. This concentration of monuments enhances these landscapes' exceptional historical and cultural value.

Acknowledgements This work was supported by Shota Rustaveli National Science Foundation of Georgia (SRNSFG) [grant number FR-22-3989].

References

Bibliographical sources

Battaini-Dragoni G (2008) The rural vernacular habitat, a heritage in our landscape. Futuropa, for a New Vision of landscape and Territory 1(3)

Berdzenishvili D (1988) Vardzia and Tmogvi Fortress-Country. Monument's Friend 06(02):46–55 (In Georgian)

Braudel F, Wallerstein I (2009) History and the social sciences: the longue du-rée. Review (Fernand Braudel Center) Jan 1:171–203

Charter A (1931) The Athens Charter for the restoration of historic monuments. In: 1st International congress of architects and technicians of historic monuments, Athens

Charter V (1964) International charter for the conservation and restoration of monuments and sites. Venice, Italy

Gaprindashvili G (1966) Monument's Friend 09:41–47. (In Georgian)

Khoshtaria T, Chachava N, Vardosanidze V, Tsilosani S (2019) Main factors of spatial landscape planning in Samtskhe-Javakheti region. In: The international multidisciplinary conference APLSENSOPO, Tbilisi, pp 9–13

Maisuradze R, Khardziani T, Eradze T (2021) Landscapes, settlements and traditional housing in Samtskhe-Javakheti, Georgia. Quaest Geogr 40(1):85-95

Mann ME (2002) Little ice age. Encycl Glob Environ Change 1(504):e509

Matthews JA, Briffa KR (2005) The Little ice age: revaluation of an evolving concept. Geogr Ann Ser A, Phys Geogr, 87(1):17–36

Nakhutsrishvili G (2012) The vegetation of Georgia (South Caucasus). Springer SBM

Narimanishvili D (2019) Late Bronze-Iron Age fortification complexes of the historical Javakheti region, Georgia. Antiquity 93(367):e5

Narimanishvili G, Shanshashvili N, Narimanishvili D (2022) The megalithic monuments of Georgia. Tbilisi, Georgian National Museum, Centre for Preservation of Antiquities "Trialeti"

National Statistics Office of Georgia Census of the Republic of Georgia. Tbilisi (2014) (in Georgian)

Sangiorgi F (2008) The vernacular rural heritage: from the past to the future. Futuropa 1(1)

Sauer CO (1948) Environment and culture during the last deglaciation. Proc Am Philos Soc 92(1):65–77

Tatishvili T (1970) Mitsisqvesha nagebobani mtkvris zemo tselze (Underground buildings in the upper reaches of the Mtkvari). Monument's Friend 21:18–22 (In Georgian)

Cartographic sources

Beruchashvili N (1979) Landscape map of Caucasus. Georgia Soviet republic, USSR, Tbilisi (in Russian)

Gamkrelidze P (1940) Map of geological structure and minerals, scale 1: 200 000. Tbilisi: s.n. (in Russian)

Gamkrelidze P (1945) Map of geological structure and minerals, scale 1: 200 000. Tbilisi: s.n. (in Russian)

Gamkrelidze P (1952) Map of geological structure and minerals, scale 1: 200 000. Tbilisi: s.n. (in Russian)

Gavasheli S (1964) Average annual absolute maximums, map. Main Division of Geodesy and Cartography of the State Geological Committee USSR, Tbilisi-Moscow, p 57 (in Georgian)

Gogishvili K (1964) Precipitation. Year. Largest, smallest and average monthly points; map. Main Division of Geodesy and Cartography of the State Geological Committee USSR, Tbilisi-Moscow, pp 75–76 (in Georgian)

Jakeli C, Saneblidze M, Ukleba D (1970) Landscape map of Georgia. Tbilisi, Georgia Soviet republic, USSR: Tbilisi 8th Cartographic Factory (in Georgian)

Javakhisvili A (1947) Scheme of geomorphological landscapes of the Georgian SSR. Moscow-Leningrad, USSR Academy of Sciences Publishing House (in Russian), Geomorphological regions of Georgia

Javakhishvili Z (2012) Earthquakes, with inclusion—seismic zones. Georgian National Atlas. Cartography, Tbilisi

Kandelaki N (1995) Map of geological structure and minerals, scale 1: 200 000. Tbilisi: s.n. (in Russian)

Matcharashvili T, Chelidze T, Javakhishvili Z (2000) Nonlinear analysis of magnitude and interevent time interval sequences for earthquakes of the Caucasian region. Nonlinear Process Geophys 7(1/2):9–20

Saneblidze M, Ukleba D, Javakhishvili A, Jakeli C (1964) Landscape map in Georgian SSR. USSR. Main Division of Geodesy and Cartography of the State Geological Committee, Atlas, Tbilisi-Moscow, pp 153–154 (in Geiorgian)

Compromised Riparian Landscapes: A Landscape Ecological Assessment of the Riparian Forests in the Lower Kartli Lowland, Georgia

Tamar Khardziani

Abstract The world's forest systems, including riparian forests, have long been altered by natural events and human activities, intensifying impacts over the past century. Riparian landscape fragmentation poses a significant challenge, disrupting ecosystem resilience, altering ecological processes, and reducing biodiversity. Protected areas and landscape restoration are key strategies for addressing this issue. This chapter examines riparian forest fragmentation in Georgia using case studies from protected and unprotected areas. Remote sensing, landscape ecology, and geospatial analysis were employed to assess current fragmentation, while historical fragmentation was analysed through old literary and cartographic sources. Findings indicate that severe fragmentation has affected riparian forests and surrounding landscapes since the early eighteenth century, significantly impacting biodiversity and even causing regional extinctions of particular fauna. Although fragmentation has slowed in recent years and some reversal has been observed over the past eight years, this positive trend is primarily confined to protected areas. Riparian forests remain compromised outside these zones, highlighting the ongoing need for restoration.

Keywords Riparian forest · Landscape fragmentation · Landscape statistics · Mtkvari river · Ktsia river

Introduction

For millennia, the Earth's forest systems have been modified by natural events and human impacts. Global forest cover has declined by one-third in the approximately 10,000 years since the last great ice age, with half of that loss occurring in the past century (Ritchie 2021). Globally, 178 million hectares of forest were lost between 1990 and 2020, although the annual net rate of forest loss has decreased compared to

T. Khardziani (✉)
Faculty of Exact and Natural Sciences, Department of Geography, Ivane Javakhishvili Tbilisi State University, Tbilisi, Georgia
e-mail: tamar.khardziani@tsu.ge

L. Soares et al. (eds.), *Compromised Geography*, Key Challenges in Geography,
https://doi.org/10.1007/978-3-032-05201-8_18

previous decades (FAO 2020). Riparian forest landscapes have also historically experienced decline. A riparian landscape is an ecotone developed on the shores of water bodies formed by the interaction of water and land. Malanson (1993) characterises the riparian zone as an environment of unique configuration, the main factor in its formation being the interaction and mutual influence of water and land. Protecting, maintaining, and restoring riparian forests are essential for biodiversity and adaptation to climate change. They have functions such as preventing floods, mitigating droughts, providing shade and cooling, filtering pollutants and thus preventing eutrophication, and acting as a natural connection between habitats and species (EEA 2023).

Due to the abovementioned characteristics, riparian forests are often a priority for conservation and restoration. Due to their unusual species diversity and the multitude of ecological and physical processes occurring, Riparian forests were actively considered a habitat of high conservation importance in the early 1990s. As the most diverse, dynamic, and complex biophysical habitat on Earth, it should play an essential role in water and landscape planning and restoring aquatic systems (Naiman et al. 1993). Emphasis was placed on maintaining connectivity and a continuous riparian corridor (Ness and Harris 1990). During the same period, adopting the UN Convention on Biological Diversity, which emphasised the need to create a system of protected areas to protect ecosystems and natural habitats and conserve biological diversity, provided an impetus for creating protected areas worldwide, including Georgia. In the early 1990s, shortly after the dissolution of the Soviet Union, the country regained its independence. Soon after adopting the Convention, many protected areas were established in Georgia, including several created to preserve riparian forest ecosystems. One such protected area, the Gradabni Managed Reserve, established in 1996, is in our study area.

Nowadays, riparian forest restoration is considered an important climate change adaptation action. It is regarded as an effective means of coping with the adverse effects of climate change and a form of blue-green infrastructure (Nakamura 2022). If creating protected areas is an effective measure to prevent riparian forest fragmentation, blue-green infrastructure is a way to restore an already fragmented and degraded landscape, even outside the protected area's boundaries. Landscape fragmentation is the breakdown of more extensive, entire habitats or ecosystems into smaller, isolated patches, often caused by anthropogenic factors. Fragmentation negatively impacts biodiversity, ecosystem resilience, and local climate, so its assessment is important in biodiversity conservation, climate mitigation and adaptation, landscape restoration, and planning. The use of remote sensing techniques and geoinformation technologies plays a significant role in the study of landscape fragmentation. The availability of satellite images and various landscape fragmentation indicators has been developed to simplify the analysis and facilitate recent and current fragmentation. However, analysing old cartographic and literary materials is important to overview historical fragmentation. Therefore, in the study presented below, which concerns the fragmentation of riparian landscapes, we have reviewed both current and historical fragmentation in the Mtkvari River and Ktsia River floodplains of the Kvemo Kartli Lowland, Georgia. We studied the current fragmentation using satellite images from 2016 to 2024, analysing landscape metrics and spatial autocorrelation. To review historical fragmentation, we used old maps (from 1735) and geographical texts (from 1745), which contained information about the riparian forest landscapes of the study area and its surroundings. This paper is the first attempt to assess the fragmentation of the riparian forest landscape in the study area, aiming to complement the existing knowledge of the region.

Study Area, Materials, and Methods

Study Area

The study area is in southeast Georgia, on the Kvemo Kartli Lowland (Fig. 1). Its geomorphological environment includes alluvial and pluvial plains, which are geologically distinguished by porous Quaternary sediments. Groundwater plays a crucial role in sustaining rivers. A moderately humid, subtropical steppe climate is prevalent here, characterised by hot summers and two minimum precipitation periods per year. The average annual precipitation is 400–600 mm. The average annual temperature is 12–13 °C. The average temperature in January is 0 °C, and July 24 °C. The annual amplitude of air temperature is 24–26 °C. Alluvial carbonate soils are in the floodplains (Academy of Science of Georgia 2000; Kharadze 2019). Since riparian forests have been disturbed in large parts of Georgia, the study area has a unique function and importance. The woody species common in the Mtkvari floodplain forests of the Gardabani district include: white poplar (*Populus alba var. canescens*), black poplar (*Populus nigra*), willow (*Salix alba, Salix excelsa*), *and* oak (*Quercus robur ssp. Pedunculiflora*), elm (*Ulmus minor*) *and* common walnut (*Juglans regia*) (Kavtaradze et al. 2023). The following species are found in the undergrowth: *Tamarix ramosissima, Hippophae rhamnoides, Crataegus caucasica,* and from lianas, it is widespread *Clematis vitalba, Humulus lupulus, Hedera helix and Periploca graeca* (Kavtaradze et al. 2023). The species common in the Ktsia floodplain are: weeping and white willows (*Salix alba* L., *Salix babylonica* L.), white poplar (*Populus alba* L.), black poplar (*Populus nigra* L.), mulberry (*Morus alba* L.), elm (*Ulmus minor* Mill.) and others (Gulisashvili 1977) (Fig. 2).

Mtkvari River Floodplain

The study area includes the section of the Mtkvari River flowing through the Kvemo Kartli Lowland, from the city of Rustavi south to the Georgian-Azerbaijan border. Within the study area, the border between the two municipalities runs along the left bank of the Mtkvari River. Within the study area, the border between the two municipalities runs along the Mtkvari River floodplain. The Gardabani Nature Reserve, which is included in the study area, is bordered by the villages and agricultural fields of the Gardabani Municipality, which are connected by an irrigation system

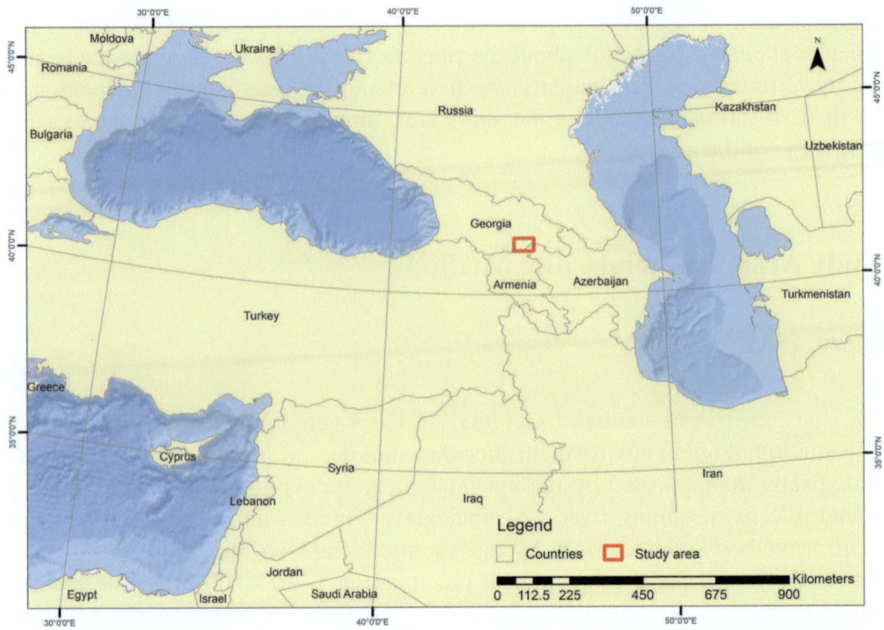

Fig. 1 Location of the study area. *Source* Author's Map

Fig. 2 *Populus alba* var. *canescens* in the forests of Gardabani Managed Reserve (41°22′49.62″N, 45° 5′23.50″E) and Ktsia River riparian zone (41°20′15.84″N, 45° 3′41.88″E). *Source* Author's photos (2023)

drawn from the Mtkvari River. These villages are Ruisbolo, Vakhtangisi, Nazarlo and Kesalo. The study area is bordered in the northeastern part by the Gardabani thermal power plant. The study area is bordered to the west of Mtkvari River, by rural settlements and agricultural fields of Marneuli Municipality. From north to south, the villages of the Marneuli Municipality encompass Ambarovka, Keshalo, Ilmazlo, Kapanakhchi, First Kesalo, and Second Kesalo.

Due to the relief features in the study area, the floodplain forest zone is asymmetrically distributed on the right and left banks of the river. To the left, within the Gardabani Municipality, the river is bordered by a flat plain, allowing the riparian forest to spread far from the riverbed. A cliff runs along the right bank, sharply separating the riverbed from the flat, plateau-like relief of the Marneuli plain. The Mtkvari River is a transboundary river that originates in Turkey, flows through Georgia, and then crosses into Azerbaijan, emptying into the Caspian Sea.

The riparian forest landscape in the Gardabani Managed Reserve extends along the Mtkvari River into Azerbaijan, where the Garayazi State Nature Reserve is located.

Ktsia River Floodplain

The study area includes the Ktsia River floodplain in the Kvemo Kartli lowland, extending southeast from the village of Nakhiduri to the Georgian-Azerbaijani border. The main section of the mentioned floodplain belongs to the Marneuli Municipality and a small part to the Bolnisi Municipality. The Ktsia River originates on the southwestern slope of the Trialeti mountain range. It will pass through a different terrain environment. Ktsia River first flows through the plateau, then passes through the Khrami Canyon and emerges onto the Kvemo Kartli Lowland. After crossing the Georgian-Azerbaijani state border, it joins the Mtkvari River in Azerbaijan.

The Ktsia floodplain is of varying widths, which are determined by the relief factor. On both the right and left sides, the floodplain is bordered by an elevated plain separated by cliffs. Almost no settlements are located directly in the floodplain, except for a small part of the village of Nakhiduri. The study area is bordered on both sides by rural settlements and farmland. The left bank is bordered by the villages of the Marneuli Municipality: Tamarisi, Beglari, Baidari, Didi Mughanlo, and Lezhbadini. The riverbed is also bordered to the south by the villages of Imiri, Kulari, Khanji-Gazlo, and Kirach Mughanlo. The study area is bordered by the following villages of the Bolnisi Municipality: Tsurtavi, Nakhiduri, Khataveti, and Arkevani.

Materials and Methods

The research presented in this paper analyses both current and historical fragmentation. We used remote sensing and landscape ecological methods to assess current fragmentation, while old cartographic and literary sources informed our review of historical fragmentation.

The primary data used in this study to assess current fragmentation are Sentinel-2 multispectral satellite images (European Space Agency 2025). We processed the Sentinel-2 MSI data using the ESA SNAP (Sentinel Application Platform) software (ESA 2025). Within the same program, we performed a supervised classification of the images using the Random Forest Classifier method (Breiman 2001; ESA 2025). We selected Sentinel-2 MSI data acquired in August and early September of 2016 and 2024, with a 10-m resolution and a maximum cloud cover of 10%. During resampling, we retained the 10-m resolution, ensuring the classified image maintained the exact resolution.

Postprocessing of classified images included validation through the calculation of the Kappa coefficient (Cohen 1960; Congedo 2021), computation of landscape class areas and fragmentation statistics using (McGarigal et al. 2012; Jung 2022), spatial autocorrelation analysis (Getis-Ord Gi*), and statistical significance assessment (Getis and Ord 1992; Rossy 2019). These analyses were conducted within the QGIS geoinformation environment (QGIS Development Team 2023). Thus, we used the following programs and plugins in the study: ESA SNAP, QGIS 3.34 Prizren, Semi-Automatic Classification Plugin (SCP), LecoS, and Visualist. In addition, we used the World Database on Protected Areas (WDPA) database (UNEP-WCMC 2025) as a source of shapefiles for protected area boundaries. We used the World Bank data catalogue (World Bank 2025) for the political map depicted on the study area map. To review historical fragmentation, we analysed several cartographic sources (1735–1982) and historical texts written about the geography of the study area (1745–1787). The sources include:

- Vakhushti Bagrationi's Geographical Atlas of Georgia, 1735–1745, republished in Tbilisi in 1997 by the Georgian Academy of Sciences (Bagrationi 1997);
- Description of the Kingdom of Georgia by Vakhushti Bagrationi, from 1745, which was published in 1842 in two languages, French and Georgian, by the French historian and researcher Marie-Félicité Brosset (1842);
- Travels through Russia and the Caucasian Mountains [Reisen durch Rußland und im Caucasischen Gebirge], published in 1798 by the eighteenth-century German naturalist and explorer Johann Anton Güldenstädt, was published in 1962 in two volumes and two languages, German and Georgian, by the Academy of Sciences of the Georgian SSR;
- The so-called "5-version" maps of the Caucasus, published in 1890–1892, during the reign of Tsarist Russia, at a scale of 1:210,000;
- Soviet topographic maps of 1:100,000 scale from 1949, 1:50,000 from the early 1950s, and 1:100,000 from 1982.

Thus, we quantified the ongoing fragmentation using satellite data, GIS programs, and landscape ecological methods. Based on historical sources, we conducted a qualitative analysis of historical fragmentation, which provided us with a context for interpreting the results of the quantitative analysis.

Land Use and Land Cover (2016–2024)

The land use and land cover (LULC) areas presented below should be interpreted in the context of classification accuracy (Table 1). Methodological limitations of the LULC layers stem primarily from the 10-m spatial resolution of Sentinel-2 MSI, as a single pixel may encompass multiple land cover types (e.g., a tree and part of a building), leading to spectral mixing and reduced classification accuracy. Thus, future research would benefit from higher-resolution satellite imagery (e.g., 1 to 5-m resolution).

Land Use and Land Cover in the Mtkvari Riparian Zone

We identified the Mtkvari River riparian forest zone within the Kvemo Kartli Lowland as 50.4 km^2. The study area almost entirely included the Gadabani Managed Reserve, which has a total area of 37.4 km^2 and was created in 1996 precisely to preserve, protect, and restore the riparian forest. Accordingly, most remaining forest is located within the protected area's boundaries.

Analysis of a 1:50,000 scale topographic map from the 1950s showed that the forest at that time was more fragmented and covered a smaller area than it does today. The forest extended to the southern part of the Mtkvari floodplain, and its area was 10.8 km^2, 21.5% of the study area and 28.9% of the current protected area.

According to satellite images, the riparian forest area was 23.3 km^2 in 2016, or 46.2% of the study area (Fig. 3). By 2024, it had increased to 29.6 km^2, which accounts for 58.7% of the study area. The forest area has increased mainly within the boundaries of the protected area. By 2016, 62% of the Gardabani protected area was covered with woodland; by 2024, 79% (Table 1). Forest growth occurred along the entire river, particularly near water.

In addition to forests, the area of meadows has also increased, which has occurred partly at the expense of agricultural land. In the 2016 image, cultivated land plots are

Table 1 Land use and land cover classes (2016–2024)

Class	Mtkvari, area (km^2)				Ktsia, area (km^2)			
	2016	Kappa (κ)	2024	Kappa (κ)	2016	Kappa (κ)	2024	Kappa (κ)
Woodland	23.3	0.99	29.6	0.96	2.5	0.88	4.3	0.99
Meadow	11.8	0.69	14.0	0.69	9.3	0.71	8.8	0.88
Agricultural land	8.8	0.80	1.4	0.97	12.0	0.94	7.8	0.97
Water bodies	2.9	0.94	2.5	0.99	1.3	0.89	2.2	0.93
Artificial surfaces and open spaces	3.5	0.83	2.8	0.75	3.7	0.91	5.5	0.93

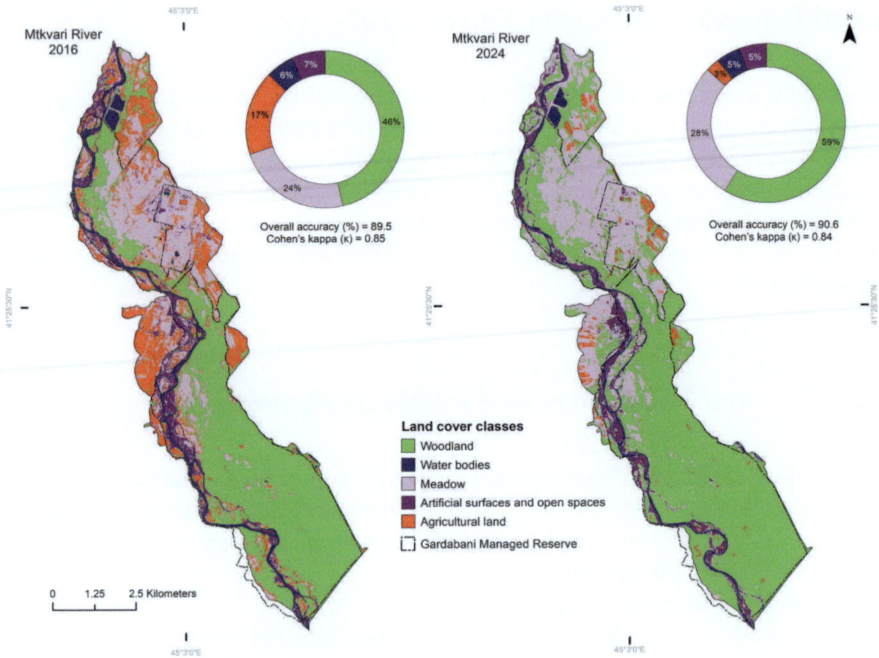

Fig. 3 Land use and land cover classes in Mtkvari River riparian zone

depicted in the northern and northeastern parts of the study area, while grassland is shown in the exact location by 2024. The area of water bodies has also been reduced, which is related to constructing two ponds in the agricultural fields northeast of the study area by 2016. Also, a pond was visible in the water treatment plant's territory in 2016. The mentioned ponds were dry in the picture of 2024. "From 2016 to 2024, the category of artificial surfaces and open spaces also decreased. This was due to forest succession on the bare rocks along the riverbanks and the spread of forest-shrubland, which caused these areas to be reclassified into the forest category.

The Mtkvari River, flowing through the Kvemo Kartli Lowland, meanders and often changes its course. In the 2016 satellite image, the river was entirely within the protected area. However, by 2024, its course had shifted, and it had moved outside the protected area near the village of Kapanakhchi (41°24′28.88″N, 45° 2′24.28″E and 41°25′6.45″N, 45° 2′56.11″E). Outside the protected area's boundaries, there is a remaining riparian zone in the exact location, which appears to be a trace of river sediment extraction. The forest succession process has begun in this zone. In this section, the remaining riparian zone outside the boundaries of protected areas, which could be restored, covers approximately 1 km^2.

Land Use and Land Cover in the Ktsia Riparian Zone

The area of the Ktsia River's riparian zone is 28.7 km^2. Based on the 1950s 1:50,000 scale map, the forest area comprised 2.33 km^2 or 8.12% of the riparian zone. Already at that time, the forest was heavily fragmented. The surviving groves extended along the left bank of the river, from the village of Lezhbadini southeast to the "Red Bridge."

The situation was nearly the same as in the 1950s, as the 2016 data indicated that floodplain forests covered 8.7% of the study area (Table 1). It has increased slightly by 2024 but still occupies a negligible portion of the total riparian zone. Between 2016 and 2024, floodplain forest groves were found in the same area where they were present in the 1950s.

The most significant change in the riparian zone of Ktsia occurred in the artificial surfaces and open spaces category, which accounted for 12.8% of the study area in 2016 and reached 19.2% in 2024 (Fig. 4). Open spaces include sedimentary materials in the riverbed and their extraction sites. A topographic map from the 1950s shows that the areas mentioned above were occupied by river sediment, sand, and gravel, with scattered scrub or meadows. Currently, intensive work is ongoing to extract river sediment.

Thus, in the Kvemo Kartli Lowland, the riparian forest area is significantly increasing in the Mtkvari River floodplain, which is associated with a protected area. Despite this slight increase, forest groves still occupy a negligible portion of the Ktsia riparian zone.

Fig. 4 Land use and land cover classes in Ktsia River riparian zone

Riparian Forest Fragmentation

Fragmentation of Mtkvari River Riparian Forest

An analysis of landscape fragmentation indices (landscape metrics) in the Mtkvari River floodplain revealed a decrease in woodland fragmentation levels. In particular, from 2016 to 2024, the Effective Mesh Size increased, the Number of Patches decreased, and the Mean Patch Area increased (Table 2). However, the fragmentation of the forests has decreased mainly within the Gardabani Managed Reserve. Fragmented sections were also identified within the study area.

Spatial autocorrelation analysis of the 2016 Effective Mesh Size (MESH) data revealed one hotspot and two coldspots (Fig. 5). Similar clusters emerged from the study of MESH data for 2024, although some changes were observed in all three clusters. In particular, compared to 2016, the area of statistically significant ($p < 0.05$) high positive values (hotspots) increased by 1.5 km^2 by 2024. In addition, the area of statistically significant ($p < 0.05$) low, negative values (coldspots) decreased by 3.1 km^2. This indicates a decrease in the level of fragmentation of the Mtkvari riparian forests. The MESH hotspot, or low fragmentation area, is located in the Gardabani protected area. One coldspot, or area of high fragmentation, is also located in the protected northeastern part. This part of the protected area is influenced by the adjacent industrial area, which includes a water treatment plant and a thermal power plant. The second coldspot is located near the village of Kafanakhchi and borders the protected area to the west.

In addition to the above, the Edge Length indicator of the forest landscape class increased in the study area from 2016 to 2024, which is related to the increase in the woodland class. Notably, the study period showed an increase in landscape homogenisation, which was reflected in a decrease in Shannon's Diversity Index (Table 2).

Overall, in the study period, the fragmentation of the riparian forests in the Kvemo Kartli Lowland section of the Mtkvari River floodplain has decreased. High fragmentation is more pronounced in the non-protected part of the study area. The northeastern part of the protected area has a high level of fragmentation, although a decrease in

Table 2 Landscape fragmentation indices

Landscape metrics	Mtkvari riparian forest		Ktsia riparian forest	
	2016	2024	2016	2024
Effective mesh size (m^2)	6,743,474	10,078,455	7129	17,531
Number of patches	913	638	1473	936
Mean patch area (m^2)	25,557	46,423	1698	4610
Edge length (m)	413,260	480,080	293,460	297,060
Shannon's Diversity Index	1.35	1.08	1.35	1.52

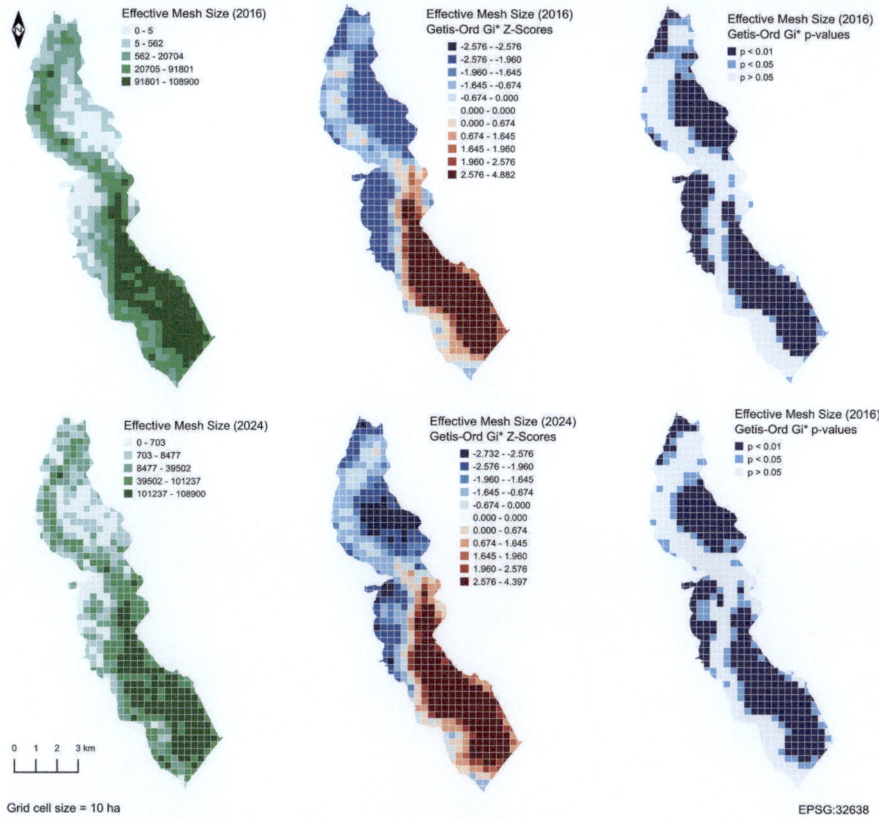

Fig. 5 Fragmentation of Mtkvari River riparian forests: effective mesh size, hotspots and coldspots

fragmentation was observed here during the study period. Analysis of landscape fragmentation indices showed that protected areas positively impact the ecological condition of riparian forests.

Fragmentation of Ktsia River Riparian Forest

The fragmentation of the forest landscape's remaining groves in the Ktsia River floodplain decreased during the study period (2016–2024) (Fig. 6). The Effective Mesh Size and Mean Patch Area increased, while the Number of Patches decreased (Table 2). However, the general context is that the riparian forest in the Ktsia floodplain has historically been fragmented and occurs in small groves. Edge Length has increased slightly, indicating an increase in woodland area.

During this period, the Ktsia floodplain landscape became more heterogeneous, as indicated by the increase in Shannon's Diversity Index (Table 2). The change

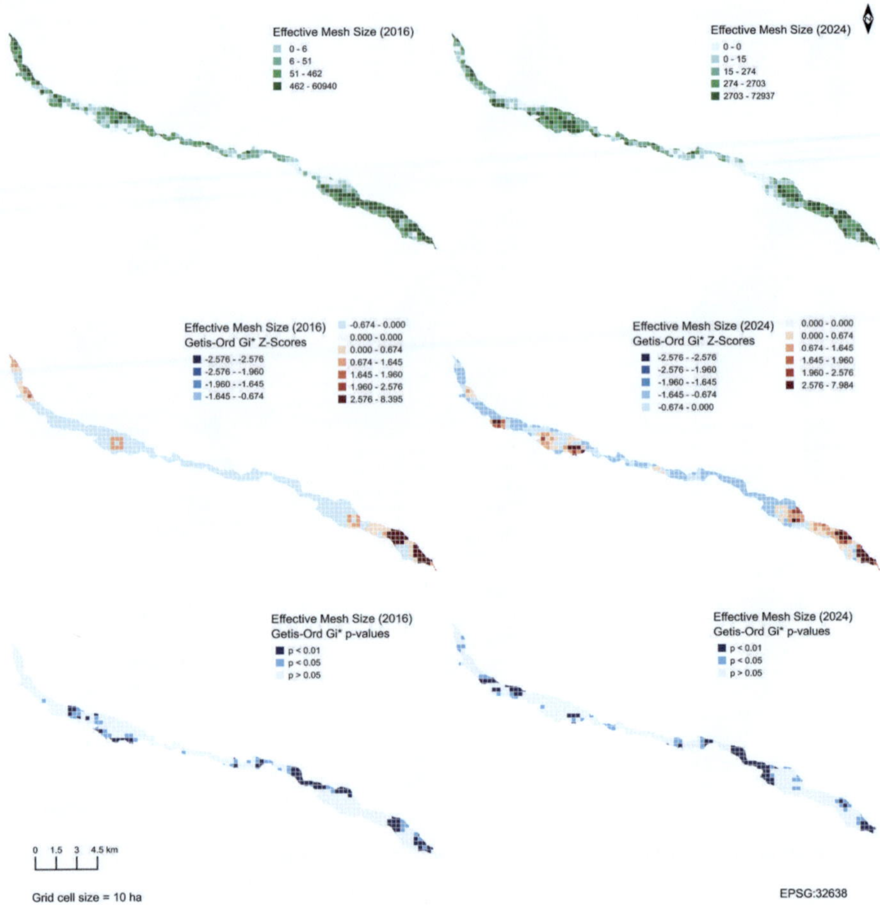

Fig. 6 Fragmentation of Ktsia River riparian forests: effective mesh size, hotspots and coldspots

in SHDI is likely related to the most changed class, open spaces (Table 1). Spatial autocorrelation analysis of the 2016 Effective Mesh Size indicator revealed two hotspots ($p < 0.01$) in the extreme southeast of the study area. Based on the 2024 values, the same space was identified as a cluster of high positive values. The hotspot begins in the marshes southeast of the village of Lezhbadini. It extends to the "Red Bridge" area on the left bank of the Ktsia River. As a result of the analysis of the 2016 data, statistically significant ($p < 0.01$) floodplain forest groves were identified northwest of the village of Tamarisi and near the village of Kvemo Arkevani. In addition, the 2024 data revealed a statistically significant ($p < 0.05$) grove in the floodplain adjacent to the Tsurtavi (Kolagiri) fortress in the northwestern part of the study area.

A statistically significant ($p < 0.01$) coldspot was found in the floodplain adjacent to the village of Lezhbadini, which borders a forest grove and where sand and gravel extraction activities are underway. This section was a coldspot in both 2016 and 2024. In addition, in both periods, coldspots were identified in the floodplain surrounding the villages of Kurtlari ($p < 0.01$), Imiri ($p < 0.05$), and Kvemo Arkevan ($p < 0.05$).

The fragmentation of forest stands in the floodplain has decreased; however, because forests occupy such a small area within the entire zone, improving the fragmentation indices of the forest landscape class cannot significantly change the ecological condition of the floodplain's riparian forests. Here, the number and extent of the cold spots outweigh the hot spots. The main factor affecting the riparian landscape is large-scale sand and gravel extraction, which has been identified as coldspots.

Land Use History

For the interpretation of the historical maps, a simple verbal description method was used based on the identification of geographical features depicted on them, which means that no georeferencing, projection, or digitising techniques were applied (except for the topographic maps). Therefore, some assumptions may be subject to revision if further research is conducted using additional sources of information.

Land Use History of the Mtkvari River Riparian Zone

The Mtkvari River and its right tributary, the Ktsia River, have historically played an important role in local culture, as reflected in old geographical texts and maps. Vakhushti Bagrationi, in his 1745 Description of the Kingdom of Georgia, mentions the existence of a large floodplain forest in the Mtkvari River basin, the Kvemo Kartli Lowland section, which was a hunting ground for kings (Brosset 1842). According to Vakhushti, to the east and south of Nagebi was the Great Karaia Plain, where the gazelle (*Gazella subgutturosa*) lived, and the Mtkvari floodplain area was full of wild boar, deer, bear, wolf, and small animals. Deer periodically appeared here, and pheasants were abundant (Brosset 1842). The Mtkvari riparian forest is also depicted on the map by Vakhushti Bagrationi (1997), representing the situation in the first half of the eighteenth century (Fig. 7).

An eighteenth-century map (1735) shows that the floodplain forests extended continuously from the territories of Rustavi and Nagebi southward to the territory of modern Azerbaijan. Nowadays, protected areas have been created in both states to protect this natural system. The same map (Bagrationi 1997) shows that the floodplain forest was widespread between Tbilisi and Rustavi. The map, created about 145 years ago (1890–1892), clearly shows the fragmentation of the Mtkvari River forest. According to the same map, the Karaia Plain was crisscrossed with irrigation canals by then, and the gazelle's habitat had been converted into arable land. According to the Soviet-era "Red Book" of Georgia, which included endangered and

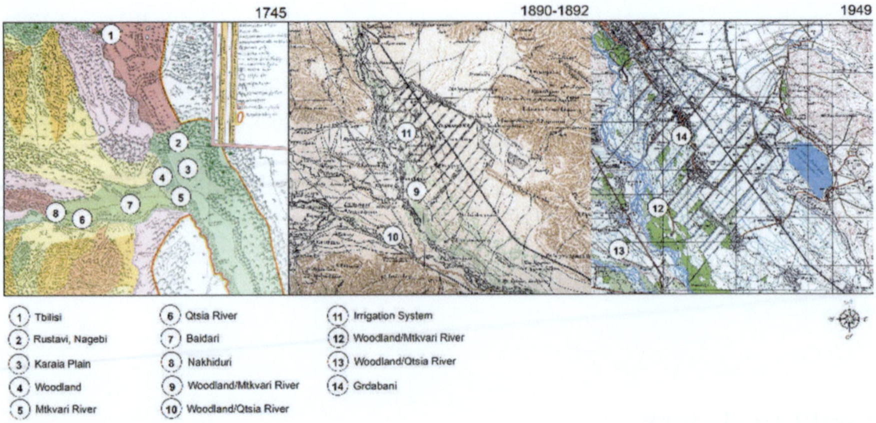

Fig. 7 Map by Vakhushti Bagrationi (1745), the so-called "5 Verst" map (1890–1892), and a 1:100,000 scale topographic map (1945)

rare species of flora and fauna, the gazelle had already disappeared from Georgia by the 1980s. The Red Book notes that the decline in gazelle was attributed to sheep grazing. During the Soviet period, "less attention was paid to the protection of biotopes of rare and endangered animals" (State Committee for Nature Protection of the Georgian SSR, Academy of Sciences SSR, Tbilisi State University, 1982). *Gazella subgutturosa* is listed as a regionally extinct (RE) species in the Georgian Red List, which was approved in 2014 (Red List of Georgia 2014). According to the IUCN assessment, *Gazella subgutturosa* is considered a Vulnerable (VU) species (IUCN SSC Antelope Specialist Group 2017).

A mid-twentieth-century (1949) topographic map shows that the riparian forest was fragmented and remained a small grove in the extreme south of the study area. To the east, in the Karaia Valley bordering it, the town of Gardabani and several villages have been developed. A network of roads also covers the territory. Later, in the early 1960s, a thermal power plant was built near Gardabani, northeast of the study area. The study area in this section is currently bordered by a water treatment plant and the new thermal power plants (2014). This location borders one of the most fragmented clusters of the protected area.

Land Use History of the Ktsia River Riparian Zone

On an eighteenth-century map (Bagrationi 1997), unlike the Mtkvari, the Ktsia floodplain is not depicted as forested (Fig. 7). However, in his work of the same period—Description of the Kingdom of Georgia—Vakhushti Bagrationi describes this area as distinguished by its biodiversity and fertile agriculture. "Below the Nakhiduri, on a wide-open field, flows the Ktsia River, which is notable for its abundance of fish. Here, one can see grazing cattle. Villages are located on both sides of the river. The villages on the plateau's edge do not have vineyards and gardens. Instead, they produce silk. This place is notable for its abundance of both domestic and wild birds. Among them is the rock partridge" (Brosset 1842). He also notes that the Ktsia floodplain was known as a hunting ground for wild birds. The riparian forest is a distinctive landscape precisely because it provides a habitat for both aquatic and terrestrial species. Vakhushti Bagrationi's description of the abundance of fish and wild birds in the early eighteenth century indicates the health of the riparian landscape. In the quote, an important emphasis related to riparian forests is on silk production instead of vineyards and orchards, which is likely related to the presence of mulberry forests in the floodplain. Mulberry trees are widespread in the Ktsia floodplain. During the expedition (May 2023), mulberry trees (*Morus alba* L.) were encountered both in the forest edge, in the transitional zone in semi-arid plains (41°20′49.19″N, 45° 3′11.14″E) in the form of groves, and also in the depths of the forest (41°20′14.32″N, 45° 3′43.06″E), mixed with poplar (*Populus alba* var. *canescens*).

As can be seen from the sources, the villages surrounding the Ktsia River were considered fertile agricultural lands. In the first half of the eighteenth century, Vakhushti Bagrationi notes that one of the villages "was previously called Tskaltashua, but now it is called Baidari, a very fertile place. All kinds of grains grow here. Many fruit orchards, vegetable gardens, and silk and cotton plants exist. You will find all kinds of livestock here, from camels to goats" (Brosset 1842). At the end of the eighteenth century, Güldenstädt also noted that on the lowland of Kvemo Kartli, the Ktsia floodplain "was fertile and covered with barley, wheat, rice, and, for the most part, cotton" (Güldenstädt 1787).

At the end of the nineteenth century, the so-called "5 Verst" map (Fig. 7) depicted the riparian forests from Beglari-Imir to the "Red Bridge." It included the floodplains surrounding Baidar and Kurtlar, which covered a much larger area than its current extent. These areas were identified as clusters of statistically significant negative indicators based on spatial autocorrelation analysis of the fragmentation indicator (MESH) for 2016–2024. Currently, abandoned or active sand and gravel mining sites can be found here. The map (1890–1892) also shows that the current forest groves in the eastern part of the study area existed at the end of the nineteenth century.

According to a 1949 topographic map at a 1:100,000 scale, forests were almost completely gone in the lower reaches of the Ktsia River. A small grove was depicted in the extreme east of the study area. In the 1950s-1980s, orchards and vineyards were planted in the Ktsia floodplains. Thus, analysing geographical texts and maps revealed that the Ktsia riparian forests almost disappeared in the first half of the twentieth century.

Conclusion

The forests of the Mtkvari and Ktsia floodplains in the Kvemo Kartli Lowland area have been historically compromised, which has been reflected in their reduction in area and historical fragmentation. By the mid-twentieth century, riparian forests in both river floodplains were already fragmented, covering a smaller area than today and were shrunk into current hotspot areas. At the beginning of the eighteenth century, the Mtkvari floodplain was densely forested, extending continuously along the river and characterised by high biodiversity, encompassing both the floodplain forest and the adjacent steppe habitat. During the same period, the Ktsia floodplain was also rich in biodiversity. By the end of the nineteenth century, the forest was already highly fragmented, although individual groves were much more widespread than in the mid-twentieth century or today. By the nineteenth century, the steppe habitat, once neighbouring riparian forests, had already been transformed, mainly for agricultural purposes, which during the Soviet period was crisscrossed with roads and covered with urban and industrial spaces. During the Soviet era, habitat protection was not practiced for wildlife preservation. The prohibition of hunting was a measure to protect species of fauna. Therefore, the habitats distinguished by their biodiversity of the study area, such as the Mtkvari and Ktsia riparian forests and the Karaia Plain steppes, were transformed into agricultural, urban, and industrial spaces, which resulted in the extinction of some species in the region.

As the study shows, the creation of a protected area, the Gardabani Managed Reserve, in the 1990s to preserve the Mtkvari riparian forests played a positive role in improving the condition of the landscape. In recent years (2016–2024), forest fragmentation has decreased in the Ktsia and Mtkvari floodplains, although it has mostly improved within the protected area's boundaries. The reduction in fragmentation was reflected in changes in several indicators. Specifically, the Effective Mesh Size and Mean Patch increased, while the Number of Patches decreased. Furthermore, in the case of the Mtkvari riparian forest, the spatial autocorrelation analysis of Effective Mesh Size showed that the area of statistically significant ($p < 0.05$) high, positive values (hotspots) increased. At the same time, the area of statistically significant ($p < 0.05$) low, negative values (coldspots) decreased. In the case of the Ktsia River, even though in 2021–2024, the fragmentation of the remaining groves decreased, the traces of historical fragmentation here are so lasting that this circumstance cannot improve the overall ecological state of the riparian forests. Here, the coldspots are spatially overlapping with river-borne sand and gravel extraction sites. At the same time, areas of statistically significant ($p < 0.05$) low, negative values in the Ktsia Valley significantly exceed hotspots, which poses a challenge to the ecological state of the Ktsia riparian forests.

Thus, the Mtkvari riparian forests' example demonstrates that a protected area effectively prevents fragmentation, although the floodplain forest landscape remains compromised outside the reserve's boundaries.

Acknowledgements This research was supported by the Shota Rustaveli National Science Foundation of Georgia (SRNSFG) [grant number YS-22-415].

References

Academy of Science of Georgia (2000) Geography of Georgia, Part I. Tbilisi: Metsniereba
Bagrationi V (1997) Atlas of Georgia. Academy of Science Georgia, Tbilisi
Breiman L (2001) Random forests. Mach Learn 45(1):5–32. https://doi.org/10.1023/A:1010933404324
Brosset M (1842) Description of the Kingdom of Georgia. Typography of the Academy, Saint Petersburg
Cohen J (1960) A coefficient of agreement for nominal scales. Educ Psychol Measur 20(1):37–46. https://doi.org/10.1177/001316446002000104
Congedo L (2021) Semi-Automatic classification plugin (SCP)—documentation. https://fromgistors.blogspot.com/p/semi-automatic-classification-plugin.html
European Environment Agency (2023) Establishment and restoration of riparian buffers. Climate-ADAPT. https://climate-adapt.eea.europa.eu/en/metadata/adaptation-options/establishment-and-restoration-of-riparian-buffer-s
European Space Agency (2025) Copernicus data space ecosystem. https://dataspace.copernicus.eu
European Space Agency (ESA) (2025) Sentinel application platform (SNAP) [Software]. https://step.esa.int/main/download/
Food and Agriculture Organization (FAO) (2020) Global forest resources assessment 2020—key findings. FAO, Rome. https://doi.org/10.4060/ca8753en
Getis A, Ord JK (1992) The analysis of spatial association by use of distance statistics. Geogr Anal 24(3):189–206
Güldenstädt JA (1787) Travels through Russia and the Caucasian Mountains, vol 1 (Reisen durch Rußland und im Caucasischen Gebirge). Imperial Academy of Sciences, St. Petersburg
Gulisashvili V (1977) Nature and natural zones of Georgia. Sabtchota Sakartvelo, Tbilisi
IUCN SSC Antelope Specialist Group (2017) Gazella subgutturosa. The IUCN Red List of Threatened Species. e.T8976A50187422. https://doi.org/10.2305/IUCN.UK.2017-2.RLTS.T8976A50187422.en
Jung M (2022) LecoS—landscape ecology statistics. https://conservationecology.wordpress.com/qgis-plugins-and-scripts/lecos-land-cover-statistics/
Kavtaradze G, Basilidze L, Pergl J, Avoiani E, Vahalik P, Aptsiauri B, Goginashvili N, Bachilava M, Manvelidze Z, Kobakhidze N (2023) Distribution and environmental impact of alien woody species in lowland riparian forest habitats: case study in the protected areas of Georgia, South Caucasus. J Sci 69:401–414
Kharadze K (2019) Marneuli municipality: nature, past, present. Sveti, Tbilisi
Malanson GP (1993) Riparian landscapes. Cambridge University Press, Cambridge
McGarigal K, Cushman SA, Ene E (2012) FRAGSTATS v4: spatial pattern analysis program for categorical map patterns. https://fragstats.org/
Naiman RJ, Decamps H, Pollock M (1993) The role of riparian corridors in maintaining regional biodiversity. Ecol Appl 3(2):209–212
Nakamura F (2022) Riparian forests and climate change: interactive zone of green and blue infrastructure. In: Green infrastructure and climate change adaptation: function, implementation and governance. Springer Nature Singapore, Singapore, pp 73–91
Ness RF, Harris LD (1990) Habitat connectivity and the conservation of biological diversity: Florida as a case history. In: Proceedings of the 1989 Society of American Foresters Conference, Spokane, Washington. Society of American Foresters, Bethesda, Maryland, USA, pp 131–135
QGIS Development Team. QGIS Geographic Information System (Version 3.34 Prizren). Open Source Geospatial Foundation. https://blog.qgis.org/2023/11/05/qgis-3-34-prizren-is-released/
Red List of Georgia (2014) Ordinance of the Government of Georgia. https://matsne.gov.ge/ka/document/view/2256983?publication=0
Ritchie H (2021) The world has lost one-third of its forest, but an end of deforestation is possible. Our World in Data. https://ourworldindata.org/world-lost-one-third-forests

Rossy Q (2019) Visualist: a spatial analysis plugin for crime analysts. Ecole des sciences criminelles, Lausanne. https://plugins.qgis.org/plugins/visualist/

UNEP-WCMC (2025) Protected area profile for Gardabani managed reserve from the world database on protected areas. www.protectedplanet.net

World Bank (2025) World Bank Official Boundaries Data Catalogue. https://datacatalog.worldbank.org/search/dataset/0038272/World-Bank-Official-Boundaries

Geoheritage and Tourism

Environment State and Cultural Heritage Areas of Serbia—Toward a Symbiotic Relationship and Sustainable Management

Ana Milanović Pešić, Jovana Brankov, and Željko Bjeljac

Abstract Cultural heritage is located in attractive sites that lead to tourism development; nevertheless, tourism and other economic activities can influence the degradation of cultural heritage sites and pollute the environmental surroundings. Numerous official data from different institutions can provide insight into the relationship between the tangible cultural heritage sites and the environmental settings in which it is located. In addition, research involving the local population can significantly help assess the quality of the environment and tangible cultural heritage sites at the local level. To examine the environment state and its relationship with natural resources pollution and tangible cultural heritage at the local level, a survey of respondents from 52 municipalities in Serbia was conducted. The results indicate that a third of respondents believe that environment and natural resources (air, surface water, groundwater, and soil) near their municipality's most significant cultural heritage sites are moderately polluted. Most respondents singled out air as a heavily polluted natural resource. Additionally, more than half of the respondents believed intensive construction endangers their municipality's tangible cultural heritage sites. A more detailed analysis in this study included the correlation of residents' responses and socio-economic variables. The results obtained can serve to improve the management of tangible cultural heritage sites and environmental protection.

Keywords Environment state · Natural resources pollution · Cultural heritage · Residents' perceptions · Serbia

A. Milanović Pešić (✉) · J. Brankov · Ž. Bjeljac
Geographical Institute "Jovan Cvijić" Serbian Academy of Sciences and Arts, Belgrade, Serbia
e-mail: a.milanovic@gi.sanu.ac.rs

J. Brankov
e-mail: j.brankov@gi.sanu.ac.rs

Ž. Bjeljac
e-mail: z.bjeljac@gi.sanu.ac.rs

A. Milanović Pešić
L. N. Gumilyov Eurasian National University, Faculty of Natural Sciences, Department of Physical and Economic Geography, Astana, Kazakhstan

L. Soares et al. (eds.), *Compromised Geography*, Key Challenges in Geography,
https://doi.org/10.1007/978-3-032-05201-8_19

Introduction

Sustainable natural resource management and environmental and cultural heritage protection are important for sustainable development. The twenty-first century is characterized by significant environmental degradation, resource depletion, increased poverty, and cultural homogenization (Parr 2016), and it is a great challenge to balance and integrate ecological, economic, and social aspects in sustainable development (Govindan et al. 2016). Golinelli (2016) pointed to the cultural value of the environment as a testament to civilization and as a result of the dynamic evolution of the relationship between humans and nature, and that it is possible to develop models for the enhancement of cultural heritage in a sustainable way.

Cultural heritage is an important part of the cultural identity of every nation. Various declarations, strategies, and conventions have particularly emphasized the importance of environmental and cultural heritage protection in the last few decades. According to the Convention for the Protection of Cultural Property in the Event of Armed Conflict with Regulations for the Execution of the Convention (UNESCO 1954), cultural property represents "movable or immovable property of great importance to the cultural heritage of every people," while the 1972 World Heritage Convention (UNESCO 1972) recognizes the synergies and connections between natural and cultural heritage. According to this Convention, cultural heritage includes, among other things, monuments that are works of man or combined works of nature and man. In addition, the European Union Directive on Environmental Impact Assessment (EIA) requires consideration of: human beings; fauna and flora; soil; water; air; climate; landscape; material assets and cultural heritage in environmental assessments of plans and programs (Council of the European Union 1997). Cultural heritage is the expression of a way of life developed by a community and transmitted from generation to generation, including "customs, traditions, places, objects, artistic expressions, and values." (ICOMOS 2002, p. 21). The Council of Europe Framework Convention on the Value of Cultural Heritage for Society (Faro Convention) defines cultural heritage as the group of resources inherited from the past that people identify, regardless of ownership, as a reflection and expression of their values, beliefs, knowledge, and constantly evolving traditions. It includes all aspects of the environment that result from the interaction between people and places over time (Council of Europe 2005).

Cultural heritage can become a tourist attraction that will increase the number of visitors to areas that have such heritage and significant financial resources (Popescu and Corbos 2010), as well as providing economic opportunities in creating highly qualified jobs and building the basis for regional identity (Žarnić et al. 2015). It encompasses both visits to tangible cultural assets and numerous manifestations related to the preservation of intangible cultural heritage, such as traditional sports and games (Bjeljac et al. 2021; Ćurčić et al. 2021; Lavega et al. 2023; Terzić et al. 2015).

Although it contributes to economic benefits, tangible cultural heritage is at constant risk due to environmental and man-induced impacts. Climate and vegetation change, air pollution, natural disasters, rapid urbanization and population growth, public neglect, and institutional or legal restrictions are the most common causes of cultural heritage and environmental degradation (Lin and Hou 2024; Sablier and Garrigues 2014; Van Grieken et al. 1998; Žarnić et al. 2015).

Given the importance of the environment, natural resources, and cultural heritage, several studies analyze their connections and the impacts of one element on another. Thus, Žarnić et al. (2015) explored the environmental impact of built heritage assessment, and Foster and Kreinin (2020) provide an overview of environmental impact indicators for the adaptive reuse of cultural heritage sites (ARCH) from a circular economy perspective. Bond et al. (2004) addressed how well cultural heritage is considered in environmental impact assessments within the European Union. Also, Mohaddes Khorassani et al. (2019) investigated the assessment of the environmental and social impacts of cultural heritage restorations. Agisilaos (2012) explored the impact on the environment caused by using cultural resources for tourism development and measures for their protection. In addition, Teller and Bond (2002) reviewed European environmental policy and legislation that includes cultural heritage.

This study aims to determine the interaction between local communities, cultural heritage sites, environmental assessment, and natural resource pollution. During survey research, several objectives were set to be explored:

- How do residents perceive cultural heritage in their surroundings and what are their attitudes about the interaction of the environment state and natural pollution with cultural heritage?
- What do they single out as the greatest risks/pressures on the environment in the surrounding area of tangible cultural heritage?
- Are there differences in attitudes about the environmental state and risks to tangible cultural heritage surroundings between different genders, ages, and educational categories?

Tangible Cultural Heritage of Serbia

According to the Institute for the Protection of Cultural Monuments of Serbia (2025), there are currently 2658 registered immovable cultural properties, of which 2277 monuments of culture, 97 spatial cultural-historical units, 203 archeological sites, and 81 landmarks. Of this number, 201 cultural properties are of exceptional importance, and 582 are of great importance (Table 1). In addition, eight sites from Serbia have been declared UNESCO World Heritage. Considering that many of these facilities are located in cities or settlements and present important tourist sites, examining the relationship between cultural heritage and the environment's state can be an important topic for research in Serbia.

Among the archeological sites of exceptional importance, nine are located along the Danube or near the Danube, such as the Belo Brdo—Vinča sites in the Vinča settlement near Belgrade from the Neolithic Period (Vinča culture period between 5300 and 4500 years B.C.) and Lepenski Vir (6500–4500 year B.C.; Photograph 1)

Table 1 Overview of the tangible culture heritage of Serbia

Immovable cultural properties	Total number	Cultural properties of exceptional importance	Cultural properties of great importance
Monuments of culture	2277	155	512
Archeological sites	203	18	25
Spatial cultural-historical units	97	12	28
Landmarks	81	16	17
TOTAL	2658	201	582

Source of data Institute for the Protection of Cultural Monuments of Serbia (2025), created by Milanović Pešić, A

from the Neolithic period in the Iron Gate (Đerdap Gorge) to the archeological sites from the ancient Roman period (Viminacijum, Trajan's Bridge, Diana— Karataš Fortress, etc.).

Among the cultural monuments of exceptional importance are numerous medieval Orthodox churches and monasteries, Orthodox churches and monasteries from a later period and other sacral buildings, medieval fortresses, and other facilities. It is important to emphasize that seven fortresses from different periods stand out

Photo 1 Lepenski Vir archeological site, 7th–5th c. B.C. (author: Milanović Pešić, A.)

Photo 2 Golubac Fortress (Golubački Grad), 14th c. (author: Milanović Pešić, A.)

along the Danube in Serbia, and among them is the medieval Golubac Fortress (Golubački Grad), which represents one of the monuments of exceptional importance (Photograph 2).

UNESCO World Heritage Sites in Serbia

Eight sites from Serbia have been declared UNESCO World Heritage List due to their artistic, cultural, social, and historical importance (UNESCO 2025a). These include the imperial Roman palace of Felix Romuliana in Gamzigrad, medieval Serbian Monasteries—Studenica, Stari Ras, and Sopoćani, a group of medieval monuments in Kosovo and Metohija (Visoki Dečani Monastery, Patriarchate of Peć Monastery, Gračanica Monastery and Bogorodica Ljeviška Church [Our Lady of Ljeviš Church]), and unique medieval tombstones known as Stećci.

A group of medieval Serbian monuments located in the Raška Region (southwestern part of Serbia), comprising fortresses, monasteries, and a church, was declared the first UNESCO World Heritage site in Serbia in 1979 under the common name "Stari Ras and Sopoćani" (UNESCO 2025a). It represents a cultural heritage which consists of four separate objects: Sopoćani Monastery, Đurđevi Stupovi Monastery, Petrova Crkva (Holy Apostles St Peter Church -St Peter's Church), and the archeological site of the Medieval Town of Ras. These Orthodox monasteries and churches date from the 10th to the 13th centuries and belong to the culture of the Byzantine Empire. The architectural complex Stari Ras (Old Ras) dates from the 12th to the early 14th centuries, when the ancient town was the first capital of the Serbian medieval state.

Studenica Monastery (Photograph 3) is among the most distinguished 12th-century Serbian Monasteries (1186). It was built by Grand Župan (Grand Prince) Stefan Nemanja, the founder of the Nemanjić dynasty, which formed the Serbian medieval state, first a kingdom and then an empire. Studenica Monastery is located in the Raška Region (south-western part of Serbia) near the Studenica River and represents one of the largest and wealthiest of Serbia's Orthodox monasteries. The most significant monuments in the monastery complex are the Bogorodičina Crkva (Church of the Virgin Mary) and Kraljeva Crkva (the Church of King), both built of white marble and with valuable collections of frescoes (Byzantine painting) from the 13th and 14th centuries. Studenica Monastery was declared a UNESCO World Heritage site in Serbia in 1986 (UNESCO 2025a).

The first medieval Orthodox monastery from the Autonomous Province Kosovo and Metohija, inscribed on the UNESCO World Heritage List, was the Visoki Dečani Monastery in 2004. Subsequently, in 2006, the monasteries Pećka Patrijaršija (the Patriarchate of Peć Monastery), the Gračanica Monastery and the Church of Bogorodica Ljeviška (the Church of the Virgin of Ljeviška) were also inscribed

Photo 3 Studenica Monastery, 12th c. (author: Milanović Pešić, A.)

and are on the UNESCO List of World Heritage in Danger under the joint name Medieval Monuments in Kosovo and Metohija (UNESCO World Heritage Convention 2025a). The Patriarchate of Peć Monastery was built in the 13th century, and the two other monasteries and the church in the 14th century by the royal family Nemanjić. The Patriarchate of Peć is located near the Peć town, in the valley of the Pećka Bistrica River. The frescoes of the Church of the Holy Apostles were painted in a unique, monumental style in the 13th century. The Patriarchate of Peć has been the headquarter of the Patriarch of the Serbian Orthodox Church since the 14th century to the date. Visoki Dečani Monastery (Photo 4) is located near the Peć town, along the Dečanska Bistrica River. Its construction was begun as an endowment by Serbian King Stefan Dečanski in 1327 and was completed by his son, the Serbian Emperor Stefan Dušan, in 1335. This monastery is also the mausoleum of the Stefan Dečanski. The Dečani Paintings are the most extensive preserved gallery of Serbian medieval paintings, a real treasury of Orthodox iconography comprising over a thousand compositions (Orthodox Metropolitanate of Montenegro and the Littoral 2011b).

Gračanica Monastery is located in Gračanica settlement near the Priština town, on the banks of the Gračanka River. It was built in the Byzantine-Serbian style between 1315 and 1321 as an endowment of King Milutin. In Gračanica Monastery, for the first time in old Serbian painting, the Nemanjić royal family line was depicted in an upright, upward-branching vine, with 16 portraits. Today, the monastery treasury contains many valuable icons from the 16th, 17th, and 18th centuries (Orthodox Metropolitanate of Montenegro and the Littoral 2011a; Diocese of Raška-Prizren

Photo 4 Visoki Dečani Monastery, 14th c. (author: Milivojević, M.)

2017). Church of Bogorodica Ljeviška (Church of the Holy Virgin of Ljeviška) is located in Prizren town. The Serbian King Milutin renovated it in the period 1306–1309. Its frescoes date back to the first decade of the 14th century and represent the appearance of the new so-called Palaiologian Renaissance style that combines the influences of the Eastern Orthodox Byzantine and Western Romanesque traditions. The style played a decisive role in subsequent Balkan art (UNESCO 2025a).

Felix Romuliana—Gamzigrad is the only archeological site of classical antiquity in Serbia to be inscribed on the UNESCO World Heritage List in 2007. The Roman Emperor Galerius built the imperial palace in the 3rd and 4th centuries to honor his mother, Romula. Felix Romuliana—Gamzigrad consisted of fortifications, the palace, basilicas, temples, hot baths, a memorial complex, and a tetrapylon (UNESCO 2025a). It is located near Gamzigrad village, in the City of Zaječar (north-eastern part of Serbia).

Stećci Medieval Tombstone Graveyards represent a unique and exceptional testimony to the medieval cultural tradition of the Balkan peoples. They were carved between the 12th and 16th centuries in various shapes. On 28 component sites, 4000 medieval tombstones (stećci) were included on the territory of four states: Serbia, Bosnia and Herzegovina, Montenegro, and Croatia. They were inscribed on the UNESCO World Heritage List in 2016 as a joint world cultural heritage of these countries (UNESCO 2025a).

In addition to sites inscribed on the UNESCO World Heritage List, member states may maintain a list of tentative sites that they may consider for nomination. As of 2021, Serbia has inscribed eleven sites on its tentative list, including six cultural sites. These are the following sites: archeological site Caričin Grad (Iustiniana Prima) (1st century), Frontiers of the Roman Empire—The Danube Limes (Serbia) (1st to 7th century), Fortified Manasija Monastery (15th century), Smederevo Fortress (15th century), Cultural Landscape of Bač and its surroundings (12th to 19th century) and Negotinske Pivnice, rural settlements consisting of wine cellars (18th to 19th century) (UNESCO 2025b).

Frontiers of the Roman Empire—The Danube Limes consists of 130 parts in the territory of four countries: Croatia, Serbia, Bulgaria, and Romania. It includes seven legionary fortresses (Singidunum, Viminacium, Ratiaria, Oescus, Novae, Durostorum, and Troesmis), about 120 auxiliary forts and smaller fortifications; civil settlements, cemeteries, production complexes, roads, and other elements that were all related to the functioning of the Roman frontier along the Danube (UNESCO 2025b). It is also a part of the European cultural route. In addition, seven certified Euro cultural routes and five national ones pass through Serbian territory (Terzić and Bjeljac 2016).

Data and Methodology

Data

In order to examine residents' perceptions about the environmental state, natural resources pollution, and tangible cultural heritage, as well as the relationship between these complexes at the local level, a survey was conducted as a data collection method. It includes 245 respondents from 52 cities or municipalities from all parts of Serbia (Map 1).

The questionnaire was distributed online via different social networks in 2023–2024. The questionnaire was initially posted on the website and on the Facebook page of the Geographical Institute "Jovan Cvijić" SASA. Then, in agreement with several cultural centers in Serbia (Loznica, Kladovo, Kosjerić, Smederevska Palanka, and Novi Sad), the questionnaire was posted on their social media platforms to ensure a geographically relevant sample.

The survey was conducted voluntarily and completely anonymously, and all of the respondents were adults. They were informed about the institution conducting the research and its goals, as well as that the obtained results would be used only for scientific purposes. According to socio-demographic characteristics, different categories of respondents participated in the survey, as given in Table 2.

According to age, more than half of the respondents were between 31 and 50 years old. Also, a little more women than men participated in the survey. Based on education, three categories were distinguished. The first category consists of respondents with high school education (including also seven respondents who had only attended elementary school); the second category includes respondents with university degrees, and the third category includes those with MSc and PhD education. Each of the three categories included about a third of the total number of respondents.

Methodology

The original questionnaire consisted of different thematic groups. For this chapter, one thematic group regarding environment state assessment in the cultural heritage zone of Serbia was singled out for analysis. The questions were related to residents' perceptions of cultural heritage, attitudes about the interaction of environment state and natural resources pollution with cultural heritage, assessments of the environmental state in the vicinity of tangible cultural heritage, and recognition of the greatest risks/pressures on the environment nearby tangible cultural heritage in their municipalities. Most of the questions were open/closed, with the option to rate your perceptions according to a numerical scale. One question was open-ended, with the option to write an answer. The questions set analysis was conducted according to defined socio-demographic variables to determine whether there was a significant difference in attitudes among different groups of respondents.

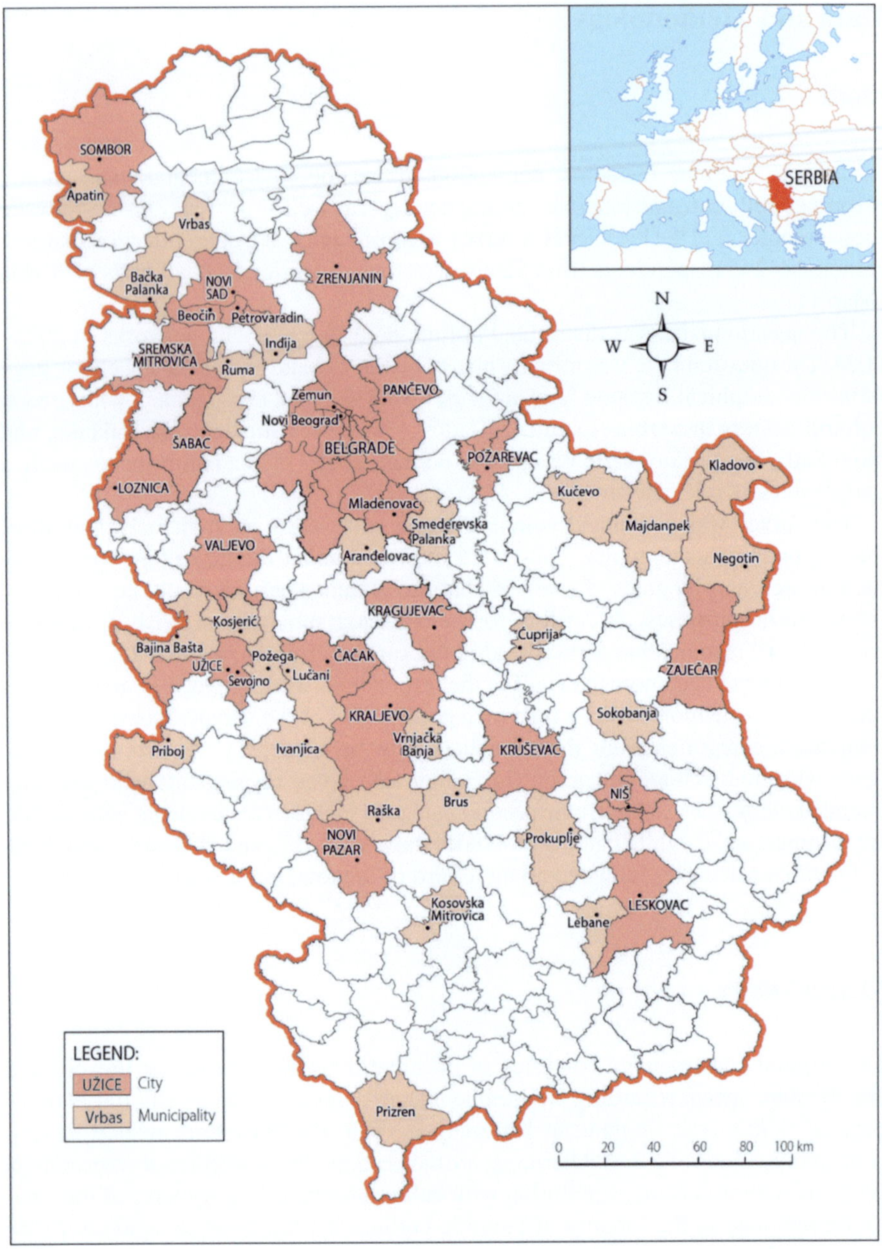

Map 1 Cities and municipalities of the survey respondents' residence

Table 2 Socio-demographic characteristics of the respondents

Categories	Number	Percentage (%)
Age		
18–30	33	13.5
31–50	134	54.7
>50	78	31.8
Gender		
Male	106	43.3
Female	139	56.7
Education		
High school (including elementary school)	86	35.1
Faculty	70	28.6
MSc and PhD	89	36.3

(created by Milanović Pešić, A.)

The data obtained from the survey were analyzed using descriptive statistics (frequency, central tendency) and bivariate analysis (T-test, ANOVA) within the SPSS version 20 software package. These statistical methods have wide application when it comes to population perceptions surveys related to various topics, such as the impact of tourism on the sustainable development of protected areas (Brankov et al. 2022; Etongo et al. 2023; Gajić et al. 2025), assessment of the state of natural resources in national parks (Abukari and Mwalyosi 2020; Brankov et al. 2021; Das and Hussain 2016; Milanović Pešić et al. 2020; Trakolis 2001) and other protected areas (Cvetković et al. 2023; Mateos et al. 2020; Milanović Pešić et al. 2023; Vodouhê et al. 2010; Xu et al. 2006), human impact on karst geoheritage (Telbisz et al. 2021), conservation of mangrove forests (Badola et al. 2012), population perception of natural disasters (Netzel et al. 2021; Qasim et al. 2015; Šiljeg et al. 2022), etc. In addition to analyzing socio-economic variables, some statistical analyses from this software package are also used to analyze trends and differences in specific quantitative data related to climate and hydrology. Thus, the T-tests are widely used in analyzing trends and differences in flow, air temperature, and precipitation between two observation periods (Cigizoglu et al. 2005; Milanović Pešić et al. 2025; Yang et al. 2014; Zhang and Yu 2021).

Results and Discussion

Residents' Perceptions of Tangible Cultural Heritage and Environment State—The Analysis of Survey Results

Given Serbia's rich material cultural heritage, many of which are sites of exceptional global and national importance, on the one hand, and the state of the environment and natural resources in their vicinity, on the other, a survey was conducted to examine residents' perceptions and attitudes related to this issue.

The first question in this part of the survey related to how respondents perceive cultural heritage and which tangible cultural heritage they highlight as the most significant in their cities or municipalities. The question had three answer options to check whether respondents knew what constitutes tangible cultural heritage. The following answers were offered with options to mark only one:

- Monument, building, or facility (archeological site, historical/artistic building, fortress, monastery, church, etc.);
- Groups of objects (groups of urban or rural objects that have ar-cheological, historical, scientific, and artistic value);
- Protected site (combined works of man and nature, as partially built areas in which both natural values and historical, archeological, artis-tic, scientific, social or technical interest can be observed).

Figure 1 presents the respondents' attitudes. It shows that the highest number of them highlight monuments, buildings, or facilities as the most significant in their municipalities.

Additionally, residents were asked to write down the most significant cultural heritage site. Considering the different numbers of respondents from different municipalities, there were similar answers in some cases. A significant number of respondents distinguished as a most important cultural heritage: Belgrade Fortress—Kalemegdan, the Petrovaradin Fortress in Novi Sad, and the Tršić Complex near the City

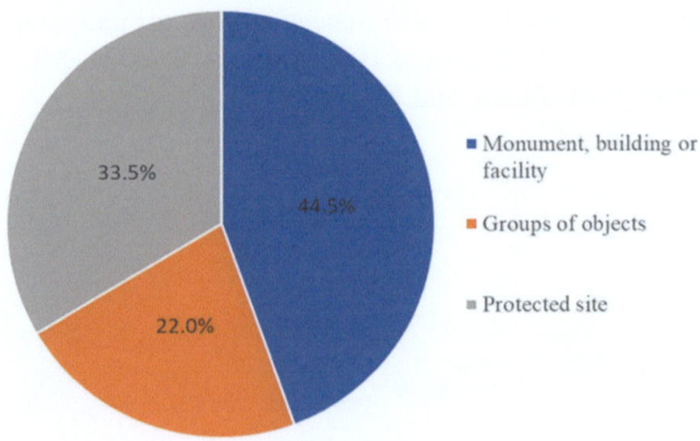

Fig. 1 Respondents' answers about the most significant tangible cultural heritage in their city or municipality (created by Milanović Pešić, A.)

of Loznica. Also, the most important monasteries in the respondents' municipalities are denoted.

The Belgrade Fortress (Photograph 5), along with Kalemegdan Park, is a cultural monument of exceptional importance at the national level located in the center of the capital, Belgrade. At the beginning of the 1st century, the first Roman military camp-castrum was built. As a defensive complex, the fortress was built on a ridge above the confluence of the Sava and Danube rivers from the 1st to the 18th century. The complex consists of the Belgrade Fortress, divided into the Upper and Lower Towns, and Kalemegdan Park.

Petrovaradin Fortress was built on a ridge along the Danube from the 1st to 18th centuries, and the current fortress was built in the 17th and 18th centuries. It is located in Novi Sad, the second largest city in Serbia, and is the most important tourist site in this city. It represents a spatial, cultural, and historical entity of great importance.

Tršić Complex consist of the house of Vuk Stefanović Karadžić, the reformer of the Serbian language in the nineteenth century. The original house was demolished during the 19th century, and the Vuk Memorial House was built in the 1930s. Together with the built objects of folk architecture, it is a cultural monument of exceptional

Photo 5 Kalemegdan Fortress, 1st–18th c. (author: Brankov, J.)

importance. It is also a popular tourist site, mainly visited by students during school trips.

Considering that the above-mentioned cultural monuments, as well as numerous others, are located in cities or in their immediate vicinity and represent significant tourist sites, it is important to examine the state of the natural environment in their surroundings. Therefore, the respondents were asked to rate their level of agreement with the statements about what factors endanger tangible cultural heritage in their municipalities (Table 3).

Regarding residents' attitudes, more than half of the respondents agree or strongly agree that intensive construction endangers tangible cultural heritage (53.5%). Almost half of them had the opinion that tangible cultural heritage is endangered by wild landfills (49.4%). About a third of the respondents believe that the tangible cultural heritage in their municipality is endangered by polluted natural resources (35.1%), and almost the same number of respondents have a neutral attitude (37.6%). Interesting results were obtained regarding tourist activity. About 55% of the respondents do not agree that tourism endangers tangible cultural heritage in their municipality, and including the respondents who strongly disagree, it is almost 70%.

The next question related to residents' perceptions of the environmental state near their municipalities' most significant cultural heritage. The obtained results indicate that a third of respondents (35.6%) believed that the environment near the most significant cultural heritage in their municipality is moderately polluted. Analyzing the pollution level of natural resources, more than a third of respondents had the opinion that surface water (38.4%), groundwater (34.7%), soil (37.1%), and air (32.2%) are moderately polluted (Fig. 2).

A similar number of respondents thought that air (24.9%), surface water (25.9%), groundwater (23.3%), and soil (24.5%) were significantly polluted. From quarter to a third of the total number of respondents believed that natural resources are very clean or slightly polluted: air (31%), surface water (25.7%), groundwater (35.1%), and soil (32.4%). In general, respondents had the perception that groundwater is the cleanest resource. Most respondents singled out the air as a very clean (16.7%) but also as a heavily polluted (11.8%) natural resource.

Table 3 Respondents' attitudes toward the factors that endanger tangible cultural heritage in their municipalities (in percentages)

	Strongly disagree	Disagree	Neutral	Agree	Strongly agree
Intensive construction in the area	5.7	13.9	26.9	27.8	25.7
Wild landfill	4.5	19.6	26.5	29.0	20.4
Polluted natural resources	4.1	23.3	37.6	20.4	14.7
Tourism	14.3	55.1	21.2	6.5	2.9

(created by Milanović Pešić, A.)

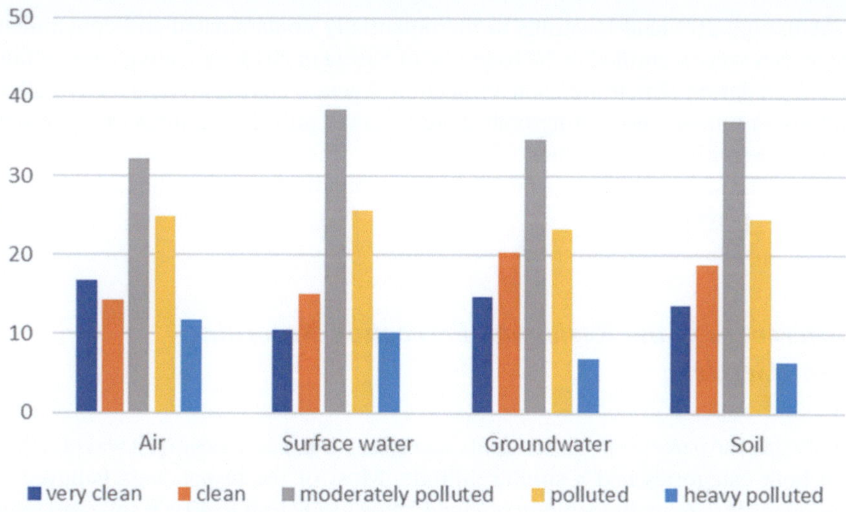

Fig. 2 Residents' perceptions of the natural resource state in the cultural heritage zones of their municipalities (created by Milanović Pešić, A.)

The obtained results indicate that the respondents in the smallest percentage had the opinion that surface waters are very clean or clean compared to other natural resources. This can be explained by the fact that numerous fortresses of exceptional and great importance (such as Petrovaradin, Belgrade, Smederevska, Golubac, etc.) and archeological sites (such as Vinča, Lepencki Vir, etc.) are located along the Danube. Numerous water quality investigations indicate moderate or significant water pollution in the middle course of the Danube in Serbia (Ismail and Robescu 2019; Jakovljević et al. 2019; Takić et al. 2017), as well as in the area of the Đerdap Gorge (Hagemann et al. 2019; Milanović Pešić et al. 2020), where the fortresses Golubac and Diana are located, and the Neolithic archeological site Lepenski Vir.

In addition, respondents often obtain information about the quality of surface waters and air on websites that publish monitoring data in national and local networks, via mobile applications, television, and visual observation. This is also confirmed by the survey on air quality in Serbia, where 27% of respondents indicated visual observation as one of the ways of obtaining information (Malinović-Milićević and Stanojević 2024).

It is also important to highlight interesting contradictory results regarding air quality, which is singled out by the largest percentage of residents as both the cleanest and most polluted natural resources. Available data on concentrations of pollutants indicate significant polluted air in Serbia (Malinović-Milićević and Stanojević 2024). The above-mentioned ways of informing the population contribute to the fact that part of the population singles out air as the most polluted resource. On the other hand, a small number of settlements in Serbia have measuring stations and continuous monitoring, and therefore, large part of the population does not know what the quality of the air they breathe (Malinović-Milićević and Stanojević 2024).

Although 309 land locations in the potentially contaminated and contaminated categories were identified in the Republic of Serbia in 2019, indicating land pollution and degradation (Environmental Protection Agency 2020), lower data availability and discontinuous monitoring contribute to lower public awareness of soil quality. Therefore, a higher percentage of respondents singled out this natural resource as cleaner. The same reasons can be named for groundwater quality monitoring and the residents' perception of it.

Differences in the Residents' Perception According to the Gender

Analyzing the perceptions of residents according to gender category, it was concluded that both categories had a similar attitude. Most of the respondents believed that natural resources are moderately polluted; male (35.8% air to 40.6% soil) and female (29.5.8% air to 38.1% surface water) (Fig. 3).

In addition, it is concluded that a higher percentage of women have the opinion that natural resources are polluted (25.9% of groundwater to 29.5% of surface water) compared to men (18.9% of soil to 22.6% air). A larger number of male respondents (14.2% of surface water to 20.8% of air) believed that natural resources were very clean compared to female respondents (7.9% of surface water to 13.7% of air). Based on all of the above mentioned, it can be concluded that a significantly lower percentage of female respondents believed that natural resources nearby cultural heritage zones are very clean.

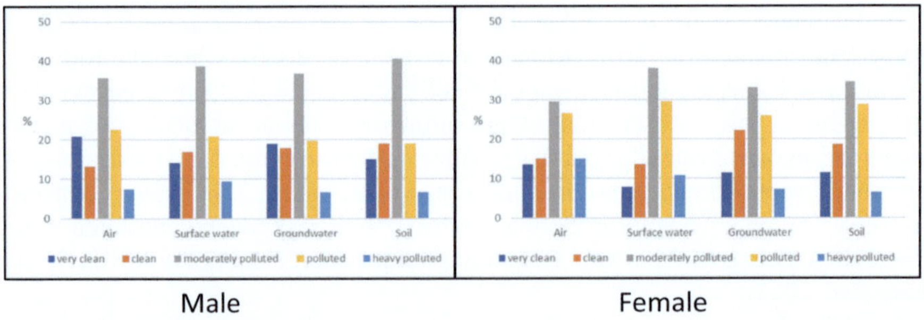

Fig. 3 Residents' perceptions according to gender about natural resource state in the cultural heritage zones (created by Milanović Pešić, A.)

Differences in the Residents' Perception According to the Age

In order to examine the differences in attitudes by age, the respondents were divided into three categories: the young group (18–30 years of age), the middle group (31–50 years of age), and the elder group (over 51 years of age). Regarding age categories, a similar attitude can also be observed; most respondents believed that natural resources are moderately polluted. Based on the obtained results, it can be concluded that more than a third of respondents in all three categories perceived that air, soil, and surface water are moderately polluted: young group (30% soil to 45.5% surface water), middle group (30.9% air to 35.3% soil), and elder group (2.9% air to 43.4% surface water and soil) (Fig. 4).

Regarding a clean natural resource, the largest percentage of respondents marked groundwater and soil: young people ((groundwater and soil per 33.3%), middle-aged people (groundwater, 19.1%), and older people (soil, 18.4%). The air was rated as the cleanest source in all three age categories (12–18%).

Residents of different age categories have divided opinions about polluted natural resources. As polluted natural resources, the young category denoted air (24.2%) and soil (24.2%), the middle-aged category surface water (27.2%), and the elder category air (27.6%). When it comes to the most polluted natural resource, young people (9.1%) and older people (10.5%) marked surface water, while people between 31 and 50 singled out air (14.7%).

Differences in the Residents' Perception According to the Education

Respondents from different education categories generally have similar attitudes regarding natural resource quality (Fig. 5). Most respondents believed that natural resources are moderately polluted, with several exceptions; namely, most people with faculty degrees believe that surface water is polluted (31.4%), and most respondents with MSc or PhD degrees considered the air as a polluted natural resource (29.2%).

Regarding clean natural resources, the largest percentage of respondents marked groundwater (about 20%) in the categories with faculty and MSc /PhD degrees and soil (24.4%) in the category of high school degree. As the very clean resource, most people with high school (12.1%) and faculty (15.7%) singled out air, while respondents with Msc/PhD degrees marked groundwater (18%).

The largest percentage of people with high school (7%) and MSc/PhD degrees (14.6%) thought that the most polluted natural resource is air, while the largest percentage of people with faculty believed that surface water is the most polluted (15.7%). In general, respondents with a high school education believed that natural resources are clean or slightly polluted, while people with faculty and MSc/PhD degrees in a higher percentage had the opinion that natural resources are polluted.

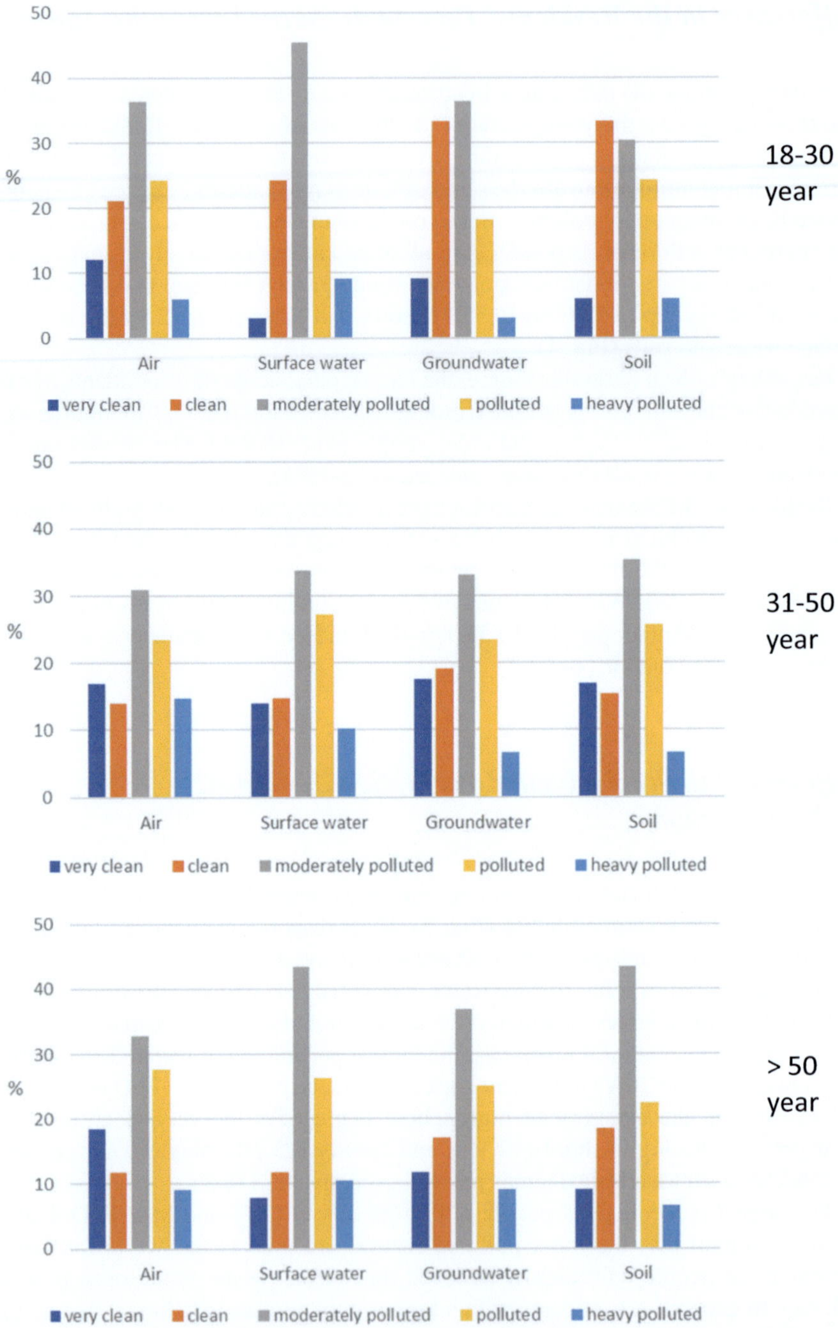

Fig. 4 Residents' perceptions according to age about natural resource state in the cultural heritage zones (created by Milanović Pešić, A.)

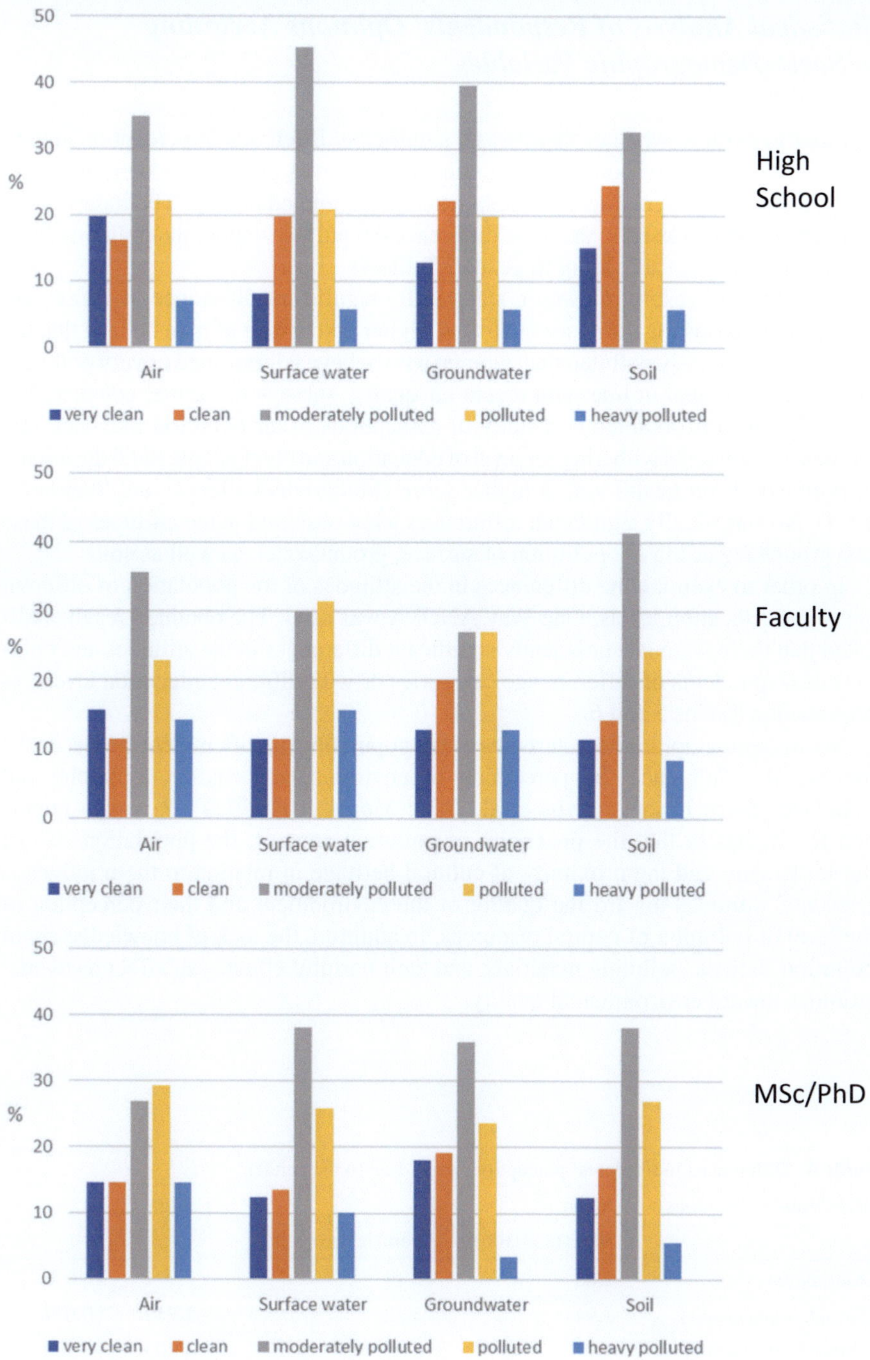

Fig. 5 Residents' perceptions according to education about natural resource state in the cultural heritage zones (created by Milanović Pešić, A.)

Statistical Analysis of Respondents' Opinions According to Socio-Demographic Variables

In order to explore whether there was a significant difference in perception among different groups of respondents, the questionnaire was analyzed according to the specific socio-demographic variables: gender, age, and education. The independent samples T-test was used to examine differences in attitudes about the pollution level of natural resources according to gender (Table 4).

The obtained results showed a statistically significant difference in mens' and womens' opinions regarding air quality. A higher percentage of men believe that the air is clean or slightly polluted than women who believe it is polluted or very polluted. Our results are also in line with recent air quality surveys in Serbia, conducted in 2023. The results obtained in a study of perceptions of air pollution indicated that women, respondents with a higher level of education and elder people rated the impact of polluted air on health with a higher score (Malinović-Milićević and Stanojević 2024). No statistically significant differences were observed in the attitudes of these two groups regarding the pollution of surface, groundwater, as well as soil.

In order to examine the differences in the attitudes of the population of different years and education levels, One-Way ANOVA was used. The obtained results indicated that there were no statistically significant differences in the attitudes and opinions of respondents of different age categories or with different education levels, as presented in Tables 5 and 6.

Socio-demographic characteristics of the population are often identified as determinants that influence the perception of environmental quality (Costache and Sencovici 2015; Lysenko-Ryba et al. 2024; Moses et al. 2022). It is also important to emphasize that the proximity of pollution sources, the population size of the settlement, and the proximity of cultural heritage in relation to them influence residents' attitudes toward the quality of the environment and their perception of the level of pollution of natural resources. In addition, the lack of knowledge about pollution sources, polluting materials, and their harmful effects can affect residents' attitudes toward environmental quality.

Table 4 Differences in residents' perceptions according to the gender

Residents' perceptions	Mean		*t*-value	*p*
	Male ($n = 106$)	Female ($n = 139$)		
Air quality	2.83	3.14	−1.971	0.050
Surface water quality	2.94	3.22	−1.910	0.057
Groundwater quality	2.77	2.95	−1.203	0.230
Soil quality	2.83	3.00	−1.195	0.233

(created by Milanović Pešić, A.)

Table 5 Differences in residents' perceptions according to the age

Residents' perceptions	Mean			F-value	p
	Young (18 –30) ($n = 33$)	Middle-aged (31–50) ($n = 134$)	Elderly (51+) ($n = 78$)		
Air quality	2.91	3.07	2.95	0.344	0.709
Surface water quality	3.06	3.07	3.17	0.218	0.805
Groundwater quality	2.73	2.84	3.00	0.829	0.438
Soil quality	2.91	2.90	2.97	0.107	0.898

(created by Milanović Pešić, A.)

Table 6 Differences in residents' perceptions according to the education

Residents' perceptions	Mean			F-value	p
	High school ($n = 86$)	Faculty ($n = 70$)	MSc and PhD ($n = 89$)		
Air quality	2.80	3.09	3.15	1.882	0.155
Surface water quality	2.97	3.29	3.08	1.634	0.197
Groundwater quality	2.84	3.07	2.75	1.616	0.201
Soil quality	2.79	3.04	2.97	1.101	0.334

(created by Milanović Pešić, A.)

Conclusion

Tangible cultural heritage represents a significant wealth of any country, both for preserving the identity of a particular nation and as important tourist sites for economic development. Considering the above, it is very important to examine the environment in their vicinity because polluted natural resources can contribute to their degradation. This study analyzes the state of the environment in the cultural heritage zones of Serbia based on the residents' perceptions.

The survey research results indicated that a third of respondents believed that the environment near their municipality's most significant cultural heritage is moderately polluted. Analyzing the residents' perceptions toward the pollution level of natural resources, it is concluded that more than a third of respondents have the opinion that surface water, groundwater, soil, and air are moderately polluted. A quarter of respondents thought that these four natural resources are significantly polluted. Most respondents singled out air as a heavily polluted natural resource. In addition, more than half of the respondents believe that intensive construction endangers tangible cultural heritage in their municipality, and almost half believed that wild landfills endanger it.

The obtained results can serve as an additional source of data on the environment state at the local level, recognized sources of pollution, and dangers of cultural heritage degradation. This can serve stakeholders and decision-makers in the sustainable management of cultural heritage, as well as environmental protection. Some of the further research on this topic will include further survey research with more socio-economic variables (such as income and profession).

Acknowledgements The research was supported by the Ministry of Science, Technological Development and Innovation of the Republic of Serbia (Grant No. 451-03-136/2025-03/200172) and the Ministry of Culture of the Republic of Serbia (Grant No. 451-04-1564/2023-02).

The authors would like to thank colleague Dr. Milovan Milivojević from the Geographical Institute "Jovan Cvijić" SASA for his valuable advice and assistance in creating the map, as well as for providing a photograph of the Visoki Dečani Monastery.

References

Abukari H, Mwalyosi RB (2020) Local communities' perceptions about the impact of protected areas on livelihoods and community development. Global Ecolol Conserv 22:e00909. https://doi.org/10.1016/j.gecco.2020.e00909

Agisilaos E (2012) Cultural heritage, tourism and the environment. Sustain Dev Cult Traditions J 1A:70–78. https://doi.org/10.26341/issn.2241-4002-2012-1a-6

Badola R, Barthwal S, Hussain SA (2012) Attitudes of local communities towards conservation of mangrove forests: a case study from the east coast of India. Estuarine, Coastal and Shelf Sciences 96:188–196. https://doi.org/10.1016/j.ecss.2011.11.016

Bjeljac Ž, Brankov J, Ćurčić N (2021) Traditional sports and games as an element of intangible cultural heritage. Issues Ethnol Anthropol 16(4):1257–1283. https://doi.org/10.21301/eap.v16i4.12

Bond A, Langstaff L, Baxter R, Kofoed HGWJ, Lisitzin K, Lundström S (2004) Dealing with the cultural heritage aspect of environmental impact assessment in Europe. Impact Assess Proj Apprais 22(1):37–45. https://doi.org/10.3152/147154604781766085

Brankov J, Micić J, Ćalić J, Kovačević-Majkić J, Milanović R, Telbisz T (2022) Stakeholders' attitudes toward protected areas: the case of Tara National Park (Serbia). Land 11(4):468. https://doi.org/10.3390/land11040468

Brankov J, Milanović Pešić A, Milijašević Joksimović D, Radovanović MM, Petrović DM (2021) Water quality estimation and population's attitudes: a multi-disciplinary perspective of environmental implications in Tara National Park (Serbia). Sustainability 13(1):241. https://doi.org/10.3390/su13010241

Cigizoglu HK, Bayazit M, Önöz B (2005) Trends in the maximum, mean, and low flows of Turkish rivers. J Hydrometeorol 6(3):280–290. https://doi.org/10.1175/JHM412.1

Costache AM, Sencovici M (2015) Influence of the socio-demographic variables on environmental perception. case study: Targoviste (Dambovita County, Romania). In: SGEM2015 conference proceedings, Book5 vol 1, 18–24 June 2015; Albena, Bulgaria International Multidisciplinary Scientific GeoConference, pp 431–438

Council of the European Union (1997) Council Directive 97/11/EC of 3 March 1997 amending Directive 85/337/EEC on the assessment of the effects of certain public and private projects on the environment. Official J European Communities 40(L73):5–14. https://op.europa.eu/en/publication-detail/-/publication/d0772bfd-51a4-4437-a5f2-752f46212a58

Council of Europe (2005) Council of Europe framework convention on the value of cultural heritage for society. Council of Europe Treaty Series—No 199. https://rm.coe.int/1680083746

Ćurčić N, Mirković Svitlica A, Brankov J, Bjeljac Ž, Pavlović S, Jandžiković B (2021) The role of rural tourism in strengthening the sustainability of rural areas: the case of Zlakusa Village. Sustainability 13(12):6747. https://doi.org/10.3390/su13126747

Cvetković M, Brankov J, Ćurčić N, Pavlović S, Dobričić M, Tretiakova TN (2023) Protected natural areas and ecotourism—priority strategies for future development in selected serbian case studies. Sustainability 15(21):15621. https://doi.org/10.3390/su152115621

Das D, Hussain I (2016) Does ecotourism affect economic welfare? Evidence from Kaziranga National Park, India. J Ecotourism 15(3):241–260. https://doi.org/10.1080/14724049.2016.1192180

Diocese of Raška-Prizren (2017) Manastir Gračanica [Gračanica Monastery]. Sisterhood of the Gračanica Monastery

Environmental Protection Agency (2020) Ministry of Environmental Protection Republic of Serbia. Izveštaj o stanju zemljišta u Republici Srbiji 2018–2019. [Land state report in the Republic of Serbia 2018–2019]. https://degradacijazemljista.sepa.gov.rs/wp-content/uploads/2023/04/Zemljiste_18_19.pdf

Etongo D, Lafleur H, Vel T (2023) Community perceptions and attitudes towards the management of protected areas in Seychelles with Morne Seychellois National Park as case study. World Dev Sustain 3:100091. https://doi.org/10.1016/j.wds.2023.100091

Foster G, Kreinin H (2020) A review of environmental impact indicators of cultural heritage buildings: a circular economy perspective. Environ Res Lett 15:043003. https://doi.org/10.1088/1748-9326/ab751e

Gajić T, Vukolić D, Spasojević A, Blešić I, Petrović MD, Bugarčić J, Bugarčić M, Drašković BD, Milivojević M (2025) Exploring attitudes on the sustainable balance between nature conservation and economic development through ecotourism—lessons from EU and Non-EU Countries. Land 14:395. https://doi.org/10.3390/land14020395

Golinelli GM (2016) Patrimonio Culturale e Creazione di Valore. La Componente Naturalistica. CEDAM

Govindan K, Garg K, Gupta S, Jha PC (2016) Effect of product recovery and sustainability enhancing indicators on the location selection of manufacturing facility. Ecol Indic J 67:517–532. https://doi.org/10.1016/j.ecolind.2016.01.035

Hagemann L, Kašanin-Grubin M, Gajica G, Štrbac S, Šajnović A, Jovančićević B, Vasić N, Schwarzbauer J (2019) Four decades of organic anthropogenic pollution: a compilation for Djerdap Lake sediments, Serbia. Water Air Soil Pollution 230:246. https://doi.org/10.1007/s11270-019-4277-8

ICOMOS (2002) Principles and guidelines for managing tourism at places of cultural and heritage significance. In: International council on monuments and sites. International Cultural Tourism Committee. https://www.icomos.org/en/89-english-categories/home/118410-the-new-icomos-international-charter-for-cultural-heritage-tourism-supports-a-more-responsible-and-sustainable-tourism-management

Institute for the Protection of Cultural Monuments of Serbia (2025) Central Catalogue. https://www.heritage.gov.rs/english/nepokretna_kulturna_dobra.php

Ismail AH, Robescu D (2019) Assessment of water quality of the Danube river using water quality indices technique. Environ Eng Manag J 18(8):1727–1737. https://doi.org/10.30638/eemj.2019.163

Jakovljević D, Milanović Pešić A, Milijašević JD (2019) Water resources management issues in the Danube River Basin District—examples from Serbia. Water Resour 46(2):286–295. https://doi.org/10.1134/S0097807819020039

Lavega P, Alcaraz Munoz V, Mallen Lacambra C, Pic M (2023) Roles, relationships, and motor aggressions: keys to unveiling the emotions of a traditional sporting game. Front Psychol 14:1127602. https://doi.org/10.3389/fpsyg.2023.1127602

Lin Q, Hou P (2024) Spatio-temporal evolution of cultural heritage and its response to the geographical environment: a case study of Fujian Province. Environ Res: Infrastruct Sustain 4(4):2–17. https://doi.org/10.1088/2634-4505/ad92a6

Lysenko-Ryba K, Zimon D, Zatwarnicka-Madura B (2024) The influence of socio-demographic characteristics on environmental awareness. Int J Q Res 18(4):1259–1268. https://doi.org/10.24874/IJQR18.04-22

Malinović-Milićević S, Stanojević G (2024) Zagađenje vazduha u Srbiji. [Air pollution in Serbia] Geographical Institute "Jovan Cvijić" SASA

Mateos AB, Leco F, Pérez A (2020) Visitors' perception of the overcrowding of a protected natural area: a case applied to the natural ("Garganta de los Infiernos" (Caceres, Spain). Sustainability 12:9503. https://doi.org/10.3390/su12229503

Milanović Pešić A, Brankov J, Milijašević JD (2020) Water quality assessment and populations' perceptions in the National Park Djerdap (Serbia): key factors affecting the environment. Environ Dev Sustain 22:2365–2383. https://doi.org/10.1007/s10668-018-0295-8

Milanović Pešić A, Jakovljević D, Rajčević V, Gnjato S (2025) Assessment of hydroclimatic trends in Southeast Europe—examples from two adjacent countries (Bosnia & Herzegovina and Serbia). IDŐJÁRÁS—Q J Hungarian Meteorol Service 129(1):69–87. https://doi.org/10.28974/idojaras.2025.1.5

Milanović Pešić A, Jojić Glavonjić T, Denda S, Jakovljević D (2023) Sustainable tourism development and Ramsar Sites in Serbia: exploring residents' attitudes and water quality assessment in the Vlasina protected area. Sustainability 15(21):15391. https://doi.org/10.3390/su152115391

Mohaddes Khorassani S, Ferrari AM, Pini M, Blundo DS, García Muiña FE, García JF (2019) Environmental and social impact assessment of cultural heritage restoration and its application to the Uncastillo Fortress. Int J Life Cycle Assess 24:1297–1318. https://doi.org/10.1007/s11367-018-1493-1

Moses A, McLain JET, Kilungo A, Root RA, Abrell L, Buxner S, Sandoval F, Foley T, Jones M, Ramírez-Andreotta M (2022) Minding the gap: socio-demographic factors linked to the perception of environmental pollution, water harvesting infrastructure, and gardening characteristics. J Environ Stud Sci 12:594–610. https://doi.org/10.1007/s13412-022-00769-7

Netzel LM, Heldt S, Engler S, Denecke M (2021) The importance of public risk perception for the effective management of pluvial floods in urban areas: a case study from Germany. J. Flood Risk Manag 14(2):e12688. https://doi.org/10.1111/jfr3.12688

Orthodox Metropolitanate of Montenegro and the Littoral (2011a) Manastir Gračanica [Gračanica Monastery]. Group of authors, IIU Svetigora

Orthodox Metropolitanate of Montenegro and the Littoral (2011b) Manastir Visoki Dečani [Visoki Dečani Monastery]. Group of authors, IIU Svetigora

Parr A (2016) Capital, environmental degradation, and economic externalization. The Oxford handbook of environmental political theory. Oxford University Press

Popescu RI, Corbos RA (2010) The role of urban tourism in the strategic development of Brasov area. Theor Empir Res Urban Manag 7(16):69–85

Qasim S, Khan AN, Shrestha RP, Qasim M (2015) Risk perception of the people in the flood prone Khyber Pukhthunkhwa province of Pakistan. Int J Disaster Risk Reduction 14(4):373–378. https://doi.org/10.1016/j.ijdrr.2015.09.001

Sablier M, Garrigues P (2014) Cultural heritage and its environment: an issue of interest for environmental science and pollution research, editorial. Environ Sci Pollut Res 21:5769–5773. https://doi.org/10.1007/s11356-013-2458-3

Šiljeg S, Milošević R, Panđa L (2022) Public perception of the urban pluvial floods risk—case study of Poreč (Croatia). J Geogr Inst Jovan Cvijić SASA 72(2):147–158. https://doi.org/10.2298/IJGI2202147S

Takić L, Mladenović-Ranisavljević I, Vasović D, Lj Đ (2017) The assessment of the danube river water pollution in Serbia. Water Air Soil Pollution 228:380. https://doi.org/10.1007/s11270-017-3551-x

Telbisz T, Ćalić J, Kovačević-Majkić J, Milanović R, Brankov J, Micić J (2021) Karst Geoheritage of Tara National Park (Serbia) and its geotouristic potential. Geoheritage 13:88. https://doi.org/10.1007/s12371-021-00612-5

Teller J, Bond A (2002) Review of present European environmental policies and legislation involving cultural heritage. Environ Impact Assess Rev 22(6):611–632. https://doi.org/10.1016/S0195-9255(02)00009-4

Terzić A, Bjeljac Ž (2016) Cultural routes—cross-border tourist destinations within Southeastern Europe. Forum Geogr. 15(2):180–188. https://doi.org/10.5775/fg.2067-4635.2015.041.d

Terzić A, Bjeljac Ž, Ćurčić N (2015) Common histories, constructed identities: intangible cultural heritage and the rebranding of Serbia. Int J Intang Herit 10:101–120

Trakolis D (2001) Local people's perceptions of planning and management issues in Prespes Lakes National Park, Greece. J Environ Manag 61(3):227–241. https://doi.org/10.1006/jema.2000.0410

Van Grieken R, Delalieux F, Gysels K (1998) Cultural heritage and the environment. Pure Appl Chem 70(12):2327–2331. https://doi.org/10.1351/pac199870122327

UNESCO (1954) Convention for the protection of cultural property in the event of armed conflict with regulations for the execution of the convention. https://www.unesco.org/en/legal-affairs/convention-protection-cultural-property-event-armed-conflict-regulations-execution-convention

UNESCO (1972) World heritage convention. https://whc.unesco.org/en/convention/

UNESCO (2025a) World heritage convention. properties inscribed on the World Heritage List in Serbia. https://whc.unesco.org/en/statesparties/rs

UNESCO (2025b) World heritage convention. Tentative List, Serbia. https://whc.unesco.org/en/tentativelists/?action=listtentative&state=rs&order=states

Vodouhê FG, Coulibaly O, Adégbidi A, Sinsin B (2010) Community perception of biodiversity conservation within protected areas in Benin. Forum Policy Economics 12(7):505–512. https://doi.org/10.1016/j.forpol.2010.06.008

Xu J, Chen L, Lu Y, Fu B (2006) Local people's perceptions as decision support for protected area management in Wolong Biosphere Reserve, China. J Environ Manage 78(4):362–372. https://doi.org/10.1016/j.jenvman.2005.05.003

Yang D, Shi X, Marsh P (2014) Variability and extreme of Mackenzie River daily discharge during 1973–2011. Quatern Int 380:159–168. https://doi.org/10.1016/j.quaint.2014.09.023

Zhang J, Yu X (2021) Analysis of land use change and its influence on runoff in the Puhe River Basin. Environ Sci Pollut Res 28:40116–40125. https://doi.org/10.1007/s11356-020-09798-7

Žarnić R, Rajčić V, Skordaki N (2015) A contribution to the built heritage environmental impact assessment. ISPRS Ann Photogramm Remote Sensing Spatial Inf Sci II-5/W3:389–394, https://doi.org/10.5194/isprsannals-II-5-W3-389-2015

Can They Endure? The Role of Tourism and Community Engagement in Preserving Traditional Sports and Games

Jovana Brankov, Ana Milanović Pešić, and Željko Bjeljac

Abstract Traditional sports and games (TSGs) represent a heritage from the past that has been preserved into the modern era. After being included on the UNESCO Intangible Cultural Heritage (ICH) List, TSGs were formally acknowledged as an important aspect of cultural heritage. TSGs are an integral part of Serbian heritage and identity, and also of other peoples of the former SFR Yugoslavia. Unlike some other countries in the region, TSGs in Serbia are not listed in the National Register of ICH, indicating a lack of recognition of their importance. The promotion of TSGs in Serbia today is primarily achieved through event tourism. This research analyzes the results of a national project aimed at preserving and revitalizing TSGs while highlighting their endangered status. The survey results confirmed that only a third of survey respondents (35%) attended an event that promotes TSGs, highlighting the need to promote such events and boost community interest. The vast majority of respondents (over 80%) agree that TSGs can be marketed through tourism. Further analysis confirmed that women and respondents with lower education had more positive attitudes. The focus group identified the lack of knowledge transfer and weak support from the local community as major obstacles to the affirmation of TSGs.

Keywords Traditional sports and games · Event tourism · Community involvement · Survey · Focus group

J. Brankov (✉) · A. Milanović Pešić · Ž. Bjeljac
Geographical Institute "Jovan Cvijić" of the Serbian Academy of Sciences and Arts, Belgrade, Serbia
e-mail: j.brankov@gi.sanu.ac.rs

A. Milanović Pešić
e-mail: a.milanovic@gi.sanu.ac.rs

Ž. Bjeljac
e-mail: z.bjeljac@gi.sanu.ac.rs

A. Milanović Pešić
Department of Physical and Economic Geography, Faculty of Natural Sciences, L. N. Gumilyov Eurasian National University, Astana, Kazakhstan

L. Soares et al. (eds.), *Compromised Geography*, Key Challenges in Geography,
https://doi.org/10.1007/978-3-032-05201-8_20

Introduction

In the middle of the first decade of the XXI century, new definitions of cultural tourism appeared, based on the growing interest of tourists in destinations where the main tourist product was cultural heritage. As a result, a specific type of cultural tourism—cultural heritage tourism is gaining importance. According to Richards (2021), the expanding scope of cultural tourism, its expanding impact on both people and places, and its increasing connections to issues of cultural heritage and identity have resulted in a shift in research focus. This shift moved away from traditional management and marketing approaches toward greater emphasis on cultural studies and anthropology. Cultural heritage can be identified as part of the past, which is presented in the present, all together through cultural, socio-political, and economic aspects (Graham et al. 2000; Terzić and Bjeljac 2014).

At the end of the 80 s of the XX century, TSGs became increasingly popular, both in practice (by organizing various competitions, establishing numerous sports associations), and as part of various scientific and professional research. In the beginning, sports and recreational competitions were organized, and TSGs were studied through sociological, anthropological, ethnological research, and through the disciplines of sports science. However, since the '90 s of the XX century, and especially during the first two decades of the XXI century, TSGs in the world are seen as part of the cultural heritage and significant tourist potential.

This term is widely used to refer to the culturally significant forms of play and movement of specific cultural groups. However, using it as a broad term can overlook the unique differences in how these games are structured, what they mean, and how they function in different cultures (Edwards 2012).

Cultural and sports tourism in many aspects are in a complex, causal connection, not only as industries, but also as world phenomena. Gammon and Ramshaw (2013) emphasize that the convergence of heritage, sport, and tourism involves many different and diverse fields, including sport tourism, heritage tourism, sport management, and sports geography. All these segments have a significant impact on the socio-cultural and political life on the globe. Depending on the size and importance of sports event, sport has a different role in the context of promoting tourist destination, as well as different values. In addition to professional sports (competition) events, sports and recreational events are also organized. They are characterized by large mass attendance, enjoyment, and socializing, testing one's own strength and health preservation (Bjeljac et al. 2017). Some sports and games have been held for several thousand years and are the backbone of various social communities. They represent a significant tourist potential by their authenticity and qualitative programs through the development of tourism and presentation of cultural heritage. TSGs range from simple, spontaneous activities to more organized, sport-like competitions often held at larger social events (Edwards 2012).

Tourism is a powerful sphere of action for showcasing the identity and cultural diversity inherent in TSGs. When presented effectively, TSGs in tourism can play a key role in preserving and promoting cultural heritage, helping to safeguard these valuable traditions for future generations (Lavega 2019). TSGs have found their place in the tourist offer as a segment of complex tourist products. Araya-Góchez and associates (Araya-Góchez et al. 2018) highlight the strong connection in the traditional sports–cultural heritage–tourism triangle. Ramshaw (2014) highlighted a strong connection between sport, tourism, and heritage. He also suggested that sport offers insight into cultural identity and serves as a means of remembering past conflicts and deepening our understanding of different cultures. It is also important to point out that sports allow to publicize the culture and heritage of a certain place. Eichberg (2009) emphasize how relevant body culture is for the understanding of society and cultural diversity.

This chapter focuses on TSGs present in Serbia and on the territory of former SFR Yugoslavia. Due to the limited contemporary research on this subject and the endangered status of these sports in Serbia, the chapter aims to broaden understanding of TSGs, facilitate their proper presentation, and analyze the initial findings from the field research. The authors provided an overview of the concept and significance of TSGs within a global context, highlighting their importance in the former SFR Yugoslavia and, specifically, in Serbia. Additionally, the chapter presents the outcomes of a project centered on the education, knowledge transfer of TSGs, and their tourist promotion.

Literature Review and Chronology of Development

In the scientific literature worldwide, there are different definitions of TSGs. The one given by UNESCO and adopted at the Conference in Tehran in 2009 is generally accepted and most often used (Bronikowska and Groll 2015). According to this definition: "*Traditional games are motor activities of leisure and recreation which can have a ritual character. They are part of the universal heritage diversity. They are practiced in an individual or collective manner, deriving from regional or local identity; they are based on rules accepted by a group that organizes competitive or non-competitive activities. Traditional games dispose of a popular character in their practice and in their organization, yet if turned into sport tend to be uniform and institutionalized. The practice of traditional games promotes global health*" (UNESCO 2017a, pp. 1–2).

The most commonly used term in scientific and professional research worldwide, and characteristic of the English-speaking world, is "traditional sports and games". Also, in different parts of the world, other terms are used, such as: "ethnosports" (Kylasov and Gavrov 2020), "ethnic sports" (Sogawa 2006), "sport heritage" (Ramshaw and Gammon 2017), "indigenous games" (Vaczi and Bairner 2024), and "old folk sports, games, and competitions" (Bjeljac and Brankov 2024). The literature emphasizes that a common element of this diverse group of sports and games is their "popular" status or close association with folk culture. TSGs include particularly traditional, ethnic, or indigenous sports and games, but also new activities that are based on traditional practices (Eichberg 2012).

There are numerous classifications of TSGs based on various criteria, of which some are highlighted in this text: types of sports, types of materials used (sports equipment), calendar of events, geographical area of events, complexity of performing certain games, static/dynamic nature, venue (outdoor/indoor), historical–geographical and socio-political circumstances, socio-demographic characteristics, locomotor abilities, etc. (Bale 2022; Bjeljac et al. 2021a; Bronikowska and Groll 2015; Bronikowska et al. 2015; Nefil et al. 2021; Lavega et al. 2023). The following classification by European Association of Traditional Sports and Games (AEJeST) which was adopted by the Council of Europe in 2003, is often used. According to this classification, there are a total of eight categories of TSGs: ball games with interaction between the players; ball (or bowl) and pins games; throwing games; shooting games; fighting and jousting games; animal games; locomotion games, and acrobatics (Bronikowska and Groll 2015). This classification was developed based on various strategies and policies aimed at understanding cultural diversity and sports. However, its limitations are evident in practice, which also encompasses numerous other concepts and classifications of TSGs.

The development, role, and importance of TSGs as part of cultural heritage can be observed chronologically through numerous strategies, documents, and the formation of organizations. First, in 1978, the UNESCO General Assembly adopted *UNESCO's International Charter for Physical Education, Physical Activity and Sport*. In following period, various seminars and gatherings were held in several European countries, with the aim of promoting sports activities and traditional games. After that, in 1990, *The Association for International Sport for All (TAFISA)* was founded, within which TSGs programs were later launched.

During the '90 s of the XX century, the practice of holding festivals, gatherings, and symposiums continued. The meetings revealed a growing desire to establish a European Platform representing both the networks of TSGs confederations, federations, and associations, as well as academic researchers and educators focused on using these sports for educational purposes. This vision was implemented in 2001 with the creation of the *European Traditional Sports and Games Association (AEJeST)* (Jaouen 2022). A total of 18 countries, 53 sports associations and federations, 20 cultural associations and museums, 12 universities and 24 respectable professors and researchers from TSGs area are members of this organization (European Association for Traditional Sports and Games, n.d.).

Within the framework of the Third International Conference of Ministers and Senior Officials Responsible for Physical Education and Sport (MINEPS), the *Declaration of Puenta del Este* (Uruguay) was adopted, which supports the policy of preserving and promoting traditional sports and indigenous sports and games based on the cultural heritage of regions and nations, and encourages the realization of regional and world festivals (UNESCO 1999).

In 2003, UNESCO adopted the *Convention for the Protection of Intangible Cultural Heritage*, highlighting that globalization and social transformation pose threats to ICH, including its deterioration and loss, due to limited preservation resources. The Convention emphasizes the important role of communities, especially indigenous groups, in creating, preserving, and revitalizing such heritage, thereby enriching cultural diversity and human creativity (UNESCO 2003).

One of the objectives of the 2005 *Preliminary Report on the Desirability and Scope of an International Charter on Traditional Games and Sports,* within the framework of the General Conference UNESCO, 33rd session, was to integrate the socio-cultural dimension into physical education and sport. It was highlighted that in the modern world of globalization, cultural identities are under threat and that the general public has very little, if any, knowledge on the scope and richness of cultural heritage in terms of TSGs (UNESCO 2005). In 2006, UNESCO held a *Collective Consultation* in Paris, bringing together numerous experts from diverse institutions and scientific fields to establish an international platform for the promotion and development of TSGs. During this meeting, key objectives were outlined, including the cataloging of TSGs as living cultural heritage and exploring their applications in various contexts such as education, culture, environmental conservation, and sports (Lavega 2006; UNESCO 2006).

The year 2009 is significant for the chronology of the development of TSGs for several reasons. First, during the second *Collective Consultation* organized by UNESCO in Tehran (Iran), an *Advisory Committee* was formed to provide a platform for the exchange of knowledge and capacity-building on TSGs (UNESCO 2017a). Also, in 2009, the first TSG was included on the UNESCO List of ICH: Dragon Boat festival (China), and by 2025, 39 more TSGs will be included on this list (Table 1).

This was followed by the adoption of a *European Parliament resolution of February 2, 2012, on the European dimension in sport* by the European Parliament in 2012, which emphasized the recognition of TSGs as ICH and called on the European Commission to support TSGs (European Parliament 2012). It should be emphasized that despite this resolution, TSGs have been excluded from EU policy papers since the sport competence was established in 2007, raising concerns about the future and sustainable support for TSG within the EU (European Platform for Traditional Sports and Games 2017).

In 2014, TAFISA established regional sport centers to create a global network promoting TSGs through research and policy (TAFISA n.d.). Two years later, a mapping of TSGs in Europe identified 174 active organizations in this field and analyzed key initiatives across the EU (European Commission 2016). The *2017 Kazan Action Plan* aligned international policies with the UN's 2030 Agenda for sustainable development in physical education and sports (UNESCO 2017b). In 2019, the Foundation of the Institute for the Development of Sports and Education in Poland launched an online encyclopedia of TSGs, alongside an annual competition calendar featuring over 800 TSGs (Traditional Sports, n.d.-a). In 2022, a worldwide network of teachers and researchers in TSGs was formed, and the ITSGA established August 14 as the "International Day of Traditional Sports and Games."

Table 1 Traditional sports and games on the UNESCO List of ICH

Name	State	Entry year
Dragon boat festival	China	2009
Naadam, traditional Mongolian festival	Mongolia	2010
Pahlevani and Zoorkhanei rituals	Iran	2010
Kirkpinar, oil wrestling festival	Turkey	2010
Sinjska alka, a knights' tournament in Sinj	Croatia	2010
Taekkyen, traditional Korean martial art	Republic of Korea	2011
Jultagi Tightrope walking	Republic of Korea	2011
Program of cultivating ludodiversity: safeguarding traditional games in Flanders	Belgium	2011
Equitation in the French tradition	France	2011
Chovgan, a traditional Karabakh horse-riding game	Azerbaijan	2013
Mongolian knuckle-bone shooting	Mongolia	2014
Capoeira circle	Brazil	2014
Tugging rituals and games	Cambodia, Philippines, Republic of Korea, Vietnam	2015
Classical horsemanship and the High School of the Spanish Riding School Vienna	Austria	2015
Kuresi, wrestling	Kazakhstan	2016
Tahteeb, stick game	Egypt	2016
La charrería, tradición ecuestre en México	Mexico	2016
Kok boru, traditional horse game	Kyrgyzstan	2017
Kazakh traditional Assyk game	Kazakhstan	2017
Chogān, a horse-riding game accompanied by music and storytelling	Iran	2017
Taskiwin, martial dance of the western High Atlas	Morocco	2017
Traditional Korean wrestling (Ssirum/Ssireum)	Republic of Korea, Democratic People Republic of Korea	2018
Hurling	Ireland	2018

(continued)

Table 1 (continued)

Name	State	Entry year
Horse and camel Ardhah	Oman	2018
Chidaoba wrestling in Georgia	Georgia	2018
Alpinism	Italy, France, Switzerland	2019
Traditional Turkish archery	Turkey	2019
Silat	Malaysia	2019
Traditions of Pencak Silat	Indonesia	2019
Traditional intelligence and strategy game: Togyzqumalaq, Toguz Korgool, Mangala/Göçürme	Kazakhstan, Kyrgyzstan, Turkey	2020
Camel racing, a social practice and festive heritage associated with camels	Oman, United Arab Emirates	2020
Wine horses	Spain	2020
Taijiquan	China	2020
Nomad games, rediscovering heritage, celebrating diversity	Kyrgyzstan	2021
Falconry	Austria, Belgium, Croatia, Czechia, France, Germany, Hungary, Ireland, Italy, Kazakhstan, Kyrgyzstan, Republic of Korea, Mongolia, Morocco, Nederland, Pakistan, Poland, Portugal, Qatar, Saudi Arabia, Slovakia, Spain, Syria	2021
Lipizzan horse breeding traditions	Austria, Bosnia and Herzegovina, Croatia, Hungary, Italy, Romania, Slovakia, Slovenia	2022
Pehlevanliq culture: traditional zorkhana games, sports, and wrestling	Azerbaijan	2022
Tocatì, a shared program for the safeguarding of traditional games and sports	Belgium, France, Croatia, Italy, Cyprus	2023
Traditional folk games of Uzbekistan and Tajikistan	Uzbekistan, Tajikistan	2024
Equestrian art in Portugal	Portugal	2024
Nawrouz (Novruz, Nowrouz, Nowrouz, Nawrouz, Nauryz, Nooruz, Nowruz, Navruz, Nevruz, Nowruz, Navruz) rituals, and games	Afghanistan, Azerbaijan, India, Iran, Iraq, Kazakhstan, Kyrgyzstan, Mongolia, Turkey, Uzbekistan, Pakistan, Tajikistan, Turkmenistan	2024

Source UNESCO (2024)

TSGs as World Intangible Cultural Heritage

Starting from the first half of the XX century, ICH is gaining more importance and becoming a significant tourist resource, i.e., the local cultural heritage is turned into new products for trade in the global tourism market (Bjeljac et al. 2014; Kim et al. 2019). ICH is what reflects national identity in the best possible way as something unique and common to a nation, and is a particularly attractive element of a country's cultural and tourist offer. The promotion of ethnic and ethnological elements, which become clear icons of the destination, is highly desirable from the aspect of tourism development. UNESCO emphasizes that ICH can be recognized in oral traditions and language, landscape as a part of the visual arts, social practices, rituals and festivities, traditional knowledge and applied knowledge about nature and the universe, and traditional arts and crafts (UNESCO 2003). The recognition of TSGs as specific practices is linked to the importance of the country and its local culture. Aligned with tourism strategies of recent decades, these intangible cultural goods are seen as valuable assets and play a key role in regional or national tourism promotion and development policies (Alonso et al. 2020).

TSGs were officially recognized as a significant segment of cultural heritage and local identity by UNESCO at the beginning of the XXI century. During this period, the inclusion of TSGs on the UNESCO List of ICH begins, with the first sports entered the list in 2009. So far, a total of 41 traditional sports and games from different parts of the world have been included (Table 1).

Historical migrations and political, social, and religious events have caused a mixture of different peoples and ethnic groups. The area's geography and environment shaped settlement types and influenced lifestyles, with many people traditionally living nomadically. In such settings, folk competitions became important to showcase martial and military abilities. As a result, certain TSGs included in the previous List, highlight warrior skills, often combining elements of fighting, dance, and music. They are often held in the form of festivals, preserving oral traditions, religious customs and folklore, promoting national cuisine and crafts, with special specificities for each country or people. Also, Masaeli and colleagues (Masaeli et al. 2020) emphasize the spiritual dimension of certain TSGs, as a special segment of sports tourism, which should receive more attention in research on tourism as an attraction.

Animals played a major role in the nomadic/semi-nomadic way of life, and, depending on the area, they form an integral part of the sports and games included on the UNESCO List. In addition, certain games and sports are closely linked to specific rituals and customs (Table 1). In general, most of the listed sports also represent tourist attractions, visited by millions, since they have not only sports, but also cultural and tourist significance. For example, the Dragon boat festival in China was visited by 110 million domestic tourists in 2024 (The State Council of the People's Republic of China 2024).

In addition to the listed sports of world importance as ICH, it is worth mentioning TSGs that represent an element of national, ethnic, and continental cultural heritage, such as traditional aboriginal games in Australia ("*bubberah*" (boomerang)); *lacrosse* (and other traditional sports and games of Native Americans in USA), *kho-kho* (India), *field hockey* (Canada), *Scottish Highland games* (Scotland), *laamb* (West Africa), *bocce* (Italy), *huka huka* (Brazil), *ayo* (Nigeria), *ulama* (Mexico), *pato* (Argentina), and *garrote* (Spain) (Traditional Sports, n.d.-b). There are various studies dealing with the topic of the previously listed TSGs (Brewster et al. 2009; Edwards 2012; Wiser 2014; Aguilar-Moreno2015; Watson et al. 2024; Bandyopadhyay 2021; Giambartolomei 2023).

TSGs on the Territory of the Former SFR Yugoslavia

One of the most significant migration movements during the early Middle Ages in Europe was the settlement of South Slavic tribes, particularly the Serbs and Croats, on the Balkan Peninsula. They brought with them their culture, customs, and various forms of sports and competitions. Additionally, the cultural landscape was further shaped by the numerous other peoples who, alongside the South Slavs, settled in the territory of the former SFR Yugoslavia in later periods—such as the Turks, Germans, Hungarians, Romanians, Albanians, and Slovaks—each contributing their own traditions, customs, and sports. The coexistence and cultural mixture in this region resulted in a wide range of shared or similar games and competitions. These practices became not only part of the cultural heritage but also an essential aspect of the everyday life of the population across the former SFR Yugoslavia.

At the end of the first decade of the XXI century, TSGs in the states created on the territory of the SFR Yugoslavia began to be interpreted as ICH. Some countries (Bosnia and Herzegovina, Croatia, Slovenia, and North Macedonia) have recognized the importance of these sports and included them on their national lists of intangible heritage (Table 2), while in others (Montenegro, Serbia) this is still not the case.

Given the nomadic and semi-nomadic way of life of the South Slavs who moved to the territory of the former SFR Yugoslavia, primarily in the hilly and mountainous regions, but also in the lowlands, the main role in the development of TSGs was initially played by shepherds. From their creation of rules, shepherds' games were born, which later grew into sports competitions for young people during numerous church celebrations and other social gatherings, most often in rural areas. These games were transmitted orally and became part of the folk tradition. Such contests are also known as folk all-around. In addition to this, similarly to Western and Central Europe, nobles organized numerous knightly games in these areas, which also influenced the later development of TSGs (Bjeljac and Brankov 2024). As part of the oral tradition in the context of TSGs, heroic epic poems that included a romanticized depiction of heroes, fighters against the occupation of numerous empires that ruled this area should also be mentioned (Bjeljac and Brankov 2024; Bjeljac et al. 2021b).

The promotion of TSGs today is mainly carried out through tourist events within the offer of rural and sports-recreational tourism, together with other elements of cultural heritage. In this way, contributions are made to the current recognition of

Table 2 TSGs as elements of ICH in the territory of the former SFR Yugoslavia

Name	State
Sava Valley štehvanje, equestrian tournament	Slovenia
Pandolo, social, and sporting game	Slovenia
Walk on the path of remembrance and comradeship	Slovenia
Traditional breeding and raising of Lipizzan horses	Slovenia, Croatia, Bosnia, and Herzegovina
Tocatì, a shared program for the safeguarding of TSGs (Škuljanje, Pljočkanje)	Slovenia, Croatia
Sinjska Alka, a knights' tournament in Sinj	Croatia
Falconry	Croatia, Slovenia
Trka na prstenac, equestrian game	Croatia
Moreška, battle fencing dance	Croatia
Kumpanija on the island of Korčula, fencing dance	Croatia
Picigin, social game with ball	Croatia
The custom of Easter gun shooting in Kostel	Croatia
The art of sailing with lateen and lug sails along the Croatian coast	Croatia
Traditional and shepherds' games in Lika	Croatia
Nevesinje Olympiad	Bosnia and Herzegovina
Vidovdan Olympiad	Bosnia and Herzegovina
Dajak boat building skills	Bosnia and Herzegovina
Jumps from the Old Bridge in Mostar	Bosnia and Herzegovina
Traditional oil wrestling competition	Republic of North Macedonia
Rusalii processions in the Republic of Macedonia, ritual dance with stick	Republic of North Macedonia

Sources Koordinator varstva nesnovne kulturne dediščine Republika Slovenija (2024); Ministarstvo kulture i medija Republike Hrvatske (n.d.); Vlada Federacije Bih, Federalno Ministarstvo kulture i sporta (n.d); Republika Severna Makedonija, Uprava za zaštita na kulturnoto nasledstvo (n.d.)

similarities and the building of new sports and cultural cooperation and togetherness. According to the author's research, around 100 TSGs events are held on the territory of the former SFR Yugoslavia today. In addition to those recognized on the national lists of ICH, the following events particularly stand out: the Ancient Sports Olympiad (Brođanci, Croatia), the Roman Games of Poetovio (Ptuj, Slovenia), the Noblegames (Nova Gorica, Slovenia), the Uskok Battle for Klis (Klis, Croatia), and the Bocce Olympics (Tivat, Montenegro). Also significant are the village Olympics, which primarily feature competitions in folk all-around disciplines, as well as events dedicated to heroic sports and competitions practiced by the hajduks and uskoks.

Some of these events are also a place where knowledge is transferred between experts of different profiles. Such an example is the International Pljočke Festival (Svetvinčenat, Croatia), which includes experts' conferences and panel discussions on the role of TSGs in the contemporary life of local communities, formal education, etc.

TSGs as Intangible Cultural Heritage in Serbia

In the context of TSGs in Serbia, it is emphasized that these practices are integral to the national tradition and identity, with their origins and underlying circumstances mirroring those of neighboring countries within the former SFR Yugoslavia. The factors contributing to this shared heritage are multifaceted, including geographical location, historical conditions, and cultural interconnections. However, unlike other countries of the former SFR Yugoslavia, which have acknowledged the significance of these sports by incorporating them into their lists of national ICH, Serbia has yet to do so.

Over the past two decades, these games have gained prominence in Serbia through event tourism, with various events organized to honor the knightly tradition associated with the hajduks' way of life, promote Serbia's cultural heritage, or explore other thematic interests. In some events, TSGs serve as the central element, while in others, they complement the broader program. Such events are typically organized by citizens' associations, sports federations, tourist organizations, and local governments (municipal offices). These events usually take place between May and September, with August being the most common month for their occurrence (Bjeljac and Brankov 2024) (Photo 1).

The author's research so far has confirmed that 42 TSGs are held on the territory of Serbia today, within 39 events. The following particularly stand out: The Olympics of Old Sports of Vojvodina (Bačko Gradište, municipality of Bečej), "World Championship in Old Shepard's Game Činga" (Kumane, municipality of Novi Bečej), Lika Olympiad (Junaković Spa, municipality of Apatin), Belmužijada event (Svrljig), Prayer under Midžor (Balta Berilovac, municipality of Knjaževac), Hajduk's Evening (Crna Bara, municipality of Bogatić), Old national sports games of Bosnians (Novi Pazar), Shepard's Days (Kosjerić), etc. In some cases, TSGs are an integral part of the events promoting national ICH in Serbia (trumpet playing, oral tradition and gusle playing, Đurđevdan festivities, etc.).

Photo 1 Stone put, a competition within the event "Shepherd's Days" Kosjerić, Serbia. *Source* The authors

The Project: Old Folk Sports, Games, and Competitions—Preservation, Transfer of Knowledge, and Presentation

Considering the importance of TSG is insufficiently recognized in Serbia, the Geographical Institute "Jovan Cvijić" SASA, together with the Culture Center "Vuk Karadžić" (Loznica City), implemented a project called *Old folk sports, games, and competitions—preservation, transfer of knowledge, and presentation.* The project was funded by the national Ministry of Culture of the Republic of Serbia and realized in the period July 2023–February 2024. The main goal of the project was to revitalize and "revive" specific sports and games through the education of the local population. Project activities included three basic groups of tasks: implementation of a series of workshops, survey research, and organization of a focus group (Fig. 1).

Workshops in elementary schools were implemented with the aim of transferring knowledge, demonstrating skills and organizing competitions on the field. One-day workshops were attended by students from higher grades (V-VIII), and were organized in three elementary schools in City of Loznica.

The workshops consisted of two parts. The first part included the implementation of lectures by experts of various profiles (geographers, ethnologists, and sports scientists), and was realized at the beginning of the workshops. The geographer gave an overview of the TSGs concept and historical background, the ethnologist introduced the workshop participants to traditional children's games, while students

Fig. 1 Basic activities of the project

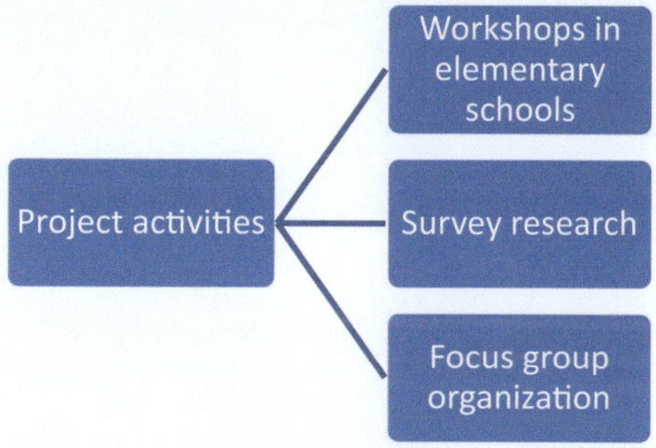

from the University of Belgrade—Faculty of Sport and Physical Education—demonstrated various TSGs using props. The second part of the workshops included the competition of participants in certain disciplines of TSGs (Photography 2 and 3). The workshops had an interactive character, with the participation of children in conversation and exercises. It was noticed that knowledge about TSGs was mainly transmitted to children orally by family members, but to a limited extent, meaning TSGs are generally not practiced today.

Photo 2 Workshops in elementary schools—demonstration of TSGs, City of Loznica. *Source* The authors, 2023

Photo 3 Workshops in elementary schools—TSGs competition, City of Loznica. *Source* The authors, 2023

Opinions on TSGs—The Analysis of Survey Results

The survey research was carried out to examine attitudes, previous experience, and knowledge about TSGs, as well the connection to tourism.

The survey method is frequently used to examine the population's opinions and attitudes on various topics. Surveys can be designed to address different topics within social geography (Brankov et al. 2019, 2022; Niemczewska 2021; Ghabouli et al. 2023), but they are also a useful tool in multidisciplinary geographical research (Moreno 2010; Lee and Lee 2015; Milanović Pešić et al. 2020, 2023; Brankov et al. 2021).

Online survey was used as data collection method, and the survey was distributed through different social networks. The questionnaire was initially posted on the Facebook page of the Geographical Institute "Jovan Cvijić" SASA. Subsequently, various cultural centers across Serbia (Loznica, Kladovo, Kosjerić, Smederevska Palanka, and Novi Sad) agreed to share the survey on their social media platforms to ensure a geographically diverse sample. The sample included 245 adult respondents, of whom 43.3% were male and 56.6% were female.

The questionnaire consisted of 18 questions, divided into thematic groups. Some questions were open-ended, while others used a five-point opinion scale. The first group reflected the socio-demographic and socio-economic structure of the respondents. The second group of questions was related to the knowledge on TSGs, as well as the previous experiences that the respondents had with these sports. The third group related to general attitudes about the benefits and presence of TSGs, while the fourth category referred to the knowledge of intangible cultural heritage, the promotion of TSGs through tourism, as well as the presence of marketing. For the purpose of this chapter, the focus will be on questions related to knowledge and tourism aspect of TSGs. Descriptive statistics (frequency, central tendency), as well as bivariate analysis (t-tests, ANOVA) were used for the data analysis. Questionnaire analysis was processed according to the specific socio-demographic variables, in order to determine whether there was a significant difference in attitudes among different groups of respondents.

First, the respondents were offered a list with the names of over 30 TSGs, and they were asked to mark which ones they knew from past experience or everyday life. As shown, only seven sports or games were identified by survey participants at a percentage greater than 35%. More than 60% TSGs on the offered list were entirely unknown to respondents (Fig. 2).

The research confirmed that the respondents are mostly familiar with certain children's games ("trule kobile," "janjine," sling shooting, etc.), as well as certain individual disciplines, such as arm wrestling and shepherd's game "klis." The participants in the survey research had never played most of the sports and games. When it comes to the ones they played, most of the respondents singled out the disciplines they recognized from everyday life, primarily certain children's games and folk all-around disciplines.

To determine whether the participants of the survey had the opportunity to attend a certain event or manifestation where TSGs are presented, or competitions in these sports are held, they were asked a question related to this topic. Results confirmed that most of the respondents (64.8%) did not have the opportunity to attend such an event, while only one third stated that they had attended (35.2%), which imposes the

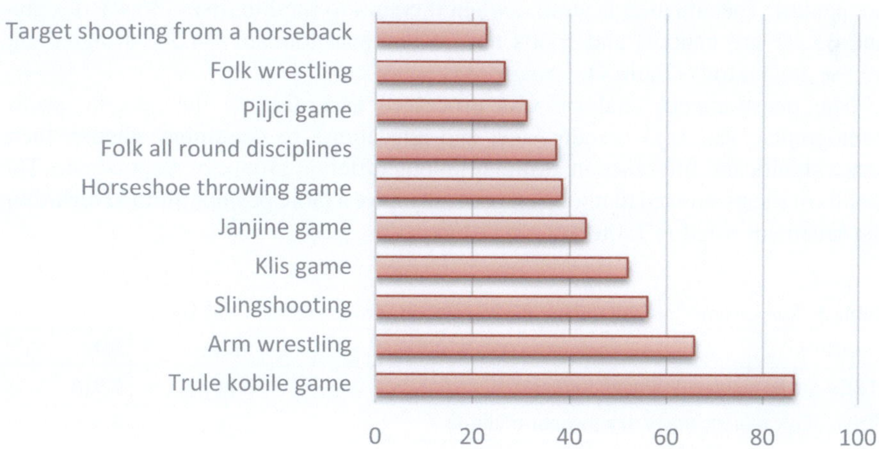

Fig. 2 Respondents' answers (in percentages) to the question: "Have you heard of the following sports/games?"

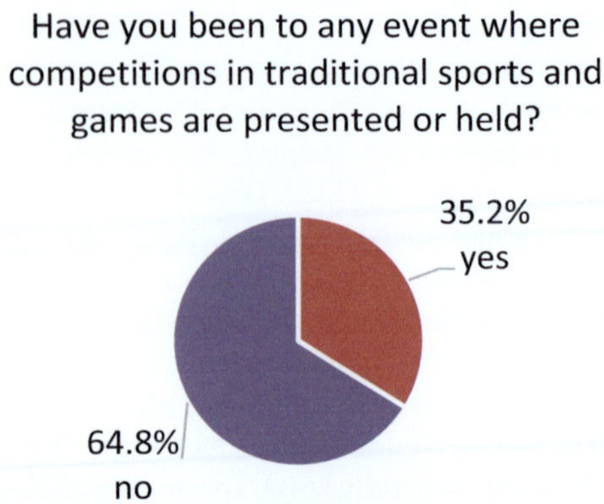

Fig. 3 Respondents' answers regarding visiting an event where TSGs are practiced and promoted

need for both a more comprehensive promotion of such events and for raising the awareness and interest of the population in this form of cultural heritage (Fig. 3).

A specific group of questions referred to the tourism aspect of TSGs. Respondents were given a series of statements and asked to indicate their degree of agreement using a five-point scale. The vast majority of respondents agreed that TSGs can be marketed through tourism (82%) and they make it possible for tourists to get to know our tradition (88%). When it comes to the ratio of TSGs values for domestic and foreign tourists, the respondents did not have a clear opinion on it, a large number had a neutral attitude (34.6%) (Table 3).

The next type of question was related to different ways of TSGs promotion. Respondents believe that TSGs are not adequately promoted through different types of tourism activities. Specifically, 39.6% of respondents disagree with the statement that TSGs are adequately promoted through the cultural tourism offer, while 38.1% are neutral. The situation is similar when it comes to rural tourism (36.4% disagree and 35.9% are neutral) and sports and recreational tourism (44.5% disagree and 33.5% are neutral) (Table 4).

The questionnaire analysis was processed according to the specific socio-demographic variables (gender, age, and education), to determine whether there was a significant difference in attitudes among different groups of respondents. The results of t-tests showed that women generally have a more positive attitude regarding the statement listed in Table 5.

Table 3 Respondents' attitudes about the connection between tourism and TSGs

Variables	Mean	SD
TSGs can be marketed through tourism	3.93	0.916
TSGs allow tourists to get to know our tradition	3.99	0.859
TSGs have more value for foreign than for domestic tourists	3.06	1.004

Scale 1—Strongly disagree; 2—Disagree; 3—Neutral; 4—Agree; 5—Strongly agree

Table 4 Respondents' attitudes toward the promotion of TSGs through tourism offer

TSGs are adequately promoted	Mean	SD
Through the offer of cultural tourism	2.78	0.950
Through the offer of rural tourism	2.87	0.998
Through the offer of sports and recreational tourism	2.73	0.991

Scale 1—Strongly disagree; 2—Disagree; 3—Neutral; 4—Agree; 5—Strongly agree

Table 5 Differences in attitudes by gender

Variables	M ($n = 106$)	W ($n = 139$)	t-value	p
TSGs allow tourists to get to know our tradition	3.84	4.11	−2.447	0.015

Table 6 Differences in attitudes by education

TSGs are adequately promoted	Elementary/secondary school ($n = 86$)	University and higher ($n = 159$)	t-value	p
Through the offer of cultural tourism	3.02	2.65	3.082	0.002
Through the offer of rural tourism	3.06	2.77	2.238	0.026
Through the offer of sports and recreational tourism	2.95	2.60	2.698	0.008

When it comes to education, statistically significant differences were registered in the statements related to the ways of promoting TSGs through different types of tourism. It was noted that respondents with secondary and elementary education have a more positive attitude than those with higher levels of education, which can be explained by the fact that the latter have greater knowledge about this topic, and therefore a more developed critical opinion (Table 6).

Results of the Focus Group

As part of the implementation of the project, a focus group was held, which included the participation of experts of various profiles, with the aim of collecting relevant opinions and answers regarding several key research problems, which were previously defined. The research topic included the following aspects: (a) the status of TSGs as an endangered segment of ICH, with special emphasis on their challenges and ways of protection; (b) the necessity of education about TSGs within different educational levels, in order to raise awareness and interest among new generations; (c) the affirmation of TSGs, as well as their preservation within the folk tradition, with the aim of revitalization and preservation of cultural identity. This research method enabled a deeper understanding of the problem and provided the foundations for further work on the protection and promotion of TSGs.

Regarding the first topic, the participants emphasized that it is necessary to improve the status of TSGs, which can be achieved by the implementation of national and international projects focusing on this topic. Given the endangered status of TSGs, the goal is the restoration of what is preserved in the old written sources or exists only in the memory of older people, so that current generations can become acquainted with the heritage of the past. Placing TSGs on the National list of ICH, as is the case with countries in the region, would also be one of the useful steps.

The need for education across various educational levels represents a crucial research issue that the focus group participants paid particular attention to. It was highlighted that UNESCO is actively working to enhance both formal and informal education, which should serve as a key framework for future initiatives. The school system could serve as an effective means of introducing and preserving TSGs, giving students a deeper understanding of the richness and tradition of national cultural heritage. This approach could encompass areas such as Serbian language, geography, history, physical education, and others. The importance of interaction with children was also highlighted, particularly in terms of fostering their engagement and participation in the practical application of these sports. It was emphasized that hands-on experiences, where children can actively participate in these traditional activities, play a crucial role in helping them better understand the cultural significance and historical context of TSGs. Engaging children in this way not only preserves the heritage but also helps them develop important skills such as teamwork, physical coordination, and an appreciation for cultural diversity. Additionally, such interactions create a more dynamic and memorable learning experience, ensuring that the knowledge and practice of these traditions are passed down to future generations. All of the above recommendations are in line with numerous international guidelines for the preservation and promotion of TSGs (UNESCO 2006, 2017a; Lavega et al. 2023). Given that such sports have their roots in rural environments, their promotion could be interpreted as a form of strategic planning for the development of these areas, based on sustainability (Ćurčić et al. 2021).

When it comes to affirmation, it was emphasized that it is inextricably linked to education. Several modern ways of promoting TSGs were highlighted, which should be more widely practiced; first, organization of the "team building" for companies, with the implementation of competitions in individual disciplines of TSGs, which, in addition to promoting sports, also serves as a unique and engaging way to foster teamwork, communication, and cooperation among employees; second, the hosting of sports-themed birthday parties for children, where competitions in TSGs are integrated into the celebration (Bjeljac and Brankov 2024). Other ideas included organizing community festivals, outdoor events, and recreational activities that incorporate TSGs, encouraging more widespread participation. Such initiatives help raise awareness and appreciation for TSGs in a modern context, ensuring they are both preserved and made relevant to contemporary audiences.

Conclusion

During the '90 s of the twentieth century, TSGs gained more importance in the world as a part of tradition and cultural heritage. Official UNESCO recognition started at 2009, and today TSGs represent an important element of cultural heritage and the overall tourist offer of the regions where they are held.

TSGs in the territory of the former SFR Yugoslavia (including Serbia) have a long tradition of being held. In Serbia, they are not registered in the National Register of ICH. As in the case of the countries in the region, their promotion in modern times is mainly done through different types of events of local character and the visitors are mostly from the hosting municipalities.

Survey research as part of the project realized in 2023–2024 indicated the necessity of preserving and revitalizing TSGs. The results confirmed that a large number of sports and games are completely unknown to the surveyed participants, which proves that they are largely forgotten and raise a need for stronger transfer of knowledge in the future. Only one-third of respondents attended an event that promotes TSGs which imposes the need for more comprehensive promotion of such events, as well as for raising awareness and interest of the population in this form of cultural heritage.

The results also confirmed that the status of TSGs should be improved by conducting future scientific research and implementing projects with the aim of identifying the social and economic mechanisms of the revival of these sports. Also, future research should explore the potential of integrating TSGs into formal and informal education systems, tourism development strategies, and community-based heritage programs. Investigating the role of TSGs in strengthening cultural identity, intergenerational exchange, and social cohesion could provide valuable insights for their sustainable preservation. Additionally, digital documentation and innovative approaches—such as digital storytelling—could be examined as tools for engaging younger generations and ensuring the continued transmission of this intangible cultural heritage.

This study has certain limitations that should be addressed in future research. It is important to note that this current research is primarily quantitative, focusing on examining the relationships between variables that shape attitudes toward different segments of knowledge on TSGs. However, as it captures perceptions without fully explaining their underlying causes, there is a clear need for complementary qualitative research. In this context, conducting in-depth interviews could serve as a valuable complement to the questionnaire by offering additional insights.

This research is also limited to a case study of respondents from Serbia. However, given the shared historical roots of TSGs in Serbia and other republics of the former Yugoslavia, the findings may hold broader relevance if similar studies are conducted across the region. In this context, comparative research within the former Yugoslav territory could highlight both regional similarities and differences, contributing to a deeper understanding of TSGs as a shared yet diverse cultural heritage.

Funding Ministry of Science, Technological Development and Innovation, Republic of Serbia (Grant No. 451-03-136/2025-03/ 200172). The research is also part of the project *Old Folk Sports,*

Games and Competitions—Preservation, Transfer of Knowledge and Presentation (Ministry of Culture of the Republic of Serbia, Grant No. 451-04-1564/2023-02).

References

Aguilar-Moreno M (2015) Ulama: pasado, presente y futuro del juego de pelota mesoamericano. Anales De Antropología 49(1):73–112. https://doi.org/10.1016/S0185-1225(15)71645-0

Alonso VL, Medina FX, Londoño MD (2020) Traditional games and sports as unesco's intangible cultural heritage facing tourist strategies. J Tour Heritage Res 3(1):94–106. http://www.jthr.es/index.php/journal/article/view/130

Araya-Góchez A, Castro-Pérez M, Morales-Ramírez M (2018) Intergenerational perspective of the game as cultural heritage. Latin Am J Human Rights 28(2):119–138. https://doi.org/10.15359/rldh.28-2.5

Bale J (2022) Sports geography. Routledge

Bandyopadhyay N (2021) The status and future prospect of Indian traditional game Kho-Kho in West Bengal. Int J Fitness Health Phys Educ Iron Games 8(2):73–80

Bjeljac Ž, Brankov J (2024) Stari narodni sportovi, igre i nadmetanja—očuvanje, prenošenje znanja i prezentacija. Geografski institut Jovan Cvijić, SANU

Bjeljac Ž, Terzić, A, Ćurčić N (2014) Tourismological valorization of intangible cultural heritage of Serbia according to the Hilary du Cros method. Issues Ethnol Anthropol 9(1):195–217. https://doi.org/10.21301/EAP.v9i1.11

Bjeljac Ž, Ćurčić N, Ivolga A (2017) Tourismological classification of sporting events. Jour Geograph Inst "Jovan Cvijić" SASA 67(1):44–57. https://doi.org/10.2298/IJGI1701053B

Bjeljac Ž, Brankov J, Ćurčić N (2021a) Traditional sports and games as an element of intangible cultural heritage. Issues Ethnol Anthropol 16(4):1257–1283. https://doi.org/10.21301/eap.v16i4.12

Bjeljac Ž, Terzić A, Brankov J, Vujović S (2021b) Image of Hajduks and Uskoks and its role in formation of traditional sports and games as intangible heritage of Ex-Yugoslav Area. Forum Geogr 20(2):224–237. https://doi.org/10.5775/fg.2021.017.d

Brankov J, Milanović Pešić A, Milijašević Joksimović D, Radovanović MM, Petrović DM (2021) Water quality estimation and population's attitudes: a multi-disciplinary perspective of environmental implications in Tara National Park (Serbia). Sustainability 13(1):241. https://doi.org/10.3390/su13010241

Brankov J, Micić J, Ćalić J, Kovačević-Majkić J, Milanović R, Telbisz T (2022) Stakeholders' attitudes toward protected areas: the case of Tara National Park (Serbia). Land 11(4):468. https://doi.org/10.3390/land11040468

Brankov J, Penjišević I, B. Ćurčić N, Živanović B (2019) Tourism as a factor of regional development: community perceptions and potential bank support in the Kopaonik National Park (Serbia). Sustainability 11(22):6507. https://doi.org/10.3390/su11226507

Brewster M, Connell J, Page SJ (2009) The Scottish highland games: evolution, development and role as a community event. Curr Issues Tourism 12(3):271–293. https://doi.org/10.1080/13683500802389730

Bronikowska M, Groll, M (2015) Definition, classification, preservation and dissemination of traditional sports & games in Europe. Tafisa Recall: Games of the Past—Sports for Today 1–10. https://api.semanticscholar.org/CorpusID:198903381

Bronikowska M, Petrović L, Horvath R, Hazelton L, Ojaniemi A, Joae A, Filomena Silva C (2015) History and cultural context of traditional sports and games in selected European countries. Tafisa Recall, Games of the Past, Sports for Today 1–19. https://www.researchgate.net/publication/281784480_History_and_Cultural_Context_of_Traditional_Sports_and_Games_in_Selected_European_Countries

Ćurčić N, Mirković Svitlica A, Brankov J, Bjeljac Ž, Pavlović S, Jandžiković B (2021) The role of rural tourism in strengthening the sustainability of rural areas: the case of Zlakusa Village. Sustainability 13(12):6747. https://doi.org/10.3390/su13126747

Edwards K (2012) A typology of the traditional games of Australian Aboriginal and Torres Strait Islander peoples. Ram Skulls Press

Eichberg H (2009) Body culture. In: Pope SW, Nauright J (eds) Routledge companion to sports history, pp 178–197

Eichberg H (2012) Folk sports and traditional games in Europe. In: Nauriht J, Parrish C (eds) Sports around the world: history, culture, and practice, pp 304–312

European Association for Traditional Sports and Games. Members. https://aejest.com/es/miembros/

European Commission (2016) Directorate-general for education, youth, sport and culture. Mapping of traditional sports and games (TSG) in Europe—Final report. https://doi.org/10.2766/00810

European Parliament (2012) European Parliament resolution of 2 February 2012 on the European dimension in sport. https://www.europarl.europa.eu/doceo/document/TA-7-2012-0025_EN.pdf

European Platform for Traditional Sports and Games (2017). European traditional sports and games state of play. www.sportanddev.org/sites/default/files/downloads/eptsg_state_of_play.pdf

Gammon S, Ramshaw G (eds) (2013) Heritage, sport and tourism: sporting pasts–tourist futures. Routledge

Ghabouli E, Soltani A, Ranjbar E (2023) Heritage and the regeneration of urban brownfields: insights on public perception in Tehran, Iran. Heritage 6(5):4451–4471. https://doi.org/10.3390/heritage6050235

Giambartolomei L (2023) ¿Qué tiene el Pato Ñato? En busca de un Pato Ñato renovado. In: Actas 2023. Universidad Nacional de La Plata. Facultad de Humanidades y Ciencias de la Educación. https://www.memoria.fahce.unlp.edu.ar/trab_eventos/ev.16820/ev.16820.pdf

Graham B, Ashworth GJ, Tunbridge JE (2000) Geography of heritage. Power, culture and economy. Routledge

Jaouen G (2022) Educational stakes for the 21st century society through inheritance, diffusion, and transmission of traditional sports & games. In: Jaouen G, Guibert J (eds) Indigenous sports & games which leisure sports for tomorrow's society? Inheritance, transmission and diffusion, pp 9–12

Kim S, Whitford M, Arcodia C (2019) Development of intangible cultural heritage as a sustainable tourism resource: the intangible cultural heritage practitioners' perspectives. J Heritage Tourisme 14(5–6):422–435. https://doi.org/10.1080/1743873X.2018.1561703

Koordinator varstva nesnovne kulturne dediščine Republika Slovenija (2024) Register of the Intangible Cultural Heritage on the webpage of the Ministry of Culture of the Republic of Slovenia. Accessed 10 Jan 2025. http://www.nesnovnadediscina.si/en/register-of-intangible-cultural-heritage

Kylasov AV, Gavrov SN (2020) Ethnocultural diversity of sport. In: Georgescu L, Kharkin VN (eds) Sport science and physical education, vol. II, pp 266–306

Lavega P (2006) Traditional sports and games in 21st century Europe: future challenges. In: Lavega P (ed) Games and society in Europe, pp 281–282

Lavega P (2019) Safeguarding traditional sports and games: a holistic approach. In: Collected papers nr. 1, International conference: the importance of the UNESCO register of good practices in the safeguarding of the traditional games and sports. Svetvinčenat, Croatia, Istarski pljočkarski savez, pp 57–66

Lavega P, Alcaraz Munoz V, Mallen Lacambra C, Pic M (2023) Roles, relationships, and motor aggressions: keys to unveiling the emotions of a traditional sporting game. Front Psychol 14:1127602. https://doi.org/10.3389/fpsyg.2023.1127602

Lee LH, Lee YD (2015) The impact of water quality on the visual and olfactory satisfaction of tourists. Ocean Coast Manag 105:92–99

Masaeli M, Farsani T, Mortazavi M (2020) A study on tourist demand for Pahlevani and Zoorkhanei rituals as ancient Iranian sport. J Sport Tourism 24(1):1–11. https://doi.org/10.1080/14775085.2020.1728364

Milanović Pešić A, Brankov J, Milijašević JD (2020) Water quality assessment and populations' perceptions in the National Park Djerdap (Serbia): key factors affecting the environment. Environ Dev Sustain 22:2365–2383. https://doi.org/10.1007/s10668-018-0295-8
Milanović Pešić A, Jojić Glavonjić T, Denda S, Jakovljević D (2023) Sustainable tourism development and Ramsar Sites in Serbia: exploring residents' attitudes and water quality assessment in the Vlasina protected area. Sustainability 15(21):15391. https://doi.org/10.3390/su152115391
Ministarstvo kulture i medija Republike Hrvatske. Registar kulturnih dobara Republike Hrvatske. Accessed 10 Jan 2025. https://registar.kulturnadobra.hr/
Moreno A (2010) Mediterranean tourism and climate (change): a survey-based study. Tour Hosp Plann Dev 7(3):253–265. https://doi.org/10.1080/1479053X.2010.502384
Nefil I, Laaoudad-dodo S, Bordes P, Torki A (2021) Traditional games and sports of the women in the Kabylie. Front Psychol 11:614746. https://doi.org/10.3389/fpsyg.2020.614746
Niemczewska ZE (2021) The sociocultural impact of adaptive reuse of immovable cultural heritage from the perspective of direct users and the local community. J Cult Herit Manag Sustain Dev 11(3):240–261. https://doi.org/10.1108/JCHMSD-07-2019-0093
Ramshaw G (2014) Sport, heritage, and tourism. J Herit Tour 9(3):191–196. https://doi.org/10.1080/1743873X.2014.904320
Ramshaw G, Gammon SJ (2017) Towards a critical sport heritage: implications for sport tourism. J Sport Tourism 21(2):115–131. https://doi.org/10.1080/14775085.2016.1262275
Republika Severna Makedonija, Uprava za zaštita na kulturnoto nasledstvo. Accessed 1 Feb 2025. http://uzkn.gov.mk/
Richards G (2021) Rethinking cultural tourism. Edvard Elgar Publishing
Sogawa T (2006) Ethnic sport, its concept and research perspectives. Int J Sport Health Sci 4:96–102. https://doi.org/10.5432/ijshs.4.96
Traditional Sports. About traditional sports foundation (n.d.-a) https://www.traditionalsports.org/about-us/about-irsie-foundation.html
Traditional Sports. About traditional sports foundation (n.d.-b). https://www.traditionalsports.org/traditional-sports.html
TAFISA. TAFISA Regional TSG Centers. http://www.tafisa.org/tafisa-regional-tsg-centres
Terzić A, Bjeljac Ž (2014) The extermination of Jewish population and heritage in backa region of AP Vojvodina (Serbia). Trames J Human Soc Sci 18(68/63)1:39–55
The State Council of the People's Republic of China. China sees 110 mln domestic tourist trips during Dragon Boat Festival holiday (2024). https://english.www.gov.cn/archive/statistics/202406/10/content_WS6666f824c6d0868f4e8e7f7d.html
UNESCO (1999) Third international conference of ministers and senior officials responsible for physical education and sport (MINEPS III), Final Report. https://unesdoc.unesco.org/ark:/48223/pf0000119812
UNESCO (2003) Convention for the safeguarding of the intangible cultural heritage. https://ich.unesco.org/en/convention
UNESCO (2005) Preliminary report on the desirability and scope of an international charter on traditional games and sports. https://unesdoc.unesco.org/ark:/48223/pf0000141146
UNESCO (2006) Collective consultation in view of proposing an international platform to promote and develop the traditional sport and games/TSG: reference document. https://unesdoc.unesco.org/ark:/48223/pf0000146400
UNESCO (2017a) Traditional sports and games, challenge for the future. Concept note on traditional sports and games. https://unesdoc.unesco.org/ark:/48223/pf0000252837
UNESCO (2017b) Kazan Action Plan. https://unesdoc.unesco.org/ark:/48223/pf0000252725
UNESCO (2024) The lists of intangible cultural heritage and the register of good safeguarding practices. https://ich.unesco.org/en/lists
Vaczi M, Bairner A (2024) Introduction. In: Vaczi M, Barnier A (eds) Indigenous, traditional, and folk sports—contesting modernities
Vlada Federacije BiH, Federalno Ministarstvo kulture i sporta. Preliminarna otvorena lista nematerijalnog kulturnog naslijedja bastine Federacije Bosne i Hercegovine. Accessed 12 Jan

2025. https://fmks.gov.ba/hr/preliminarna-otvorena-lista-nematerijalnog-kulturnog-naslijedja-bastine-federacije-bosne-i-hercegovine/

Watson G, Dichter HL, Hess R (2024) Local, national and global histories of field hockey: an overview. Int J History Sport 41(10–11):927–938. https://doi.org/10.1080/09523367.2024.2433364

Wiser MC (2014) Lacrosse history, a history of one sport or two? A comparative analysis of men's lacrosse and women's lacrosse in the United States. Int J History Sport 31(13):1656–1676. https://doi.org/10.1080/09523367.2014.930709

The Geoheritage of Romania. From Theory to Practice—Legislation, Inventory, and Management

Laura Comănescu and Alexandru Nedelea

Abstract Global initiatives are underway to catalogue geoheritage across various nations, including those in Europe and elsewhere. These efforts occur within a legislative context, where geoheritage is acknowledged, assessed, and informs relevant policies, or through projects designed to enhance its visibility via geotourism and protect it through geoconservation. This study aims to inventory geoheritage at the national level, assess its status within Romanian specialised literature and legislation, and propose measures for its enhanced management, both structurally and regarding official status. The legislative framework and activities concerning inventory and valorisation in Romania are in their early stages, lacking a cohesive national effort. Local initiatives have emerged at the level of relief units or administrative-territorial units through various studies. Romanian legislation does not include the terms geoheritage, geosite, or geomorphosite; the only term utilised is natural heritage, as defined in Law no. 49/2011 and Law no. 5/2000, which pertains to protected natural areas. Regarding the management of geoheritage at the national level, it is necessary to protect valuable elements that hold scientific, educational, cultural, aesthetic, and touristic importance by establishing legislative, educational, and structural measures.

Keywords Geoheritage · Legislation · Inventory · Management · Romania

Geoheritage—Theoretical and Methodological Background

According to Pereira et al. (2005), geoheritage is made up of all geosites and geomorphosites, which are essential parts of natural heritage, as well as valuable components of the biogeographical envelope (eco-heritage) (Fig. 1). The natural heritage, together with the man-made (cultural-historical) constitutes the total heritage.

L. Comănescu (✉) · A. Nedelea
Faculty of Geography, University of Bucharest, Balcescu 1, Bucharest, Romania
e-mail: laura.comanescu@geo.unibuc.ro

A. Nedelea
e-mail: alexandru.nedelea@geo.unibuc.ro

L. Soares et al. (eds.), *Compromised Geography*, Key Challenges in Geography,
https://doi.org/10.1007/978-3-032-05201-8_21

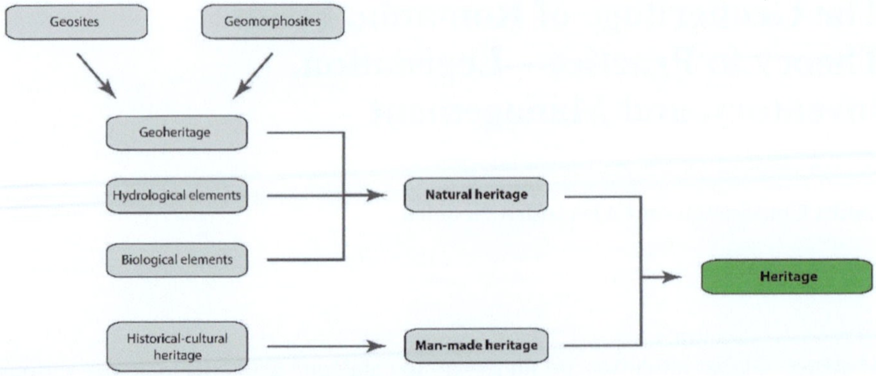

Fig.1 The general framework of the (geo)heritage concept

Geoheritage should obviously not be separated from the other components of heritage because the interrelationships established over time are evident.

Thus, the evolution of the biotic layer is closely related to the evolution of the relief and tectonic movements, as well as to the paleoclimate, and it supports various habitats through its morphographic, morphodynamic, and morphogenetic characteristics. On the other hand, geomorphological characteristics are of great importance for the inclusion of man-built elements in the cultural-historical heritage of an area or even for recognition at the level of UNESCO World Heritage.

According to Panizza and Piacente (2003), the cultural heritage includes natural sites with historical, religious, architectural, and scientific value, alongside monuments and isolated or grouped constructions with historical, artistic, architectural, and religious value.

Coratza and Hoblea (2018) consider heritage from a triple perspective: the legacy from the past that is used in the present and will be passed on to future generations.

Although the use of the notion of geoheritage is relatively recent, the concept has been utilised since the 1980s—Earth's scientific heritage or Earth's heritage (Gonggrijp and Boekschoten 1981; Andersen et al. 1990; quoted Brilha 2021).

Reynard (2009) considers that the "*concept of heritage refers to what is transmitted from one generation to another; the geomorphological heritage may, therefore, be considered as the set of landforms worthy of being protected and transmitted to future generations*".

Brilha (2021) considers that geoheritage encompasses those geological features that represent a legacy of the past, and their value makes them worthy of preservation for the future. It includes minerals, rocks, fossil soils, geological processes, landforms, and their genetic processes. Geoheritage has a series of values (identical to those of geosites and geomorphosites), namely scientific/educational, aesthetic, cultural, functional, and economic/touristic.

Brilha (2018) appreciates the importance of geoheritage for education and the development of geosciences.

The International Union of Geological Sciences considers that a geosite, a component of geoheritage, is a key point where elements or geological processes of international scientific importance are present, which can provide information regarding the evolution of geosciences over time (https://iugs-geoheritage.org).

The definition, importance of geoheritage, and the necessity of its protection are encompassed within two UN framework documents: the Convention Concerning the Protection of the World Cultural and Natural Heritage (1972) (https://whc.unesco.org/en/conventiontext/) and the International Geoscience and Geoparks Programme (2015) (https://www.unesco.org/en/iggp).

Thus, in accordance with the Convention Concerning the Protection of the World Cultural and Natural Heritage, the following are considered to be part of the natural heritage:

> Natural monuments consist of physical and biological formations or groups of such formations that have outstanding universal value from an aesthetic or scientific point of view; geological and physiographic formations and strictly delimited areas constitute the habitat of threatened animal and plant species, which have outstanding universal value from the point of view of value or conservation; natural sites or strictly delimited natural areas have outstanding universal value from the point of view of science, conservation, or natural beauty" (UNESCO 2011)

Geoheritage, through its scientific value, provides highly important information regarding the evolution of the Earth as a whole, as well as its layers and the changes occurring in the geographical environment. The geological-geomorphological heritage is also related to the cultural heritage when connections between relief and human society are established over time (Coratza and Hoblea 2018). This feature is closely linked to its perception and use by human society.

Legislation Related to Geoheritage

The International Union for Conservation of Nature (IUCN) was founded in 1948 and initially focused on the protection of various species and ecosystems (biodiversity). After 2008, the area of interest expanded to include geoheritage as a component of the natural heritage (Crofts et al. 2015), with resolutions adopted in 2008, 2012, and 2016 reconsidering the role of geoheritage in nature conservation and establishing the Geoheritage Specialist Group under the World Commission on Protected Areas (Brilha 2021).

The year 1972 is of great importance regarding the valorisation of heritage on a global scale through the adoption of the *Convention concerning the Protection of the World Cultural and Natural Heritage*, which was subsequently implemented through the *Operational Guidelines for the Implementation of the World Heritage*.

In 1988, the European Working Group for the Conservation of Earth Sciences was created, which in 1993 transformed into the European Association for the Conservation of the Geological Heritage (ProGEO). Today, it is affiliated with IUGS and IUCN. The primary action directions are represented by creating inventories of geoheritage at the regional, national, and European levels; organising symposia, courses, and other categories of educational activities; publishing a journal and various promotional materials; and representing in various international organisations (https://www.progeo.ngo).

In 2001, on the occasion of the Tokyo Conference of Geomorphology, a working group focused on geomorphosites and geomorphological heritage was established within the International Association of Geomorphologists. In 2007, within the International Union of Geography, the Geoheritage Commission was established.

In 2014, within the WCPA (World Commission on Protected Areas), a working group of geoheritage specialists was established. The main objectives of this group have included and continue to include the creation of a good practices guide related to geoheritage management, the revision of elements related to the study and value of volcanic geosites, and the establishment of objective criteria for the inclusion of geosites in the UNESCO heritage list (Woo 2017).

In 1991, in Digne-les-Bains, within the Haute-Provence Geological Reserve, the first symposium dedicated to nature conservation was organised under the auspices of UNESCO, where participants from 30 countries adopted the *International Declaration of the Rights of the Memory of the Earth.* Within this declaration, the relationship between humans and the Earth is paramount, and it is the first to define geological heritage. The statement, structured in 9 points, has been translated into 50 languages and is the foundation stone for national conservation policies (https://www.geoparchauteprovence.com).

The existence of the geological-geomorphological heritage has been recognised at the European level through the Council of Europe Recommendation no. 3 of May 5, 2004—*On the conservation of geological heritage and areas of special geological interest*. In this Act, member states are requested to inventory the geological heritage, develop strategies and programs to determine its protection and management, create legal instruments to ensure the conservation of geoheritage, implement information and educational programs, strengthen international cooperation in the field, and allocate financial resources for the conservation of geoheritage (Brilha 2021).

In Romania, the first association for tourism and environmental protection was established at the beginning of the nineteenth century, and the first publication dedicated to nature conservation was *Nature Protection in Romania* (Borza 1924). In 1900, the establishment of the first national park in the Bucegi Mountains was proposed, and later, in 1935, the first national park was declared—Retezat National Park—and in 1938, the first protected geological site, the Detunatele in the Apuseni Mountains (Smaranda 2008). In 1938, there were 30 protected natural areas in Romania, mostly with biogeographic or landscape characteristics, and by 1965, their number had increased to 130 (Stanciu and Florescu 2009).

At the legislative level, in 1930, the first law for the protection of nature in Romania was adopted (Law no. 213), which was considerably improved in 1950 when a new law was approved. In this law, numerous sites were proposed as geological monuments and reserves.[1] Through the 1973 law, new protected areas were declared—The law of Environment Protection (no. 9/1973).[2]

After 1989, the first law on this topic was the Environmental Protection Law no. 137/1995, which presented a new vision regarding nature conservation in general and protected areas in particular. In Emergency Ordinance no. 236 of November 24, 2000,

> (...) regarding the regime of protected natural areas and the conservation of natural habitats, reference is made to elements of geoheritage within a subchapter that addresses measures for the protection and conservation of endangered, vulnerable, endemic, and/or rare wild animal and plant species, as well as those for the protection of geomorphological and landscape formations of ecological, scientific, aesthetic, cultural-historical, and other interests, natural goods of speleological, paleontological, geological, anthropological interest, and other natural goods with natural heritage value, existing within the perimeters of protected natural areas and/or outside of them". In Chapter 5—Conservation of Other Natural Heritage Assets, the terms geomorphological interest natural areas, geomorphological heritage assets, and geological heritage, are used.[3]

Law no. 5/2000—on the approval of the National Territory Planning Plan—Section III—protected areas provides a comprehensive definition for protected areas, referring to *(…) natural or constructed areas, geographically and/or topographically delimited, which include natural and/or cultural heritage assets.* In this law, Sect. 3 is dedicated to the inventory of protected natural areas of national interest and natural monuments (including geological and geomorphological), as well as their geographical location.

In current Romanian legislation, the terms geoheritage, geosite, and geomorphosite are missing, the only one used being natural heritage. It is precisely defined in Law no. 49/2011 for the approval of Government Emergency Ordinance no. 57/2007 regarding the regime of protected natural areas, the conservation of natural habitats, flora, and wildlife as being

> (…) the set of components and structures of the physical-geographical, floral, faunal, and biocenotic environment, whose ecological, economic, scientific, biogenic, sanogenic, landscape, and recreational importance and value have a relevant significance in terms of conserving biological, floral, and faunal diversity, the functional integrity of ecosystems, and the conservation of genetic, plant, and animal heritage, as well as for meeting the life, welfare, culture, and civilization needs of present and future generations.

In contrast to the geological and geomorphological legacy, which is restricted to physical-geographical components and structures, as the definition makes clear, the focus is on biotic aspects, even though some of its values—such as ecological, economic, scientific, and landscape—are highlighted.

In the same law, the significance of the notion of a natural heritage asset is also explained, which represents the component of natural heritage that requires a special

[1] https://acad.ro/institutia/misiune_istoric_cmn.html.

[2] https://legislatie.just.ro/Public/DetaliiDocument/311.

[3] https://www.cdep.ro/pls/legis.

regime of protection, conservation, and sustainable use in order to be maintained for the benefit of present and future generations. The importance of some of these natural heritage assets is already recognised through their inclusion in the category of protected natural areas.

Law no. 49/2011, art. 4(18), defines a protected natural area as

> (…) the terrestrial and/or aquatic zone where wild plant and animal species, biogeographical, landscape, geological, paleontological, speleological, or other elements and formations with exceptional ecological, scientific, or cultural value exist, which has a special protection and conservation regime established in accordance with legal provisions.

Thus, it follows that geosites and geomorphosites are included in the category of landscape, geological, paleontological, speleological, or other natural elements and formations, but some of their values are mentioned: ecological, scientific, and cultural. In the same legislative act (art. 8), a classification of protected natural areas is made:

> (…) **of national interest** (scientific reserves, national parks, natural monuments, natural reserves, natural parks); **of international interest** (natural sites of universal natural heritage, geoparks, wetlands of international importance, biosphere reserves); of community interest or 'Natura 2000' sites (sites of community importance, special conservation areas, special avifauna protection areas); of county or local interest.

In conclusion, the legislation regarding geoheritage in Romania is present only in the broader context of protected natural areas and geoparks or European provisions in the field.

Inventory of Geoheritage

The inventory of high-value geoheritage needs to be carried out at different scales, from global to national or regional and local levels. The level of detail will increase with the size of the reference area. Regardless of the area where the inventory will be conducted, four defining elements must be considered: the topic, the value, the aim, and the scale (Brilha 2018; Díaz-Martínez et al. 2016).

In scientific literature, there are numerous methods for the selection and inventory of geoheritage, all aimed primarily at reducing subjectivity. To carry out a complete, accurate, and relatively uniform inventory, it is necessary to have a sheet with clear selection criteria. According to Brilha (2018), scientific value (representativeness, integrity, rarity, scientific knowledge), educational value/geoeducation (didactic potential, variety of geological elements, accessibility, lack of risks) and value for use through geotourism (scenery, interpretation, safety, and accessibility) must be considered (Bollati et al. 2012; Pica et al. 2013).

In 1978, the first stage of registering sites in the World Cultural and Natural Heritage takes place (12 properties). These are inscribed in three categories: natural, cultural, and mixed based, on the criteria presented in Table 1 (Migon 2018). Since

Table 1 The selection criteria for inclusion in the UNESCO world heritage list (https://whc.unesco.org/en)

(vii)	to contain superlative natural phenomena or areas of exceptional natural beauty and aesthetic importance;
(viii)	to be outstanding examples representing major stages of earth's history, including the record of life, significant on-going geological processes in the development of landforms, or significant geomorphic or physiographic features;
(ix)	to be outstanding examples representing significant on-going ecological and biological processes in the evolution and development of terrestrial, fresh water, coastal and marine ecosystems and communities of plants and animals;
(x)	to contain the most important and significant natural habitats for in-situ conservation of biological diversity, including those containing threatened species of outstanding universal value from the point of view of science or conservation

1992, *(…) significant interactions between people and the natural environment have been recognised as cultural landscapes.*[4]

In 2025, the World Heritage List includes 1,223 sites from 168 countries (952 cultural, 231 natural, and 40 mixed). Most of the natural sites are in Australia, Canada, the United States, China, Italy, and France. Among the natural sites, 90 of them have been inscribed only for their geological or geomorphological features (https://whc.unesco.org/en/list).

In this list, Romania has 11 sites included of which 2 are natural, but none of them belong to geoheritage.[5] Currently, on the Tentative Lists of UNESCO, Romania has proposed 17sites. Among these, in the category of geological and geomorphological sites, are present Movile Cave, Retezat Mountains, Pietrosul Rodnei Peak, and the Paleontological Site of Sânpetru.[6]

In numerous countries, especially European ones, geological heritage is inventoried at the regional or national level, most frequently within the framework of unified projects. In this regard, it can be mention countries such as Canada, Brazil, New Zealand, Australia, South Africa, Iceland, Spain, the United Kingdom, Ireland, the Netherlands, Switzerland, Italy, Germany, Denmark, Finland, Estonia, Lithuania, Poland, Slovenia, the Czech Republic, and Slovakia, have national/regional inventories with extensive and coherent databases.

Some instances of best practices in the development of national inventories are as follows: Spain's inventory of geosites at the end of the 1970s included 144 geosites of national and international interest (Carcavilla Urqui et al. 2007); Switzerland recognised more than 300 geosites of national interest starting in 1990 (Reynard et al. 2012); and the United Kingdom currently has over 3000 sites of scientific interest registered, where the inventory of sites began in the 1950s (Wimbledon et al. 1995) and was divided into 7 main categories and 100 subclasses.

---

[4] https://whc.unesco.org/en.

[5] https://www.cultura.ro/unesco.

[6] https://www.whc.unesco.org/en/list.

In Portugal, an inventory of geosites with scientific importance was created as part of a community-involved research project (Brilha et al. 2005), and a strong school related to geoconservation has developed, mainly at the University of Minho (Brilha et al. 2012).

In Slovenia, Erhartic (2010) developed a new methodology for the inventory and mapping of geomorphosites and geosites, important for the conservation of geological heritage. *In the Slovenian Nature Conservation Act* (1999, 2004), several categories of valuable elements were defined, namely geomorphological, geological, hydrological, botanical, dendrological, zoological, ecosystems, and landscapes.[7] At the time the law was enacted, 114 natural sites were inventoried, for which the relevant provisions applied.

Spain is also known for its contributions to the development of geoconservation programs and inventories in Africa and South America (Díaz-Martínez et al. 2019). In Brazil, in the state of São Paulo, the first complete inventory of geosites was created, while in the rest of the country there are scattered contributions, mostly related to geoconservation (Garcia et al. 2017). A total of 142 representative geosites from 11 geological frameworks were inventoried and evaluated, with an emphasis on their scientific value and the risk of natural and anthropogenic degradation, as a basis for future management (Garcia et al. 2017).

In 2018, 7267 geosites (3154 geological, 3935 geomorphological, and 178 geological disaster categories) were inventoried in China. These were categorised into four levels: local, provincial, national, and global, with thirteen classes and forty-six subclasses (Song et al. 2023).

The IUGS Global Geosite Working Group (GGWG) is attempting to unify and generalise inventories from different countries to develop a global database (IUGS GEOSITES) based on a methodology and principles valid at the international level.

The most important ProGeo project within IUGS began in 1995, during which numerous national inventories were created (https://iugs-geoheritage.org).

The ProGeo-RO project—Enhancing climate resilience and environmental sustainability through the knowledge and monitoring of Romania's internationally significant geological heritage (https://igr.ro/pn-23-39-03-01)—established criteria for the analysis and classification of sites in Romania possessing national and global heritage value and developed a database and digital application accordingly. Building upon the analogous initiative of the IUGS, which identified, characterised, and promoted the initial 100 sites of global geological heritage across various disciplines (including history of geosciences, stratigraphy and sedimentology, paleontology, igneous and metamorphic petrology, volcanology, tectonics, mineralogy, geomorphology and active geological processes, impact structures, and extraterrestrial rocks), the first 101 geosites in Romania of international significance will be sealed. It is requested that several of them be incorporated into the global geosite map in the forthcoming edition of the IUGS list.

[7] https://www.ecolex.org/details/legislation/nature-conservation-act-lex.

Through geointerpretation, stories will be created about these, grouped into five categories: GeoLife (BioStones), GeoFire (Fire and Stones), GeoLand (Bare Stones), GeoAir (Air, Stones, and Metal), or GeoWater (Water and Stones).[8]

In Romania, although the theoretical and methodological results related to the inventory and evaluation of geoheritage are remarkable, their applicability to different territories is limited. Thus, there is no national inventory of geosites and/or geomorphosites, nor has one been initiated. Local initiatives (Bucegi Mountains, Făgăraş Mountains, Lotrului Valley, Retezat Mountains, Karstic Plateau Anina, Trascău Mountains, Transylvanian Plateau, Călimani Mountains, Ceahlău Mountains, Ciucaş Mountains, Ponoare Karstic Area, Curvature Subcarpathians, Southern Dobrogea Plateau, Măcin Mountains, etc.) have been launched for small areas within various administrative-territorial or physical-geographical units (predominantly within doctoral theses or research projects).

As mentioned earlier, geosites and geomorphosites were partially inventoried in Law 5/2000 in the broader category of protected natural areas and natural monuments and in Annex 1 to Law no. 49/2011 for the approval of Government Emergency Ordinance no. 57/2007 regarding the regime of protected natural areas, the conservation of natural habitats, flora, and wildlife.

With a focus on how geological and geomorphological heritage elements are reflected, Table 2 offers a synthetic correspondence between protected natural areas of national interest in relation to the IUCN categories and legal documents (Annex 1 to Law no. 49/2011 for the approval of Government Emergency Ordinance no. 57/2007 regarding the regime of protected natural areas, the conservation of natural habitats, flora, and wildlife). The most important geoheritage elements have been declared geomorphological, geological, and speleological natural areas. Thus, in Fig. 2, the share of these protected natural areas at the county level is represented. It can be observed that the highest percentages overlap with the Carpathian arch (the counties of Bihor, Braşov, Alba, Hunedoara, Caraş Severin, Gorj, Argeş, Braşov, and Prahova), while the lowest percentages are found in the counties located in the plain regions (Timiş, Brăila, Ialomiţa, Ilfov, Călăraşi, Giurgiu, Teleorman, and Olt).

In Table 3, the relationship between internationally protected natural areas and the IUCN category, as well as the heritage elements included within them, is presented. In the category of biosphere reserves (Fig. 3) are included the Danube Delta Biosphere Reserve (where geoheritage elements are not protected and there are no scientific reserves of this kind), the Retezat Mountains, and the Rodnei Mountains, where valuable geoheritage predominates (with varying degrees of protection and partially inventoried). The two geoparks declared on the territory of Romania (Ţara Haţegului Geopark and Ţinutul Buzăului Geopark) hold numerous geoheritage elements, inventoried and properly managed.

Within national and natural parks (Table 2, Fig. 4), the situation is differentiated: on one hand, there are numerous natural and national parks (e.g., Retezat, Cozia, Iron Gate, Grădiştea de Munte-Cioclovina, etc.) where many geosites and geomorphosites are inventoried, partially evaluated, and subsequently conserved; on the

[8] https://igr.ro/pn-23-39-03-01.

Table 2 The correspondence between protected natural areas of national interest and the IUCN category (developed in accordance with the documents of Annex 1 to Law no. 49/2011; Stanciu and Florescu 2009)

Law 49/ 2011	IUCN categories	IUCN description	Management	Geoheritage	Examples
Scientific reserves	I (a) Protected area managed for scientific purposes	Strictly protected areas for the conservation of biodiversity and geological features. To preserve the valuable elements, tourist activities and use by human society are limited	Scientific research Environmental services	Elements of geoheritage with exceptional value/ geodiversity	Gemenele— Retezat, Pietrosul Mare Peak
National parks	II Protected area managed for the protection of ecosystems/ other elements	Protected areas with a high degree of natural nessare declared to protect large-scale ecosystems, which hold the potential for spiritual, scientific, educational, and recreational experiences compatible ecologically and culturally with the respective space	Scientific research Environmental protection/ biodiversity/ geodiversity Environmental services Tourism and recreation	They contain components of geoheritage	Domogled—Valea Cernei, Retezat, Nerei Gorges—Beuşniţa, Rodnei Mountains, Bicazului Gorges—Hăşmaş, Ceahlău, Călimani, Cozia, Piatra Craiului, Semenic—Caraşului Gorges, Măcinului Mountains, Buila—Vânturariţa, Jiului Gorges
Natural monuments	III Protected area managed especially for the preservation of specific natural characteristics	Areas with reduced surfaces were declared to protect geological or ecological natural elements/ monuments. They hold exceptional value, especially educational or recreational for visitors	Scientific research Conservation of natural and cultural elements Environmental services Tourism and recreation	Numerous geomorphosites/ geosites of high value	Piatra Albă La Grunj, Răteiului Cave, Sfinxul Bănăţean, Râpa Neagră, Liliecilor Cave, Moara Dracului Waterfall, Mihăileni Canyon, Duruitoarea Waterfall, Râpa Roşie, Groapa Ruginoasă, Spinx, Prăpăstiile Zărnesţiului, Muddy volcanoes of Hăşag, Old Ladies

(continued)

Table 2 (continued)

Law 49/2011	IUCN categories	IUCN description	Management	Geoheritage	Examples
Nature reserves	IV Area with active habitat or species management	Areas declared for the protection of valuable species or habitats	Conservation through various types of interventions Scientific research Education Protection of biodiversity/geodiversity/cultural elements	Nature reserves of scientific, geological, speological and landscape types	Izverna Cave, Beştepe Hills, Rusca Montană, Detunata, Dealul cu Melci, Golul alpin Moldoveanu—Capra, Golul alpin Podragu—Suru, Karst Plateau Padiş, Tătarului Gorges, Cocora Cave, Urşilor Gorges, Turzii Gorges, Scărişoara Cave
Natural parks	V Protected area managed for landscape conservation	Protected areas where the interaction between humans and nature over time has determined significant ecological, scientific, cultural, and aesthetic value. It is important to preserve this traditional interaction for the protection of the area and the maintenance of nature conservation values	Scientific research Education Conservation of landscapes/biodiversity/geodiversity/cultural elements Tourism and recreation Traditional activities for the use of natural resources carried out by local communities Preserving traditional elements	Elements of geoheritage included within them often set the prominent characteristic	Apuseni, Bucegi, Porţile de Fier, Grădiştea—Cioclovina, Maramureşului Mountains, Putna—Vrancea

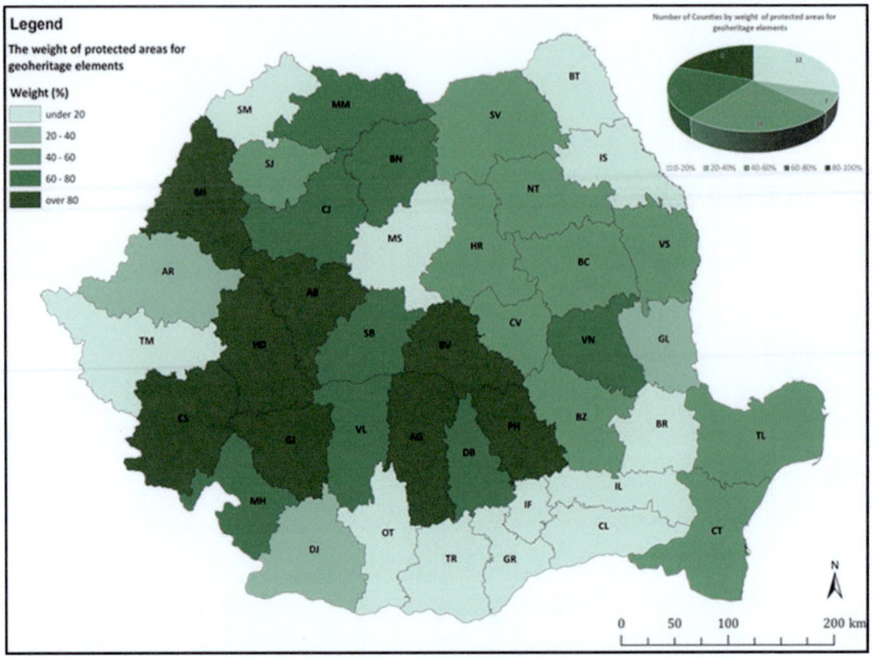

Fig. 2 The weight of geoheritage components in protected natural areas

Table 3 The correspondence between internationally protected natural areas and the IUCN category (developed in accordance with the provisions of Annex 1 to Law no. 49/2011; Stanciu and Florescu 2009)

Law 49/2011	IUCN categories	Management	Geoheritage	Examples
Biosphere Reserves—UNESCO	No correspondence	Protection of World cultural and natural heritage, UNESCO	Elements of high-value geoheritage	Danube Delta Biosphere Reserve, Retezat Mountains, Rodna Mountains
Geoparks—UNESCO	II or V, depending on the situation	Regulations established by UNESCO and the Charter of the European Geoparks Network	Elements of high-value geoheritage	Haţeg Country Dinosaurs Geopark, Buzău Land Geopark

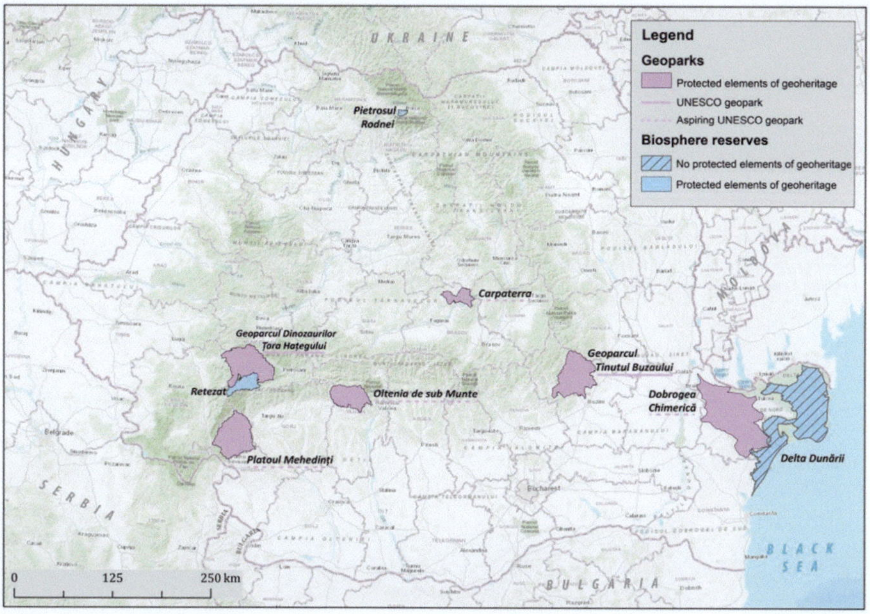

Fig. 3 The geographical distribution of geoparks and biosphere reserves in Romania (https://ananp.gov.ro)

other hand, especially in the category of natural parks, there are examples where only biogeographic elements are protected.

Geoheritage Management

Geoheritage is of particular value, including social value, but it cannot be regarded as a renewable resource. Thus, it needs to be conserved through policies at international, national, and local levels (currently biodiversity and ecosystem policies predominate)—geoconservation, but also efficiently managed through the adoption and implementation of coherent policies and strategies that contribute to sustainable scientific, educational, and tourism use—geotourism (Brilha 2021).

The inventory, management, and conservation of geoheritage require solid scientific knowledge; thus, they must be carried out with the support of experts in geosciences.

Two types of management are required, according to Stanciu and Florescu (2009): participatory (including the public, scientific community, and authorities in a variety of activities) and adaptive (consisting of ongoing adaptation to legal or environmental changes).

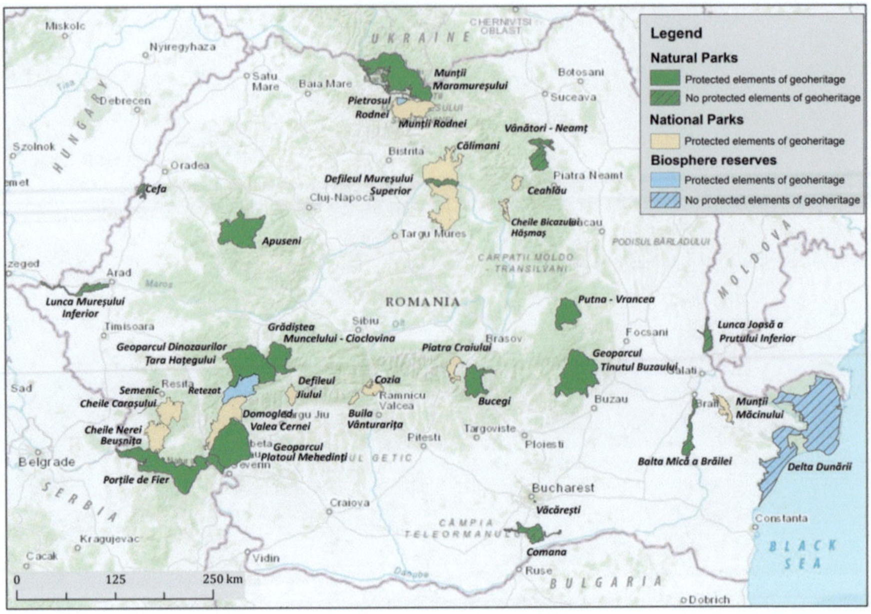

Fig. 4 The geographical distribution of natural parks and national parks in Romania (https://ananp.gov.ro)

At the level of Romania, there are no coherent national or regional plans or projects for the management of valuable geological and geomorphological elements; however, a series of geosites and geomorphosites can be identified that are correctly managed and geotouristically valued (Fig. 5).

In the case of many of the geoheritage elements, their state of conservation and valorisation is disastrous (geosites completely destroyed, some have become garbage dumps or quarries for various rocks; lack of indicator and/or informative elements or rusty panels; places covered by vegetation or destroyed due to natural hazards—earthquakes, landslides, collapses; destruction or theft of fossils or other valuable elements, hydrotechnical works, various types of infrastructure).

Geosites and geomorphosites located within geoparks, biosphere reserves, and natural or national parks that have management plans, which include general measures that can also be applied to geoheritage, are exceptions.

The geoheritage management measures that can contribute to the preservation of its values over time are of three types: legislative, educational (Reynard et al. 2018), and structural (Table 4).

The correct management of geoheritage represents the first step towards the declaration of new protected natural areas or geoparks. The latter successfully integrates geotourism, geoconservation, and geoeducation.

Fig. 5 Values of Romanian geoheritage: **a** Geomorphological natural monument—Franz Josef Rock; **b** Geomorphological natural monument—The Sphinx; **c** Geomorphological natural monument—The Old Ladies; **d** Detunatele Natural Reserve; **e** Scărişoara Glacier Reserve (the first speleological reserve in Romania)

Table 4 The measures for geoheritage management

Legislative	Educational	Structural
Improving their management	Aims at raising the awareness of the population and the authorities at different levels regarding the value and importance of geoheritage	Establishing restrictions on land use
Increasing the level of its conservation	Promoting the implementation of projects for the training of people	The implementation of protection systems, fences, walkways, drainage, etc.
Establishing an updated legislative framework aligned with international trends in the field	Developing geotourism—routes, geo-interpretation, creating legends, tourist information centre	
The involvement of local authorities in both conservation and valorisation	Working in the field to ensure the sustainable exploitation of geosites and geomorphosites	
The implementation of strong partnerships between local communities, scientists, authorities, and entrepreneurs	Making informational panels and points, brochures, specialised websites, etc.	
	Integrated scientific research through the participation of multidisciplinary teams	

Conclusion

At the level of Romania, the inventorying of geoheritage has been done partially, only for certain areas (less so within the framework of national projects, but rather within scientific research). Furthermore, these inventories are not conducted based on a unified methodology, at the same scale, and for the same purpose.

National legislation is still patchy and incomplete in this area (the most important being Law no. 5/2000), with existing provisions being part of a broader framework of protected natural areas or geoparks. Additionally, Romania complies with European legislation in this field and is a signatory to all international conventions related to heritage.

Regarding geoheritage management, the main objective at the level of Romania is to protect elements of geological and geomorphological value, which hold scientific, educational, cultural, aesthetic, and touristic importance.

For the promotion of scientific knowledge, natural heritage (especially geological and geomorphological, as well as geoparks) are great value elements. Within the Haţeg Country Geopark and Buzău Land Geopark, both part of the UNESCO Global Geoparks network, important steps are being taken with the aim of promoting geotourism, geoeducation, and geoconservation.

What are the challenges for the future?

Geoheritage must be recognised at the national legislative level, and it is necessary to create a national inventory, which should start from existing international methodologies and projects.

To achieve the objectives (geotourism and geoconservation), it is important for national and local authorities, as well as environmental organisations, to develop adequate infrastructure for tourists (informational panels, geotourism itineraries, local guides, and interpreters). Additionally, in this period when tourism has become a mass phenomenon, geotourism must ensure a balance for the rational use of sites and their protection.

One of the main challenges is related to ensuring the long-term protection of geological sites in the face of degradation that can occur as a result of industrial activities, various types of construction, and intensive use by human society.

It is necessary to identify the most important levers for raising public awareness, primarily among local communities, regarding the value of geoheritage. Additionally, the scientific community must open up to the public and present basic geological and geomorphological knowledge in an understandable manner and organise field trips for students in areas with diverse and valuable geosites and geomorphosites.

Research and higher educational institutions, together with central and local authorities, must collaborate to efficiently manage the geological heritage and to organise courses, projects, and educational programs focused on its knowledge, valorisation, and protection.

Romania has a valuable and diverse geoheritage, and its recognition, valorisation, and protection represent a priority for economic development (especially tourism) in the coming period.

Acknowledgements The authors wish to thank Dr. George Cosmin Păunescu for the assistance provided in completing the graphic part of the work.

References

Andersen S, Black G, Duff K, Erikstad L, Gonggrijp G, Kontturi O, Schonlaub H, Wimbledon W (1990) Earth-science conservation. An absolute need for science and education. Jahrbuch Der Geologischen Bundesanstalt 133(4):653–669

Bollati I, Pelfini M, Pellegrini L (2012) A geomorphosites selection method for educational purposes: a case study in Trebbia Valley (Emilia Romagna, Italy). Geogr Fis Din Quat 35(1):23–35

Borza A (1924) Protecţiunea naturii în România. Institutul de arte grafice Ardealul, Cluj Napoca

Brilha J (2018) Geoheritage: inventories and evaluation. In: Reynard E, Brilha J (eds) Geoheritage assessment, protection and management. Elsevier, pp 69–86

Brilha J (2021) Geoheritage. In: Encyclopedia of geology, 2nd ed. Elsevier, pp 569–578. https://doi.org/10.1016/B978-0-12-409548-9.12106-2

Brilha J, Andrade C, Azerêdo A, Barriga FJ, Cachão M, Couto H, Cunha P, Crispim JA, Dantas P, Duarte LV, Freitas Granja MC, Henriques MH, Henriques P, Lopes L, Madeira J, Matos JM, Noronha F, Pais J, Piçarra J, Ramalho M, Relvas JM, Ribeiro A, Santos A, Santos V, Terrinha P (2005) Definition of the Portuguese frameworks with international relevance as an input for the European geological heritage characterization. Episodes 28(3):177–186

Brilha J, Pereira D, Pereira P (2012) Geoconservation education: the leading role of the University of Minho (Portugal). Geophysical research abstracts, vol. 14, EGU General Assembly, Vienna

Carcavilla Urqui L, Lopez Martinez J, Duran Valsero J (2007) Patrimonio Geologico y Geodiversidad: investigacion, conservacion, gestion y relacion con los espacios naturales protegidos. Instituto Geologico y Minero de Espana, Madrid

Coratza P, Hoblea F (2018) The specifities of geomorphological and management. In: Reynard E, Brilha J (eds) Geoheritage assessment, protection and management. Elsevier, pp 87–106

Crofts R, Gordon JE, Santucci V (2015) Geoconservation in protected areas. In: Worboys GL, Lockwood M, Kothari A, Feary S, Pulsford I (eds) Protected area governance and management. ANU Press, Canberra, pp 531–568

Díaz-Martínez E, Brilha J, Brocx M, Erikstad L, García-Cortés A, Wimbledon WA (2016) Global geosites: an active and partially achieved geoheritage inventory initiative, waiting to regain official recognition. In: Cornée A, Egoroff G, de Wever P, Lalanne A, Duranthon F (eds) Actes du congrès international Les inventaires du géopatrimoine, 22–26 septembre 2015, Toulouse. Mémoire hors-série de la Société géologique de France 16, pp 103–108

Díaz-Martínez E, García-Cortés Á, Vegas J, Carcavilla L, Charles N (2019) Cooperación europea Para la promoción de la geoconservación en África. In: Martín-González E, Coello Bravo JJ, Vegas J (eds) Actas de la XIII Reunión Nacional de la Comisión de Patrimonio Geológico, Cuadernos del Museo Geominero, Instituto Geológico y Minero de España, 30, 27–33

Erhatic B (2010) Geomorphosite assessment. Acta Geogr Slov 50(2):296–309

Garcia GM, Brilha J, Lima F, Vargas JC, Pérez-Aguilar A, Alves A, Campanha GA, Duleba W, Faleiros FM, Fernandes LA, Fierz MS, Garcia MJ, Janasi VA, Martins L, Raposo MI, Ricardi-Branco F, Ross JL, Sallum FW, Souza CR, Bernardes-De-Oliveira ME, Neves BB, Neto MC, Christofoletti SR, Henrique-Pinto R, Lobo HA, Machado R, Passarelli CR, Perinotto JA, Ribeiro

RR, Shimada H (2017) The inventory of geological heritage of the state of São Paulo, Brazil: methodological basis, results and perspectives. Geoheritage 10(2):239–258

Gonggrijp GP, Boekschoten GJ (1981) Earth-science conservation: No science without conservation. In: van Loon AJ (ed) Quaternary geology: a farewell to AJ Wiggers Geologie en Mijnbouw, 60(3):433–445

Legea nr. 49/2011 pentru aprobarea Ordonanţei de urgenţă a Guvernului nr. 57/2007 privind regimul ariilor naturale protejate, conservarea habitatelor naturale, a florei şi faunei sălbatice

Legea nr. 5/2000 privind aprobarea Planului de amenajare a teritoriului national—Secţiunea a III-a—zone protejate

Migon P (2018) Geoheritage and world heritage sites. In: Reynard E, Brilha J (eds) Geoheritage assessment, protection and management. Elsevier, pp 237–249

Panizza M, Piacente S (2003) Geomorfologia culturale. Pitagora Editrice, Bologna

Pereira P, Pereira D, Alves C, Brilha J (2005) Geology, landscape and geomorphology: finding the place of geomorphological heritage, IV International Symposium PROGEO, Braga, Portugal. https://hdl.handle.net/1822/4823

Pica A, Del Monte M, Fredi P (2013) Geotouristic itineraries in the Lazio region, Italy. Geomorphosites in the millenarian urban environment of Rome and in the natural environment of the Ernici mountains. Collection EDYTEM. Cahiers de géographie-Gestion des géosites dans les espaces protégés, 15(1):155–162

Reynard E (2009) Geomorphosites: definitions and characteristics. In: Reynard E, Coratza P, Regolini Bissi G (eds) Geomorphosites, pp 9–20

Reynard E, Perret A, Grangier L, Kozlik L (2012) Methodological approach for the assessment, protection, promotion and management of geoheritage. Geophys Res Abstr 14, EGU 3740

Reynard E, Coratza P, Cayla N, Clivaz M, Comănescu L, Darbellay L, Giusti C, Grecu F, Hobléa F, Pereira P (2018) InterGEO: a digital platform for university education on geomorphological heritage. Geoheritage 10(4):645–657

Smaranda JS (2008) Managementul turismului in ariile naturale protejate. Edit. Risoprint, Bucureşti

Song Q, Dong Y, Miao S, Hu X, Chen F, Ji Y, Wang J, Huang Z, Li H (2023) The types, grades and distribution features of geoheritage in China: an inventory. Int J Geoheritage Parks 11(3):514–526

Stanciu E, Florescu F (2009) Ariile protejate din România. Notiuni introductive. Edit. Green Steps, Brasov

UNESCO (2011) The world heritage resource manual preparing world heritage nominations, 2nd ed

Wimbledon WA, Benton MJ, Bevind RE, Black GP, Bridgland DR, Cleal CJ, Cooper RG, May VJ (1995) The development of a methodology for the selection of British geological sites for geoconservation: part 1. Mod Geol 20:159–202

Woo KS (2017) Role of IUCN WCPA geoheritage specialist group for geoheritage conservation and recognition of World Heritage Sites, Global Geoparks and other protected areas, Geophysical Research Abstracts 19, EGU2017-1137

Links

https://ananp.gov.ro. Accessed Feb 2025
https://igr.ro/pn-23-39-03-01. Accessed Feb 2025
https://iugs-geoheritage.org. Accessed Feb 2025
https://legislatie.just.ro/Public. Accessed Feb 2025
https://whc.unesco.org/en/list. Accessed Feb 2025
https://www.cdep.ro/pls/legis, Accessed February 2025
https://www.cultura.ro/unesco, Accessed February 2025

https://www.ecolex.org/details/legislation/nature-conservation-act-lex-faoc061725/?q=4. Accessed Feb 2025
https://www.geoparchauteprovence.com. Accessed Feb 2025
https://www.progeo.ngo. Accessed Feb 2025
https://whc.unesco.org/en/conventiontext. Accessed May 2025
https://www.unesco.org/en/iggp. Accessed May 2025
https://acad.ro/institutia/Misiune_Istoric_cmn.html. Accessed May 2025
https://legislatie.just.ro/Public/DetaliiDocument/311. Accessed May 2025

Determination of Tourism Priorities in the Sustainable Development of Mountainous Ajara: Challenges and Perspectives

Merab Putkaradze and Landa Putkaradze

Abstract Sustainable development of mountainous regions and their social and economic growth is one of the significant challenges of modernity. The mentioned problem constitutes one of quite urgent issues of Georgia, in particular for mountainous Ajara region, as the study area is subject to demographical and economic crisis during recent years. Touristic and recreational resources of mountainous Ajara is evaluated via application of traditional and modern methods and the index of fitness of area usage for touristic and recreational purposes is determined. Geographical characteristics of tourism development dynamics and perspectives of operation of some particular types of tourism are found out. On the basis of theories of economic development and taking into consideration of environmental sustainability, originalities of regional economy fields are found out and the place of tourism in local economy structure is shaped. By finding out functioning of state municipal programs, business interests and attitude of local population in line with sustainable development of the region—there are determined a tourism's priorities, existing challenges and the perspectives of their further development via complex approaches are set.

Keywords Highlands · Resource · Economy · Tourism · Sustainable development

Introduction

Sustainable development of mountainous regions constitutes an important challenge of modernity, which needs setting particular events in line with identification and settlement of existing problems. Therefore, the studies conducted in line with settlement of specific problems in mountainous regions of the world (Dornier and Mauri

M. Putkaradze (✉)
Batumi Shota Rustaveli State University, Batumi, Georgia
e-mail: merab.putkaradze@bsu.edu.ge

L. Putkaradze
Ivane Javakhishvili Tbilisi State University, Tbilisi, Georgia
e-mail: landaputkaradze1997@gmail.com

L. Soares et al. (eds.), *Compromised Geography*, Key Challenges in Geography,
https://doi.org/10.1007/978-3-032-05201-8_22

2018; Magadán-Díaz, Rivas-García 2019; Saarinen 2015) gives theoretical leadings to rise their social and economic development level via rational application of the existing resources thus, creating the normal living conditions in the long run (Krevs et al. 2023; Milićević et al. 2021; Minguez 2012; Valentine et al. 2005; Vučetić 2018). Similar problems are quite urgent in almost every mountainous region of Georgia, especially in mountainous Ajara (Applis 2022; Gogitidze et al. 2022; Salukvadze and Backhaus 2020), where geodynamic processes were activated and ecological condition was worsened due to irrational natural management during recent year. On the other hand, agriculture (economies leading field) under the land scarcity conditions is unable to ensure economic development, which is led to fall of living standards and decrease of population in villages. Therefore, emphasis on the sustainable development of mountainous Ajara should be placed on local resources' usage at their maximum. In this regard, the application of touristic and recreational resources (on the study territory being in a less sustainable environment) has a huge perspective, which may be the most rational approach in sustainable development of the region. Therefore, determination of tourism priorities in the economic sectoral structure of study region will have a special importance, both in scientific-practical and application directions, which constitutes quite an urgent topic.

The main goal of research is to establish tourism priorities in sustainable development of Ajara, determine the challenges existing in front of it and set the perspectives for further development. The following tasks should be solved in order to reach the goal of research:

- Establish sectoral structure of the economy of mountainous Ajara, shape the opportunities of development of some particular field and set the tourism's further development priorities;
- Evaluate the basic properties of natural resources of mountainous Ajara and via foreseeing the environmental sustainability, shape the perspectives for application of touristic and recreational resources;
- Establish the geographical characteristics of touristic and recreational resources, determine the coefficient of fitness for tourism development and evaluate the landscape-recreational assimilation index;
- Find out the features of dynamics of tourism and determine the tourists' transmittance potential;
- Evaluate the tourism industry development level and determine the tourists' simultaneous visitor capacity in resorts areas during the season;
- Determine the influence of different types of tourism on environmental sustainability;
- Determine the attitude of locals towards the tourism development;
- Find out tourism's economic and geographical characteristics and its impact on social and economic development of the region;
- Determine the role of interested parties in development of tourism via tourism's cluster development approach;
- Establish the challenges that exist in front of tourism and set the solutions for existing problems;

- Determine the tourism priorities in the sustainable development of mountainous Ajara and set the perspectives for its further perfection.

The subject of research is a mountainous region of Autonomous Republic of Ajara, which unites Keda, Shuakhevi and Khulo Municipalities. The area of the region equals to 1750 km^2, with 57,600 population by 2023.

Literature Review

Theoretical approaches about sustainable development of mountainous regions (Amerta et al. 2018; Bramwell and Line 2011; Dwyer 2023; Sakhno et al. 2019) cover not only effective application for economic increase with existing resources, but if how to overcome the existing challenges towards sustainable development in particular region via using an international experience. In this regard firstly, there should be separated the prioritized fields of some particular territory from sectoral structure of the economy and (coming out of an existing environmental sustainability) determine the tourism's role together with all the remaining sectors of the economy, that was established by theory of balanced economic development and agreed operation of other sectors together with tourism was found out by means of coordination failure theory (Cooper 2008).

As a result of study of existing theoretical approaches on sustainable development (Bošković et al. 2020; Bugdol et al. 2019; Kuščer et al. 2017; Weaver et al. 2021). It is found that tourism economy will reach the higher level of sustainability if each and every interested party will take part in its development. Therefore, foreseeing the ideas of interested parties and satisfaction of their requests should meet the common interests, by means of which an economic effect should be acquired, and environment should be protected at the same time. Theoretical research conducted in this regard (Jelinčić and Tišma 2022; Milijic et al. 2017; Niavis et al. 2022) will become the means for achievement of main goal of the research for particular region.

It is proven that sustainable development of tourism actually requires: "tourism planning, development and functioning should be sectoral-sectional and integrated, which covers the involvement of different state departments, companies of public and private sectors, social groups and experts" (Wahab and Pigram 1998). Moreover, at the Millennium Conference of Tourism Leaders held in Osaka in 2001, there was declared that sustainable development of tourism is only possible via its support and involvement on the local level. Therefore, based on an international experience (Bugdol et al. 2019; Dodds 2007; Logar 2010; Chiodo et al. 2019; Bruyn and Meyer 2022)and local research (Khartishvili et al. 2021; Putkaradze and Abuselidze 2019; Putkaradze and Putkaradze 2022; Salukvadze et al. 2019), the state policy of tourism priorities into sustainable development of mountainous Ajara should be defined.

Materials and Methods

Sustainable development of mountainous regions constitutes one of the most important challenges of modernity, which needs determination of such trends of economy, which will be oriented on rational application of local natural resources and at a time will serve as a guarantor of environmental protection. In this regard (on the research territory with complex environment), development of tourism economy should be prioritized via application of touristic and recreational resources in mountainous Ajara. In order to achieve the goals set, it is necessary to apply traditional and modern (contemporary) research methods as well.

To determine sectoral structure of the economy of mountainous Ajara and determine the perspectives of tourism development, the theory of balanced economic development and coordination failure theory will be applied (Cooper 2008). The coefficient of fitness of the territory of mountainous Ajara regarding its usage for tourism purposes will be determined via method of mathematical research (Aleksandrov 2009) and landscape-recreational assimilation index will be evaluated therein (Vedenin and Miroshichenko 1969). By application of methodology of tourists' ecological transmittance potential perspectives (Birjakov 2012) of some particular type of tourism and touristic units were defined. In order to find out in detail the topic related to tourism development of local population, it became necessary to apply the method of sociological surveys and in-depth interviews (Lu and Nepal 2009). Via using the modern geographic information systems of research.

GIS methods, characteristics of spatial organization of touristic and recreational resources of mountainous Ajara were presented, and quantitative and qualitative properties of tourism development in the region was expressed by charts and graphic images. During the research there was also used statistical, field observation, space–time analysis and other methods.

Results and Discussion

Mountainous Ajara was quite promoted region of Georgia in line with demographic and economic topics, but changes implemented in the sectoral structure of the economy during recent years have created serious issues as a result of which the living standards fell, rural population decreased, and the villages become desolated. The problem cannot be settled via municipal programs and notwithstanding the fact that law of Georgia "on the Development of High Mountainous Regions" is applicable, according to which the salaries and pensions in mountainous regions are higher rather in plain/flatland regions. On the other hand, irrational natural management is in action during economic activity, as a result of which geodynamic processes are being activated, and ecological condition worsens (Putkaradze et al. 2019).

Therefore, the main solution for region's sustainable development is rational application of existing resources and change of sectoral structure of the economy on its basis. Complex approach to the topic firstly needs discussion of sectoral structure of regional economy and determination of perspectives of some particular field. In this regard and according to theory of balanced economic development (Cooper 2008)—to determine the connections between the economic sectors of the region, it is necessary to discuss the development level of some particular sector (field) and the perspectives of its further development.

Nowadays, agriculture is the leading field of the region and 80% and more of material productions falls upon it. Because of land scarcity there is no perspectives for field's further development. From the other side, effort of increasing agricultural lands is being held in an irrational natural management manner that worsens ecological condition as a result. Hydroenergetics is developed amongst the industrial sectors of the region, but there are no opportunities for its further development, because hydroenergetical potential is assimilated at its maximum. Besides, the food industry is weakly developed and perspectives for its further increase don't exist due to inexistence of huge amounts of raw material. characteristics, according to balanced theory approach and based on analysis of basic sectors of economy of mountainous Ajara there is being determined that development of traditional fields (sectors) of the region has not big perspectives.

By evaluation of natural and resources' potential of mountainous Ajara and overview of economic sectors existing therein, there is found out that on the basis of coordination failure theory (Cooper 2008) priority for regional economy's development should be granted to tourism (on the basis of diverse touristic and recreational resources) and it should become one of the perspective fields (sectors) of the economy. Coming out of the above-mentioned and based on research, tourism development priorities should be grounded for sustainable development of the region.

It is known that development of tourism industry firstly is based on touristic and recreational resources. In order to substantiate tourism development in the region, first of all touristic and recreational potential should be evaluated and opportunities for development of some particular types of tourism should be determined.

Mountainous Ajara is a region quite rich with touristic and recreational resources, where we have amazing mountain landscapes, natural and historical monuments, local culture, ski resorts and recreational areas, mineral waters, etc. Touristic and recreational resources located on the research territory is more or less presented in each municipality (see Fig. 1), but most of their part is not involved in the tourism industry yet.

In order to discuss the determination of tourism priorities in sustainable development of research territory, firstly it is necessary to find out the coefficient of fitness of the region regarding its usage for tourism purposes, which may be determined by mathematical method (Salukvadze et al. 2019) by using the following formula:

$$C = \frac{\text{Srf}}{\text{Saf}}$$

where

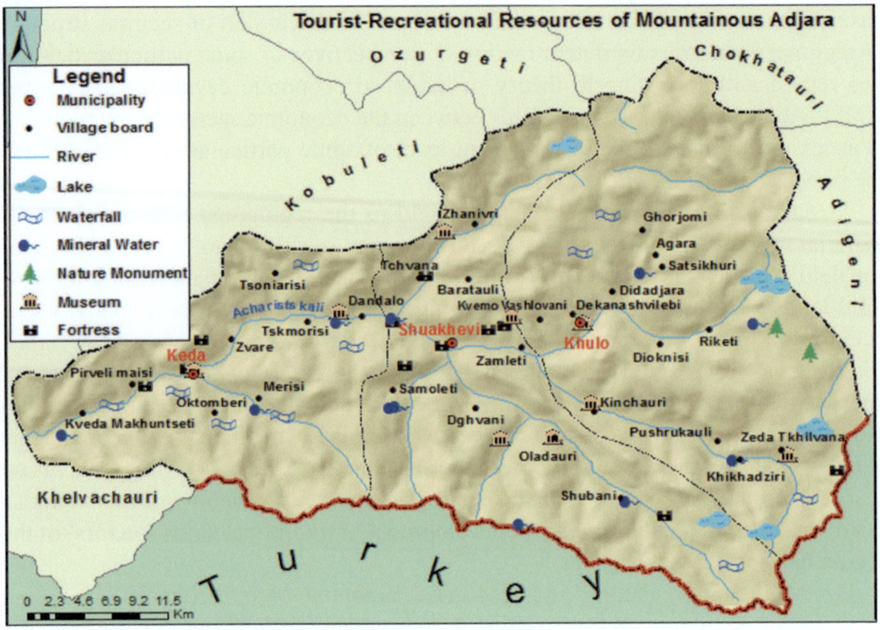

Fig. 1 Tourist-recreational resources of mountainous Ajara

C coefficient of fitness;

Srf sum of regional functions (unity of tourism types), we have determined their number as 14;

Saf sum of area functions (region covers three municipalities or the place, where more or less is developed a number of tourism types, but their number is not less than 12 in any of units)

As a result, the formula will be as follows: $C = \frac{14}{12} = 1.17$.

If the coefficient of fitness is above 1, then it will be considered that research territory is fit for tourism development purposes. Therefore, region of mountainous Ajara (where different types of tourism operate) is fit for tourism development.

For topic's complex study it will be important to determine the landscape-recreational assimilation index (Vedenin and Miroshichenko 1969), which will be calculated according to the following formula:

$$I = \frac{P_m}{K_x P_p}$$

where

I landscape-recreational assimilation index of territory;

P_m touristic units density in natural-territorial complex (NTC); up to 74 touristic units (visited by tourists) are spread within 1750 sq. km of the region;

K_x recreational fitness complex of natural conditions (which is being evaluated with 4 coefficients: very favorable—4; favorable—3; comparatively favorable—2; less favorable or almost unfavorable—1; according to the mentioned methods and on the basis of the research held the coefficient of the region was determined by 3);

P_p population density on the given territory (average population density in the region is 33 people per sq. km)

If we place the mentioned quantitative indicators, the formula will be as follows:

$$I = \frac{23.6}{3 \times 33} = \frac{23.6}{99} = 0.24$$

According to the mentioned methods there is found that landscape-recreational assimilation index of mountainous Ajara region is only 24%.

The research makes clear that mountainous Ajara possesses quite important touristic and recreational resources for tourism development and its application and creation of tourism infrastructure will be dependent on working out and implementation of the correct tourism policy as well.

In order to find out the tourism development perspectives in the region, it is necessary to determine the quantitative indicators of visitors and tourists, by means of which there will be possible to discuss the qualitative properties of the sector (see Table 1).

Table 1 shows us that the number of visitors and tourists gradually rises in 2000–2023 years, but number of tourists compared to visitors is almost 12-times less. The mentioned statistical data clearly shows that many visitors are interested in the region, thus giving the perspective to increase the number of tourists far more in case of implementation of effective tourism policy and development of tourism industry. Here we can also mention that such a high indicator of number of visitors is caused with close location (20–100 km) of most significant tourist center in Georgia—Batumi City. Most Batumi Sea tourists visit mountainous Ajara during the midday or non-swimming weather. We have studied touristic and recreational potential of Makhuntseti Waterfall—most important place of mountainous Ajara receiving the visitors and have set the conditions for creation of new resort area (Putkaradze and Putkaradze 2022).

On the other hand, big number of tourists by itself gives the perspective for development of different types of tourism and the diversity of touristic and recreational

Table 1 Dynamics of visitors and tourists in mountainous Ajara in 2000–2023 years

	2000 year	2010 year	2015 year	2019 year	2021 year	2022 year	2023 year
Visitor	15,550	85,750	135,850	160,150	134,350	175,550	182,470
Tourist	5750	695	8350	12,150	9650	14,650	15,460

Source Table is made based on materials from Department of Tourist and Resorts of Autonomous Republic of Ajara and the field research

Fig. 2 Types of tourism of mountainous Ajara

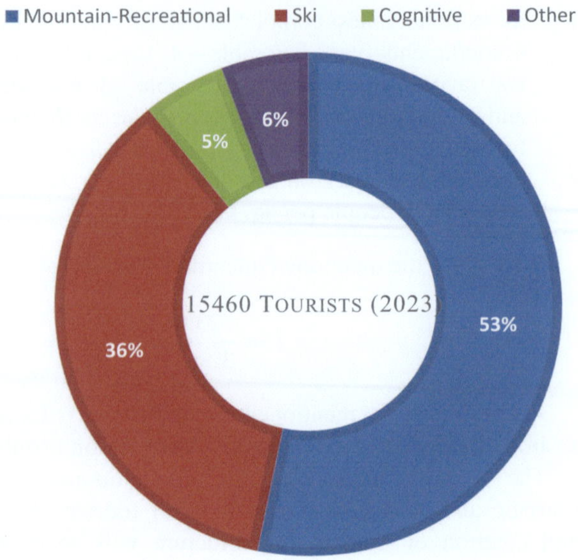

resources allows this. Therefore, to reach the main aim of research it is necessary to evaluate the quantitative indicators of development of some particular type of tourism and determine the tourism's priority trends on its basis. As a result of the research, we have determined 14 types of tourism in mountainous Ajara; the number of tourists quite differs according to some particular types of tourism, which is clearly observed on the chart (see Fig. 2).

Figure 2 shows us that a large part of tourists (53.3%) in mountainous Ajara falls upon mountain recreation (and 50.1% of it comes on ski resort Beshumi), 35.8%—ski, 5.3%—cognitive and 5,6% on other different types of tourism. It is obvious that leading tourism directions of the region constitutes mountain-recreational and ski tourism, but other tourism types also develop during the recent years, especially cognitive, agricultural and eco-tourism. It is natural to have a question, which from the listed types of tourism possess the perspectives for its further development?

Main type of tourism in mountainous Ajara is mountain-recreational, which is mainly developed on the basis of region's sole resort—Beshumi, the tourist number wherein gradually increases. We have studied in detail the touristic and recreational resources of resort Beshumi, development level of tourism industry, ecological transmittance potential of tourists, problems and development perspectives (Putkaradze et al. 2023). It is true that the resort mainly serves internal tourists, but foreign tourists also became interested in resort during recent years. Therefore, tendencies of increased mountain-recreational tourists on the basis of tourists' dynamics of recent years are visibly shaped, moreover – receipt of such type of tourist product partly is being implemented via seasonal alpine settlements.

The second significant type of regional tourism is ski tourism, which is quite quickly developing on the basis of Goderdzi and Gomarduli recreational areas. It's obvious that the mentioned areas are not assigned with resort statuses but their useful location from Batumi (80–105 km) is prioritized for tourists in line with reachability via transport (main ski resorts of Georgia like Gudauri and Bakuriani are 360–390 km away from Batumi). In this regard, special potential possesses Goderdzi, where international sport events are being held coming out of its touristic and recreational resources. Resort Goderdzi is located 1950–2025 m above sea level. It is known for high snow cover, fresh air, natural landscapes with spectacular open horizons. The length of its ski tracks are up to 9 km, starting from 2364 m above sea level. The longest downhill ski track is 4.5 km long. Therefore, the infrastructure of resort meets international standards. Therefore, due to its best properties and close location to Batumi, the resort in the nearest future has an opportunity to transform into international level resort. No less perspectives have ski resort Gomarduli area with its diverse natural properties and close (80 km) location from Batumi City.

Huge perspectives have cognitive tourism, for implementation of which a significant touristic and recreational resources are mountain landscapes, well-known natural monuments, historical and architectural buildings and other resources. Besides, there are developed so vine tourism, agricultural tourism, adventure tourism, eco-tourism, cultural tourism, gastrotourism and other types of tourism in the region.

The research makes clear that the study region is known for various touristic and recreational resources on the basis of which different types of tourism is developed and the indicators of the number of visitors and tourists show that there exist quite big perspectives for development of this field. Therefore, challenges existing in front of tourism and tourism's further development priorities should be discussed in detail.

One of the significant challenges of tourism in the research territory is that tourism potential of specific touristic and recreational resources, also its social and economic value and place in sectoral structure of the economy is not completely studied yet. Region's tourism potential is imperfect regarding its awareness relating to development of internal and international tourism as well. We have such touristic and recreational resources for now, which are unknown to large society. Tourism's role and place are not determined in the economy regarding sustainable development of related municipality of the region. Actually, there do not exist subsidies for tourism development in municipal programs and they completely depend on central finances. Tourism infrastructure is not perfect, there are issues in transportation availability for tourists, there are no internet connection, Wi-Fi available everywhere, etc. Except objective reasons, there are subjective factors in the tourism development—in particular, local population is less informed about tourism priorities. From the other side, private companies (at the creation of tourism infrastructure) take less account of basic principles of rational natural management. As a result, geodynamic processes are being activated, and ecological condition worsens. Besides, there are no intense works held in line with involvement of diverse touristic and recreational resources (especially those of exotic mountain landscapes) in tourism economy and arrangement of new touristic routes.

In order to reach the basic aim of the study, first of all it is necessary to substantiate the priorities of tourism development in the region compared to other sectors of the economy. The main topic is that agriculture stays as the leading sector in the region with quite a high demographic load, which is unable to ensure the region's economic development in a less sustainable environment and under the irrational natural management conditions and the ecological condition worsens at the same time. Therefore, priority in increase of population's welfare should be placed on sustainable development, which should be implemented in line with economic growth of the region and in environmental protection's direction. Thus, determination of tourism development priorities in the region (except the natural and geographical properties) needs definition of tourism place in sectoral structure of the economy and setting the perspectives of its further development. For this purpose, firstly the tourism place in gross domestic product of the region should be defined.

According to the research held by us (Putkaradze and Abuselidze 2019), by 2017, the leading sector of the economy on the research territory was agriculture and tourism had only 3%. According to evaluations by 2023, this indicator was somehow changed in favor of tourism, and it was increased up to 6%. Coming out of region's environmental sustainability, priority should be granted to tourism, but this does not imply weakening the development of traditional sectors. Moreover, agriculture should be the leading sector still and the tourists catering system (including tasting the local cuisine and development of gourmet tourism) should develop on the basis of natural, ecologically pure agricultural products.

The most important perspective for tourism development in mountainous Ajara is that touristic and recreational potential existing in the region is not completely assimilated (used). The research proved that coefficient of fitness of the regional territory regarding its usage for tourism purposes (1.17 coefficient) and the landscape-recreational assimilation index (only 24% is assimilated) clarifies the perspectives of sector's further development. In order to clear up the topic in detail, firstly we should discuss the number of tourists in the region and about receipt of tourists under existing tourism industry and the tourists' transmittance potential. We have studied in detail the resort Beshumi and found out that under existing infrastructure, average seasonal load is only 1.6% (Putkaradze et al. 2023). Actually, resort's load is at its minimum, but this indicator during the season drastically differs and its average load during the peak (first week of August) period increases up to 45.3%. Almost similar situation we have on other resort areas and recreational places. We have studied potential of tourists' simultaneous receipt, number of tourists and maximum potential of tourists' receipt under the development of the tourism industry at four resort places (see Fig. 3).

Figure 3 shows us that the number of tourists at all resort units is far less, compared to possibilities of tourists' receipt during the season under existing potential. Actually, existing quantitative indicators prove that tourists' transmittance potential in the region is far more, allowing the possibility for perspectives of tourism's further development in the region. According to this, further development of tourism in the region (to use existing potential at its maximum) needs implementation of effective management and marketing on the basis of working out the tourism policy.

Coming out of basic purposes of research and for the determination of tourism priorities in sustainable development of the region, there should be shaped the views of interested parties (Damian et al. 2023; Duglio et al. 2019; Richins 2016). Complex

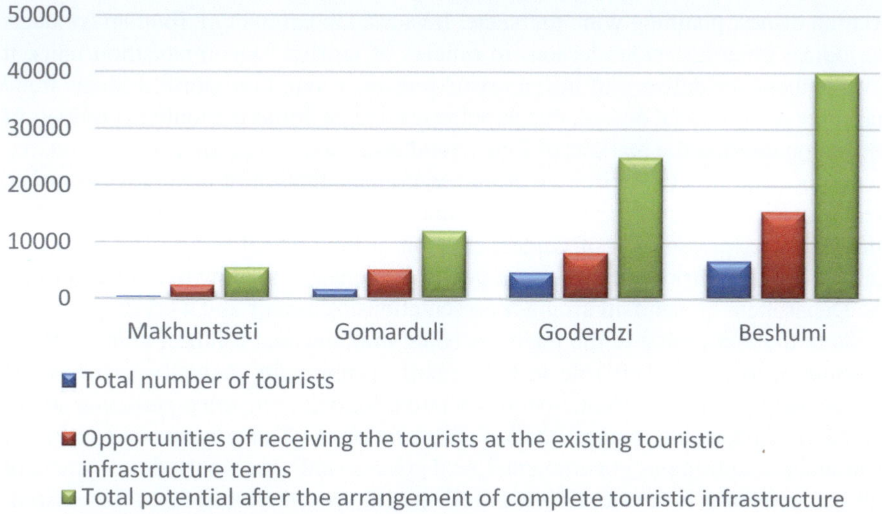

Fig. 3 Tourism transmittance potential of the resorts

evaluation of the topic needs to find out the interests (related to tourism in mountainous Ajara) according to cluster approach. Exactly on the basis of the requirements of interested parties, the tourism development policy of the region should be determined. Not less important is the local population's attitude regarding the creation of tourism industry and resort areas. To find out the topic, we have conducted sociological surveys and in-depth interviews. 247 respondents were asked with following question: "which sector of the economy do you prefer (agriculture, industry, tourism) order them according to priorities") and 64% of them set agriculture on the first place, 33% -if them—tourism and only 3% of them—industry. These respondents also were asked with question: "which touristic and recreational resource should be used more?" the answers were as follows: 41% out of 247 respondents prioritized mountain-climatic, 32% of them—ski, 22%—beautiful natural landscapes and 5% of them—the rest of the resources. According to local populations' view, there is found that tourism development should be prioritized in order to maintain the environment and increase the living standards. Moreover, the population is interested to grant the special priority to ski tourism development thus, employing some particular part of labor resources freed during winter season and application of such resources would have minimum impact on environment's ecological condition.

The state policy of tourism development in region is conducted by Department of Tourism of Autonomous Republic of Ajara. To exercise the tourism policy in the region the department has short-term tactical and long-term strategic programs. Main trends of tactical action programs are collection of information about and advertisement of touristic and recreational resources of the region, showing up local tourism products, spreading the information about new touristic directions, making banners, performing advertising works, arranging culinary expeditions for popularization of

Ajarian dishes, planning wine tours, etc. Besides, Department of Tourism conducts periodical trainings, video-lessons to officers of tourists' accommodation units, it awards those for delivery of best touristic resources, etc. Long-term strategic trend has more international design, that popularizes and positions the tourism products of the region through the markets of Europe and Asia, actively participating in international tourism fairs (exhibitions), promotes implementation of novelties in tourism field, attraction of tourists, attraction of financial resources for creation of tourism infrastructure, preparation of guides, creation of the manual of touristic routes for guides. For the purpose of increasing public awareness regarding tourism priorities, the Department of Tourism arranges presentations, seminars and workshops.

State tourism policy in the region is being implemented by local municipalities at some degree, but their role in this regard is minimum. Actually, there are no independent services in municipalities relating tourism, but some particular works are fulfilled regarding implementation of resolutions by Departments of Tourism of central and autonomous government. Local government's role in the development of tourism economy is developed more in line with preventive measures for environmental protection and ecological condition's changes. Some works in this regard are fulfilled by teaching university and general-educational schools existing in the region. They play a huge role in increasing local public awareness about the rational application of touristic and recreational resources and attitude towards tourism. Students and pupils are actively involved in protection and restoration of green cover.

According to general trends, the decisive role in creation of tourism industry and infrastructure is connected to private sector or business interests. As is known, any tourism activity is profit-oriented, which at the same time is less interested in rational natural management and often their activity develops in the direction of worsening the ecological condition. The mentioned problem is observed on the research territory to some degree, but it has no urgent nature, as there are no large touristic units in the region (except Goderdzi ski recreational place). Some particular issues in this regard are observed in touristic units with smaller size, where due to lack of land—hostels are being built without taking into consideration the sustainability of natural conditions. Moreover—private sector has a state support for tourism development in the region and most of the part of full package of tourism infrastructure development falls upon them.

Public sector and non-governmental organizations consist in the cluster group determining the tourism priorities into sustainable development of research territory. Their role in tourism development is insignificant, which (by financing foreign foundations) are also involved in social and economic development studies of the region. Besides, regions are being presented via mass media with their participation.

Financial institutions or private banks are involved in the cluster chain of development of regional tourism. Different banks operating in the country had financed several tourism products. Actually, banks present financial sources for the development of regional tourism and many tourism units are being built and operated on the basis of correctly projected business projects.

Scientific-research organizations serve the biggest part in the evaluation of tourism potential of the region, establish the quantitative properties of tourism's geographical characteristics, find out the existing challenges towards the tourism and setting its perspectives. It is known that development of the economy in lagging regions is impossible due to interdisciplinary research. Economists, research specialists from tourism and environmental protection fields are involved in the study region in this regard (Gogitidze et al. 2022; Putkaradze et al. 2019, 2023;Putkaradze and Putkaradze 2022; Abuselidze and Devadze 2021). Due to the complexity of the topic, it is desirable to involve research specialists from political (governance) and sociological fields as well, though their role at this stage remains insignificant. Thus, integral connections should be implemented via joint coordinated research involving public services and representatives of wide public. As a result, perspectives of sustainable tourism's further development will be better defined via innovative approaches. As for the regional layout to ensure the sustainable tourism, strategic development programs are developed on the basis of recommendations defined in research (Gogitidze et al. 2022; Salukvadze and Backhaus 2020; Salukvadze et al. 2019; Abuselidze and Devadze 2021; Gogitidze and Nadareishvili 2021) held in this regard, taking into consideration the environment in local municipalities and according to interests of local population.

It is obvious that tourism is an economic sector, which has relatively less influence on the environment, but because of regional specifics, tourism development bears some particular risks towards the environment. According to research conducted by us (Putkaradze et al. 2019, 2022, 2023) it is confirmed that works performed at the several locations for tourism infrastructure creation purposes caused activation of exodynamic processes. In particular, construction of hostels on highly inclined slopes of middle-sized mountains is related to some risks. The situation is different in high-mountain alpine zone (here operates Beshumi mountain and climate resort and Goderdzi ski resort), where tourism's impact on the environment is at its minimum. Therefore, alpine meadows of mountainous Ajara (average elevation 1800–2300 m above the sea level) should be turned into the main region for mountain tourism development. Here should also be noted that up to 50 seasonal settlements exist throughout the alpine meadows (which have recreational functions together with development of cattle breeding), where exodynamic processes are observed rarely. It can be said that the development of tourism has huge perspectives in the case of taking into consideration the main principles of rational natural management in mountainous Ajara.

Teaching and research institutes also implement significant works in increasing local public awareness about rational application of touristic and recreational resources and attitude towards tourism. We have determined the population's friendly attitude towards tourism development in recent five years by means of in-depth interview method of qualitative research (Putkaradze and Abuselidze 2019; Putkaradze and Putkaradze 2022; Putkaradze et al. 2023). As a result, the population's positive mood regarding tourism development was determined.

Figure 4 reflects the role and place of parties (or subjects those involved in cluster chain) those interested in the tourism development in mountainous Ajara.

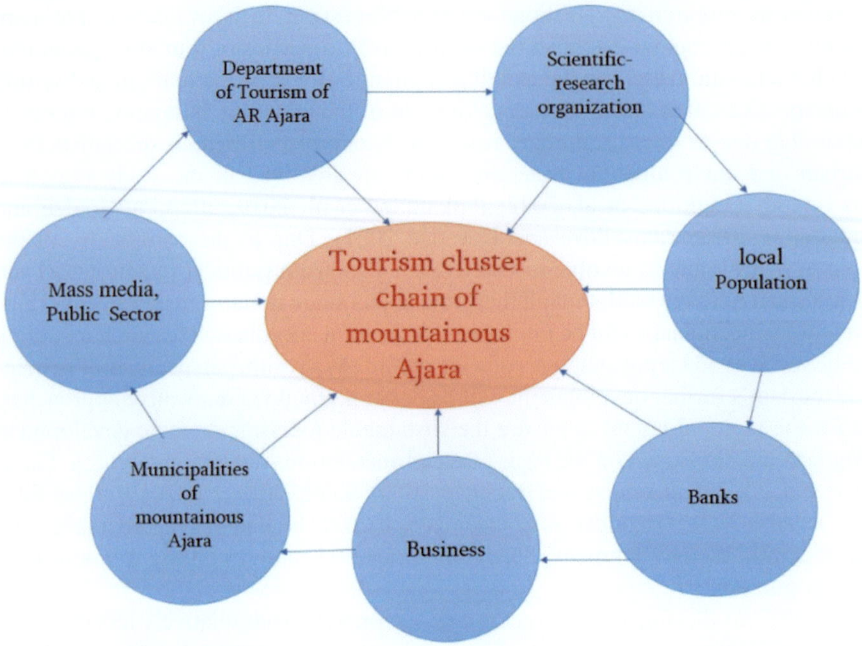

Fig. 4 The cluster chain of tourism development of mountainous Ajara

Thus, tourism holds the special place in sectoral structure of the economy of mountainous Ajara, the perspectives of further development of which will be depended on rational usage of existing touristic and recreational resources and mutually agreed activity of the parties, those interested in tourism.

Conclusion

On the basis of analysis of the theory of balanced economic development there was determined that under the land scarcity and environment's less sustainability conditions in mountainous Ajara—there are no opportunities for further increase of traditional sector of the economy—agricultural development. According to evaluation of natural resources of the region, there was determined that coefficient of fitness for usage for recreational purposes and landscape-recreational assimilation index gives the opportunity to grant tourism with prioritized direction into sustainable development of the research territory.

Diversity of touristic and recreational resources of the region, opportunities for development of different types of tourism, potential of accommodation places and dynamics of tourists makes clear that the number of tourists in the region quite lags the total potential, which is one of the main challenges. There exists another problem in tourism development—transportation availability for tourists, completion of tourism service infrastructure, low awareness of tourism potential of the region, inefficient management and marketing of tourism.

In order to determine the tourism priorities in the sustainable development of mountainous Ajara, the role of cluster subjects interested in tourism industry was studied in detail and there was found out that all subjects (including the local population) are more or less involved in their development.

Overcome of challenges of tourism in mountainous Ajara and setting the perspectives for its further development is largely depended on the usage of applicable law of Georgia "on the Development of High Mountainous Regions", working out flexible marketing and management for the development of new tourism types coming out of environmental sustainability.

References

Abuselidze G, Devadze N (2021) Stakeholders perceptions and model of engagement for sustainable tourism development in Adjara autonomous republic. Int J Innov Technol Econ 3(35):1–6. https://doi.org/10.31435/gsglobalijite/30092021/7638

Aleksandrov A (ed) (2009) Geography of tourism. Moscow

Amerta IMS, Sara IM, Bagiada K (2018) Sustainable tourism development. Int Res J Manage IT Soc Sci 5(2):248–254

Applis S (2022) Crises around concepts of hospitality in the mountainous region of Svaneti in the North of Georgia. Tour Hospitality 3(22):416–434. https://doi.org/10.3390/tourhosp3020027

Birjakov M (2012) Theory of tourism. Tbilisi

Bošković N, Vujičić M, Ristić L (2020) Sustainable tourism development indicators for mountain destinations in the Republic of Serbia. Curr Issues Tour 23(22):2766–2778. https://doi.org/10.1080/13683500.2019.1666807

Bramwell B, Lane L (2011) Critical research on the governance of tourism and sustainability. J Sustain Tour 19(4–5):411–421. https://doi.org/10.1080/09669582.2011.580586

Bruyn C, Meyer N (2022) An assessment of the viewpoints of local stakeholders on tourism sustainability and local economic development (LED). Forum Scientiae Oeconomia 10(1):82–96. https://ojs.wsb.edu.pl/index.php/fso/article/view/487

Bugdol M, Puciato D, Borys T (2019) Development of tourism—barriers to shaping and recommendations for further research. Probl Sustain Dev 14(2):157–170. https://core.ac.uk/downloar/pdf/288119312.pdf

Chiodo E, Fantini A, Dickes L et al (2019) Agritourism in mountainous regions—insights from an international perspective. Sustainability 11(13):3715. https://doi.org/10.3390/su11133715

Cooper C (2008) Tourism: principles and practice. Pearson Education, Harlow

Damian I, Navarro-Jurado E, Ruiz F (2023) Involving stakeholders in the evaluation of the sustainability of a tourist destination: a novel comprehensive approach. J Sustain Tour 31(7):1631–1650. https://doi.org/10.1080/09669582.2021.1919687

Dodds R (2007) Sustainable tourism and policy implementation: lessons from the case of Calvia, Spain. Curr Issues Tour 10(1):296–322. https://doi.org/10.2167/cit278.0

Dornier R, Mauri C (2028) Overview: tourism sustainability in the Alpine region: the major trends and challenges. Worldw Hosp Tour Themes 10:136–139

Duglio S, Bonadonna A, Letey M, Peira G, Zavattaro L, Lombardi G (2019) Tourism development in inner mountain areas-the local stakeholders' point of view through a mixed method approach. Sustainability 11:5997

Dwyer L (2023) Resident well-being and sustainable tourism development: the 'capitals approach'. J Sustain Tour 31(9):2119–2135. http://hdl.handle.net/10453/176353

Gogitidze G, Nadareishvili N (2021) Access to information about the tourist products of mountainous Adjara for tourists and visitor to the Adjara region. In: Proceedings of the SADAB 8th international online conference on social researches and behavioral sciences, Tuzla, Bosnia and Herzegovina, pp 174–183

Gogitidze G, Nadareishvili N, Harun R, Lrien L, Muresan L (2022) Exploring residents' perceptions towards tourism development—a case study of the Adjara mountain area. Sustanability 15(1):492. https://doi.org/10.3390/su15010492

Jelinčić DA, Tišma S (2022) Tourizm innovation in the Adriatic-Ionan region: questioning the understanding of innovation. Eur J Geogr 13(5): 97–114. https://doi.org/10.48088/ejg.d.jel.13.5.097.114

Khartishvili L, Muhar A, Dax T, Khelashvili I (2021) Rural tourism in Georgia in transition: challenges for regional sustainability. Sustainability 11:410

Krevs M, Repe B, Mazej T (2023) Reconsidering the basics of mountain trail categori-sation: case stude in Slovenia. Eur J Geogr 14(2):44–63. https://doi.org/10.48088/ejg.m.kre.14.2.044.063

Kuščer K, Mihalič T, Pechlane H (2017) Innovation, sustainable tourism and environments in mountain destination development: a comparative analysis of Austria, Slovenia and Switzerland. J Sustain Tour 25(4):489–504. https://doi.org/10.1080/09669582.2016.1223086

Logar I (2010) Sustainable tourism management in Crikvenica, Croatia: an assessment of policy instruments. Tour Manage 31(1):125–135. https://doi.org/10.1016/j.tourman.2009.02.005

Lu J, Nepal K (2009) Sustainable tourism research: an analysis of the papers published. J Sustain Tour 17(1):5–16. https://doi.org/10.1080/09669580802582480

Magadán-Díaz M, Rivas-García JI (2019). Sustainable tourism in mountain protected areas of Asturias: an analysis from tourists' perceptions and profiles. Sustain Dev Mt Territ 4(11):459–474. https://doi.org/10.21177/1998-4502-2019-11-4-459-474

Milićević S, Bošković N, Lakićević M (2021) Sustainable tourism development in mountain areas in Šumadija and Western Serbia. J Mountain Sci 18(3):735–748. https://doi.org/10.1007/s11629-020-6239-4

Milijic S, Micic S, Maksin M (2017) Retrospective of and prospects for the development and strategic planning of tourism in the mountain regions of Serbia. Spatium (Belgrade) 37:42–48. https://doi.org/10.2298/SPAT1737042M

Minguez C (2012) The management of cultural resources in the creation of Spanish tourist destinations. Eur J Geog 3(1):68–82. https://doi.org/10.12795/rea.212.j41.03

Niavis S, Papatheochari T, Koutsopoulou T, Coccossis H, Psycharis Y (2022) Considering regional challenges when prioritizing tourism policy interventions: evidence from a Mediterranean community of projects. J Sustain Tour 30(4):663–684. https://doi.org/10.1080/09669582.2021.1876074

Putkaradze M, Abuselidze G (2019) The impact tourism on business development in mountain regions: a case study. Int J Econ Bus Admin 7(3):70–83. https://ijeba.com/journal/308

Putkaradze M, Putkaradze L (2022) Touristic and recreational potential of Makhuntseti waterfall: application and perspectives (Georgia). Int J Innov Technol Econ 2(38):1–5. https://doi.org/10.31435/rsglobal_ijite/30062022/7826

Putkaradze M, Phagava N, Kamadadze C, Chichileishvili K, Ananidze M (2019) Ecological issues of buziness activity in mountains of Ajara (Georgia). Int Multidiscip Scient GeoConference: SGEM; Sofia 5(2):307–312. https://www.caucasus-mt.net/Publications/ecological-issues-of-business.html

Putkaradze M, Michalski T, Abuselidze G (2022) Tourism and rekreational resources usage perspectives through according to mountainous Adjaria S Example. J Econ Manage 19(2):27–39. https://doi.org/10.31435/rsglobaljjjte/30122023/8077

Putkaradze M, Putkaradze L, Duadze S (2023) Tourism challenges and perspectives in the mountainous Ajara: on the Beshumi resort example. Int J Innov Technol Econ 44(4):1–7. https://doi.org/10.31435/rsglobal_ijite/30122023/8077

Richins H (2016) Experience provision in mountain tourism: overview, contextual development and emphasis. In: Richins H, Hull J (eds) Mountain tourism: experiences, communities, environments and sustainable futures. CABI, UK, pp 13–24. https://doi.org/10.1079/9781780644608.0013

Saarinen J (2015) Conflicting limits to growth in sustainable tourism. Curr Issues Tour 18(10):903–907. https://doi.org/10.1080/13683500.2014.972344

Sakhno A, Hryvkivska O, Salkova I, Kucher L (2019) Evaluation of the efficiency of enterprises by the method of analysis of functioning environment. J Environ Manage Tour 10(3):499–507. https://doi.org/10.14505/jemt.v10.3(35).04

Salukvadze G, Backhaus N (2020) Is tourizm the beginning or the end? Livelihoods of Georgian mountain people at stake. Mountain Res Dev 40(1):28–36. https://doi.org/10.1659/MRD-JOURNAL-D-19-00078.1

Salukvadze G, Gugushvili T, Salukvadze J (2019) Spatial pecularities of local tourisms supply chains in high-mountainous Georgia: challenges and perspectives. Eur J Geogr 10(3):173–188. https://eurogeojournal.eu/index.php/egj/article/view/203

Valentine KM, Knopf RC, Vogt CA (2005) Residents' perceptions of community tourism impacts. Ann Tour Res 32(4):1056–1076. https://doi.org/10.1016/j.annals.2005.03.001

Vedenin Y, Miroshichenko N (1969) Evaluation of environmental conditions for organization of recreation. Issue Geogr 4

Vučetić A (2018) Importance of environmental indicators of sustainable development in the transitional selective tourism destination. Int J Tour Res 20:317–325

Wahab S, Pigram JJ (1998) Tourism and sustainability: policy considerations. In: Wahab S, Pigram JJ (eds) Tourism, development and growth. The challenge of sustainability. Routledge, London, pp 277–290

Weaver D, Moyle M, Mclennan C (2021) The citizen within: positioning local residents for sustainable tourism. J Sustain Tour 30(2):1–18. https://doi.org/10.1080/09669582.2021.1903017

MIX
Papier aus verantwortungsvollen Quellen
Paper from responsible sources
FSC® C105338

FSC
www.fsc.org

If you have any concerns about our products,
you can contact us on
ProductSafety@springernature.com

In case Publisher is established outside the EU,
the EU authorized representative is:
Springer Nature Customer Service Center GmbH
Europaplatz 3, 69115 Heidelberg, Germany

Printed by Libri Plureos GmbH
in Hamburg, Germany